# Springer Series in Materials Science

Volume 209





More information about this series at http://www.springer.com/series/856

The Springer Series in Materials Science covers the complete spectrum of materials physics, including fundamental principles, physical properties, materials theory and design. Recognizing the increasing importance of materials science in future device technologies, the book titles in this series reflect the state-of-the-art in understanding and controlling the structure and properties of all important classes of materials.

Hisao Ishii • Kazuhiro Kudo
Takashi Nakayama • Nobuo Ueno
Editors

# Electronic Processes in Organic Electronics

## Bridging Nanostructure, Electronic States and Device Properties

Springer

*Editors*
Hisao Ishii
Chiba University
Chiba, Japan

Kazuhiro Kudo
Chiba University
Chiba, Japan

Takashi Nakayama
Chiba University
Chiba, Japan

Nobuo Ueno
Chiba University
Chiba, Japan

ISSN 0933-033X ISSN 2196-2812 (electronic)
ISBN 978-4-431-56154-5 ISBN 978-4-431-55206-2 (eBook)
DOI 10.1007/978-4-431-55206-2
Springer Tokyo Heidelberg New York Dordrecht London



Printed on acid-free paper

Springer is part of Springer Science+Business Media (www.springer.com)

# Preface

This book is intended to serve as a textbook for a Ph.D.-level course or as reference material for researchers in organic electronics and related nanoscale science and technology, from fundamental science to technological applications. It includes the newest findings and ideas. Accordingly, it does not duplicate the books on organic electronics, but it focuses mainly on electronic properties in relation to the structure of organic films and device applications. We also intend to contribute to an understanding of the nature of organic semiconductors that arises from universal features of molecular crystals. The nature of organic semiconductors is considered to appear rather commonly not only in so-called molecular crystals but also in various systems in a vast number of organic systems including bio-related material systems. The new experimental methods introduced in this book are applicable to various materials. Thus the book will be useful for experts working in physics, chemistry, materials, and related engineering and industrial fields as well.

We selected research topics that were studied at the Global Center of Excellence (Global COE) program (MEXT: FY2008–FY2012, G-03), "Advanced School for Organic Electronics", operated at Chiba University. Some projects have been carried out according to the results obtained at the twenty-first Century COE program (MEXT: FY2003–FY2007, G-04), "Frontiers of Super-Functionality Organic Devices", also operated at Chiba University. Therefore this book covers a variety of studies from fundamental electronic states to device applications covering theoretical studies and applications to information tags and displays. Furthermore, innovative experimental techniques in molecular systems, e.g., (1) ultrahigh sensitivity photoelectron spectroscopy, (2) photoelectron yield spectroscopy, (3) spin-resolved scanning tunneling microscopy (STM), and (4) a new material processing method, with optical-vortex and polarization-vortex lasers, are introduced. The former two spectroscopic techniques, (1) and (2), enable us to directly correlate electronic states and electrical conductivity of organic semiconductors, and the third one (3) to study magnetic properties of organic molecules. The fourth technique, utilizing optical-vortex (4), offers a novel method for materials processing and a new possibility to probe spin-related properties, which had never been expected. Furthermore, some related research on molecule- and carbon-based systems are described, namely, functions of conjugated $\pi$-electronic carbon-walled nanospaces, tuned by molecular tiling, and of ionic liquids. Sciences in these studies are related to those in organic semiconductor systems through the nature of organic semiconductors as summarized in Chap. 1. For instance the interaction between curved graphene surfaces and $\pi$-conjugated molecules introduces the possibility of discovering new functions.

Because many of the young researchers who contributed to this book have moved to other universities or institutes, we prefer to list their current institutions in this book.

Chiba, Japan
November 30, 2013

Hisao Ishii
Kazuhiro Kudo
Takashi Nakayama
Nobuo Ueno

# Contents

**Part I Bridging Structure, Electronic States and Electrical Conductivity**

1 **Fundamental Aspects and the Nature of Organic Semiconductor** . . . . . 3
Nobuo Ueno

2 **Ultraviolet Photoelectron Spectroscopy (UPS) I: Band Dispersion Measurements of "Insulating" Organic Single Crystals** . . . . . 11
Yasuo Nakayama, Steffen Duhm, Qian Xin, Satoshi Kera, Hisao Ishii, and Nobuo Ueno

3 **Ultraviolet Photoelectron Spectroscopy (UPS) II: Electron–Phonon Coupling and Hopping Mobility** . . . . . 27
Satoshi Kera, Hiroyuki Yamane, and Nobuo Ueno

4 **Ultraviolet Photoelectron Spectroscopy (UPS) III: Direct Study of "Invisible" Band Gap States by Ultrahigh-Sensitivity UPS** . . . . . 51
Nobuo Ueno, Tomoki Sueyoshi, Fabio Bussolotti, and Satoshi Kera

5 **Pentacene Becomes Mott–Hubbard Insulator by Potassium Doping** . . . . . 69
Fabio Bussolotti, Satoshi Kera, and Nobuo Ueno

6 **Vertical Bonding Distances Impact Organic-Metal Interface Energetics** . . . . . 89
Steffen Duhm, Christoph Bürker, Takuya Hosokai, and Alexander Gerlach

**7 Structure Matters: Combining X-Ray Scattering and Ultraviolet Photoelectron Spectroscopy for Studying Organic Thin Films** . . . . . 109
Alexander Hinderhofer, Keiichirou Yonezawa, Kengo Kato, and Frank Schreiber

**8 Photoelectron Yield Spectroscopy for Organic Materials and Interfaces** . . . . . 131
Hisao Ishii, Hiroumi Kinjo, Tomoya Sato, Shin-ichi Machida, and Yasuo Nakayama

**Part II Organic Devices and Their Properties**

**9 Fabrication and Characterization of Organic Devices** . . . . . 159
Kazuhiro Kudo and Masatoshi Sakai

**10 Mobility Limiting Factors in Practical Polycrystalline Organic Thin Films** . . . . . 185
Ryosuke Matsubara, Noboru Ohashi, Shi-Guang Li, and Masakazu Nakamura

**11 Materials for Organic Light Emitting Diode (OLED)** . . . . . 227
Takashi Karatsu

**12 DNA Electronics and Photonics** . . . . . 253
Norihisa Kobayashi and Kazuki Nakamura

**Part III Theoretical Study**

**13 Theory of Photoelectron Spectroscopy** . . . . . 285
Takashi Fujikawa and Kaori Niki

**14 Theory of Metal-Atom Diffusion in Organic Systems** . . . . . 303
Yoko Tomita and Takashi Nakayama

**15 Numerical Approach to Charge Transport Problems on Organic Molecular Crystals** . . . . . 319
Hiroyuki Ishii

**Part IV Bridging Different Fields: Challenges for the Future**

**16 Function of Conjugated π-Electronic Carbon Walled Nanospaces Tuned by Molecular Tiling** . . . . . 351
Toshihiko Fujimori, Fitri Khoerunnisa, Tomonori Ohba, Suzana Gotovac-Atlagic, Hideki Tanaka, and Katsumi Kaneko

**17 Understanding of Unique Thermal Phase Behavior of Room Temperature Ionic Liquids: 1-Butyl-3-Methylimdiazolium Hexafluorophosphate as a Great Example** . . . . . 379
Takatsugu Endo and Keiko Nishikawa

**18 Single Molecular Spintronics** . . . . . 403
Toyo Kazu Yamada

**19 Vortex Lasers Twist Materials to Form Chiral Nanostructures** . . . . . 417
Takashige Omatsu, Nobuyuki Aoki, and Katsuhiko Miyamoto

**Index** . . . . . 429

# Contributors

**Nobuyuki Aoki** Graduate School of Advanced Integration Science, Chiba University, Chiba, Japan

**Christoph Bürker** Institut für Angewandte Physik, Universität Tübingen, Tübingen, Germany

**Fabio Bussolotti** Department of Nanomaterial Science, Graduate School of Advanced Integration Science, Chiba University, Chiba, Japan

**Steffen Duhm** Institute of Functional Nano and Soft Materials (FUNSOM), Soochow University, Suzhou, China

**Takatsugu Endo** University of California, Davis, CA, USA

**Takashi Fujikawa** Graduate School of Advanced Integration Science, Chiba University, Chiba, Japan

**Toshihiko Fujimori** Research Center for Exotic Nanocarbons (JST), Shinshu University, Nagano-City, Japan

**Alexander Gerlach** Institut für Angewandte Physik, Universität Tübingen, Tübingen, Germany

**Suzana Gotovac-Atlagic** Public Health Institute, Banja Luka, Bosnia and Herzegovina

**Alexander Hinderhofer** Department of Nanomaterial Science, Graduate School of Advanced Integration Science, Chiba University, Chiba, Japan

**Takuya Hosokai** Research Institute of Instrumentation Frontier (RIIF), National Institute of Advanced Industrial Science and Technology (AIST), Ibaraki, Japan

**Hisao Ishii** Center for Frontier Science, Chiba University, Chiba, Japan

**Hiroyuki Ishii** JST-PRESTO, University of Tsukuba, Ibaraki, Japan

Division of Applied Physics, Faculty of Pure and Applied Sciences, University of Tsukuba, Ibaraki, Japan

**Katsumi Kaneko** Center for Energy and Environmental Science, Shinshu University, Nagano, Japan

**Takashi Karatsu** Department of Applied Chemistry and Biotechnology, Graduate School of Engineering, Chiba University, Chiba, Japan

**Kengo Kato** Department of Nanomaterial Science, Graduate School of Advanced Integration Science, Chiba University, Chiba, Japan

**Satoshi Kera** Department of Nanomaterial Science, Graduate School of Advanced Integration Science, Chiba University, Chiba, Japan

**Fitri Khoerunnisa** Department of Chemistry, Indonesia University of Education, Bandung, Indonesia

**Hiroumi Kinjo** Department of Nanomaterials science, Graduate School of Advanced Integration Science, Chiba University, Chiba, Japan

**Norihisa Kobayashi** Graduate School of Advanced Integration Science, Chiba University, Chiba, Japan

**Kazuhiro Kudo** Graduate School of Engineering, Chiba University, Chiba, Japan

**Shi-Guang Li** Graduate School of Engineering, Chiba University, Chiba, Japan

Faculty of Automation and Information Engineering, Xi'an University of Technology, Shaanxi, China

**Shin-ichi Machida** Department of Nanomaterials Science, Graduate School of Advanced Integration Science, Chiba University, Chiba, Japan

**Ryosuke Matsubara** Graduate School of Materials Science, Nara Institute of Science and Technology, Nara, Japan

Graduate School of Engineering, Chiba University, Chiba, Japan

**Katsuhiko Miyamoto** Graduate School of Advanced Integration Science, Chiba University, Chiba, Japan

**Masakazu Nakamura** Graduate School of Materials Science, Nara Institute of Science and Technology, Nara, Japan

Graduate School of Engineering, Chiba University, Chiba, Japan

**Kazuki Nakamura** Graduate School of Advanced Integration Science, Chiba University, Chiba, Japan

**Takashi Nakayama** Department of Physics, Chiba University, Chiba, Japan

**Yasuo Nakayama** Department of Nanomaterials science, Graduate School of Advanced Integration Science, Chiba University, Chiba, Japan

**Kaori Niki** Graduate School of Advanced Integration Science, Chiba University, Chiba, Japan

**Keiko Nishikawa** Graduate School of Advanced Integration Science, Chiba University, Chiba, Japan

**Noboru Ohashi** Graduate School of Engineering, Chiba University, Chiba, Japan

Department of Applied Chemistry School of Engineering, Tohoku University, Miyagi, Japan

**Tomonori Ohba** Department of Chemistry, Graduate School of Science, Chiba University, Chiba, Japan

**Takashige Omatsu** Graduate School of Advanced Integration Science, Chiba University, Chiba, Japan

CREST Japan Science and Technology Agency, Tokyo, Japan

**Masatoshi Sakai** Graduate School of Engineering, Chiba University, Chiba, Japan

**Tomoya Sato** Department of Nanomaterials Science, Graduate School of Advanced Integration Science, Chiba University, Chiba, Japan

**Frank Schreiber** Institute for Applied Physics, University of Tübingen, Tübingen, Germany

**Tomoki Sueyoshi** Department of Nanomaterial Science, Graduate School of Advanced Integration Science, Chiba University, Chiba, Japan

**Hideki Tanaka** Department of Chemical Engineering, Kyoto University, Kyoto, Japan

**Yoko Tomita** Graduate School of Pure and Applied Sciences, University of Tsukuba, Ibaraki, Japan

Japan Science and Technology Agency, CREST, Tokyo, Japan

**Nobuo Ueno** Department of Nanomaterial Science, Graduate School of Advanced Integration Science, Chiba University, Chiba, Japan

**Qian Xin** School of Physics, Shandong University, Jinan, Japan

**Toyo Kazu Yamada** Graduate School of Advanced Integration Science, Chiba University, Chiba, Japan

**Hiroyuki Yamane** Institute for Molecular Science, Okazaki, Japan

**Keiichirou Yonezawa** Department of Nanomaterial Science, Graduate School of Advanced Integration Science, Chiba University, Chiba, Japan

# Part I
# Bridging Structure, Electronic States and Electrical Conductivity

# Chapter 1
# Fundamental Aspects and the Nature of Organic Semiconductor

**Nobuo Ueno**

## 1.1 Brief History and Technological Applications

In 1954 the term *organic semiconductor* was used for polycyclic aromatic compounds with a molecular structure similar to fragments of a graphite sheet (graphene) by Hiroo Inokuchi, when he confirmed the notion that such organic materials are electrically conductive [1] through a number of careful experiments by himself and other pioneers [2–5]; For history before 1988 see [6, 7]. As such, it is generally accepted that organic semiconductors were discovered in the mid-twentieth century. Following to their pioneering work in this period, much of the research concentrated on revealing the nature of the electrical conduction in molecular single crystals, which exhibited charge carrier mobility values of a few $cm^2/Vs$ at room temperature, and even much higher values at low temperature, as shown in the work of Karl et al. (see for example, [8]) So far, the highest mobility (40 $cm^2/Vs$) was reported for rubrene single crystals in organic field effect transistor (OFET) [9]. For development of organic devises pioneering works have also been performed on organic light emitting diode (OLED) [10], organic solar cell (OSC) [11] and OFET [12]. For practical device applications, however, organic semiconductor thin films, comprising evaporated small-molecule compounds [13] or polymers processed from solution, are more viable.

Since organic semiconductors were widely used as photoconductors in copiers and laser printers, they have recently gained increasing attention because of their potential applications in electronic and opto-electronic devices, such as OLED [13–17], OSC [18, 19] and OFETs [16, 17, 20–23]. OLEDs are already used in displays of mobile phones and TV, and are entering the commercial lighting market. As a result of the continuous drive to fabricate organic electronic devices

N. Ueno (✉)
Department of Nanomaterial Science, Graduate School of Advanced Integration Science, Chiba University, Inage-ku, Chiba 263-8522, Japan
e-mail: uenon@faculty.chiba-u.jp



H. Ishii et al. (eds.), *Electronic Processes in Organic Electronics*,
Springer Series in Materials Science 209, DOI 10.1007/978-4-431-55206-2_1

on lightweight, large-area plastic substrates by low-cost processing techniques, organic electronics is fast-tracked for applications that can overcome general energy problems and global warming. Following this trend, OSCs and OTFTs have developed rapidly over the past decade. Many potential applications of OFETs have been demonstrated, ranging from flexible displays [24] and sensor systems [23, 25] to radio frequency identification tags [26], and some of these systems are now close to commercialization.

Organic semiconductors have other unique physical properties that offer numerous advantages compared to their inorganic counterparts: (i) The extremely high absorption coefficient of many organic molecules in the visible wavelength range offer the possibility of very thin, and therefore resource-efficient, photodetectors and solar cells. (ii) Many fluorescent molecules emit light efficiently, where wavelength of the light depends on spatial spread of highest occupied molecular orbital (HOMO) and lowest unoccupied molecular orbital (LUMO) states. However, charge transport in organic semiconductors is often limited by low intrinsic carrier density and mobility. Therefore, controlled and stable doping, in analogy to doping of inorganic semiconductors for increasing carrier density, is desirable for reaching higher efficiency of many organic-based devices [17, 27, 28]. In addition, if one succeeds in shifting the Fermi level toward the transport states by doping or control of the interface structure, this could reduce Ohmic losses at contacts, improve carrier injection from electrodes, and increase the built-in potential of Schottky or p-n junctions.

To comprehend the recent progress and expansion of organic electronics we show the progress of efficiency of OSCs compared with different solar cell technologies in Fig. 1.1 (http://commons.wikimedia.org/wiki/File:PVeff(rev130528).jpg). Although the device performance is still not enough, the very rapid improvement of the performance can be seen in Fig. 1.1, which is undeniably related to the progress in synthesis of new functional molecules with desired electronic states, the science of interfaces and the availability of techniques to control film structure and interfaces.

Despite of the rapid progress of organic electronics, however, it is not easy to clearly understand essential difference between organic and inorganic semiconductors. In the following sections we summarize universal features of organic crystals/solids and discuss the nature of organic semiconductor. This summary would help to understand what we should study for revealing science of organic semiconductor that is different from inorganic semiconductor.

## 1.2 Universal Features of Molecular Crystals as the Nature of Organic Semiconductor

For years, it has been very difficult to answer the question, "what is key difference between organic semiconductor and inorganic counterpart?" As described in Sect. 1.1 there are clear answers from technological points and application use

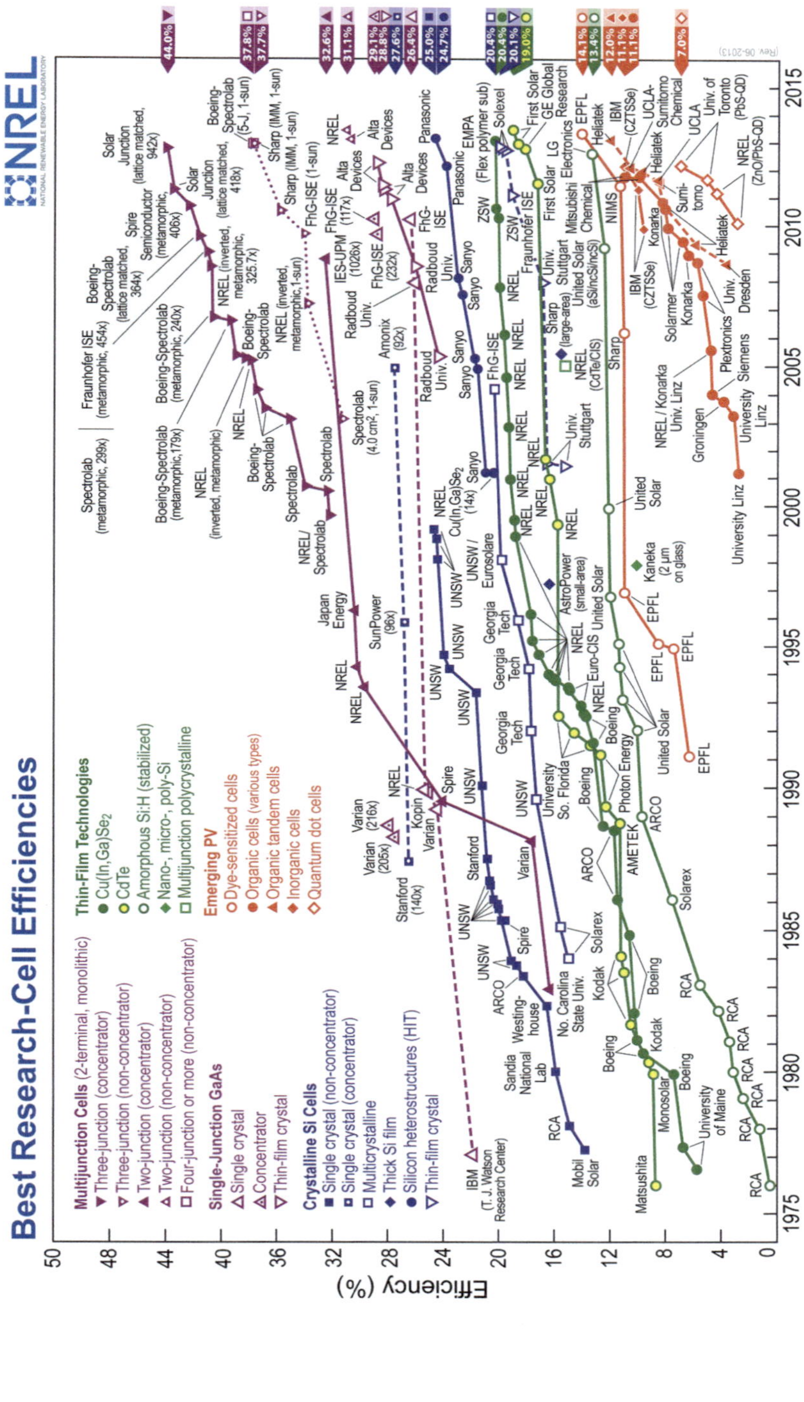


**Fig. 1.1** Efficiency progress of different solar cell technologies (May, 2013: *Source*: L.L. Kazmerski, National Renewable Energy Laboratory (NREL), Golden, CO) http://commons.wikimedia.org/wiki/File:PVeff(rev130528).jpg

for overcoming global energy and warming problems, which are well known advantages of organic devices comparing with inorganic ones. However, the difficulty of giving the answer exists in scientific points. We think that we are currently on the right way to finding the answer by the recent progress of both of experimental and theoretical studies. Here we summarize characteristics of organic semiconductors, namely those of molecular crystals, to find some peculiar scientific features that cannot be expected in inorganic semiconductors. Molecular crystals, which are in general electrical insulators but have semi-conductive function if they are used as active materials in various organic devices, have following universal features that are very different from inorganic solids. These can be considered as the nature of organic semiconductors and thus offer various new possibilities.

1. Intermolecular interaction in molecular crystals is very weak.
2. Molecular size is very large.
3. Symmetry of the molecular structure is very low (the structure is planar or complicated 3-dimensional structure etc.).
4. Each molecule consists of light elements (mainly H, C, O, N), but the molecular weight is very large [i.e. pentacene ($C_{22}H_{14}$) MW = 278; $C_{60}$ MW = 720; metal free phthalocyanine ($C_{32}H_{18}N_8$) MW = 514]. All of these molecules are much heavier than Si (atomic weight = 28) and available heaviest elements (atomic weight ~100)].

These characteristics may result in following properties (for brief summary see Fig. 1.2):

- Bandwidth is very narrow, wave function is localized and spatial distribution of frontier orbitals, HOMO and LUMO, are very bumpy. These features make the electronic states very sensitive to tiny changes in the packing structure of molecules and the crystal structure. (see for example Chaps. 2, 3, 4, 7 and 10),
- There are many local electric dipoles in each molecule, which are related to local chemical bonds and spatial distribution of relevant molecular orbitals. This gives molecular orientation dependent electronic properties [29],
- Organic semiconductor cannot be approximated by a continuum medium that is widely used in considering the energy levels at the interfaces and band bending in inorganic semiconductors,
- A weak perturbation (i.e. existence of interfaces with other materials, impurity or charge doping, gas exposure, heating, electric field etc.) may easily mediate imperfectness of molecular packing structure or structural disorder, yielding band gap states (see for example Chaps. 4, 7 and 10). This could be the most important origin of the Fermi level pinning (better to write as quasi Fermi level pinning),
- Tiny amount of excess changes injected into the packing structure can produce larger changes in the electronic structure than supposed (see for example Chap. 4),
- A weak perturbation (i.e. doping) may in some cases enhance a wave function localization to result in Mott-Hubbard insulator even at room temperature,

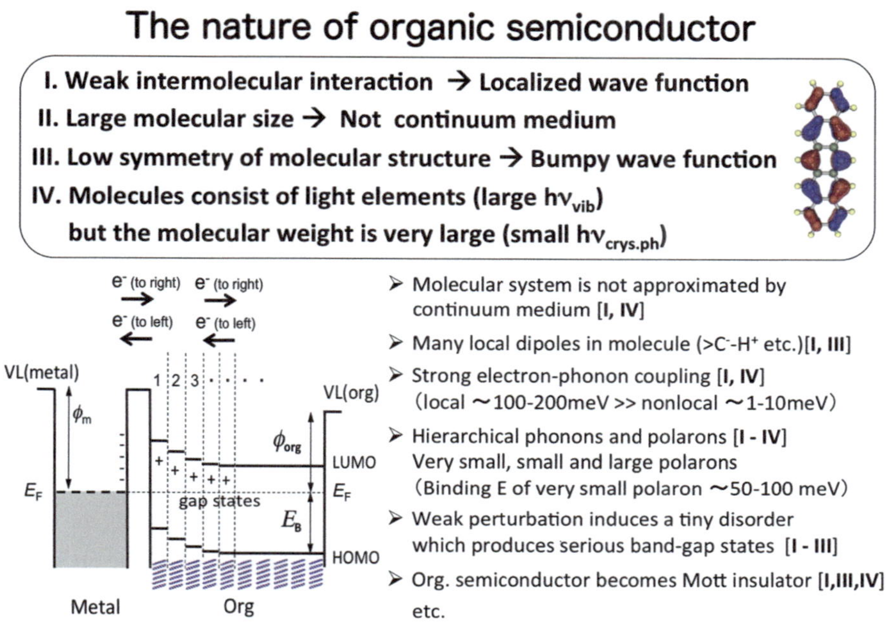


**Fig. 1.2** The nature of organic semiconductor from universal features of organic molecular crystals [I–IV]. Complicated but important properties of organic semiconductors come out from each of them and/or their interplay. Some of examples are written (see text). The energy diagram illustrates a weakly interacting metal/organic film system, where the interface dipole due to metal-molecule electronic interaction is assumed to be negligible. The organic film is not approximated by a continuum medium, which result in "band bending" like shift due to an energy-level jump at quasi interfaces in the molecular layers after achieving thermal equilibrium of the electron system (see the *inset* figure and Chap. 4, Fig. 4.1). The narrow bandwidth of HOMO and LUMO levels reflects weak intermolecular interaction (localization of wave functions). The HOMO of pentacene molecule is shown as an example of "bumpy" distribution of molecular wave function

i.e. hydrocarbon based conventional organic semiconductors may change into strongly correlated systems (Hubbard U becomes larger than the band width W) (see for example Chap. 5),

- Energies of local phonons (molecular-vibration related phonons) are much larger than those of nonlocal phonons (crystal phonons). This property is related also to formation of hierarchical polarons, electronic polaron and polalons related to geometrical structure changes [very small polaron/related to molecular vibration, small polaron (size<~ unit cell and related to crystal phonons) and large polaron (size>~ unit cell and related to crystal phonons)]. These polarons have different time scales of polarization/screening effects (see for example Chaps. 3 and 4),
- Coupling between the molecular-vibration-related phonons and HOMO/LUMO wave functions are very large (see for example Chap. 3).

- As the transfer integral ($t$) is so small that electron-local phonon (including molecular vibrations) coupling as well as electron-nonlocal phonon coupling may seriously affect $t$ (see for example Chap. 15),etc.

In addition to the above-described nature, there is another important advantage of using organic molecular systems, namely flexibility in synthesizing new functional molecules. This is the most well known advantage of organic molecules, thus has been successfully used in developing many new organic semiconductor molecules. This flexibility can also offer *self-synthesizing function,* which is not easy to realize by chemical reaction in solution so far. Evidence was reported recently [31, Chap. 6], where a new possibility to have metallic molecular layer is demonstrated experimentally and theoretically for non-interesting molecules when they adsorb on metal surfaces.

## 1.3 What Do We Need for the Next?

One can easily expect that all or some of above-described properties may appear not only in molecular crystals but in various other organic solids including polymers and bio-related molecular systems, thus study of organic semiconductor is of crucial importance not only for understanding organic semiconductor itself but also realizing high performance organic devices and study of electronic functions of polymers and bio materials etc.

Organic semiconductors commonly have strange or peculiar properties, which are originated from the features I–IV (Fig. 1.2), and thus are much different from those of inorganic counterparts. Accordingly, one observes mysterious phenomena that are not easy to understand with the knowledge from physics of inorganic semiconductors. Some of these are discussed in this book.

Also from this reason, we need to learn science of various phenomena in organic systems and to develop new experimental and theoretical methods suitable for organic semiconductors to unravel their mysterious properties and dig out new functions still hidden in organic molecule-based systems (see for example [30, 31]).

## References


1. H. Inokuchi, Bull. Chem. Soc. Jpn. **27**, 22 (1954)
2. D.D. Eley, Nature **162**, 819 (1948)
3. H. Akamatu, H. Inokuchi, J. Chem. Phys. **18**, 810 (1950)
4. H. Akamatu, H. Inokuchi, Y. Matsunaga, Nature **173**, 168 (1954)
5. H. Inokuchi, Org. Electron. **7**, 62 (2006)
6. H. Inokuchi, M. Sano, Y. Maruyama and N. Sato, Proceedings of the Oji international seminar on organic semiconductors – 40 years. Mol. Cryst. Liq. Cryst. **171**, 1–156 (1989)
7. M. Pope, C.E. Swenberg, *Electronic Processes in Organic Crystals and Polymers*, 2nd edn. (Oxford University Press, Oxford, 1999), pp. 337–340

8. N. Karl, *Defect Control in Semiconductors*, vol. 2 (North Holland: Amsterdam, 1990)
9. J. Takeya, M. Yamagishi, Y. Tominari, R. Hirahara, Y. Nakazawa, T. Nishikawa, T. Kawase, T. Shimoda, S. Ogawa, Appl. Phys. Lett. **90**, 102120 (2007)
10. C.W. Tang, S.A. Van Slyke, Appl. Phys. Lett. **51**, 913 (1987)
11. C.W. Tang, Appl. Phys. Lett. **48**, 183 (1986)
12. K. Kudo, M. Yamashita, T. Moriizumi, Jpn. J. Appl. Phys. **23**, 130 (1984)
13. S.R. Forrest, Chem. Rev. **97**, 1793 (1997)
14. L.S. Hung, C.H. Chen, Mater. Sci. Eng. R **39**, 143 (2002)
15. D. Berner, H. Houili, W. Leo, L. Zuppiroli, Phys. Status Solidi A **202**, 9 (2005)
16. Editorial, Chem. Mater. **16**(23), 4381–4846 (2004), Special issue on organic electronics
17. K. Walzer, B. Maennig, M. Pfeiffer, K. Leo, Chem. Rev. **107**, 1233 (2007)
18. P. Peumans, A. Yakimov, S.R. Forrest, J. Appl. Phys. **93**, 3693 (2003)
19. S.E. Shaheen, D.S. Ginley, G.E. Jabour (eds.), MRS Bull. **30**, 10–52 (2005), Special issue on organic based photovoltaics
20. C.D. Dimitrakopoulos, D.J. Mascaro, IBM J. Res. Dev. **45**, 11 (2001)
21. C.D. Dimitrakopoulos, P.R.L. Malenfant, Adv. Mater. **14**, 99 (2002)
22. Y. Wen, Y. Liu, Y. Guo, G. Yu, W. Hu, Chem. Rev. **111**, 3358 (2011)
23. Y.L. Guo, G. Yu, Y.Q. Liu, Adv. Mater. **22**, 4427 (2010)
24. Y. Nakajima, T. Takei, T. Tsuzuki, M. Suzuki, H. Fukagawa, T. Yamamoto, S. Tokito, J. Soc. Inf. Disp. **17**, 629 (2009)
25. T. Sekitani, Y. Noguchi, K. Hata, T. Fukushima, T. Aida, T. Someya, Science **321**, 1468 (2008)
26. R. Rotzoll, S. Mohapatra, V. Olariu, R. Wenz, M. Grigas, K. Dimmler, O. Shchekin, A. Dodabalapur, Appl. Phys. Lett. **88**, 123502 (2006)
27. A. Kahn, N. Koch, W.Y. Gao, J. Polym. Sci. B **41**, 2529 (2003)
28. Y. Gao, Mater. Sci. Eng. R **68**, 39 (2010)
29. S. Duhm, G. Heimel, I. Salzmann, H. Glowatzki, R.L. Johnson, A. Vollmer, J.P. Rabe, N. Koch, Nat. Mater. **7**, 326 (2008)
30. G. Heimel, S. Duhm, I. Salzmann, A. Gerlach, A. Strozecka, J. Niederhausen, C. Bürker, T. Hosokai, I. Fernandez-Torrente, G. Schulze, S. Winkler, A. Wilke, R. Schlesinger, J. Frisch, B. Bröker, A. Vollmer, B. Detlefs, J. Pflaum, S. Kera, K.J. Franke, N. Ueno, J.I. Pascual, F. Schreiber, N. Koch, Nat. Chem. **5**, 187 (2013). and see also Chap. 6
31. N. Koch, N. Ueno, A.T.S. Wee (eds.), *The Molecule-Metal Interface* (Wiley, 2013, Weinheim)

# Chapter 2
# Ultraviolet Photoelectron Spectroscopy (UPS) I: Band Dispersion Measurements of "Insulating" Organic Single Crystals

**Yasuo Nakayama, Steffen Duhm, Qian Xin, Satoshi Kera, Hisao Ishii, and Nobuo Ueno**

## 2.1 Introduction

Organic electronic devices owe their functionalities to the motion of the charge carriers in solid state organic semiconducting materials; in other words, development of the organic electronics is nothing but exploration of *smarter* ways to manipulate the electrons inside the organic solids as we wish. Generally speaking, behaviors of charge carriers inside a solid are dominated by its electronic structure, that is the molecular orbital energy distribution for ordinary organic materials or the electronic band structure for crystalline solids, and thus deep insight about the electronic structures of the functional organic materials has been not only of *academic* interests but also perpetually desired by the *practical* research and development field.

Photoelectron spectroscopy is an established methodology to demonstrate the electronic structures of matters. Particularly a sophisticated version of this technique, angle-resolved ultraviolet photoelectron spectroscopy (ARUPS), enables one to directly measure the valence band structures of crystalline samples. In order to adopt this useful method for organic molecular crystals, however, a serious technical drawback so called "sample charging" has to be overcome. In the present chapter, we explain fundamental aspects of this problem based on elemental

Y. Nakayama (✉) • S. Kera • N. Ueno
Graduate School of Advanced Integration Science, Chiba University, Chiba, Japan
e-mail: nkym@restaff.chiba-u.jp

S. Duhm
Institute of Functional Nano and Soft Materials (FUNSOM), Soochow University, Suzhou, China

Q. Xin
School of Physics, Shandong University, Jinan, Japan

H. Ishii
Center for Frontier Science, Chiba University, Chiba, Japan



H. Ishii et al. (eds.), *Electronic Processes in Organic Electronics*,
Springer Series in Materials Science 209, DOI 10.1007/978-4-431-55206-2_2

principles of the photoelectron spectroscopy. In addition, several examples to access the 'valence bands' of the organic semiconductor crystals through successful settlement of the sample charging are introduced.

## 2.2 How to Measure the Band Dispersion in Organic Solids

In this section, we describe primary essences of an experimental technique, photoelectron spectroscopy, to directly probe the electronic structures of matters. The aim of this article is to provide a concise guide of the ways how to demonstrate the 'valence bands' of the organic solids. Thus we mostly spotlight some special technical requirements to conduct reliable measurement on crystalline organic samples. More general and accurate theoretical instruction on this technique will be given in the later chapters.

Photoelectron spectroscopy is an experimental technique to measure the kinetic energy and momentum of "photoelectrons" ejected from the sample caused by a photo-excitation event. The primary requirement of this technique is conservation of the energy and momentum throughout this event. Since the *observable* values for us are the kinetic energy and momentum of the photoelectron at an analyzer position, that requirement has to be fulfilled through an overall measurement system. This principle is illustrated in Fig. 2.1a. In the ordinary measurement condition, the photoelectron analyzer is electrically in contact to the sample; this means that the Fermi level ($E_F$) of each side aligns. The vacuum levels of the sample and analyzer therefore do not agree with each other because the work function of the sample ($\varphi_s$) is generally not equal to that of the analyzer ($\varphi_a$). The contact potential difference ($\varphi_s - \varphi_a$) accelerates the photoelectrons while they are travelling from the sample surface toward the analyzer. The overall energy conservation can therefore be written as follows;

$$E_b = h\nu - E_k{}^{out} - \varphi_s = h\nu - E_k{}^{obs} - \varphi_a, \tag{2.1}$$

where $E_b$ is the binding energy of the target electronic state with respect to the Fermi level, $h\nu$ is the excitation photon energy, $E_k{}^{out}$ is the kinetic energy of the photoelectrons just out of the sample surface, and $E_k{}^{obs}$ is its *observed* kinetic energy at the analyzer. Since $\varphi_a$ is a fixed value independent of the sample properties, $E_b$ can be uniquely traced back from $E_k{}^{obs}$ by substituting the known $h\nu$ into Eq. (2.1).

In a practical point of view, Eq. (2.1) implicitly manifests the following two technical restrictions. Firstly, the UPS results obtained in the ordinary condition do not give us any knowledge about the sample work function $\varphi_s$. The second point is leveling of the "Fermi energy" throughout the sample.

In order to determine $\varphi_s$ from the UPS measurements, a standard method is to apply a negative voltage to the sample to lift its vacuum level position sufficiently above that of the analyzer. In this condition, the low kinetic energy end of a spectra give a massive inflation mainly caused by secondary electron emission which is terminated by a steep cut-off at a certain kinetic energy. As schematically shown in

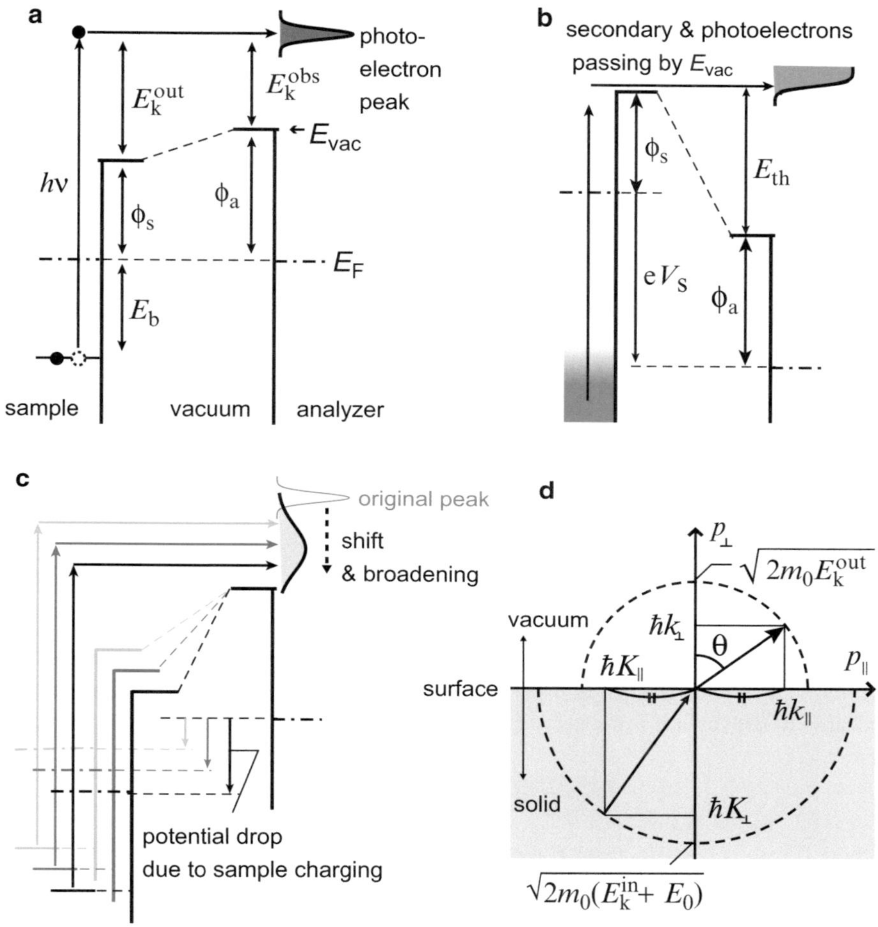


**Fig. 2.1** (**a**) A schematic diagram representing the energy conservation relationship throughout the photoelectron spectroscopy measurements. (**b**) An energy level diagram of the measurement condition for work function determination by UPS. (**c**) The substance of the sample charging problem. (**d**) A schematic diagram of the momentum conservation relation during the external photoemission event

Fig. 2.1b, $\varphi_s$ can be determined from this secondary electron cut-off (SECO) position ($E_{th}$) as follows;

$$\varphi_s = \varphi_a + E_{th} + eV_s, \tag{2.2}$$

where $e$ is the elementary charge and $V_s$ is the sample bias voltage (usually $V_s < 0$). This evaluation is in principle independent of the excitation energy.

The second requirement is automatically fulfilled for well-conductive samples such as metals. However, in the case of most organic solids, the situation depends on a balance of the total photoemission amount and electric conductance between the *front* surface and backside of the sample. If the former is too large and/or the latter is insufficient, a potential drop of notable magnitude across the sample to its thickness direction is induced by photoemission. This potential variation is commonly inhomogeneous over the surface because of a local fluctuation of the sample condition and thus causes broadening and distortion of the spectra (Fig. 2.1c). This technical drawback is known as the sample charging problem. Available solutions on this issue will be given in the later part of this section.

The photoelectrons ejected out of the sample satisfy an energy-momentum (*E*-*k*) dispersion relation of the free electron,

$$E_{\mathrm{k}}{}^{\mathrm{out}} = \hbar^2 |\boldsymbol{k}|^2 / 2m_0, \tag{2.3}$$

where $\boldsymbol{k}$ is the momentum of a photoelectron outside the sample and $m_0$ is the free electron mass. Inside the solids, in case the excitation energy is sufficiently large ($E_{\mathrm{k}} > 10$ eV), the *E*-*k* dispersion relation can also be approximated as free electron-like;

$$E_{\mathrm{k}}{}^{\mathrm{in}} - E_0 = \hbar^2 |\boldsymbol{K}|^2 / 2m^*, \tag{2.4}$$

where $\boldsymbol{K}$ and $m^*$ are the momentum and effective mass of a photoelectron inside the sample, respectively, and $E_0$ is an *adequate* energy standard that corresponds to the bottom energy of the *pseudo*-parabolic dispersion. The momentum component parallel to the surface is conserved when the photoelectrons pass across the surface;

$$\hbar K_{||} = \hbar k_{||} = \left(2m_0 E_{\mathrm{k}}{}^{\mathrm{out}}\right)^{1/2} \sin\theta, \tag{2.5}$$

where θ is the photoelectron take off angle from the surface normal. In contrast, the surface normal component is modulated by an effective potential barrier which is called as the inner potential ($U_0 = E_0 + \varphi_{\mathrm{s}}$). The relation can be written as;

$$\hbar K_{\perp} = \left(\hbar^2 k_{\perp}{}^2 + U_0\right)^{1/2} = \left(2m_0 E_{\mathrm{k}}{}^{\mathrm{out}} \cos^2\theta + U_0\right)^{1/2}. \tag{2.6}$$

The above relations between $\boldsymbol{k}$, $\boldsymbol{K}$, $E_{\mathrm{k}}{}^{\mathrm{out}}$, and θ is geometrically illustrated in Fig. 2.1d. For the UPS case ($h\nu \ll 1$ keV), $\boldsymbol{K}$ can be regarded to be identical to the momentum of the electrons before the photoexcitation event, which we want to know, because the momentum of the excitation photons is minor enough in comparison to $\boldsymbol{K}$. Hence the *E*-*k* dispersion relation can be traced back from the observable kinetic energy of the photoelectrons detected at a certain detection angle[1]; this methodology is generally called as angle-resolved ultraviolet

[1] Strictly speaking, the *E*-*k* set of a photoelectron is slightly modulated during its flight from the sample toward the analyzer due to the contact potential difference. This modulation is generally sufficiently small and thus can be neglected in practice.

photoelectron spectroscopy (ARUPS). According to Eqs. (2.1) and (2.5), the parallel component can be derived by simply varying θ at an adequate $h\nu$ through the following relationship;

$$K_{||} = [2m_0(h\nu - E_b - \varphi_s)]^{1/2}/\hbar \sin\theta \approx 0.511\,(h\nu - E_b - \varphi_s)^{1/2}\sin\theta, \quad (2.7)$$

where the unit of the wave number is taken in $Å^{-1}$ hereafter in this chapter. On the other hand, $K_\perp$ is ordinary measured by changing $h\nu$ at the normal emission geometry ($\theta = 0$). In this case, Eq. (2.6) is rewritten as;

$$K_\perp = [2m_0(h\nu - E_b - \varphi_s) + U_0]^{1/2}/\hbar. \quad (2.8)$$

The explication given in last paragraph also requires the charge neutrality throughout the specimen; otherwise the momentum of a photoelectron is also disturbed by an unknown electrostatic field of *un-canceled* charge and does not convey any usable information of the initial state. In a viewpoint to regard the whole measurement system to be a closed electric circuit, the potential drop $\Delta V_s$ across the thickness direction of a specimen can be *phenomenologically* approximated as;

$$\Delta V_s \sim j_s \rho d, \quad (2.9)$$

where $j_s$ is the current density caused by the photoemission event, ρ is the electric resistivity of the material, and $d$ is the vertical distance between the sample surface and metallic 'substrate'.[2] The electrostatic potential difference across the sample has to be negligibly small (practically less than the energy resolution of the measurement system) for reliable photoelectron spectroscopy measurements. Hence, in order to resolve the sample charging problem for *insulating* materials of insufficient ρ, additional tactics to reduce at least one of these three parameters in the right term of Eq. (2.9) are required.

The simplest way to suppress $j_s$ is a reduction of the excitation photon flux, which is commonly taken care of but on the other hand is in a trade-off relationship with a diminution of the signal intensity. A widely adopted way in this direction is substantial cancelation of $j_s$ by a reverse electric current; compensation of the photoemission-induced positive charge by using a low-energy electron beam from a "flood gun". This technique is, however, generally not suitable for organic samples because these materials are easily damaged by the electron irradiation via chemical decomposition or polymerization. Another tactic that has been most commonly taken for organic materials is shrinkage of $d$, that is preparing an ultra-thin film thinner than a few ten nanometer-thick as a specimen. Although it

[2] This approximation is valid only for estimating a necessary condition to fulfill the sufficiently small $\Delta V_s$ without any sample charging. A 'charged' insulating sample has to be modeled as a parallel component of a resistor and capacitor rather than a single resistor if one wants to guess the magnitude of $\Delta V_s$ for a certain measurement condition.

is not always possible to produce well-ordered crystalline films, this approach has succeeded in demonstration of the valence band electronic structures of several organic semiconductors by means of ARUPS. In the present article, we briefly introduce several examples of successful valence band measurement on crystalline thin-films of π-conjugated organic molecules. On the other hand, in the case of organic single crystals, it is not easy to control $d$ of the sample as intended, accordingly reduction of $\rho$ has been challenged as the only possible solution. Our group corroborated that enhancement of the photoconductivity is an applicable way for this purpose. The ARUPS results on the *bulk* single crystal of rubrene (5,6,11,12-tetraphenyltetracene) obtained under the illumination of a continuous-wave (cw) laser light is described as an example.

## 2.3 From Thin Film to Single Crystal

The valence band dispersion of organic materials was initially investigated on monolayer films of aliphatic molecules with a polyethylene-like backbone structure [1–3], where each single molecule can be regarded as a one-dimensional 'micro-crystal' consisting of repeating ethylene groups. Since these 'crystalline units' are bound each other by the covalent bonding, the *intra*-molecular band dispersion of such molecules is much wider (typically ~5 eV) than the *inter*-molecular ones in van-der-Waals molecular crystals and thus is technically easier to be probed by the photoemission.

The first successful demonstration of the *inter*-molecular band dispersion of single-component organic semiconductors was accomplished on a few nanometer-thick film of bis-(1,2,5-thiadiazolo)-p-quinobis(1,3-dithiole) (BTQBT) by Hasegawa and co-workers [4]. The BTQBT molecules form a well-oriented thin film on the highly-oriented pyrolitic graphite (HOPG) substrate, and they conducted the excitation energy dependent UPS measurements at the normal emission geometry to indicate the valence band dispersion along the π-stacking direction at room temperature (290 K). As shown in Fig. 2.2a, the photoemission peak derived from the highest-occupied molecular orbital (HOMO) of BTQBT exhibited notable shift depending on $h\nu$; it took the maximum and minimum ionization energy ($E_b + \varphi_s$) values at $h\nu = 20$ and 48 eV, respectively. This phenomenon is a direct evidence of the $E$-$k$ dispersion along the surface normal direction. The vertical component of the electron wave vector in the HOMO band (valence band) can be determined from this $h\nu$-$E_b$ relationship by using Eq. (2.8). They found that this $E_b$-$K\perp$ relation can be fit by a simple cosine function under an adequate assumption for the $U_0$ value. This indicates that the valence band dispersion can be expressed by the one-dimensional tight-binding (1D-TB) approximation as;

$$E_b = E_c - 2t\cos(aK), \tag{2.10}$$

where $E_c$ is the energy of the band center, and $t$ is the transfer integral; $a$ and $K$ are the lattice constant and electron wave vector component along the measured

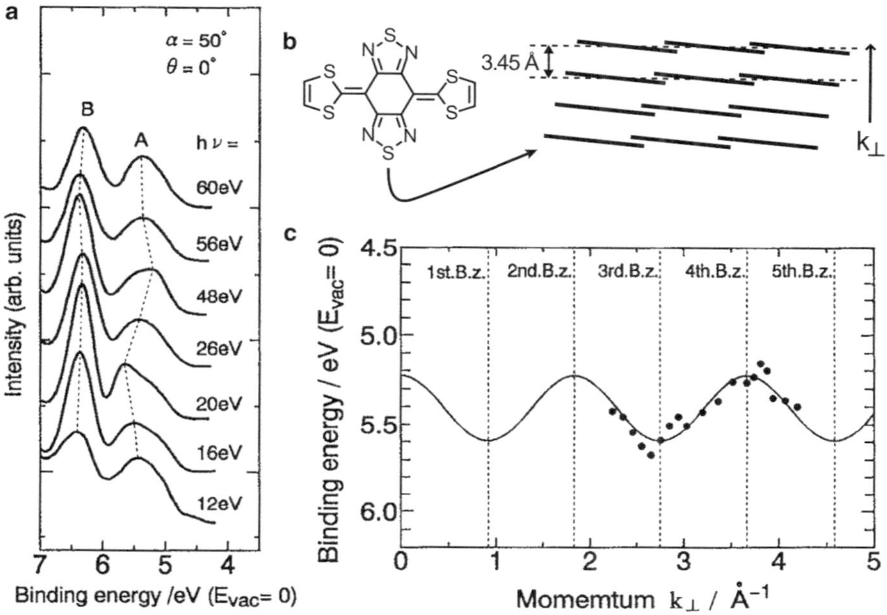


**Fig. 2.2** (**a**) Photon energy dependence of the normal emission UPS spectra for the BTQBT thin film. The abscissa is taken with respect to the vacuum level (ionization energy). (**b**) The molecular structure and a cross-sectional view of the crystalline π-stacking of BTQBT (the *bars* indicate the molecular plane). (**c**) The experimental $E_b$-$K\perp$ relationship (*filled circles*) and their best fit curve (*solid line*) of the BTQBT crystalline film. The viewgraphs (**a**) and (**c**) are reproduced from [4]

direction, respectively. The best fit of the topical data set to Eq. (2.10) was achieved as shown in Fig. 2.2b when $U_0 = 12.5$ eV, $E_c = 5.4$ eV, $t = 0.092$ eV, and $a = 0.34$ nm were input, where the last parameter $a$ agrees well to an expected value (0.345 nm) from the known crystal structure [5].

The obtained magnitude of $t$ is quite large for the van-der-Waals molecular solids. The dispersion width of the valence band, which is equal to $4t$, reaches as wide as 0.4 eV. Since this band width is much larger than the thermal energy (0.025 eV in this case), a necessary condition for the band transport is fulfilled. In this framework, the drift mobility of a conducting hole $\mu_h$ is described as;

$$\mu_h = e\tau / m_h{}^*, \tag{2.11}$$

where $\tau$ and $m_h$* are the relaxation time and effective mass, respectively, of the carrier. The latter value is derived from the curvature of the valence band at the topmost part. For a cosine-shaped $E$-$K$ dispersion as the present case, it can be written as;

$$m_h{}^* = \hbar^2 / \left(\mathrm{d}^2 E / \mathrm{d}K^2\right) = \hbar^2 / 2ta. \tag{2.12}$$

In this case, $m_h^* \approx 3.1 m_0$ was obtained. Although the accurate value of $\tau$ is not known, the time-energy uncertainty relationship gives its shortest limit as;

$$\tau > \hbar / k_B T, \tag{2.13}$$

where $T$ is the sample temperature. Substitution of Eqs. (2.12) and (2.13) into (2.11) gives the lower limit of the expected *ideal* mobility[3] of this molecular film as [6];

$$\mu_h > 20\,(m_0/m_h^*)\,(300/T) \approx 6.5\,\left[\mathrm{cm^2V^{-1}s^{-1}}\right]. \tag{2.14}$$

This value is considerably large for an organic semiconductor, and in practice this molecule is regarded as a promising material for a vertical field effect transistor.

As another example of the valence band measurements on the organic semiconductor thin films, we introduce a series of the ARUPS results on pentacene crystalline films. Pentacene is one of the most prototypical and extensively studied organic semiconductors owing to its considerably high charge carrier mobility [7, 8]. As an *electronic* origin of such an excellent transport property, its valence band structure has been attracting much interest. The first successful demonstration of the valence band dispersion of pentacene was achieved by Koch and co-workers [9]. They fabricated a crystalline stack of almost flat-lying pentacene molecules onto a HOPG substrate (Fig. 2.3a) and conducted ARUPS measurements at the normal emission geometry with various $h\nu$ to map out the valence band structure to the vertical direction. In the similar manner to the case of BTQBT described above, the band width were found to be 0.24 ($\pm$0.05) eV at 120 ($\pm$10) K as shown in Fig. 2.3, where they determined the lattice parameter perpendicular to the surface by x-ray diffraction to be 0.37 ($\pm$0.02) nm and assumed $U_0 = 5$ eV. Interestingly, the band width shrank to 0.19 ($\pm$0.05) eV at room temperature presumably owing to increased population of intra- and inter-molecular phonons. In contrast to this obvious energy dispersion toward the surface normal direction, only an insignificant energy shift of the HOMO peak depending on the electron take-off angle was observed by ARUPS at a fixed excitation energy ($h\nu = 25$ eV).[4] An absence of the energy band dispersion parallel to the surface is quite reasonable for a pentacene crystalline film of the flat-lying molecular orientation because of an insufficient intermolecular $\pi$-orbital overlap in the molecular plane. In contrast, Yamane and co-workers found a distinct *in-plane* two-dimensional energy band dispersion in the flat-lying pentacene monolayer on Cu(110) by means of ARUPS [10].

---

[3] The UPS measurements are much more insensitive to structural imperfectness (such as impurities and lattice defects) than *electric* characterizations in general. Therefore the mobility estimated from the ARUPS results has to be regarded as that of a *perfect* crystal; the *practical* mobility exhibited by an actual sample is more or less reduced with respect to this ideal value. In fact, an experimentally observed hole mobility in a BTQBT single crystal was ~4 $\mathrm{cm^2V^{-1}\,s^{-1}}$ [29] which is a little smaller than the expected value in Eq. (2.14).

[4] For systems with a clear $E$-$K_\perp$ dispersion, an energy shift in the ARUPS spectra cannot always be attributed to an $E$-$K_\parallel$ dispersion singly but also may be influenced by the $K_\perp \cos\theta$ component.

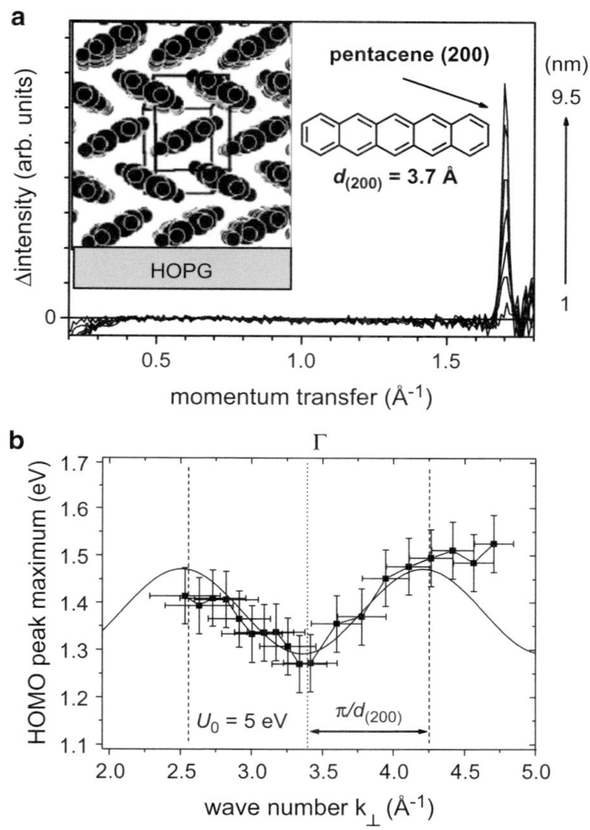


**Fig. 2.3** (**a**) X-ray diffractogram of the pentacene crystalline film on HOPG for increasing thickness. (*Inset*) A schematic *side view* of the molecular orientation and crystallographic unit cell of this film. The molecular structure of pentacene is also shown. (**b**) The experimental $E_b$-$K_\perp$ relationship at 120 K and its 1D-TB fitting curve. Reproduced from [9]

They attributed the origin of this band to an interface state caused by the hybridization between the molecular orbital and the wave function of the substrate surface. This indicates that a strong interaction between a molecular orbital and the electronic state of the substrate sometimes bring about the in-plane energy band dispersion.

Practically, pentacene exhibits several crystalline polymorphs and theoretical calculations have predicted that the electronic band structures, so as to the charge carrier transport properties, vary depending on the crystal lattice structures. So far, three crystallographic phases have been established; "thin film", "bulk", and "single crystal" phases. As a guide for an innovation of practical high-performance devices, accurate one-to-one correspondence of these three crystalline phases to the electronic band structures is keenly desired. Yoshida and Sato presented clear comparison of the expected band structures of these phases based on a band calculation, and also experimentally showed that the pentacene films of the "thin film" and "bulk" phases exhibited different HOMO peak profiles in the angle-integrated UPS spectra reflecting the respective $K$-integrated

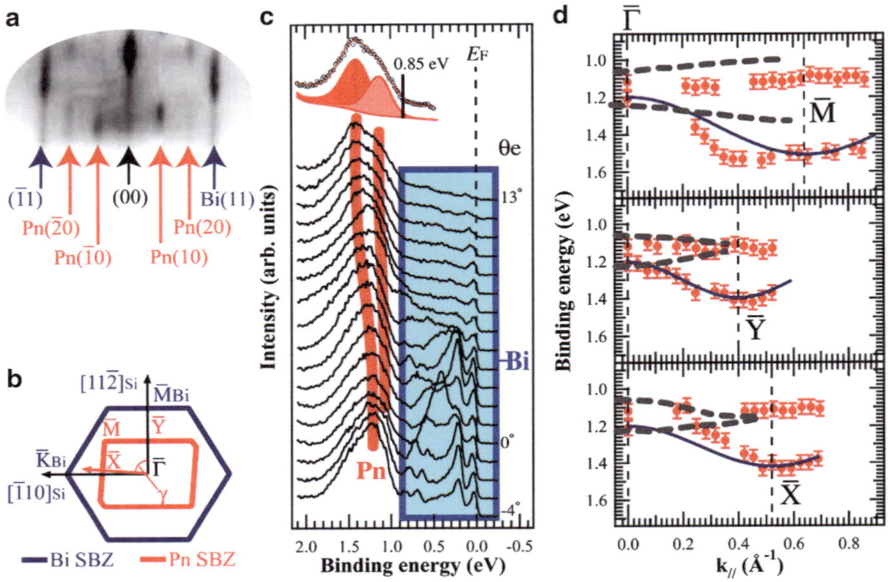


**Fig. 2.4** (**a**) RHEED pattern of the one-monolayer pentacene crystalline film on Bi/Si(111). (**b**) A schematic drawing of SBZs of the pentacene film and Bi(001) substrate. (**c**) ARUPS spectra of the pentacene film on Bi(001) measured at 140 K along the Γ-M direction. The peak separation of the low and high $E_b$ bands is displayed at the upper-most part. (**d**) The $E_b$-$K_{\|}$ relationships to three symmetric directions of the pentacene film. 1D-TB fitting results for the high $E_b$ band (*solid lines*) and band calculation results (*dashed lines*, details see the original article) are also represented. Reproduced from [13]

density-of-states (DOS) distributions [11]. Although knowledge of an angle-integrated DOS as well can provide a piece to solve the puzzle of the charge carrier transport [12], more *direct* evidence of the band structure is necessary to go into the detail of the transport phenomena.

The first *one-to-one* demonstration of the crystal lattice structure and valence band measurement of a pentacene crystalline film was achieved by Kakuta and co-workers [13]. They fabricated a crystalline film of the one-monolayer pentacene molecules onto a single crystalline film of Bi(001) pre-coating over a Si(111) substrate. In this system, contrary to the aforementioned Cu(110) case, the pentacene molecules take the upright-standing molecular orientation. The surface lattice parameters were determined by the reflection high-energy electron diffraction (RHEED), which suggested the molecular packing in this film was that of the "bulk" phase (Fig. 2.4a, b). They conducted the ARUPS measurement on this film at 140 K along three symmetric directions of the surface Brillouin zone (SBZ) to map out the two dimensional valence band structure as shown in Fig. 2.4c. The valence band was found to split into two as a result of two unequivalent pentacene molecules within the single unit cell, and the high $E_b$ one exhibited the wider energy dispersion for all symmetry directions (Fig. 2.4d). The maximum energy dispersion was 0.33 (±0.04) eV along the Γ-M direction, and $m_h$* was evaluated to be 1.26 $m_0$ under the 1D-TB approximation. As mentioned above, Eq. (2.14) gives a rough estimation for the expected in the broad band model; $\mu_h > 34\ cm^2V^{-1}\ s^{-1}$ was derived in this case. This value of the charge carrier mobility is considerably large, and the authors concluded that the band-like transport plays a role in the bulk phase of pentacene. This result also corroborated that the band-width depends on the crystalline direction. The two-dimensional valence band structure of the bulk phase pentacene film was more precisely investigated by Hatch and co-workers [14]. They illustrated a significant anisotropy of intermolecular transfer integral and hole effective mass values based on the two-dimensional (2D) TB approximation. In addition, this group also conducted the ARUPS measurements at various sample temperature to quantify the magnitude of electron-phonon coupling strength [15]. From a careful analysis for the thermal broadening of the valence band feature, they concluded that the electron-phonon coupling enhances the effective mass by 36 %, and its origin, related to the ARUPS measurements, has to be ascribed to the *inter*-molecular vibrations taking a smaller Einstein energy (~0.01 eV) in comparison to the cutoff for the intra-molecular ones into consideration.

On the other hand, pentacene mostly exhibits the "thin-film" phase at the vicinity of insulator substrates such as silicon dioxide. Therefore the valence band structure for this crystalline structure is of the highest interest in direct correlation to the field effect transistor applications. Ohtomo and co-workers succeeded in preparation of a "single domain" thin-film phase pentacene overlayer onto a bismuth-passivated Si (111) substrate with regularly bunching steps of 2–3 nm-high through a graphoepitaxial approach [16]. They carried out the ARUPS measurements on this pentacene thin film at 130 K to illustrate the 2D valence band structure. The intermolecular transfer integral values to ten molecules were evaluated under the 2D-TB approximation, and a relatively isotropic effective mass was derived, which is strikingly in contrast to the case of the bulk phase. They also observed a declination of the valence band dispersion with increasing the sample temperature to 300 K, however, interestingly, the energy dispersion did not disappeared isotropically but it survived only for one direction (Γ-X) [17]. A strong anisotropy for possible de-coherence of the wave function was proposed as a reason of this finding which was also in a relation to the anisotropic intermolecular transfer integrals.

While aforementioned achievements on the crystalline thin-films have provided a key to open a door for accessing the fundamental insights of the charge carrier behaviors in the organic semiconductors, organic *bulk* single crystals have been a fascinating target to unveil the *intrinsic* materials properties. The first successful photoemission results were achieved by Sato and co-workers in the early 1980s on *bulk* single crystal samples of several polyacenes and their derivatives [18]. Their way to avoid the sample charging was lessening the effective resistivity ρ of the sample by a combination of capping of an ultra-thin gold film of the crystal surface and enhancement of the photoconductivity by white-light illumination. Another successful work was reported by Vollmer and co-workers [19]. They shined a single

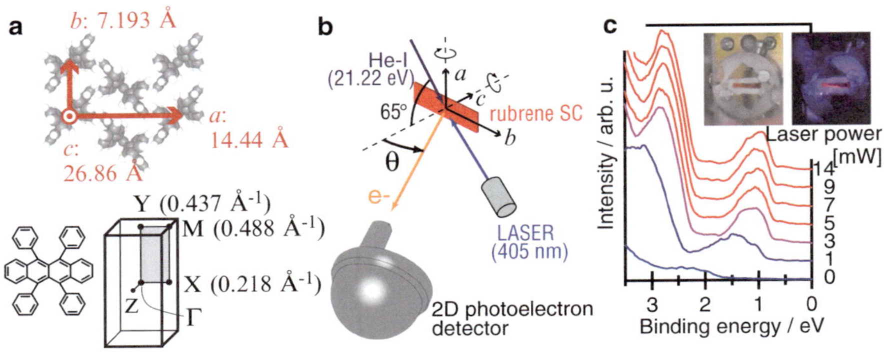


**Fig. 2.5** (**a**) The crystal lattice constants (at 293 K) [20], molecular structure, and Brillouin zone shape of rubrene. The *shaded area* corresponds the SBZ of the rubrene single crystal samples. The molecular arrangement of the rubrene single crystal surface is also shown. (**b**) A schematic drawing of the experimental setup of the ARUPS measurement on the rubrene single crystal. (**c**) Evolution of angle "integrated" UPS spectra of a rubrene single crystal depending on the additional laser power (405 nm). (*Inset*) Photographs of the sample without and with the laser illumination. The rubrene crystal (a *red* stripe at the center) was bound onto a piece of Au/Si by conductive Ag paste surrounding the crystal in order to enhance the electric contact as possible

crystal of pentacene with a cw laser for the sake of activating the electron injection into the crystal to cancel the photoemission-induced holes by the internal photoemission effect. These works were, however, still abortive for demonstrating the valence band structures by means of ARUPS.

Our group attempted to demonstrate the valence band measurement of the bulk single crystal of rubrene, whose molecular structure and crystallographic properties are presented in Fig. 2.5a [20]. Firstly, the ionization energy, which corresponds to the energy position of the valence band maximum with respect to the vacuum level, was determined by means of photoelectron yield spectroscopy (PYS) [21]. This technique, however, is an angle-integrated one and thus is not applicable to survey detailed structures of the valence band. UPS by ordinary measurement conditions was abortive because of severe sample charging. We discovered that this situation can be improved dramatically by illumination of the crystal by an additional cw laser light during the UPS measurements [22], where the experimental setup is presented in Fig. 2.5b. As shown in Fig. 2.5c, a photoemission peak derived from the HOMO of rubrene was clearly resolved under the additional light of sufficient intensity suggesting successful suppression of the sample charging, while the spectral profile exhibited apparent broadening accompanied by a rightward energy shift when the crystal was not illuminated or the laser power was not high enough.

The valence band measurement was then demonstrated by means of ARUPS in this condition with the sufficiently intense laser light illumination. Figure 2.6a shows the ARUPS spectra of the rubrene single crystal along the crystalline ***b*** axis which is identical to the Γ-Y direction of SBZ. A prominent energy shift of the

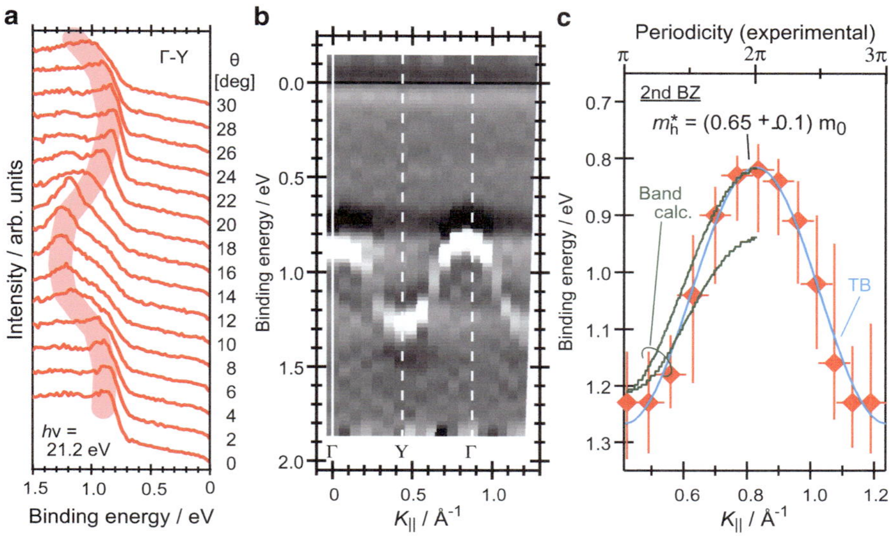


**Fig. 2.6** (**a**) ARUPS spectra of the rubrene single crystal taken along the Γ-Y direction. The *thick line* is a guide for eyes indicating the shifting trend of the HOMO peak position. (**b**) The second-derivative of the ARUPS spectra mapped onto the $E_b$-$K_{||}$ plane. The wave number corresponding the critical points (Γ and Y points) are also indicated. (**c**) The $E_b$-$K_{||}$ relationship of the valence band of rubrene along the Γ-Y direction within the second Brillouin zone. The least-square fitting result under the 1D-TB approximation is overlaid as a *blue curve*. A predicted band structure by a theoretical calculation [23] is also shown as *green curves*

HOMO peak was found depending on the electron take-off angle θ. This spectral series is re-traced on the $E_b$-$K_{||}$ plane as shown in Fig. 2.6b, where θ is transposed into $K_{||}$ by Eq. (2.7). The bright part, which corresponds to the HOMO peak, exhibited a periodic variation in the binding energy as a function of $K_{||}$, and its periodicity agreed with the SBZ size of this molecular crystal. This is a clear evidence for construction of an *inter*-molecular electronic band. The $E_b$-$K_{||}$ relation of this valence band structure was mapped out in Fig. 2.6c. This relation can be reproduced by the 1D-TB approximation, and $t = 0.11$ eV and $m_h{}^* = 0.65(\pm 0.1)m_0$ was obtained. It is noteworthy that this $m_h{}^*$ is the known lightest value for the organic semiconductors. The estimated band width is equivalent or even wider than theoretical predictions [23, 24]. On the other hand, a notable discrepancy between this result and calculations was found in particular at the Γ point; the calculations predict two bands dispersing for the same direction to degenerate at the Y point, as indicated in Fig. 2.6c, whereas this second band was not resolved in the present ARUPS spectra. This mismatch was also confirmed by another group with a different measurement technique [25]. One possible reason for this *invisible* second band can be ascribed to a thermal fluctuation to level off the inequivalency of two molecules in the unit cell. To clarify this point, low temperature ARUPS measurements are currently in progress.

Meanwhile, Eq. (2.14) gives the lower limit of the *ideal* $\mu_h$ of the rubrene single crystal from the present $m_h{}^*$ to be 29 cm$^2$V$^{-1}$ s$^{-1}$. As the largest *practical* $\mu_h$ of the rubrene single crystal at room temperature reported so far is ~40 cm$^2$V$^{-1}$ s$^{-1}$ [26], this requirement is actually fulfilled. Reversely, one can estimate the relaxation time $\tau$ from this known mobility value as; $e\tau/m_h{}^* > 40$ [cm$^2$V$^{-1}$ s$^{-1}$], then; $\tau > 15$ [fs], where this sign of inequality means that the *practical* mobility always must be smaller than the *ideal* value. Since the energy uncertainty corresponding to this lower limit of $\tau$ is still ten times smaller than the band width, a requirement for the band transport is not violated. The coherent length of the charge carrier $l_h$ can be derived from $\tau$ as;

$$l_h = u\tau, \tag{2.15}$$

where $u$ is the speed of the charge carrier. If one simply assumes the Maxwell-Boltzmann distribution for $u$, where the mean speed of the charge carriers $\bar{u}$ can be written to be $(3k_BT/m_h{}^*)$, the following value is derived for the lower limit of the *average* coherent length as; $\bar{l}_h = \bar{u}\tau > 2.1$ [nm], which is at least a few times greater than the crystalline lattice constant to this direction (0.719 nm). This strongly suggests delocalization of the charge carrier over several molecules even at room temperature.

As an extension of this accomplishment, we probed the valence band structure of the rubrene single crystal in a three-dimensional manner [27]. For the vertical direction, no energy shift of the HOMO peak was corroborated independent of the excitation energy, which is a clear evidence of an absence of the valence band dispersion. ARUPS measurements conducted to three directions within the crystalline plane evinced a clear anisotropy of the valence band dispersion as shown in Fig. 2.7a. To the Γ-X direction, which is orthogonal to the most conductive axis (Γ-Y), the valence band is substantially flat without any energy dispersion. In contrast, significant energy dispersion was observed to the diagonal direction of the SBZ (Γ-M); in comparison to the most conductive Γ-Y direction, the dispersion width was equivalent whereas the *turnaround* angle was farther from the normal emission reflecting a wider Brillouin zone size. From these ARUPS spectra, the 2D valence band *E-K* relation can instantaneously be mapped out as shown in Fig. 2.7b. The 2D-TB approximation reproduces this valence band structure satisfactory, and the *nearest-neighbor* intermolecular transfer integrals in the crystal lattice can be derived as given also in the figure. The present values agrees with the theoretical calculations fairly well. This result indicates that the intermolecular charge carrier transport efficiency, regardless of either the band or hopping transport framework, is strongly anisotropic, which has to be the *electronic* origin of known orientation-dependent $\mu_h$ discovered for the rubrene single crystal field effect transistor [28].

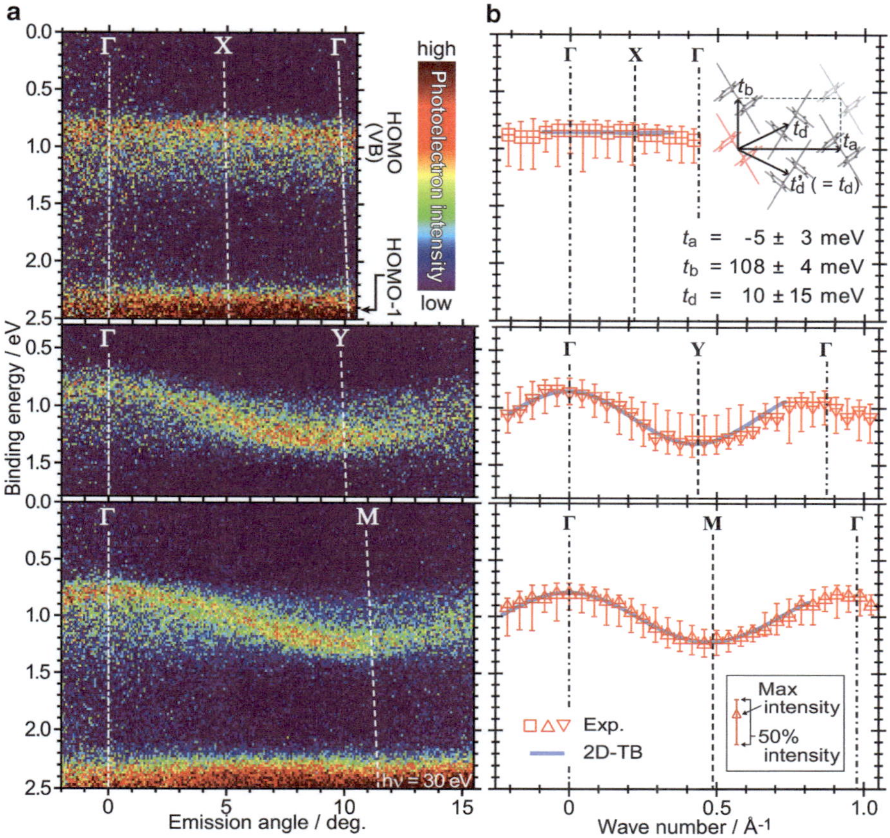


**Fig. 2.7** (**a**) ARUPS spectra of the rubrene single crystal taken along three symmetric direction of the SBZ. The excitation energy was 30 eV. The horizontal and vertical axes of these two-dimensional spectral images corresponds to the electron take-off angle and binding energy, respectively, while the *color tone* indicates the photoelectron intensity. The angles corresponding to the critical points of the SBZ are also indicated. (**b**) The $E_b$-$K_{\parallel}$ relationships to three symmetric directions of the rubrene single crystal. The best fit curves of the 2D-TB approximation are shown as *blue curves*. The vertical axes are common to (**a**). (*Inset*) A schematic view of the surface molecular arrangement of the rubrene single crystal. The inter-molecular transfer integral values to the four *nearest-neighbor* molecules derived under the 2D-TB approximation is also represented (where $t_d' = t_d$ by symmetry)

**Acknowledgement** YN would like to thank Dr. Shin'ichi Machida, Mr. Akihiro Funakoshi, Mr. Naoki Ogawa, Mr. Yuki Uragami, and Dr. Kaveenga Rasika Koswattage of Chiba University and Prof. Kazuhiko Mase of KEK (under an approval of the PF Program Advisory Committee [2011G161]) for their help during the work presented in the last part of this article. Financial supports from a Grant-in-Aid for Young Scientists B [23750209] from the Japan Society for the Promotion of Science, and G-COE Program of Chiba University (Advanced School for Organic Electronics; G-3) are also gratefully acknowledged.

## References

1. K. Seki, U. Karlsson, R. Engelhardt, E. Koch, Chem. Phys. Lett. **103**, 343 (1984)
2. N. Ueno, W. Gaedeke, E.E. Koch, R. Engelhardt, R. Dudde, L. Laxhuber, H. Moehwald, J. Mol. Electron. **1**, 19 (1985)
3. H. Fujimoto, T. Mori, H. Inokuchi, N. Ueno, K. Sugita, K. Seki, Chem. Phys. Lett. **141**, 485 (1987)
4. S. Hasegawa, T. Mori, K. Imaeda, S. Tanaka, Y. Yamashita, H. Inokuchi, H. Fujimoto, K. Seki, N. Ueno, J. Chem. Phys. **100**, 6969 (1994)
5. Y. Yamashita, S. Tanaka, K. Imaeda, H. Inokuchi, Chem. Lett. **20**, 1213 (1991)
6. H. Meier, in *Organic Semiconductors*, vol. 2, ed. by H.F. Ebel (Verlag Chemie, Weinheim, 1974)
7. O.D. Jurchescu, J. Baas, T.T.M. Palstra, Appl. Phys. Lett. **84**, 3061 (2004)
8. C.-H. Wang, C.-Y. Hsieh, J.-C. Hwang, Adv. Mater. **23**, 1630 (2011)
9. N. Koch, A. Vollmer, I. Salzmann, B. Nickel, H. Weiss, J. Rabe, Phys. Rev. Lett. **96**, 156803 (2006)
10. H. Yamane, D. Yoshimura, E. Kawabe, R. Sumii, K. Kanai, Y. Ouchi, N. Ueno, K. Seki, Phys. Rev. B **76**, 165434 (2007)
11. H. Yoshida, N. Sato, Phys. Rev. B **77**, 235205 (2008)
12. Y. Nakayama, J. Niederhausen, S. Machida, Y. Uragami, H. Kinjo, A. Vollmer, J.P. Rabe, N. Koch, H. Ishii, Org. Electron. **14**, 1825 (2013)
13. H. Kakuta, T. Hirahara, I. Matsuda, T. Nagao, S. Hasegawa, N. Ueno, K. Sakamoto, Phys. Rev. Lett. **98**, 247601 (2007)
14. R. Hatch, D. Huber, H. Höchst, Phys. Rev. B **80**, 081411 (2009)
15. R.C. Hatch, D.L. Huber, H. Höchst, Phys. Rev. Lett. **104**, 047601 (2010)
16. M. Ohtomo, T. Suzuki, T. Shimada, T. Hasegawa, Appl. Phys. Lett. **95**, 123308 (2009)
17. T. Shimada, T. Suzuki, M. Ohtomo, T. Hasegawa, Hyomen Kagaku (J. Surf. Sci. Soc. Jpn. **30**, 7 (2009)
18. N. Sato, H. Inokuchi, B. Schmid, N. Karl, J. Chem. Phys. **83**, 5413 (1985)
19. A. Vollmer, O.D. Jurchescu, I. Arfaoui, I. Salzmann, T.T.M. Palstra, P. Rudolf, J. Niemax, J. Pflaum, J.P. Rabe, N. Koch, Eur. Phys. J. E. 17339 (2005)
20. O.D. Jurchescu, A. Meetsma, T.T.M. Palstra, Acta Crystallogr. B Struct. Sci. **62**(330) (2006)
21. Y. Nakayama, S. Machida, T. Minari, K. Tsukagoshi, Y. Noguchi, H. Ishii, Appl. Phys. Lett. **93**, 173305 (2008)
22. S. Machida, Y. Nakayama, S. Duhm, Q. Xin, A. Funakoshi, N. Ogawa, S. Kera, N. Ueno, H. Ishii, Phys. Rev. Lett. **104**, 156401 (2010)
23. D.A. da Silva Filho, E.-G. Kim, J.-L. Brédas, Adv. Mater. **17**, 1072 (2005)
24. Z.Q. Li, V. Podzorov, N. Sai, M.C. Martin, M.E. Gershenson, M. Di Ventra, D.N. Basov, Phys. Rev. Lett. **99**, 016403 (2007)
25. A. Vollmer, R. Ovsyannikov, M. Gorgoi, S. Krause, M. Oehzelt, A. Lindblad, N. Mårtensson, S. Svensson, P. Karlsson, M. Lundvuist, T. Schmeiler, J. Pflaum, N. Koch, J. Electron Spectros. Relat. Phenomena **185**, 55 (2012)
26. J. Takeya, M. Yamagishi, Y. Tominari, R. Hirahara, Y. Nakazawa, T. Nishikawa, T. Kawase, T. Shimoda, S. Ogawa, Appl. Phys. Lett. **90**, 102120 (2007)
27. Y. Nakayama, Y. Uragami, S. Machida, K.R. Koswattage, D. Yoshimura, H. Setoyama, T. Okajima, K. Mase, H. Ishii, Appl. Phys. Express **5**, 111601 (2012)
28. V.C. Sundar, J. Zaumseil, V. Podzorov, E. Menard, R.L. Willett, T. Someya, M.E. Gershenson, J.A. Rogers, Science (New York, N.Y.) **303**, 1644 (2004)
29. K. Imaeda, Y. Yamashita, Y. Li, H. Inokuchi, T. Mori, J. Mater. Chem. **2**, 115 (1992)

# Chapter 3
# Ultraviolet Photoelectron Spectroscopy (UPS) II: Electron–Phonon Coupling and Hopping Mobility

**Satoshi Kera, Hiroyuki Yamane, and Nobuo Ueno**

## 3.1 Introduction

Organic semiconductors are molecular solids with specific charge transport properties, due to weak intermolecular interaction [1], and have been increasingly studied for various device applications. However, the transport properties of organic single crystals and organic thin films are far from being adequately understood [2–13]. Important subjects still to be understood are related to molecular and lattice vibrations (phonons) and their coupling to charge carrier [14–17]. The electron–phonon interaction depends on the molecular structure and their packing motif and therefore it can impact both molecular site energies (HOMO energies, for instance) and transfer integrals. First coupling mechanism is commonly referred to as the local electron–phonon coupling mechanism while the second as non-local electron–phonon coupling mechanism [2–17]. Here we focus on local electron–phonon coupling mechanism by molecular (local phonon) and lattice (non-local phonon) vibrations. The overall strength of this interaction is given by the relaxation energy of neutral and ionized states and the reorganization energy ($\lambda$) associated [3]. The relaxation energy of the ionized state in the context of small or very small (molecular) polaron theory is also referred to as the polaron binding energy ($E_{\mathrm{pol}}$) [18, 19]. In general both molecular and lattice vibrations contribute to $E_{\mathrm{pol}}$. However, the results of recent theoretical investigations indicate that in systems such as oligoacenes the main contribution to the polaron binding energy is due to the molecular vibrations [20].

S. Kera (✉) • N. Ueno
Department of Nanomaterial Science, Graduate School of Advanced Integration Science, Chiba University, Inage-ku, Chiba 263-8522, Japan
e-mail: kera@ims.ac.jp

H. Yamane
Institute for Molecular Science, Myodaiji, Okazaki 444-8585, Japan



H. Ishii et al. (eds.), *Electronic Processes in Organic Electronics*,
Springer Series in Materials Science 209, DOI 10.1007/978-4-431-55206-2_3

Theoretical studies on intramolecular contribution to the polaron binding energy were performed and compared with the ultraviolet photoelectron spectroscopy (UPS) spectra of gas-phase molecules measured at an elevated temperature. These studies provided vital information about electron–phonon interaction and contributed to the understanding of charge transport in organic solids [2–13, 16–31]. It is important to note that most theoretical and experimental studies on the local electron–phonon interaction have been performed on isolated molecules. Therefore it is to understand how these results are altered by intermolecular interactions that take place in solid state. However, direct experimental study on this issue has been neglected until recently because of experimental difficulties in resolving vibronic satellites of the HOMO state in UPS spectra of organic thin films [15]. The UPS linewidth of the thin film is broadened by intermolecular interaction, molecule-substrate interaction and insufficient uniformity of the film structure, which mask the vibronic satellites. Here after we use a term of hole-vibration coupling instead of electron–phonon coupling in phenomenological point of view in the UPS experiment. Although LUMO electron-vibration coupling in principle can be measured by inverse photoelectron spectroscopy (IPES), the measurements have not yet been realized because of poor energy resolution of IPES.

The origin of the UPS bandwidth for organic solids was discussed extensively until the end of the 1970s [32–34]. After the work done by Salaneck et al., in 1980 [34], the UPS bandwidth of the HOMO state in organic films was considered to have been mainly dominated by the dependence of relaxation (polarization) energy [33] on the site/depth, since it yielded a broadened UPS feature with a full-width-at-half maximum (FWHM) of >~ 0.4 eV. As a result, it was considered that intermolecular energy-band dispersion could not be measured, since the dispersion width may be smaller than the spectral bandwidth due to the dependence of relaxation (polarization) energy ([35] and references there in) on the site/depth. Nowadays, the measurement of band dispersion has been successfully achieved as described in Chap. 2, and very recently even for very small band dispersion was observed by Yamane et al. for phthalocyanine (Pc) films thanks to development in the measurement technique and the sample preparation [36]. Unfortunately, the measurement of hole-vibration coupling in organic thin films has also been believed to be impossible, although the possibility of molecular vibration contributing to the width of the HOMO band in UPS has been suggested by taking into consideration the excellent one-to-one correspondence of the valence-band features to those in gas-phase [32], and evidence was found for a tetracene film in an early work on UPS [37]. Therefore, gas-phase UPS spectra, where hole-vibration coupling is resolved for HOMO, have been used in theoretical studies of hole-hopping mobility [2, 3]. There is a serious problem in using the gas-phase spectrum, since the molecules are in more thermally excited states due to the evaporation of larger functional molecules. We need to measure HOMO hole-vibration coupling at lower temperatures using thin films or adsorbed molecules to discuss hopping mobility.

The first meaningful result of the vibronic satellites by high-resolution UPS measurements was reported by Kera et al. for Cu-phthalocyanine (CuPc) ultrathin films [38]. Yamane et al. extended the study to the HOMO hole-vibration coupling with for pentacene (PEN) monolayer on graphite and determined the reorganization energy ($\lambda$) [14, 15]. They pointed out that $\lambda$ in the film is larger than that in a free molecule (gas-phase). Afterward, Paramonov et al. discussed the difference of $\lambda$'s in gas and monolayer phases theoretically by considering intermolecular and molecule-substrate interactions [23].

In this study, we introduce our high-resolution UPS experiments on HOMO hole-vibration coupling in relation to hole-hopping mobility and reorganization energy estimated for various organic semiconductors. We will then discuss current problems that should be overcome for a more thorough understanding of hole-hopping mobility including polaron-like effects. Moreover, we demonstrate a possible analysis method for high-resolution UPS data for molecular films. The results are compared by single mode analysis (SMA) of UPS spectrum and those derived by means of theoretical approaches, indicating that the purely experimental method with SMA is useful for studying the reorganization energy and the hopping mobility of organic systems.

## 3.2 How to Realize the High-Resolution UPS

We summarize the possible origins of the UPS-band width (shape) in Fig. 3.1 due to solid-state effects, which led to the features broadening. They can roughly be classified into five origins: (i) a non-uniform film structure, (ii) lifetime broadening (*in shape*), (iii) vibration (phonon) coupling (*in shape*), (iv) energy-band dispersion and (v) polarization/relaxation-related phenomena. The schematics for the UPS-band shape concerning these phenomena are shown.

As organic molecules consist of light elements, the intramolecular vibration energy is much larger than that of lattice phonons, and molecular vibrations with larger energies (~100 meV) contribute more to increasing $\lambda$. This means that we do not need an ultimate-resolution electron-spectrometer system for measurement in this level of the study. The principal drawback that has hindered us from obtaining high resolution UPS is that spectral broadening is much larger than vibration energy, i.e., the inhomogeneities of molecular film dominate the observed spectral-line width. We thus need to minimize such origins of peak broadening for UPS. The non-uniform structure of the thin film introduces serious dependence of the ionization energy on the position, which originates from the dependence of the initial electronic states on the site and the intermolecular relaxation energy (polarization energy) that appears in the final state upon ionization [39, 40]. Furthermore, molecule-substrate electronic interaction should be minimized to reduce the appearance of new interface states with different binding energies. For instance the electronic coupling between PEN and graphite is evaluated negligibly small [23]. Polarization/relaxation effects in a molecular solid can be divided into two groups, static and dynamic effects. The static effect produces the dependence of ionization energy on the site for film with a non-uniform molecular-packing structure. Moreover, molecular ordering and orientation determine the ionization

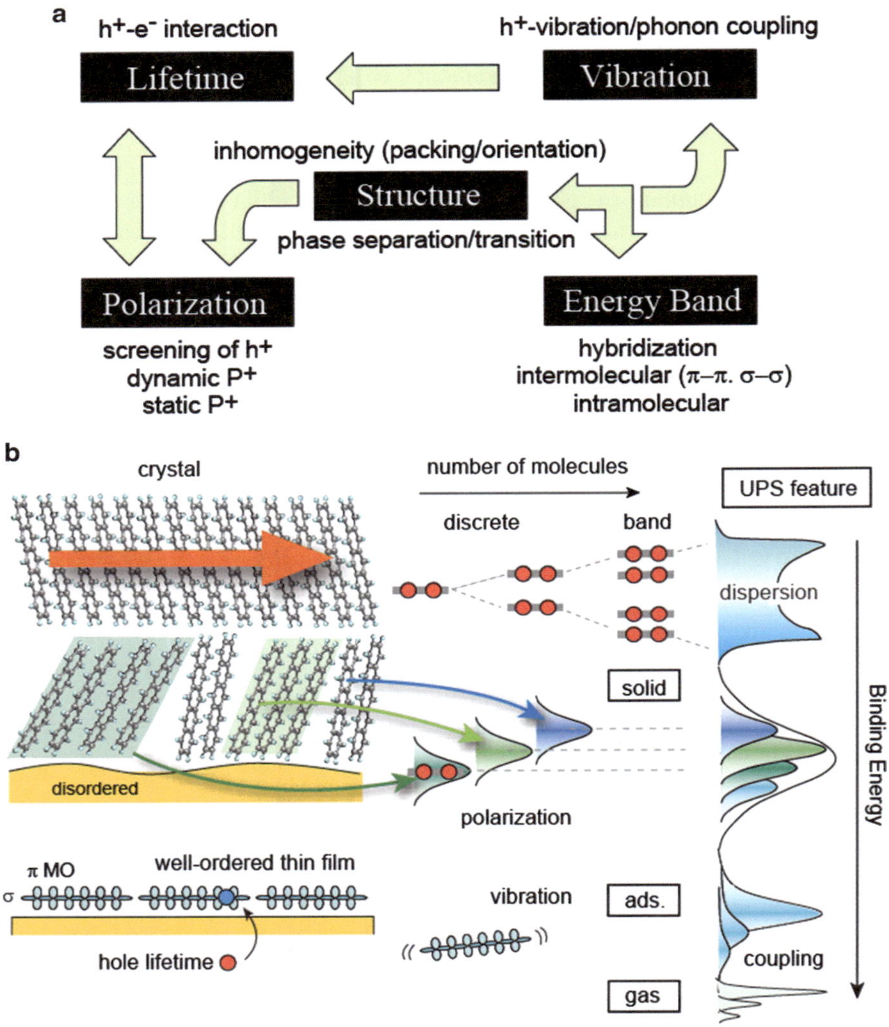


**Fig. 3.1** (**a**) Origins of UPS band broadening of organic solids. *Arrows* indicate their relationships. (**b**) Schematic illustration of UPS-band shape and energy position for various molecular packing/orientations. UPS results on pentacene are used (From [15] with permission)

potential due to the electrostatic potential distribution in all molecules [41]. Briefly, the energy shift due to such non-uniform molecular-packing is within a range of several tens of meV for planer molecules [42–44] and of several hundreds of meV for polar or bulky molecules [45] due to polarization effects dependent on the site, and several hundreds of meV for systems in the electrostatic potential model, namely orientation dependence [41].

The lifetime broadening in photoemission appears differently from that in an electronic transition to a bound state. The lifetime broadening in photoemissions from an ultrathin film of weakly interacting molecules may yield a skewed shape for the line while tailing to a higher kinetic energy, since the photoelectron kinetic energy increases by elimination of the hole potential existing outside the film [46, 47]. The lifetime of the photogenerated HOMO hole is dominated by the electronic interaction between ionized molecules and the substrate, also providing information, e.g., on the electron transfer rate from the substrate to the molecule. Note that briefly the dynamic effects are related to the time required to polarize/relax ionized molecules and neighboring molecules upon ionization, that is a time series in the various polarons (electronic, very small, small and large polaron) for molecular system (see Chap. 4, Sect. 4.3.3) [48–50]. That is, if the time for polarization/relaxation is on a time scale for a photoelectron to travel to an electron analyzer, the experimental line width (ionization energy) reflects the dependence of the polarization/relaxation energy on time. The study on the dynamic polarization, that is polaron phenomena, would open a quantum view of charge transport in molecular solids. Very recently, we have obtained some evidences on the dynamic effects of charges in the molecular solid by angle-resolved UPS measurements. The hole-vibration coupling intensity depends on the photoemission angle (not shown). It will be an interesting target to study the transport mechanism for organic molecular solids. As a result, the observed UPS spectra show various line shapes.

In Fig. 3.2, the HeI UPS taken for various phases of PEN are presented. The spectra for gas-phase, monolayer (ML) films, amorphous film, and crystal film are measured by using the same apparatus. In the gas spectrum, a very sharp HOMO band is observed and hole-vibration coupling is detected at the high-$E_b$ side with asymmetric feature. Note that the line shape of gas-phase UPS depends on the instruments and the temperature measured in which a contamination by low-vapor pressure molecules at a higher temperature experiment may cause serious problems on the system. Brédas et al., reported a spectrum which shows a larger skewed shape in the high-kinetic energy side [22]. A similar sharp HOMO band is observed for the ML film prepared on the HOPG substrate. We find some differences in the line shape of ML depending on the temperature in the detail as discussed later. A broadened HOMO band with a large tail is observed for the films prepared on the $SiO_2$ or polycrystalline metal substrates, where the band gives ~0.4 eV width and shows no fine features. PEN is known to have polymorphs depending on the film preparation [51], hence two peaks with a sharp onset features are observed as an evidence of a crystalline sample prepared on CuPc/GeS substrate [52].

To avoid strong molecule-substrate interaction, we need to use an inert substrate such as graphite (mostly we use a highly oriented pyrolytic graphite: HOPG, ZYA grade). Furthermore, organic semiconductor thin films do not grow layer-by-layer to create non-uniform films with a broadened HOMO-bandwidth in UPS. Thus, one must prepare a uniform molecular monolayer to attain high-resolution UPS. Although islands of molecules with orientations different to the monolayer region grow at defects/step bunches on the substrate surface, the islands may be selectively cleaned away by appropriate heat treatment after a thicker molecular layer has been

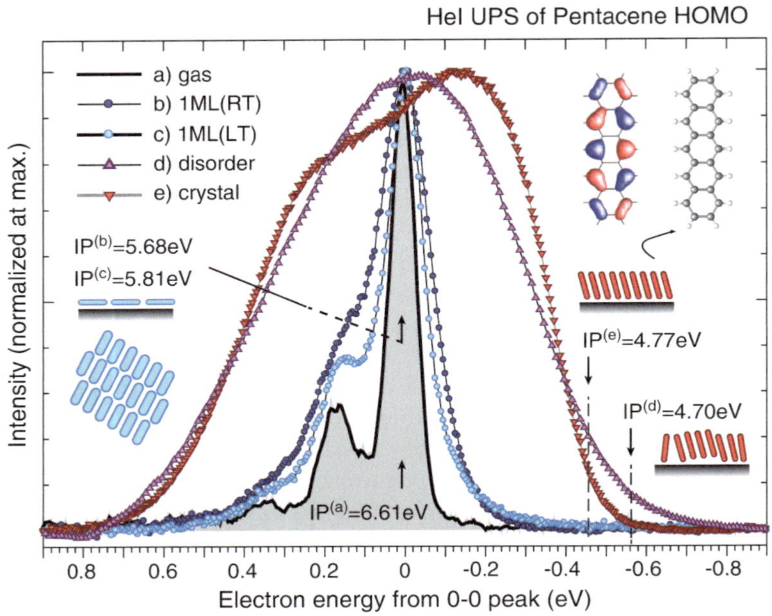


**Fig. 3.2** HeI UPS for various phases of pentacene. (**a**) gas spectrum (at 433 K) is obtained preliminary (unpublished), (**b**) lying-monolayer (ML) on HOPG (295 K, RT), (**c**) ML on HOPG (50 K, LT), (**d**) standing-disorder ML film (1 nm) on $SiO_2$ (295 K), and (**e**) standing-crystalline ML film (1.5 nm) on CuPc/GeS (295 K). All spectra are recorded by the same electron analyzer (VG-CLAM4). The background signals are subtracted. The abscissa is aligned to 0–0 transition peak in the gas-phase spectrum. Ionization potential (IP) for each film ((0–0) peak energy for sharp spectra and onset energy for broaden spectra) is described in the figure

deposited, as observed with photoemission electron microscopy (PEEM) [53, 54]. In our experience, the detection of vibration coupling with UPS, which is appeared as skewed asymmetric line shape, is a good indication of the formation of a very uniform and large domain of ultrathin organic film.

Yamane et al. studied what effect molecular orientation had on the molecular electronic structure for CuPc on a graphite (HOPG) substrate via the dependencies of metastable-atom electron spectra and UPS [55] on film thickness. Figure 3.3 shows the UPS spectra for HOMO regions and the intensity map as a function of film thickness. The position and shape of HOMO are drastically changed with increasing the thickness. These differences can be explained by spontaneous electric dipoles produced in the film by a gradient of intermolecular electronic interaction along the surface normal due to the continuous increase in the molecular angle. This indicates that the change in molecular orientation is an important origin of band-bending-like shift in the molecular electronic states even if the molecule has no permanent electric dipole. Note that for in the thin-film region only (a couple of layers), we can observe an asymmetric HOMO band with fine features.

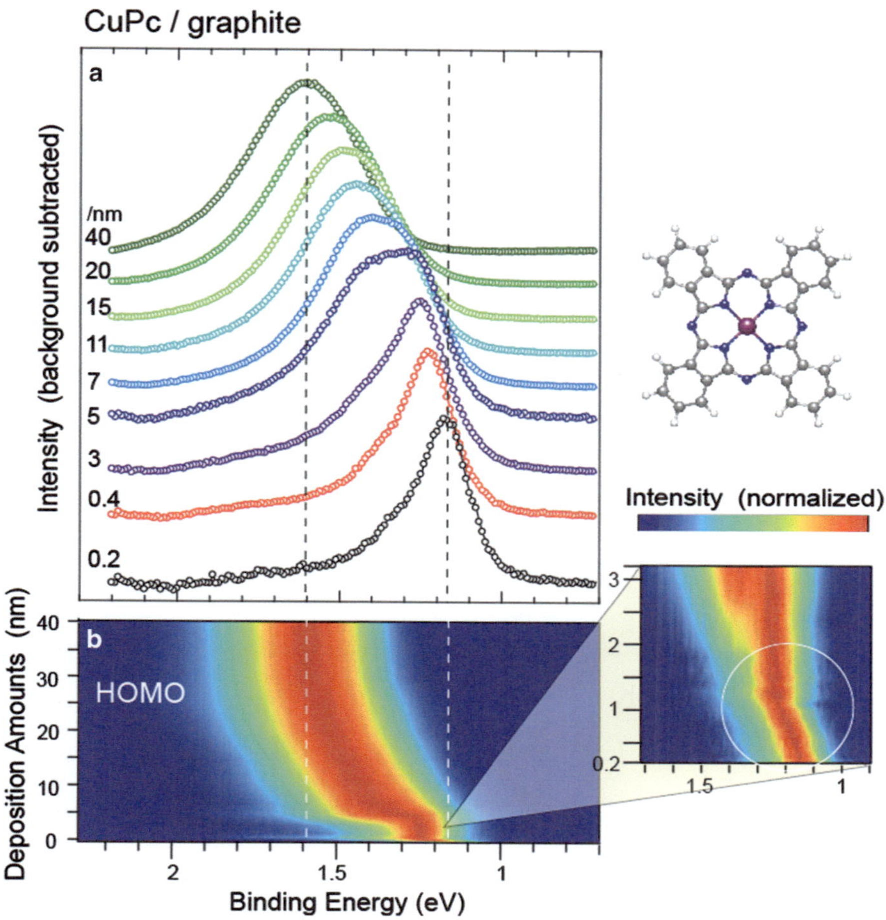


**Fig. 3.3** (**a**) Dependence of He I UPS spectra of HOMO of CuPc/HOPG system on film thickness measured at normal emission $\theta = 0°$, where binding-energy scale is referenced to Fermi level of substrate. (**b**) Intensity map of UPS spectra for HOMO regions of system as function of film thickness. Background photoelectrons from substrate have been subtracted for mapping. Thinner region is magnified at *right*. Vibration fine features are observed for thinner films marked by *circle* (From [15] with permission)

Another important phenomenon that we should remember is that molecular ultrathin film changes its packing structure depending on temperature, since intermolecular and molecule-substrate interactions depend on temperature, and this makes it very difficult to prepare a uniform film. Figure 3.4 has an example of the dependence (86–162 K) of HOMO-band UPS for pentacene (flat ML)/HOPG on temperature. Vibration satellites can be seen at all temperatures, although there is a structural transition at 130–140 K. The HOMO band at 132 K is well reproduced by convoluting two asymmetric HOMO bands at 86 and 162 K.

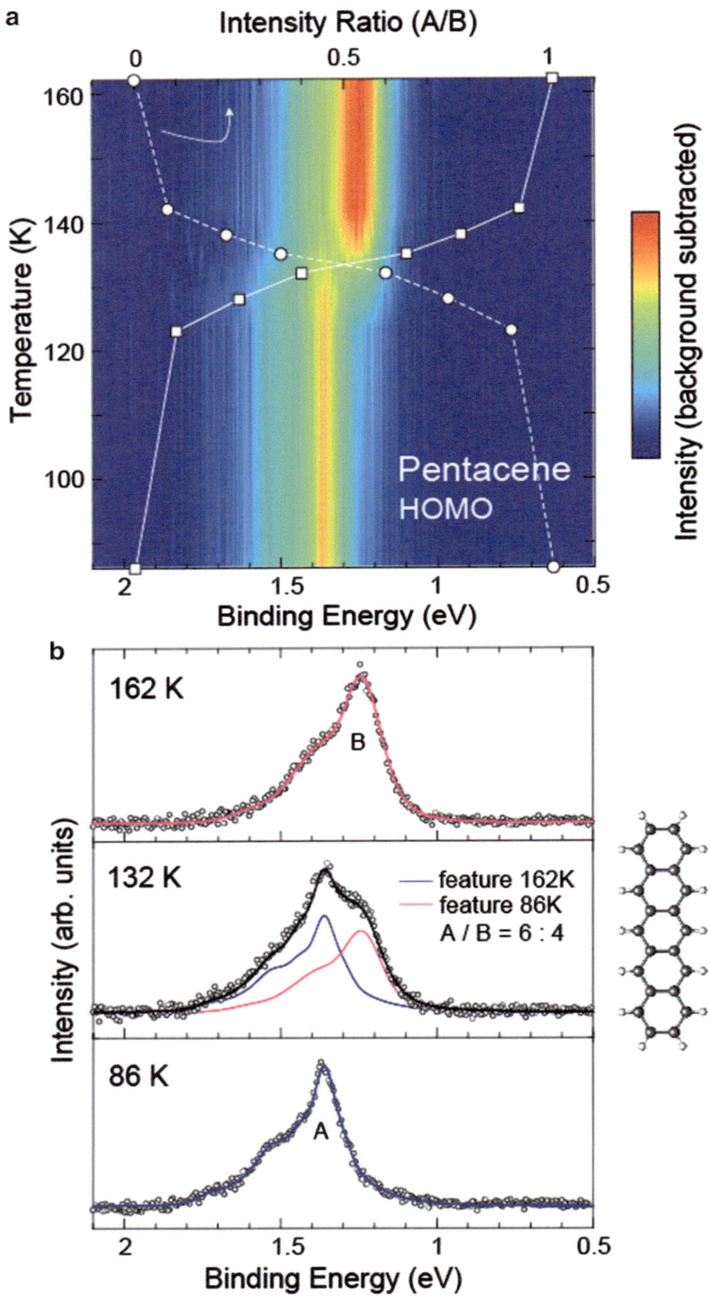


**Fig. 3.4** Intensity mapping of temperature dependence of UPS spectra of HOMO band region (panel (**a**)) and typical UPS spectra (panel (**b**)) for pentacene (1 ML)/HOPG at photoelectron

Such a transition in UPS was also observed for tetracene/HOPG (not published) and CuPc/HOPG [56]. Although Kataoka et al. found that the UPS line width for the CuPc/HOPG system suddenly changes at the transition temperature and at temperatures below the transition, the line width is larger due to the appearance of intermolecular interactions, origins for the transition in molecular monolayer is still being studied. Finding an appropriate temperature to prepare the film is also an important factor in obtaining a highly-resolved UPS spectrum. Indeed, the impact of temperature dependence on the electronic structure is more significant for ML films prepared on other substrates (e.g. even for noble metal, Cu, Ag, and Au) and becomes complicated to understand much more. The formation mechanisms of interface state and band-gap state for molecules on the metal are described briefly in Chap. 6 and elsewhere [57, 58].

The bandwidth of UPS features contains information on electron–phonon interaction, electron (hole) lifetime, and electron–defect scattering provided that final state band-structure (dispersion) effects are not taken into account. These have been the subjects of an excellent review by Matzdorf [59]. There are few examples of molecular adsorbates on metal surfaces where vibrational features can be resolved in a valence photoemission spectrum and where structural heterogeneity can be eliminated from molecule–metal interfaces to allow the bandwidth to be quantitatively determined.

The first meaningful measurement on the UPS line shape was accomplished for a CuPc sub-monolayer (ML) on HOPG, as shown in Fig. 3.5, where the molecular plane is parallel to the surface [38]. This result was obtained at room temperature (RT: 295 K), since the molecular orientation becomes tilted to yield a larger bandwidth at a lower temperature due to the appearance of a non-uniform packing structure [56, 60]. Nevertheless, we succeeded in obtaining a much sharper HOMO band for the CuPc/HOPG system by properly controlling the method of cooling afterward [15]. The HOMO (labeled A) in panel (a) appears as a sharp peak at ~1 eV below the Fermi level ($E_F$). Panel (b) shows an enlarged view of the HOMO peak after the contribution from the HOPG substrate has been subtracted. The peak is asymmetric and can be decomposed into three vibrational features (v = 0, 1, 2) with an energy separation of 170 meV (this value has been improved later on). This gives a full-width-at-half-maximum (FWHM) of 172 meV for the main component (v = 0). Considering an instrument resolution ($\Delta$R) of 80 meV, the intrinsic line width we obtained was 150 meV, which corresponds to a lifetime of 2.2 fs within the assumption of using an uncertainty relation. The tail feature at low-$E_b$ side of main peak, that is Lorentzian contribution in the Voigt function, could be a virtual feature that has originated from the tail of the main peak due to the thermal

**Fig. 3.4** (continued) emission angle $\theta = 0°$. In panel (**b**), *circles* represent observed spectrum after background has been subtracted. *Solid curve* that overlaps observed spectrum is convolution of fitting curves. Each HOMO band at measured temperature is well reproduced by changing ratio of two asymmetric curves, viz., spectra for low-temperature (A) and high-temperature phases (B). In panel (**a**), intensity ratio between phases A (*circles*) and B (*squares*) obtained by convolution curves is also superimposed on map (From [15] with permission)

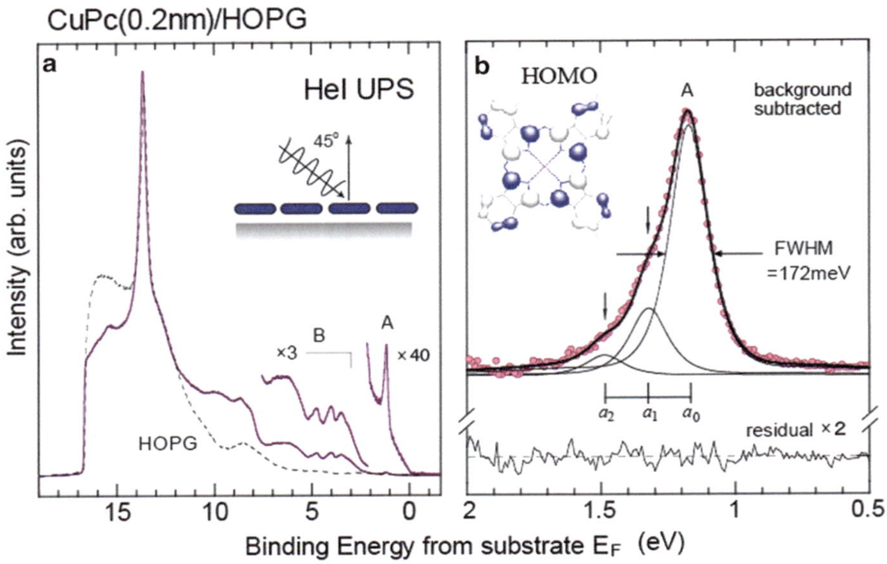


**Fig. 3.5** (**a**) He I UPS of CuPc ML (0.2 nm) on HOPG substrate (*solid curve*) and clean HOPG (*dashed curve*). Spectra were measured at sample temperature of 295 K at $\theta = 0°$. (**b**) Expanded spectrum of top band (A) (HOMO) region after background photoelectrons had been subtracted from substrate. *Circles* represent observed spectrum after subtraction of background. Three components ($a_0$–$a_2$) of pseudo-Voigt profiles used in curve fitting are indicated by *thin solid curves*. *Solid curve* that overlaps observed spectrum is convolution of three fitting curves. Residual of curve fitting is also shown at *bottom* of (**b**). Orbital pattern of HOMO of CuPc is also illustrated. Caution should be taken with vibration energy (170 meV) obtained from this result (From [15] with permission)

excitation of phonons in the initial state of photo-ionization and the finite lifetime of the HOMO hole. Later on, the lifetime effect on the lines' shape was also discussed without using the uncertainty relation [15, 38]. This value can be an upper limit for the lifetime of the hole generated by the photoemission process. The transient hole was considered to be filled by electron transfer from the HOPG substrate.

In relation to achieving a uniform film, HOMO-level splitting with an energy separation of $2t$ (transfer integral ($t$)) was observed for two interacting PbPc molecules in a bilayer (BL) on HOPG, as shown in Fig. 3.6 [61], where the HOMO (A) in the ML splits into two features ($A_1$ and $A_2$) at 295 K. A sharp peak, A, which has an asymmetric profile with a few fine features, is observed for the ML. It is convoluted with five Gaussian functions (FWHM of 147 meV), and satellite components on the high-$E_b$ side of the main peak originating from hole-vibration coupling, as observed for other Pcs [38, 56, 60, 62–65], and oligoacene (Ac) monolayers [14, 66]. The observation of the fine features indicates that large ML domains with good molecular ordering of PbPc are formed. The BL spectrum can be convoluted using three ML spectra with different $E_b$ positions. The intensities of $A_1$ and $A_2$ are nearly equal, and the energy separation is 0.35 eV. The origin

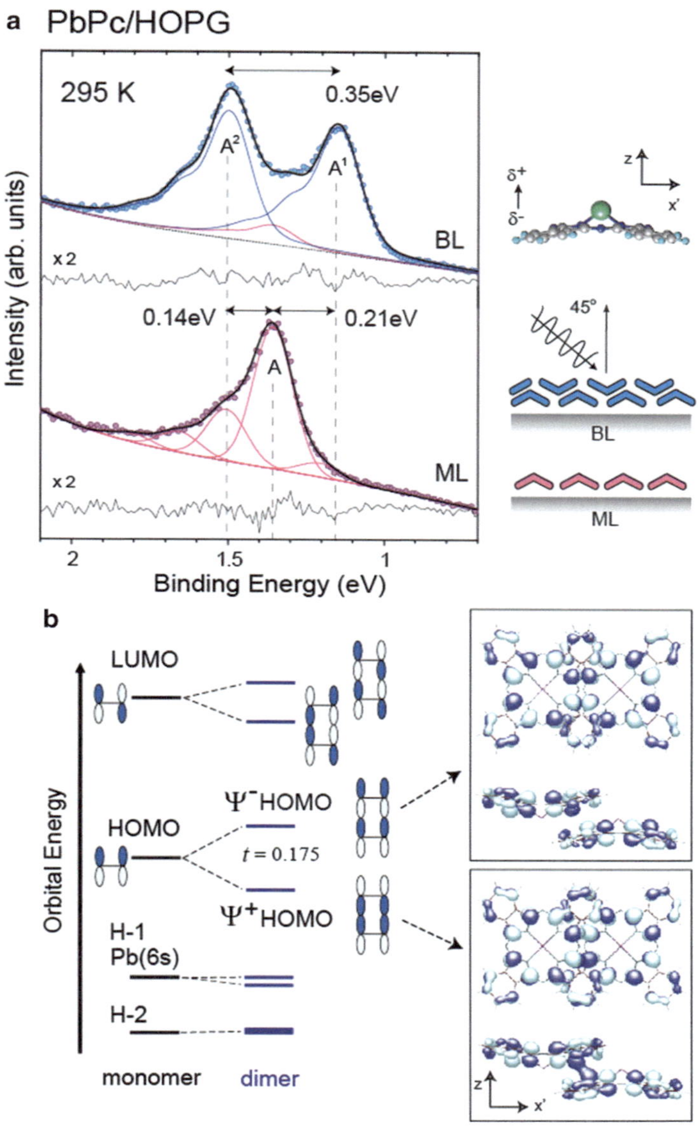


**Fig. 3.6** (**a**) Expanded spectra of HOMO region for PbPc ML and BL films on HOPG substrate measured at $\theta = 0°$ at 295 K. *Circles* represent observed spectra and *thin curves* are deconvoluted profiles. *Solid curve* that overlaps observed spectrum is convoluted spectrum of five Gaussian peaks for ML. BL spectra are convoluted with three asymmetric functions of ML spectrum. Residual curve is also shown at *bottom* of each spectrum. (**b**) Scheme of split of orbital levels on dimerization. MO patterns of bonding ($\Psi$+) and antibonding ($\Psi$−) HOMO are shown (From [15] with permission)

of the two components, $A_1$ and $A_2$, must be considered here, because the small component denoted by A may originate from the small ML area uncovered by the overlayer. Although the origins of UPS spectral features, peak shift, shape, and width for organic thin films, are generally extremely complicated, we can conclude that the marked differences observed in the HOMO peak can be ascribed to energy-level splitting by inter-orbital interaction appearing in the BL. This is also supported by the theoretical calculations [61]. A possible dimer model and the level splitting that was calculated are shown in Fig. 3.6b. The displacement between the two molecules is (x, z) = (7.21 Å, 4.0 Å) and we calculated $2t = 0.35$ eV for the HOMO and 0.1 eV for the Pb 6s state (H-1). The calculated bonding ($\Psi+$) and anti-bonding ($\Psi-$) HOMO patterns are also given in Fig. 3.6b. This $2t$ separation was observed to increase to 0.41 eV on cooling down to 47 K, which is reasonably understandable because the intermolecular distance reduces due to the decreasing thermal excitation of molecular vibrations (not shown). It is worth noting, as a coincidence, that the bilayer spectrum can be well convoluted by using the mono-layer spectrum, indicating that vibration coupling in the PbPc dimer structure is not extremely different from that in the monolayer film. However, recently we observed a huge difference in the vibration coupling for other molecular systems, where the wave-function spread would be largely modified upon intermolecular interaction. The dependence of the molecular packing structure on the reorganization energy will be studied further.

## 3.3 Estimation of Charge Reorganization Energy, Polaron Binding Energy and Hopping Mobility

### *3.3.1 Role of Hole-Vibration Coupling in Hopping Mobility*

Charge hopping dominates the mobility of systems with very narrow bandwidths and systems with larger bandwidths if the mean free path of the conduction charge is of the order of the intermolecular distance. Charge-transfer processes and carrier dynamics of organic molecules have been widely studied in various fields [2–13, 16–31, 48–50, 67]. Detailed theoretical descriptions can be found in several reviews [2–4]. According to general microscopic models, total mobility can be expressed as the sum of two contributions, i.e., (i) coherent charge tunneling that dominates transport at low temperatures and (ii) incoherent charge hopping that becomes dominant at high temperatures. The relative contributions of either mechanism depend on the film structures, where key parameters are charge (electron or hole)-phonon (molecular vibration) coupling, the electronic bandwidth, and phonon bandwidths. For ideal crystalline films, the model has been satisfactorily described in theoretical work for two limiting cases, i.e., weak and strong electronic coupling limits; however, a model that exhibits intermediate coupling with dependence on temperature has not yet been well explained.

The Holstein polaron model has been applied [18, 19] within the limits of weak intermolecular interactions. Since the electronic state is localized, charge transport occurs via hopping from one molecule to the next (local electron–phonon coupling). There is also a detailed theoretical description in a review [2, 3]. Of course, transport can be described by band theory, i.e., coherent electron-transport scenarios [4, 8, 10, 35], within the limits of strong intermolecular interactions (non-local electron–phonon coupling). Moreover, the limits in other directions need to be considered for static-disorder systems, which are often found in actual organic solids. When an electron or hole is injected into a molecular solid, electronic–nuclear coupling leads to localization and transport occurs via localized charge hopping, which is closely related to Marcus electron-transfer theory [68, 69]. The semi-classical Marcus-hopping model for self-exchange charge transfer has been widely studied. Here, we present one semi-classical approach to obtain hopping mobility. These molecular parameters have appeared in the following equations, which have also often been used in other theoretical models, and can experimentally be obtained by using high-resolution UPS measurements.

The mobility (μ) in the high-temperature regime can be approximated from the electron-transfer rates by considering the Einstein relation for diffusive motion,

$$\mu = \frac{ea^2}{k_B T} k_{\mathrm{ET}}, \tag{3.1}$$

where $a$ is the intermolecular distance and $k_{\mathrm{ET}}$ is the electron transfer/hopping probability per unit time.

For electron transfer between an initial electronic state and a final state within the weak electronic coupling limit, a common starting point for the electron-transfer rate is Fermi's Golden Rule form,

$$k_{ET} = \frac{2\pi}{\hbar} |V_{if}|^2 (\mathrm{FC}), \tag{3.2}$$

where $|V_{if}| = \langle \Psi_i | \mathrm{H} | \Psi_f \rangle$ is the electronic-coupling matrix element between the electronic states, $i$ and $f$, and FC is the thermally averaged Franck–Condon weighted density of states. It can easily be shown that, in the high-temperature regime, i.e., when all vibration modes are classical as $h\nu \ll k_B\mathrm{T}$ ($h\nu$: vibration energy), FC reduces to a standard Arrhenius equation for the non-adiabatic charge-transfer regime, $V < \lambda$, as

$$\mathrm{FC} = \sqrt{\frac{1}{4\pi\lambda k_B T}} \exp\left[ -\frac{(\lambda + \Delta G^0)^2}{4\lambda k_B T} \right], \tag{3.3}$$

where λ denotes the reorganization energy induced by the electron or the energy transfer and $\Delta G^0$ is the Gibbs free energy during the reaction. When a charge hops from an ionized molecule to an adjacent neutral molecule, the hopping process can

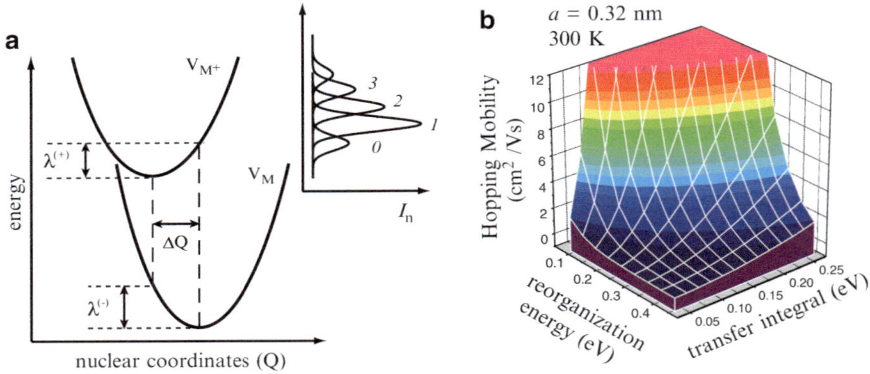


**Fig. 3.7** (**a**) Typical adiabatic energy surfaces of neutral ($V_M$) and ionized ($V_M{}^+$) states and two relaxation energies $\lambda^{(+)}$ and $\lambda^{(-)}$ at ionization ($M \rightarrow M^+$) and neutralization ($M^+ \rightarrow M$) processes. The UPS intensity of vibration satellites $I_n$ on photo-ionization is shown. $\lambda^{(+)}$ can be obtained by measuring $I_n$. (**b**) Typical simulation results of charge-hopping mobility as function of transfer integral ($t$) and reorganization energy ($\lambda = 2\lambda^{(+)}$) using Eq. (3.1) at 300 K and $a = 0.32$ nm

be viewed as a symmetric self-exchange electron-transfer reaction with $\Delta G^0 = 0$. The rate of charge hopping between two adjacent molecules can be estimated as

$$k_{ET} = \frac{2\pi}{\hbar} t^2 \sqrt{\frac{1}{4\pi\lambda k_B T}} \exp\left[-\frac{\lambda}{4k_B T}\right], \tag{3.4}$$

where $t$ corresponds to the intermolecular transfer integral that describes the strength of the electronic interaction between adjacent molecules (or electronic coupling $V$). Thus, there are two major parameters of key importance in hopping transport: (i) the electronic coupling (transfer integral ($t$)) between adjacent molecules, which needs to be large and (ii) the reorganization energy, λ, which needs to be small to obtain efficient hopping mobility (see Fig. 3.7). For hole transport, $t$ can be experimentally obtained from the HOMO-band dispersion of a molecular stacking system [35, 36] or by splitting the HOMO level of a dimer molecule [61]. The λ corresponds to the sum of the geometry-relaxation energies ($\lambda_{rel}$) when a molecule goes from a neutral to an ionized geometry and from an ionized to a neutral geometry during hole hopping through molecular aggregates. Since the vibration energies of conjugated C-C stretching modes are 1,200–1,600 $cm^{-1}$ (150–200 meV), the classical high-temperature approximation used in Eq. (3.4) is only approximate and may have problems at room temperature.

There are two $\lambda_{rel}$ components, $\lambda^{(-)}$ and $\lambda^{(+)}$, corresponding to going into an ionized state and returning to a neutral state (see Fig. 3.7). If the $\lambda^{(-)}$ does not differ too much from $\lambda^{(+)}$, λ can be written as

$$\lambda = \lambda^{(-)} + \lambda^{(+)} \approx 2\lambda^{(+)}. \tag{3.5}$$

This can be satisfied for a system where the molecular structure is not largely deformed on being ionized. The contribution of each vibration mode to the relaxation energy, $\lambda^{(+)}$, can be determined by the intensities of vibration satellites in high resolution UPS. The relaxation energy of the ionized state in the context of small (molecular) polaron theory is also referred to as the polaron binding energy ($E_{\mathrm{pol}}$). As $E_{\mathrm{pol}}$ is defined as a stabilization energy when the hole is localized on a single lattice site, $E_{\mathrm{pol}}$ is directly related to $\lambda^{(+)}$. The satellite intensities are described by the values of the Huang-Rhys factors, $S$, which in harmonic approximation are related to $\lambda^{(+)}$ by

$$E_{pol} = \lambda^{(+)} = \sum_i S_i h\nu_i. \tag{3.6}$$

It is useful to point out that $S$ is directly related to the electron–phonon coupling constant [2, 3]. When the neutral state is in the vibrational ground state, the intensities of vibrational progression are given by Poisson distribution

$$I_n = \frac{S^n}{n!} e^{-S}. \tag{3.7}$$

where $I_n$ is the intensity of the $n^{\mathrm{th}}$ vibrational satellite. These relationships mean that once the energy of vibration mode $i$ ($h\nu_{\mathrm{i}}$) and corresponding $I_{\mathrm{n}}$ are measured with high resolution UPS of HOMO, $S_{\mathrm{i}}$, $\lambda^{(+)}$, $E_{\mathrm{pol}}$ and $\lambda$ can be experimentally obtained. The $\mu$ can be directly studied by using UPS, since $t$ is also determined by UPS, making the first-principles experimental studies of $\mu$ possible. Although these molecular parameters are very helpful in designing organic semiconductor molecules, Eq. (3.4) is too simplistic when we need to calculate $\mu$ using experimental $\lambda$ obtained at low temperatures. For example, one should refer to [2, 3] to understand the limitations in using Eq. (3.4).

### 3.3.2 Analysis of HOMO Fine Feature

We report how to analyze an experimental data of the electronic structure and hole-vibration coupling of pentacene (PEN) and perfluoropentacene (PFP) ML on graphite. We compare the hole-vibration couplings in PFP and PEN monolayers and in gas phase to study the effect of perfluorination on these parameters. We observed that the UPS vibrational structure is largely different between PEN and PFP in ML phase as well as in gas phase. Furthermore although the vibration energy is decreased by perfluorination as expected from much larger mass of fluorine atom than hydrogen, the vibronic-satellite intensity due to hole-vibration coupling is significantly enhanced in PFP. We found that the latter contribution to $\lambda$ is much greater than the former one, resulting in serious increase in $\lambda$ of PFP also for the monolayer.

HeI UPS spectra of the PEN(ML)/HOPG and PFP(ML)/HOPG compared with gas-phase UPS are shown in Figs. 3.8 and 3.9, respectively. The valence-band

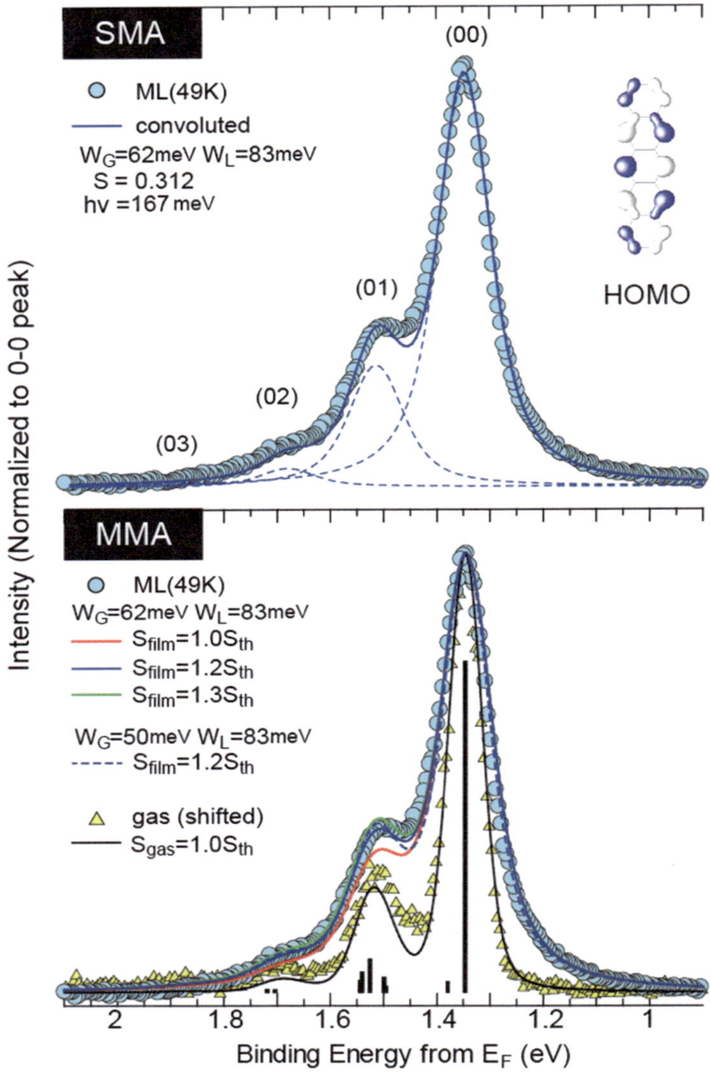


**Fig. 3.8** (*Bottom panel*) Gaseous (*triangles*) and angle-integrated-ML UPS spectra (*circles*) for pentacene (PEN), compared with convoluted curves of 18 $A_g$ vibrational modes (*solid curves*). Energy scale of gaseous is relative to 0–0 transition peak. *Vertical bars* indicate 0-n (n = 0–4) transition intensities. Convolution curve in MMA is obtained by Voigt functions ($W_G = 62$ meV and $W_L = 20$ meV for gaseous and $W_G = 62$ meV and $W_L = 83$ meV for ML) with $S_{film}$ and $S_{gas}$. The MMA for ML ($W_G = 50$ meV and $W_L = 83$ meV) is also shown by *dashed curve*. (*Top panel*) the SMA result is shown for deconvolution with four Voigt functions ($W_G = 62$ meV, $W_L = 83$ meV) with an energy separation (hν) of 167 meV and using Eq. (3.7) with $S_{film} = 0.312$ for the ML. Gas spectrum is obtained by our newly-constructed UPS apparatus (unpublished). The ML spectrum is cited from [71] with permission

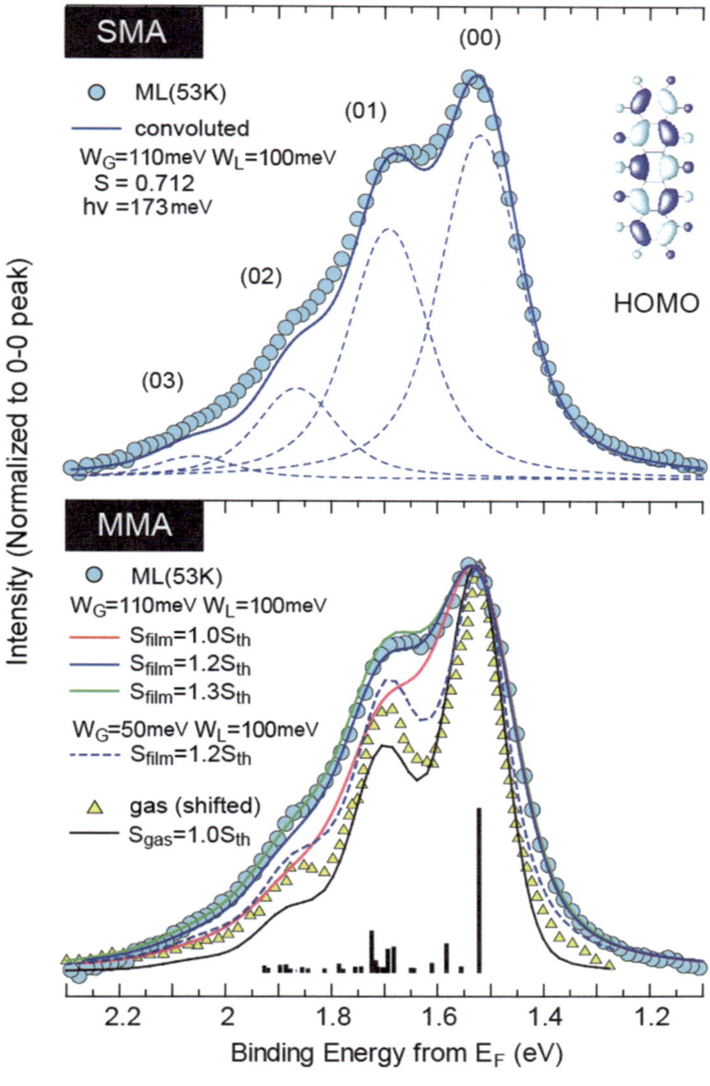


**Fig. 3.9** (*Bottom panel*) Gaseous (*triangles*) and angle-integrated-ML UPS spectra (*circles*) for perfluoropentacene (PFP), compared with convoluted curves of 18 $A_g$ vibrational modes (*solid curves*). Energy scale of gaseous is relative to 0–0 transition peak. *Vertical bars* indicate 0-n ($n = 0$–$4$) transition intensities. Convolution curve in MMA is obtained by Voigt functions ($W_G = 110$ meV and $W_L = 20$ meV for gaseous and $W_G = 110$ meV and $W_L = 100$ meV for ML) with $S_{film}$ and $S_{gas}$. The MMA for ML ($W_G = 50$ meV and $W_L = 100$ meV) is also shown by *dashed curve*. Gas spectrum is by Delgado et al. [25] (*top panel*) the SMA result is shown for deconvolution with four Voigt functions ($W_G = 110$ meV, $W_L = 100$ meV) with an energy separation (hv) of 173 meV and using Eq. (3.7) with $S_{film} = 0.712$ for the ML. The ML spectrum is cited from [71] with permission

features in the ML spectra correspond well with those in the gas phase [14, 67], indicating the electronic structure is not significantly modified by intermolecular and molecule-substrate interactions. One clearly sees that the HOMO band represents an asymmetric shape caused by a vibrational progression towards the higher $E_b$ side as in previously reported other ML systems [14, 15, 35, 38, 56, 60–66].

We compared angle-integrated UPS spectra ($\theta = 0$–$66°$ after subtracting the background and magnified at HOMO band) measured at low temperature of the PEN(ML)/HOPG (at 49 K) and the gas-phase UPS (vaporized at 433 K) (see the bottom panel). The $E_b$ of the gas-phase spectrum is shifted to align the (0–0) peak. The HOMO consists of four fine features, namely a main peak (0–0 vibronic transition) with three satellites (0–1, 0–2, 0–3 vibronic transitions) for the gas-phase and the ML spectra. Note that the intensities of the vibration satellites in the 49 K film are more intense than those in the gas phase. We found that such a fine structure shows no coverage dependence during the monolayer formation since the intermolecular interaction along the surface lateral is dominated by the negligibly weak $\sigma$–$\sigma$ interaction. These indicate that the hole-vibration coupling in PEN thin films is different from that in free molecules owing to the solid-state effect.

Brédas and co-workers reported that UPS vibrational fine structures of the gas-phase PEN and PFP were well reproduced by using all 18 totally symmetric ($A_g$) vibrational modes whose intensities satisfy the Franck–Condon principle (linear coupling model) [we call it multimode analysis (MMA)] [22, 23, 25]. Here we demonstrate that the convolution of gas spectrum, which is newly measured by us, by MMA using Voigt functions (the FWHM of $W_G$ and $W_L$ are fixed to 62 and 20 meV respectively) with considering the 0-n ($n = 0$ to 4) transitions. The width for Voigt function ($W_G$ and $W_L$) is specified for fitting the low-energy tail of (0–0) peak. We used the calculated $S$ factors ($S_{th}$) and the vibration energies, where the energies are scaled by a factor of 0.9613, for cation state obtained by Brédas et al. to simulate the intensity of vibration satellites. The theoretical simulation curve has been demonstrated by using other sets of Voigt functions for PEN, because the spectral features of gas-phase UPS depends on the instruments [22]. A different high-energy tail feature may be produced by an instrumental artifact. In the present gaseous UPS, which shows much sharper and less tailing feature, we find a lack in reproducibility by the simulated curve, even if we assume a Lorentzian contribution to adjust the tail feature. It indicates that the theoretical MMA would need to develop for an isolated molecule. The convolution of ML spectrum was performed as used for gas-phase, but by using different Voigt functions ($W_G = 62$ meV, $W_L = 83$ meV) and different coupling intensities ($S$). The difference in the spectral width and $S$ between gas-phase and ML phase would be explained by solid state effects. We tried to reproduce the ML spectrum by changing the width and $S$, and found that it shows a better agreement by assuming $S_{film} = 1.2\ S_{th}$ for each 18-vibrational modes. If we use the instrumental resolution of $W_G = 50$ meV, the MMA do not agree with the ML spectrum. This analysis yields $\lambda(\text{film}) = 108$ meV, $\lambda^{(+)}(\text{film}) = E_{pol}(\text{film}) = 54$ meV at 49 K as an example. The experimental $\lambda^{(+)}(\text{film})$ is larger than the $\lambda^{(+)}(\text{gas})$ and $\lambda^{(+)}$ by theoretical values.

Instead of the MMA, we describe purely experimental estimation of $\lambda^{(+)}$ ($=E_{pol}$), where we use only experimental values of the vibration energy and satellite intensity within the available experimental energy resolution, because the present theory cannot explain the gas-phase spectrum perfectly and the theoretical simulation by MMA is not so simplified to handle. Here we call this method as single mode analysis (SMA). Figure 3.8 (in the top panel) shows the angle-integrated UPS spectra of the PEN(ML)/HOPG together with the SMA result. The spectrum can be reasonably deconvoluted with four Voigt functions ($W_G = 62$ meV, $W_L = 83$ meV) with an energy separation (*hv*) of 167 meV and using Eq. (3.7) with $S_{film} = 0.312$. The intensity follows Poisson distribution for the vibrational progression in this evaluation which gives that an error into $\lambda^{(+)}$ is less than 5 meV. The results give $\lambda^{(+)} = E_{pol} = 52$ meV for the ML at 49 K. We found that the gas-phase spectrum can be fitted with $W_G = 62$ meV, $W_L = 20$ meV, hv = 167 meV, and $S_{gas} = 0.264$, leads to $\lambda^{(+)} = E_{pol} = 44$ meV (not shown). The reorganization energy in the film is evaluated to be ~12 % larger than that in gas phase by SMA.

We demonstrated the same analysis procedure for PFP(ML)/HOPG (at 53 K) and the gas-phase UPS (vaporized at 503–563 K) obtained by Delgado et al. [25] as shown in Fig. 3.9. We found that the gas-phase spectrum cannot be reproduced by the theoretical values and meaningful differences between ML and gas-phase UPS. We obtained $\lambda^{(+)} = E_{pol} = 129$ meV for the ML at 53 K by MMA ($W_G = 110$ meV, $W_L = 100$ meV, $S_{film} = 1.2\ S_{th}$: for 18 modes), while $\lambda^{(+)} = E_{pol} = 123$ meV for the ML ($W_G = 110$ meV, $W_L = 100$ meV, $S_{film} = 0.712$, hμ = 173 meV) and $\lambda^{(+)} = E_{pol} = 110$ meV ($W_G = 110$ meV, $W_L = 100$ meV, $S_{gas} = 0.636$, hμ = 173 meV) for the gas-phase by SMA, which represents similar trend as for PEN. The values by the simplest SMA agree fairly well with the complicated MMA due to self-convolution of vibration modes with weak coupling intensity. The $\lambda^{(+)}$ by SMA for gas and ML shows a significant difference, suggesting the theoretical MMA would be useful to describe the $\lambda^{(+)}$ for an isolated molecule, however solid-state effects must be considered to describe the electrical properties. Analogous sets of SMA and MMA analyses are described in ref [71] with slightly different values by comparing another gas-phase spectrum of PEN reported by Coropceanu et al [24]. The simplest method of SMA with UPS implies that purely experimental values of $\lambda^{(+)}$, $E_{pol}$ and $\lambda$ can be used in discussing the hopping charge mobility without using any theoretical computation. A similar agreement between the SMA and the theoretical values was recently demonstrated for rubrene by Duhm et al. [70].

The $\lambda$ value of PFP is about two times larger than those of PEN that is in agreement with the results of DFT calculations [25], indicating that there is large impact of perfluorination on the HOMO hole-vibration coupling and thus on the transport properties. Furthermore the $\lambda$(film) values are obviously larger than those estimated from gas-phase UPS and by theoretical calculation for both PFP and PEN probably due to electronic coupling to the substrate and neighbor molecules. This suggests that solid-state effects must be considered for the determination of the electron (hole)-vibration couplings, which is supported by the experimental results that weak intermolecular and molecule-substrate interaction influence the

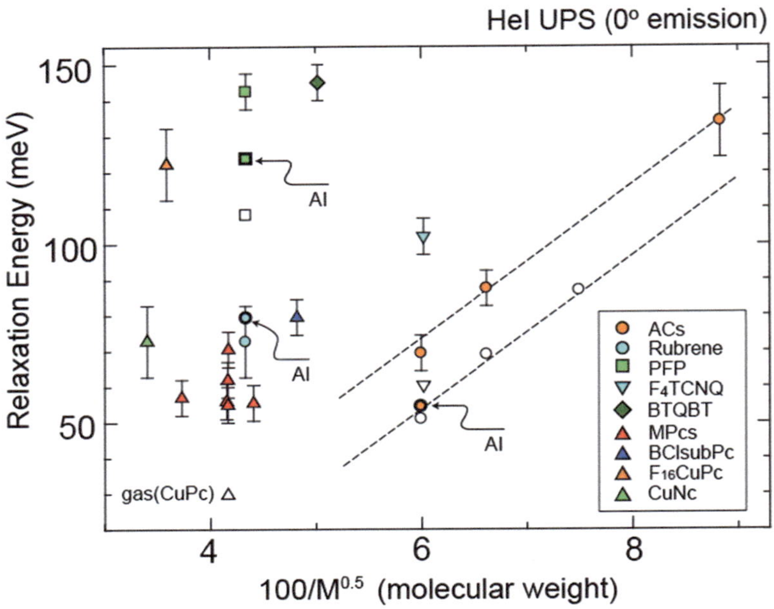


**Fig. 3.10** Experimental relaxation energy ($\lambda^{(+)} = E_{pol}$) vs. 1/M$^{0.5}$ (M: molecular weight) for various organic monolayer films. Results are for oligoacene (Ac: pentacene, tetracene, naphthalene), perfluoropentacene (PFP), rubrene, tetrafluoro-tetracyanoquinodimethane ($F_4$TCNQ), bis (1,2,5-thiadiazolo)-p-quinobis(1,3-dithiole) (BTQBT), metal-phthalocyanine (MPc; M=$H_2$, OV, OTi, ClAl, Cu, and Pb), Cu-naphthalocyanine (CuNc), Cu-hexadecafluorophthalocyanine ($F_{16}$CuPc) and chloro(subphthalocyaninato)boron (BClsubPc). Open symbols are from gaseous-Acs (pentacene, tetracene, and anthracene), PFP, CuPc, and $F_4$TCNQ. *Dashed lines* are visual guides. HOMO band measured at normal emission (acceptance angle of 12°) is fitted with Gaussian functions. Single normal mode analysis (typically 150–170 meV) was considered to determine relaxation energy. Errors are estimated from residual spectra. The SMA results for angle-integrated spectra (PEN, PFP and rubrene) are also plotted by a *thick symbol* (AI). Some data are obtained from unpublished results

vibrational spectra [72, 73]. Note that the large difference in the device mobility of these two materials is also caused by the crystal structure of the film prepared [51, 74].

The present results also clearly demonstrate that change in the wave function of by perfluorination is more critical for hole-vibration couplings and related energy parameters than change in vibration energies, although the substitution of H atoms with heavier F atoms can diminish the vibration energies and thereby contributes to reduce values of these parameters. The difference in the HOMO distributions of PEN and PFP is not very large (see insets in Figs. 3.8 and 3.9), but there is contribution of F 2p orbital to the PFP HOMO to result in slightly larger HOMO distribution over the molecule. Such a small change in the HOMO yields a significant increase in the HOMO hole-vibration coupling. This again indicates that electron–phonon coupling is more sensitive to the wave function of the responsible electronic state than the phonon energy. We observed the UPS-fine features of various organic semiconductor films. A brief summary of the experimental relaxation energies ($=E_{pol}$) are shown in Fig. 3.10, where the $\lambda^{(+)}$ estimated by SMA are plotted as a function of the inverse square root of the molecular weight (M). Here, the $\lambda^{(+)}$ for the films were estimated using the spectra recorded at $\theta = 0°$, which may give a larger error for $\lambda^{(+)}$ due to angular anisotropic dependence. Nevertheless, we can see clear a correlation between the $\lambda^{(+)}$ and the size of the $\pi$ electron system for oligoacenes (Acs) as has been described in theoretical papers [24].

Using Eq. (3.1), the upper limit of hopping mobility in a lying PEN and PFP film can be estimated as 0.07 $cm^2/Vs$ and 0.01 $cm^2/Vs$ at 295 K from $\lambda(PEN) =$ 104 meV and $\lambda(PFP) = 246$ meV, and a typical intermolecular distance of $a = 1$ nm and $T = 3$ meV for the oriented monolayer [23]. These results show that even a relatively moderate increase in $\lambda$ leads to a significant lowering of $\mu$.

**Acknowledgements** The authors are very grateful to Prof. V. Coropceanu and Dr. S. Nagamatsu for stimulating discussions on the computation as well as many students in the Graduate School of Chiba University for doing experiments.

The study is partly supported by JSPS KAKENHI (24245034, 20245039, 20685014, 20656002, 17685019), 21st Century COE program (MEXT) and Global COE program (MEXT).

## References

1. H. Inokuchi, Org. Electron. **7**, 62 (2006)
2. J.-L. Brédas, D. Beljonne, V. Coropceanu, J. Cornil, Chem. Rev. **104**, 4971 (2004)
3. V. Coropceanu, J. Cornil, D.A. da Silva Filho, Y. Oliver, R. Silbey, J.-L. Brédas, Chem. Rev. **107**, 926 (2007)
4. A. Troisi, Chem. Soc. Rev. **40**, 2347 (2011)
5. A. Troisi, Mol. Simulat. **32**, 707 (2006)
6. A. Troisi, G. Orlandi, Phys. Rev. Lett. **96**, 086601 (2006)
7. K. Mitsutake, M. Tsukada, e-J. Surf. Sci. Nanotech. **4**, 311 (2006)
8. F. Ortmann, F. Bechstedt, K. Hannewald, Phys. Status Solidi B **248**, 511 (2011)
9. F. Ortmann, F. Bechstedt, K. Hannewald, Phys. Rev. B **79**, 235206 (2009)
10. S. Ciuchi, S. Fratini, Phys. Rev. Lett. **106**, 166403 (2011)
11. L. Wang, G. Nan, X. Yang, Q. Peng, Q. Li, Z. Shuai, Chem. Soc. Rev. **39**, 423 (2010)
12. L. Wang, D. Beljonne, J. Phys. Chem. Lett. **4**, 1888 (2013)
13. H. Ishii, K. Honma, N. Kobayashi, K. Hirose, Phys. Rev. B **85**, 245206 (2012)
14. H. Yamane, S. Nagamatsu, H. Fukagawa, S. Kera, R. Friedlein, K.K. Okudaira, N. Ueno, Phys. Rev. B **72**, 153412 (2005)
15. S. Kera, H. Yamane, N. Ueno, Prog. Surf. Sci. **84**, 135 (2009)
16. A. Girlando, L. Grisanti, M. Masino, I. Bilotti, A. Brillante, R.G. Della Valle, E. Venuti, Phys. Rev. B **82**, 035208 (2010)
17. V. Coropceanu, L.Y. Zhu, Y. Li, Y.P. Yi, J.-L. Brédas, Mater. Res. Bull **38**, 57 (2012)
18. T. Holstein, Ann. Phys. (N.Y.) **8**, 325 (1959)
19. T. Holstein, Ann. Phys. (N.Y.) **8**, 343 (1959)
20. N.G. Martinelli, J. Ide, R.S. Sanchez-Carrera, V. Coropceanu, J.-L. Brédas, L. Ducasse, F. Castet, J. Cornil, D. Beljonne, J. Phys. Chem. C **114**, 20678 (2010)
21. T. Kato, T. Yamabe, J. Chem. Phys. **115**, 8592 (2001)

22. M. Malagoli, V. Coropceanu, D.A. da Silva Filho, J.-L. Brédas, J. Chem. Phys. **120**, 7490 (2004)
23. P.B. Paramonov, J.-L. Brédas, V. Coropceanu, Phys. Rev. B **78**, 041403(R) (2008)
24. V. Coropceanu, M. Malagoli, D.A. da Silva Filho, N.E. Gruhn, T.G. Bill, J.-L. Brédas, Phys. Rev. Lett. **89**, 275503 (2002)
25. M.C.R. Delgado, K.R. Pigg, D.A. da Silva Filho, N.E. Gruhn, Y. Sakamoto, T. Suzuki, R.M. Osuna, J. Casado, V. Hernández, J. Teodomiro, L. Navarrete, N.G. Martinelli, J. Cornil, R.S. Sánchez-Carrera, V. Coropceanu, J.-L. Brédas, J. Am. Chem. Soc. **131**, 1502 (2009)
26. D.A. da Silva Filho, E.-G. Kim, J.-L. Brédas, Adv. Mater. **17**, 1072 (2005)
27. G. Nan, X. Yang, L. Wang, Z. Shuai, Y. Zhao, Phys. Rev. B **79**, 115203 (2009)
28. H.-Y. Chen, I. Chao, Chem. Phys. Lett. **401**, 539 (2005)
29. W.Q. Deng, W.A. Goddard III, J. Phys. Chem. B **108**, 8614 (2004)
30. G.R. Hutchison, M.A. Ratner, T.J. Marks, J. Am. Chem. Soc. **127**, 2339 (2005)
31. F.-C. Wu, H.-L. Cheng, C.-H. Yen, J.-W. Lin, S.-J. Liu, W.-Y. Choua, F.-C. Tang, Phys. Chem. Chem. Phys. **12**, 2098 (2010)
32. K. Seki, Y. Harada, K. Ohno, H. Inokuchi, Bull. Chem. Soc. Jpn. **47**, 1608–1610 (1974)
33. W.R. Salaneck, Phys. Rev. Lett. **40**, 60–63 (1978)
34. W.R. Salaneck, C.B. Duke, W. Eberhardt, E.W. Plummer, H.J. Freund, Phys. Rev. Lett. **45**, 280 (1980)
35. N. Ueno, S. Kera, Prog. Surf. Sci. **83**, 490 (2008)
36. H. Yamane, N. Kosugi, Phys. Rev. Lett. **111**, 086602 (2013)
37. N. Ueno, Ph.D. Thesis, Tohoku University, (1976), in Japanese
38. S. Kera, H. Yamane, I. Sakuragi, K.K. Okudaira, N. Ueno, Chem. Phys. Lett. **364**, 93 (2002)
39. T. Sugiyama, T. Sasaki, S. Kera, N. Ueno, T. Munakata, Chem. Phys. Lett. **449**, 319 (2007)
40. T. Sugiyama, T. Sasaki, S. Kera, N. Ueno, T. Munakata, Appl. Phys. Lett. **89**, 202116 (2006)
41. S. Duhm, G. Heimel, I. Salzmann, H. Glowatzki, R.L. Johnson, A. Vollmer, J.R. Rabe, N. Koch, Nat. Mater. **7**, 326 (2008)
42. M.B. Casu, Y. Zou, S. Kera, D. Batchelor, T. Schmidt, E. Umbach, Phys. Rev. B **76**, 193311 (2007)
43. T. Sueyoshi, H. Kakuta, M. Ono, K. Sakamoto, S. Kera, N. Ueno, Appl. Phys. Lett. **96**, 093303 (2010)
44. F. Bussolotti, S. Kera, K. Kudo, A. Kahn, N. Ueno, Phys. Rev. Lett. **110**, 267602 (2013)
45. H. Yoshida, N. Sato, Chem. Phys. Lett. **511**, 146 (2011)
46. H. Hövel, B. Grimm, M. Pollmann, B. Reihl, Phys. Rev. Lett. **81**, 4608 (1998)
47. A. Tanaka, Y. Takeda, T. Nagasawa, S. Sato, Phys. Rev. B **67**, 033101 (2003)
48. V. Čápek, E.A. Silinsh, Chem. Phys. **200**, 309 (1995)
49. E.A. Silinsh, V. Čápek, *Organic Molecular Crystals: Interaction, Localization and Transport Phenomena* (Springer, Berlin, 1994)
50. E.A. Silinsh, *Organic Molecular Crystals. Their Electronic States* (Springer, Berlin, 1980)
51. H. Yoshida, N. Sato, Phys. Rev. B **77**, 235205 (2008)
52. H. Fukagawa, S. Kera, T. Kataoka, S. Hosoumi, Y. Watanabe, K. Kudo, N. Ueno, Adv. Mater. **19**, 665 (2007)
53. M. Shionoiri, M. Kozasa, S. Kera, K.K. Okudaira, N. Ueno, Jpn. J. Appl. Phys. **46**, 1625 (2007)
54. I. Yamamoto, N. Matsuura, M. Mikamori, R. Yamamoto, T. Yamada, K. Miyakubo, N. Ueno, T. Munakata, Surf. Sci. **602**, 2232 (2008)
55. H. Yamane, K. Ito, S. Kera, K.K. Okudaira, N. Ueno, J. Appl. Phys. **92**, 5203 (2002)
56. T. Kataoka, H. Fukagawa, S. Hosoumi, K. Nebashi, K. Sakamoto, N. Ueno, Chem. Phys. Lett. **451**, 43 (2008)
57. G. Heimel, S. Duhm, I. Salzmann, A. Gerlach, A. Strozecka, J. Niederhausen, C. Bürker, T. Hosokai, I. Fernandez-Torrente, G. Schulze, S. Winkler, A. Wilke, R. Schlesinger, J. Frisch, B. Bröker, A. Vollmer, B. Detlefs, J. Pflaum, S. Kera, K.J. Franke, N. Ueno, J.I. Pascual, F. Schreiber, N. Koch, Nat. Chem. **5**, 187 (2013)

58. Y.L. Huang, W. Chen, F. Bussolotti, T.C. Niu, A.T.S. Wee, N. Ueno, S. Kera, Phys. Rev. B **87**, 085205 (2013)
59. R. Matzdorf, Surf. Sci. Rep. **30**, 153 (1998)
60. N. Ueno, S. Kera, K. Sakamoto, K.K. Okudaira, Appl. Phys. **A92**, 495 (2008)
61. S. Kera, H. Fukagawa, T. Kataoka, S. Hosoumi, H. Yamane, N. Ueno, Phys. Rev. B **75**, 121305 (R) (2007)
62. H. Fukagawa, H. Yamane, S. Kera, K.K. Okudaira, N. Ueno, Phys. Rev. B **73**, 041302 (R) (2006)
63. H. Yamane, H. Honda, H. Fukagawa, M. Ohyama, Y. Hinuma, S. Kera, K.K. Okudaira, N. Ueno, J. Electron Spectros. Relat. Phenomena **137–140**, 223 (2004)
64. H. Yamane, H. Fukagawa, H. Honda, S. Kera, K.K. Okudaira, N. Ueno, Synth. Met. **152**, 297 (2005)
65. H. Fukagawa, S. Hosoumi, H. Yamana, S. Kera, N. Ueno, Phys. Rev. B **83**, 085304 (2011)
66. H. Yamane, S. Nagamatsu, H. Fukagawa, S. Kera, K.K. Okudaira, N. Ueno, Mol. Cryst. Liq. Cryst. **455**, 235 (2006)
67. X.-Y. Zhu, Surf. Sci. Rep. **56**, 1 (2004)
68. R.A. Marcus, J. Chem. Phys. **43**, 679 (1965)
69. R.A. Marcus, Rev. Mod. Phys. **65**, 599 (1993)
70. S. Duhm, Q. Xin, S. Hosoumi, H. Fukagawa, K. Sato, N. Ueno, S. Kera, Adv. Mater. **24**, 901 (2012)
71. S. Kera, S. Hosoumi, K. Sato, H. Fukagawa, S. Nagamatsu, Y. Sakamoto, T. Suzuki, H. Huang, W. Chen, A. T. S. Wee, V. Coropceanu, N. Ueno, J. Phys. Chem. C **117**, 22428 (2013)
72. K. Fujii, S. Kera, M. Oiwa, K.K. Okudaira, K. Sakamoto, N. Ueno, Surf. Sci. **601**, 3765 (2009)
73. K. Fujii, C. Himcinschi, M. Toader, S. Kera, D.R.T. Zahn, N. Ueno, J. Electron Spectros. Relat. Phenomena **174**, 65 (2007)
74. Y. Sakamoto, T. Suzuki, M. Kobayashi, Y. Gao, Y. Funaki, Y. Inoue, F. Sato, S. Tokito, J. Am. Chem. Soc. **126**, 8138 (2004)

# Chapter 4
# Ultraviolet Photoelectron Spectroscopy (UPS) III: Direct Study of "Invisible" Band Gap States by Ultrahigh-Sensitivity UPS

**Nobuo Ueno, Tomoki Sueyoshi, Fabio Bussolotti, and Satoshi Kera**

## 4.1 Introduction to the Role of Band Gap States and Detection by Ultrahigh Sensitivity UPS

Organic electronic devices, such as light emitting diodes, solar cells, field effect transistors or sensors, are all based on thin film architectures that comprise multiple interfaces between organic semiconductors (small molecules or polymers), organic or inorganic dielectrics, and metals or other types of conducting electrodes. Without exception, the performance of these devices is dominated by the electronic structure and electrical behavior of these interfaces. During the past two decades of development of organic electronics, organic interfaces have therefore been the object of considerable attention and extensive investigations [1–6].

The transport of charge carriers through a semiconductor interface is affected by the atomic or molecular structure of this interface, i.e., bonding or molecular orientation across the interface, as well as the electronic structure of this boundary region, i.e., the relative positions of the relevant electronic transport levels. In a simple metal/organic film/metal system, these levels are the Fermi level of the metal electrode, the lower edge of the lowest unoccupied molecular orbital (LUMO) band, the upper edge of the highest occupied molecular orbital (HOMO) band and the Fermi level of the organic film. When the system is in thermal equilibrium, the two Fermi levels coincide at the interface to result in the single Fermi level throughout the system.

For electron current through the organic system, we need electron injection from the electrode into the organic thin film structure. The injection efficiency, which determines mobile electron concentration, depends on the energy barrier defined between the electrode Fermi level and the LUMO level of the organic layer

N. Ueno (✉) • T. Sueyoshi • F. Bussolotti • S. Kera
Department of Nanomaterial Science, Graduate School of Advanced Integration Science, Chiba University, Inage-ku, Chiba 263-8522, Japan
e-mail: uenon@faculty.chiba-u.jp



H. Ishii et al. (eds.), *Electronic Processes in Organic Electronics*,
Springer Series in Materials Science 209, DOI 10.1007/978-4-431-55206-2_4

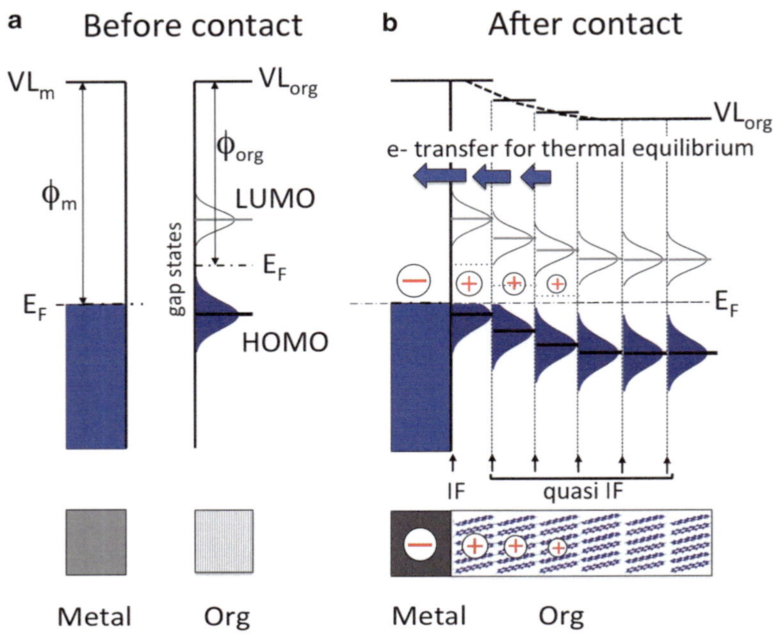


**Fig. 4.1** Energy level alignment at a weakly interacting interface of metal and organic film. (**a**) Before contact. (**b**) After contact, where the thermal equilibrium is achieved. The work function ($\phi_m$) is assumed to be larger than that of the organic film ($\phi_{org}$). The HOMO and LUMO levels (shown by *lines*) are assumed to have broadening that results in gap states due to imperfection of molecular packing structure. IF indicates the metal-organic interface, while "quasi IF" indicates the interfaces existing between the molecular layers in the film. The holes exist at the gap states produced by the packing imperfection

according to the energy level alignment. The electrons transport at the LUMO level with energy gain or loss, then finally they are collected at the other electrode. The hole transport process is very similar, involving the HOMO level of the organic film.

Without exception, therefore, the relative positions of the electrode Fermi level and the HOMO (for hole) and the LUMO (for electron) levels of the organic layer have key importance for electrical current between the electrodes, and the concept is simply called as "energy level alignment". The energy level alignment is achieved after reaching the thermal equilibrium by electron transfer through the interface as shown in Fig. 4.1. The number of transferred electrons depends on the relative position of the two Fermi levels before the contact of the electrode and the organic film, because after the contact electrons in the lower work function material flow to the higher work function material to achieve thermal equilibrium of the electron system. If the two materials do not interact strongly at the interface, we may assume "no interface electric dipole just at the interface". In this case the electron transfer creates electrostatic potential in the organic film (see also Fig. 1.2

in Chap. 1) relative to the electrode in which the potential is constant because it is electrical conductor. The Fermi level of the system is written with a line in usual energy diagram of the system and the effect of the potential is written as curved HOMO and LUMO levels with respect to the Fermi level. Thus the Fermi level position within the HOMO-LUMO gap depends on the distance from the interface. Near the interface, it is possible that the Fermi level is located closer to the HOMO even for n-type organic semiconductor, while the Fermi level is closer to the LUMO for p-type one, due to the potential produced by excess charges in the organic films (see Fig. 4.1). The Fermi level position in the HOMO–LUMO gap is controlled by the gap states, since the potential changes with the distance from the interface depending on the excess charge distribution that is related to the energy and spatial distribution of the density of occupied and unoccupied gap states.

As a result the Fermi level position in the HOMO–LUMO gap is a unique indicator of various interface phenomena and plays a crucial role for the energy level alignment and bending of the HOMO and the LUMO in the organic film. This means that the gap states control the hole and electron injection barriers thus concentration of the mobile hole and electron in organic layer. Moreover the gap states also act as charge trapping states and reduce the charge mobility significantly. Due to very anisotropic property of organic molecule the band (level) bending in an organic film must be discussed with the concept of the energy level alignment at quasi-interfaces between molecular layers even in a single component organic film (Fig. 4.1 and see Fig. 1.2 in Chap. 1).

In many cases the terminology of "interface dipole" is used also for the potential produced in the molecular layers near the interface by excess charges introduced by achieving the thermal equilibrium, but it should be distinguished from the interface dipole produced by electronic interaction just at the interface for more precise discussion on the energy level alignment [6]. The former may be specified by using a terminology, for example, *quasi-interface dipole*.

It has been known that there are following mysterious phenomena in conductor-organic interface systems:

(1) The Fermi level moves within the HOMO-LUMO gap depending on interfaces, molecules and molecular orientation, *even for weakly interacting* organic-related interfaces,
(2) In many cases pentacene shows p-type transport property, $C_{60}$ exhibits n-type transport property etc., which seem to be determined by molecular structure,
(3) We always need very high impurity concentration for doping effects (% order for organic semiconductors, while ppm order for inorganic counterparts).

An intuitive answer given by people was "there must be unknown impurities". However, we also know that these phenomena were observed insensitive to purification level and many reported experimental results were not fully consistent.

Accordingly it is necessary to know true origins of *invisible* gap states for deeper understanding of the mechanism of the energy level alignment at organic-electrode interfaces, pinning of the Fermi level, and band-bending-like phenomena in organic semiconductor films, all of which impact seriously electrical properties of various organic devices. As understood from physics of inorganic semiconductors, all these are related to electronic states in the band gap. To reveal mysteries in organic semiconductors, it was highly requested to measure directly and quantitatively the gap states in organic semiconductors. A possible method for this measurement is ultraviolet photoelectron spectroscopy (UPS), but the use of conventional UPS was impossible because of its very low sensitivity to detect the very tiny density of band-gap states in particular for organic materials due to radiation damages introduced by the photons and photoelectrons during UPS measurement. In the following we describe the ultrahigh-sensitivity UPS that enables us to observe *invisible* gap states and offer both of density of the gap states and their energy distribution, and discuss origin of gap states and their role.

## 4.2 Development of Ultralow-Background and Ultrahigh-Sensitivity UPS

UPS is well known as the most potential tool to investigate one-hole state in various materials. For solids, however, there are following principal drawback especially for UPS measurements of organic solids:

1. **Impossible to measure electrical insulator due to charging effects**
   As electrical insulator is positively charged immediately at photoelectron emission, one cannot measure kinetic energy and momentum vector of photoelectrons correctly, while can measure approximately the number of photoelectrons by applying larger negative electrical potential to the sample than the positive potential by the charging. The latter method is called as photoelectron or photoemission yield spectroscopy [see Chap. 8], but largely different from UPS. Accordingly one needs to use thin films of the insulator for UPS, where the sample charging must be negligible.
2. **Poor sensitivity**
   The sensitivity is much poorer than that of the electrical measurement of charge trapping states. This makes it difficult to measure low-density electronic states in the band gap, namely charge trapping centers and impurity levels, which play a key role in the energy level alignment and electrical properties of various organic devices.
3. **Radiation damages of organics**
   It is not easy to observe organic materials as they are, because organic materials are easily damaged by the incidence ionizing radiation. In same cases physical effects such as disorder or changes in molecular packing structure are introduced in addition to chemical changes of molecules.
4. **Too high surface sensitivity**
   Very short mean-free-path of photoelectrons [10 Å for organics (c.a. with He I radiation, $h\nu = 21.218$ eV)] makes measurements of the bulk of solids and buried interfaces impossible.

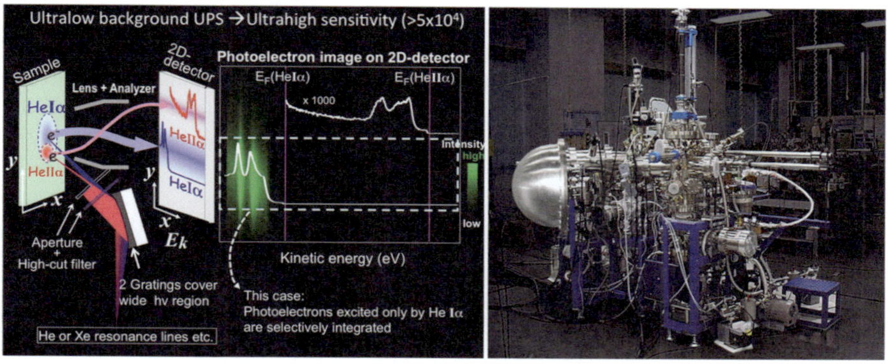


**Fig. 4.2** Method of the high sensitivity measurement (*left*) and a photograph of the UPS instrument (*right*)

As introduced in Chap. 2, the 1st drawback (I) was overcome at Chiba University by using photoconduction phenomena or using charge-trap free specimens, namely using very high quality single crystals. In this chapter, we describe the ultralow background ultrahigh sensitivity UPS developed also at Chiba University, which could overcome the drawback II-IV, and present selected results that have never been obtained.

Figure 4.2 displays the method of ultrahigh sensitivity measurement with a photograph of the apparatus. The UPS uses a hemispherical electron energy analyzer (MBS A-1) with two-dimensional detector and two monochromators (MBS M-1/modified) for the incidence lights. The first monochromator is for photon energy (h$\nu$) range of He $I_\alpha$ (h$\nu = 21.218$ eV) and He $II_\alpha$ (h$\nu = 40.814$ eV) radiations, and the second one covers lower h$\nu$ range of other gas resonance lines including Xe I (h$\nu = 8.437$ eV/$^3P_1$ emission and 9.570 eV/$^1P_1$ emission: here we conventionally use Xe $I_\alpha$ for the former line) and hydrogen $Ly_\alpha$ ( h$\nu = 10.199$ eV) and Ne I (h$\nu = 16.673$ eV and 16.850 eV). There are two discharge lamps installed to the UPS system, high-density plasma lamp (MBS L-1) for He discharge and a conventional high intensity discharge lamp (Omicron HIS 13) for the other gases. Two filters, Al thin film or LiF single crystal depending on the desired photon, are alternatively used to reduce the intensity of high/low-h$\nu$ impurity photons from the monochromator. Using a two-dimensional electron detector, moreover, photoelectrons excited by desired photons are selected for lowering secondary electron background produced by inelastic scattering of high kinetic energy photoelectrons excited by remaining high-h$\nu$ photons, since the compact monochromator is not perfect. As photons reflected by the specimen produces electron background in the measurement chamber when they hit the inner surface of the chamber, we pay attention to reduce these electrons.

The electrical wiring of the measurement system is carefully performed to minimize electrical noises by putting the electrical cables at appropriate positions. In particular connectors and coaxial cables for signals are checked frequently to minimize electrical noise level. The AC power is supplied after passing noise-cut transformer. The system is grounded with higher-class earth line. To be more careful the UPS system is installed on a floor separated mechanically from working floor to minimize vibration-induced electrical and magnetic noises. Some of the results obtained with a prototype instrument [4, 7–9] were also discussed in [10] and Chap. 5.

## 4.3 Role and Origin of Invisible Gap States

### *4.3.1 Gas Exposure Effects on Pentacene Thin Films*

The UPS measurements were performed at RT using the ultrahigh-sensitivity UPS apparatus and monochromatic Xe-I$_\alpha$ radiation. All UPS spectra were measured at normal emission with an acceptance angle of $\pm 18°$, and a bias of $-5$ V was applied to the sample in order to measure the vacuum level (VL). The energy resolution of the UPS system was set to 30 meV. The binding energy scale is relative to the substrate Fermi level. The accumulation time of the spectra was typically 40 min. Under these conditions, radiation damage effect [11] to the pentacene films could be excluded as seen in Fig. 4.3 [12]. The highest occupied molecular orbital (HOMO) band of the as-deposited pentacene thin film consists of two main components [labeled H$_1$ and H$_2$] with an energy separation of ~0.45 eV. The two components correspond to the density of states (DOS) of the dispersed HOMO

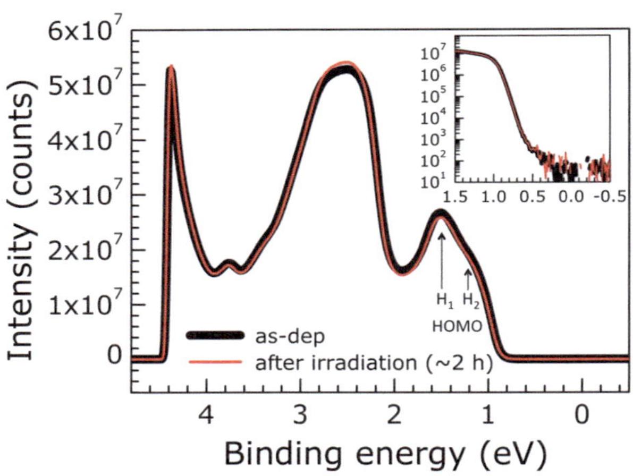


**Fig. 4.3** Full Xe I$_\alpha$-UPS spectra of the as-deposited pentacene film (thickness = 15 nm) on $SiO_2$ before (*black curve*) and after (*red curve*) prolonged XeI$_\alpha$ light irradiation (~2 h) [12]. The typical acquisition time during density of gap states (DOGS) detection is ~40 min. No spectral shift and change in DOGS (see *inset*, shown on log scale) is detected after irradiation, indicating the absence of significant effects from irradiation induced damage

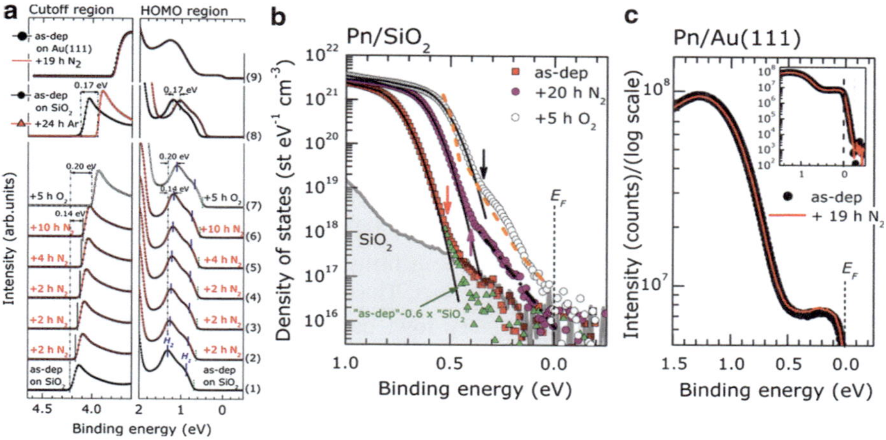


**Fig. 4.4** (**a**) XeI$_\alpha$-UPS spectra of as-deposited [spectrum (1)], $N_2$-exposed [spectra (2)–(6)], $O_2$-exposed [spectra (7)] and Ar-exposed [spectra (8)] pentacene (Pn) thin film (15 nm) on $SiO_2$ substrate. Spectrum (9) includes the UPS data of as-deposited and $N_2$-exposed pentacene/Au(111) thin film (40 nm) acquired at RT. The positions of the VL and HOMO derived band ($H_1$ and $H_2$) are indicated by *vertical* (*continuous*) *bars*. The difference between the VL (HOMO) positions of the pentacene/$SiO_2$ thin films in the "$N_2$" and "Ar" experiments reflects the difference between the initial $SiO_2$ work functions. (**b**) Density of states (DOS: log scale intensity plot) of as-deposited [*filled* (*red*) *square* symbols], $N_2$-exposed [20 h, *filled* (*magenta*) *circles*] and $O_2$-exposed (5 h, *open circles*) pentacene thin film on $SiO_2$. *Continuous* (*black*) *lines* are the cumulative fitting curve of the HOMO band. DOS values were extracted from the UPS data, as described in [8]. $SiO_2$ data are rescaled to preserve, in the DOS scale, their relative intensity with respect to the as-deposited pentacene data. *Vertical arrows* indicate the threshold energy position ($E_T$) where the DOS distribution deviates from the cumulative fitting curve and the density of gap states (DOGS) starts to appear. The *dashed line* indicates the DOGS of the pentacene thin film as determined by transport measurements. Data are adapted from [15]. (**c**) XeI$_\alpha$-UPS spectra of as-deposited pentacene thin film on Au(111) (40 nm) before and after $N_2$ exposure. The Fermi edge of the Au(111) substrate is clearly visible (see also *inset*). No spectral change was detected upon gas exposure. The figure is cited from [16] with permission

bands in pentacene thin film with an upright-standing molecular orientation and herringbone-like intermolecular packing (thin film phase crystal structure) [13, 14].

Figure 4.4a shows the Xe-I$_\alpha$ UPS spectra of the as-deposited and $N_2$- ($O_2$-) exposed pentacene film (thickness = 15 nm) in the cutoff and highest occupied molecular orbital (HOMO) regions [spectra (1)–(7)] [16]. The gas pressure of all gas ($N_2$, $O_2$ and Ar)-exposure experiments is one atmosphere (1 atm). Following each $N_2$-exposure, the $H_1$ ($H_2$) peak shifts towards low binding energies, while the vacuum level (VL) moves upwards by the same amount. This $N_2$-induced "rigid" energy shift almost saturates after a total exposure time of 20 h. Further exposure to $O_2$ (5 h) produces an additional energy shift [spectra (7)]. Interestingly, the ionization potential (IP) of the as-deposited pentacene thin film (IP = 4.90 eV) is unchanged upon the various gas exposure treatments. The IP of an organic thin film was reported to be strongly dependent on molecular packing [13] and the molecular orientation [17]. Therefore, the stability of the IP in the present case indicates that no large-scale structural rearrangement occurs. Similar results are obtained for the Ar-exposed pentacene/$SiO_2$ thin film [spectra (8)], while no spectral shift is detected for pentacene/Au(111) thin film after prolonged exposure (19 h) to 1-atm $N_2$ [spectra (9)] [12, 16].

Figure 4.4b shows the expanded UPS spectra of the HOMO and gap energy region on log scale for the as-deposited, $N_2$-exposed (total exposure time = 20 h) and $O_2$-exposed (total exposure time = 5 h following the 20 h $N_2$ exposure) pentacene thin film with the corresponding fitting curves. The UPS spectrum of the $SiO_2$ substrate is shown for comparison. The density of gap states (DOGS) of the as-deposited pentacene film is very low, but it significantly increases after $N_2$-and $O_2$-exposure. Ar-exposure also leads to a similar DOGS increase ($E_B$ shift is seen from Fig. 4.4a: spectra (8)) [12]. In the gap state binding energy region (~0–0.5 eV), the DOS of the as-deposited pentacene film and the $SiO_2$ substrate are comparable [Fig. 4.4b]. Therefore, the $SiO_2$ DOS may overlap with the DOGS of the as-deposited pentacene film in a way that depends on the electron mean free path ($\lambda_{Pn}$) in the pentacene overlayer, which was determined to be 9 nm $\leq \lambda_{Pn} \leq$ 27 nm using pentacene/Au(111) [12, 16]. Once the substrate contribution is taken into account, the DOGS of the as-deposited film (~$10^{16}$ states $eV^{-1}$ $cm^{-3}$) turns out to be comparable to that detected by electrical measurements for pentacene on $SiO_2$ [15, 18]. For the gas-exposed pentacene films, the DOGS is ~10 times larger than the contribution from the substrate, and can therefore be unambiguously related to the pentacene film. In particular, for the $N_2$-exposed sample, an exponential-type DOGS [linear on the log scale of Fig. 4.4b] is clearly visible between the HOMO threshold energy $E_T = 0.4$ eV (where the DOS starts deviating from a Gaussian line shape) and the Fermi level ($E_F$). A different energy dependence is observed for the DOGS of the $O_2$-exposed sample. Post-annealing experiments (50 °C for 18 h) on $N_2$-exposed pentacene films show a gradual recovery to the original VL and HOMO positions (i.e. before gas exposure) [12, 16]. At the same time, a decrease in the DOGS is observed. Consequently, the DOGS of the gas-exposed sample can be ascribed to a slight intermolecular packing disorder resulting from prolonged gas exposure. These gap states may in turn affect the position of $E_F$ within the energy gap [8]. Similar effect was observed for thinner pentacene films. Note that no residual $N_2$ is detected by X-ray photoemission spectroscopy in the gas-exposed film following re-introduction into UHV [12, 16]. This supports the conclusion that the DOGS is not due to $N_2$-related states but rather to intermolecular packing disorder. The increase in DOGS and its different energy distribution upon $O_2$ exposure can probably be ascribed to (i) a more effective penetration of $O_2$ molecules into the pentacene film (the penetration of a gas into organic systems strongly depends on its chemical properties and size [19, 20]) and/or (ii) chemical interaction between $O_2$ molecules and pentacene molecules [21, 22]. However, a prolonged $N_2$ exposure of pentacene films prepared at RT on single crystalline Au(111) does not induce any detectable change in the DOGS or spectral shift [Fig. 4.4a, c]. Moreover, it is worth to note that Fig. 4.4c clearly indicates that the present UPS can observe buried interface as understood

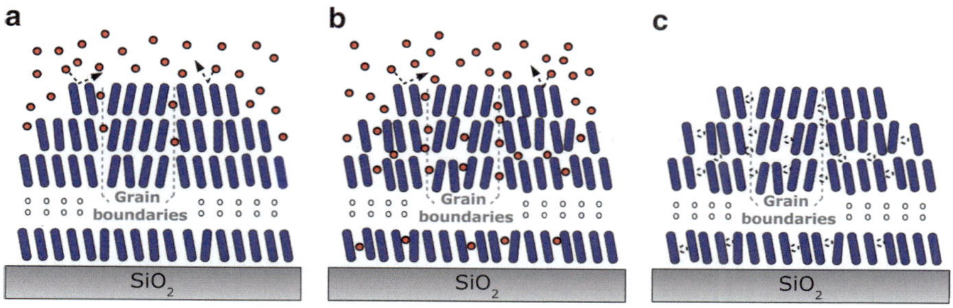


**Fig. 4.5** Schematic representation of the gas penetration and mediated imperfections in pentacene thin film on $SiO_2$. During exposure [panel (**a**)], gas molecules progressively penetrate into the pentacene film, where they *locally alter* the original intermolecular packing geometry [panel (**b**)] [16]. The gas penetration proceeds through the grain boundaries of the film. Because of the weak $N_2$-pentacene interaction, $N_2$ molecules are easily removed once the film is put back in UHV. The result is a weakly disordered film [(panel (**c**)]. The figure is cited from [16] with permission

from the intensity of the Fermi edge from the Au(111) single crystal substrate. Note that the molecular orientation in thin pentacene films is nearly standing on $SiO_2$ and nearly flat-lying on Au(111) [21, 23].

Pentacene films deposited on Au(111) was reported to exhibit larger crystallites [size ~ 200 nm] and significantly fewer grain boundaries than those on $SiO_2$ [23, 24], suggesting that grain boundaries play a crucial role in the penetration of gas molecules and in the changes of the intermolecular packing geometry as depicted in Fig. 4.5. According to the UPS data [Fig. 4.4b], the defect density is estimated to be in the range of $10^{16}$–$10^{18}$ $cm^{-3}$, corresponding to a defect-to-molecule ratio of $10^{-5}$–$10^{-3}$, which is hardly detectable via X-ray diffraction techniques at the present sensitivity [25]. The hypothesis that the DOGS is mediated by structural defects in the pentacene film is also supported by theoretical calculations [26, 27]. For pentacene the calculation showed that (i) sliding defects along the pentacene long axis create shallow gap states and (ii) the HOMO levels at the defect sites are distributed over a range of up to 100 meV from the HOMO of the unperturbed molecules [26].

It is interesting to see at which energy the DOGS increases. To verify this point, the difference and the ratio spectra between the gas-exposed and the as-deposited films are shown in Fig. 4.6. The difference spectra highlight the change of the density of states in the high intensity region [panel (b)], while the ratio spectra are sensitive to the change in very low intensity region [panel (c)]. The increase in the gap state (for $E > E_T$, spectral difference >0, and spectral ratio > 1) corresponds to a reduction in the HOMO band intensity (for $E < E_T$, spectral difference <0, and spectral ratio <1) within ~0.2 eV from the threshold position ($E_T$). In the main region of the HOMO band, the slight increase and decrease of the DOS and/or the small broadening in the HOMO lineshape, as due to tiny structural disorder induced by gas exposure [15], exist (observed at $E < E_T$). These changes are related to those of the HOMO that are observed for films without proper annealing in advance.

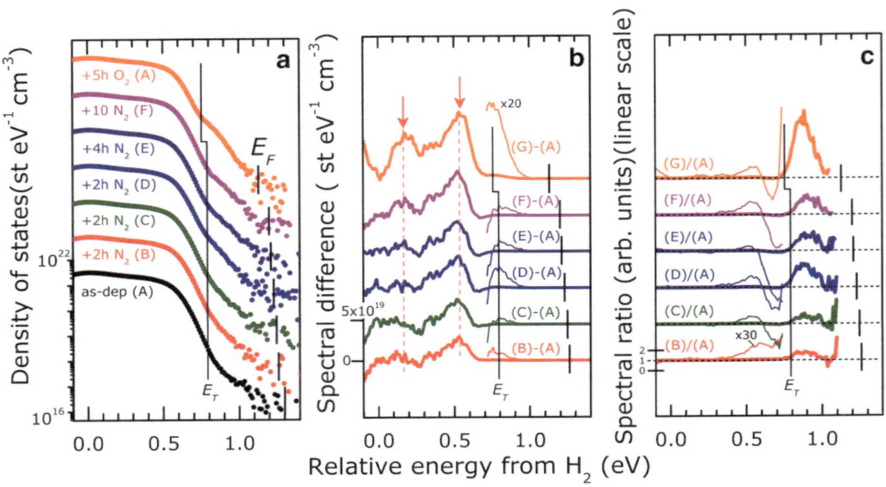


**Fig. 4.6** (**a**) $XeI_\alpha$-UPS data (log scale) of as-deposited, $N_2$-exposed and $O_2$-exposed pentacene/ $SiO_2$ thin film in the HOMO band region [16]. The energy scale is referred to the $H_2$ position [see Fig. 4.3a], where the HOMO band reaches the maximum intensity. The *vertical*, *short bars* on each plot indicate the position of the Fermi level. The $SiO_2$ substrate signal was not removed from the as-deposited sample spectra. The *continuous* (*black*) *line* indicates the position of the threshold energy ($E_T$) from which the gap state energy distribution starts to appear. (**b**) Spectral difference between the data of gas-exposed film and as-deposited film. (**c**) Spectral ratio between the data of gas-exposed film and as-deposited film [16]. Spectral ratio highlights that the gap states position is unchanged and the $E_F$ moves to the HOMO with increase in the gap states. Panels (**a**) and (**c**) are cited from [16] with permission

Finally, we comment on the impact of the gas exposure on the observed molecular level shift, i.e. the shift of the HOMO level towards $E_F$ (see Fig. 4.4a and [12]). We suggest that the lowest unoccupied molecular orbital (LUMO) of pentacene also gives rise to a distribution of (unoccupied) gap states resulting from imperfections in the intermolecular packing [9, 28]. The energy distributions of the HOMO- and LUMO-derived DOGS are not symmetric because the corresponding wave functions have different spatial spreads [9, 26, 29]. Such disorder-induced-, non-symmetric DOGS (tailing states) affect the position of $E_F$ within the gap of organic semiconductors. Depending on the DOGS and their energy distribution, $E_F$ may lie closer to the pristine LUMO or HOMO band [9]. With increasing structural disorder (as induced, for example, by gas exposure) $E_F$ is expected to lie even closer to the HOMO, as observed in the present work. The evolution of the energy level upon the $N_2$ and $O_2$ exposure is summarized in Fig. 4.7 [12], and final $E_F$ position in the band gap is shown in Fig. 4.8 using the electron affinity of pentacene film measured with radiation-damage free and higher-resolution inverse photoemission spectroscopy (IPES) [29]. Similar gas exposure effects are observed for organic pn heterojunction [30].

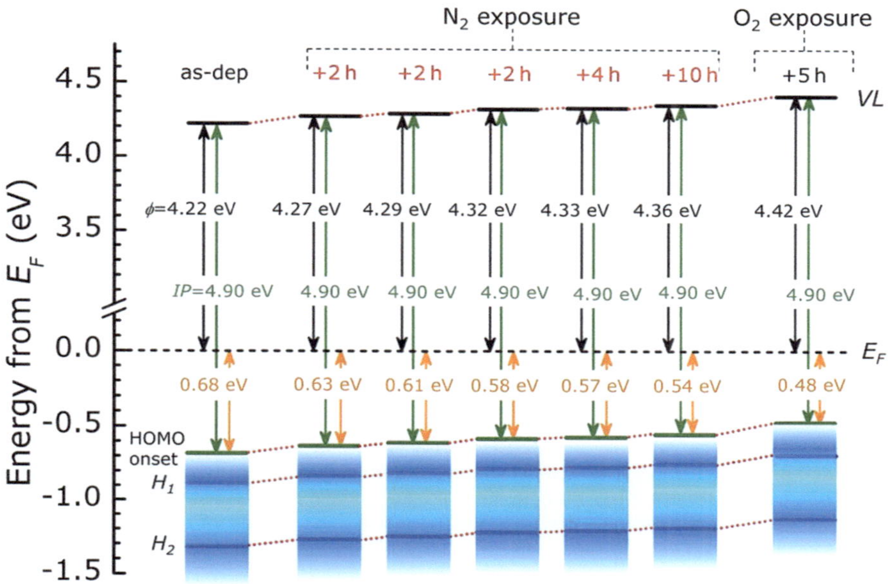


**Fig. 4.7** Energy level evolution by the gas exposure of pentacene thin film, as determined from the UPS data in Fig. 4.4a [12]. The IPs of the as-deposited and gas-exposed Pentacene film was determined from the position of the vacuum level (VL) (i.e. work function ϕ) and of the HOMO onset with respect to the Fermi level $E_F$. $H_1$ and $H_2$ indicate the HOMO sub bands. The HOMO onset position was determined by linear extrapolation of the HOMO band of the UPS data. The figure is cited from [12] with permission

### *4.3.2 Why Does Pentacene Always Show p-Type Property in Device?*

The above results indicate that pentacene becomes more p-type upon exposure to gases judging from the Fermi level position, which is an intrinsic property of bulk region of pentacene film. One should be careful in using terminology, p and n type, for organic semiconductors, since for example so-called p- or n-type conduction in organic transistors does not mean that the majority carriers are created by thermal excitation as in inorganic semiconductors, but that they are dominated by hole or electron injection efficiency from electrode with higher or lower work function than the organic film, respectively. Thus the intrinsic p-type film, which is defined by majority thermal carrier (hole), can also show n-type conduction in an organic field effect transistor when one uses an electrode with much lower work function (WF) than for example pentacene for electron injection, where the concentration of electrons under the device operation can be much larger than that of thermally created holes. It is true, however, that the intrinsic p-type property of pentacene, where the Fermi level is located closer to HOMO in the gap in air, greatly helps hole

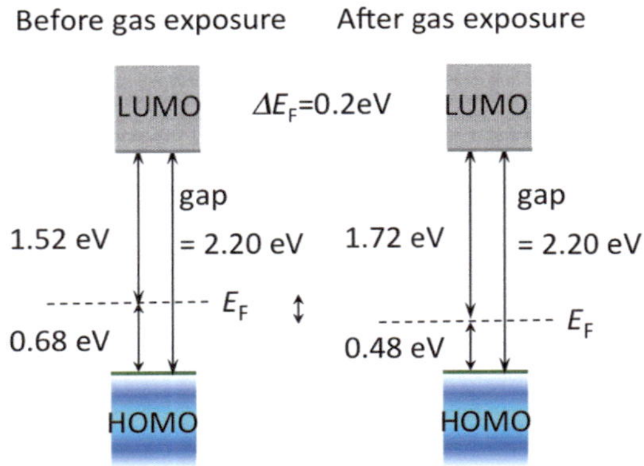


**Fig. 4.8** The Fermi level of pentacene moves by 0.2 eV to the HOMO upon the $N_2$ and $O_2$ exposure. The newest values of the electron affinity (EA = 0.27 eV) and the band gap (2.20 eV) measured with a high-energy-resolution radiation-damage-free IPES are used [29]. The threshold ionization energy (IP) is 4.90 eV for both of the pristine and the gas-exposed pentacene films (see Fig. 4.7). The figure demonstrates that the gas exposed pentacene film has intrinsically more p-type property defined from thermal carriers

injection from the electrode, thus pentacene thin film may generally show p-type property. Again as seen in Figs. 4.7 and 4.8 that even for the as-grown film the Fermi level does not locate at mid gap position but closer to HOMO, giving intrinsic p-type property. This can be due to larger tailing from LUMO into the band gap than tailing from HOMO [10, 12]. For n-type films, namely $C_{60}$, the situation is reversed. These are related to difference between spatial distributions of HOMO and LUMO orbitals and/or degeneracy of these electronic states. The larger degeneracy gives larger density of tailing states, which would push the Fermi level to the level with the lower degeneracy.

### *4.3.3 Contribution of Fully Relaxed Large Polaron*

When we introduce excess charges into organic films, various polarization (relaxation) effects appear in the film by changes in electron distribution, positions of atoms in each molecule and positions of molecules [31]. These polarization effects exist near the excess charge, and each excess charge with the polarization behaves as a quasi-particle, namely polaron. The time scale ($t$) of the polarization effects is different for the movement of electrons ($t_e$), atoms in each molecule ($t_{vib}$) and molecules in a molecular crystal ($t_{ph}$), and generally $t_e \ll t_{vib} \ll t_{ph}$ for molecular crystals since the time depends on mass of moving body. Accordingly there are various polarons, namely electronic polaron, molecular-vibration related polaron and crystal-phonon related polaron, which can be distinguished by the time required for the relaxation of the system after introducing the charges [31].

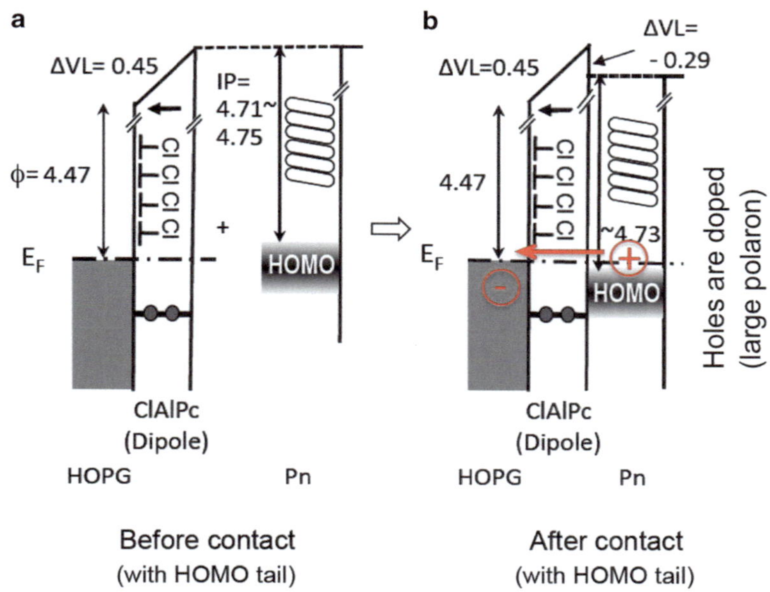


**Fig. 4.9** Hole doped pentacene (Pn) [(**a**) before Pn deposition (before hole doping), (**b**) after Pn deposition (after hole doping). The substrate is ClAl-phthalocyanine(ClAlPc) monolayer(ML)-on-HOPG. ClAlPc dipoles are aligned as shown in the figure to result in the vacuum level shift (ΔVL) [6, 32, 33]. Energy is given in eV. Large polaron exists in the Pn layer [panel (**b**)]

Here we introduce UPS results that involve large positive polaron (fully relaxed positive polaron involving all polarizations by movements of electrons, atoms in each molecule and molecules themselves), which exists in an organic film when holes are injected into the film as is realized by preparing the film on a high-work function (ϕ) substrate. Such a system is shown for pentacene in Fig. 4.9, where a high-ϕ substrate is prepared by depositing well-defined ClAl-phthalocyanine (ClAlPc) monolayer (ML) to form a two-dimensional (2-D) dipole layer on highly oriented pyrolytic graphite (HOPG) followed by deposition of ML of standing pentacene. After deposition of the pentacene ML on the ClAlPc-on-HOPG substrate, electrons are transferred from the pentacene layer to the substrate to achieve thermal equilibrium of the electron system. UPS spectra of this system thus reflect the large polaron states as well as gap states due to the structure disorder. The number of injected holes (thus electrons transferred from pentacene to the substrate) is about 0.01/pentacene molecule [6, 8].

Figure 4.10 represents ultrahigh sensitivity He-I$_{\alpha}$ spectra of the pentacene system with fully relaxed large polaron shown in Fig. 4.9b [8], where the intensity is shown with DOS on log scale. We found that the spectra have a broad parabolic (Gaussian on linear scale) and a straight line (exponential on linear scale) DOGS. The DOGS near $E_F$ for the 3-ML pentacene is weaker in intensity by an order of

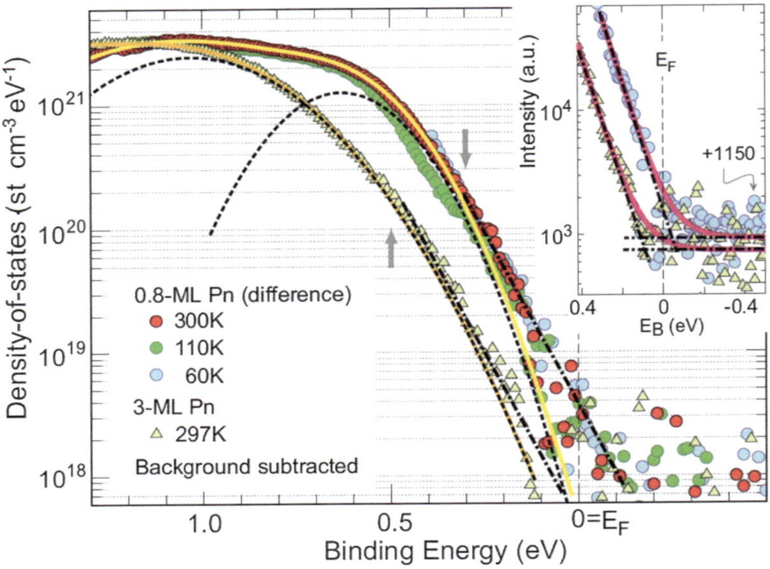


**Fig. 4.10** He $I_\alpha$ UPS difference spectra between Pn(0.8 ML)/ClAlPc(0.8 ML)/HOPG and ClAlPc (0.8 ML)/HOPG at 300, 110, and 60 K [8]. The substrate spectra were subtracted for the 0.8-ML Pn spectra. The He $I_\alpha$ UPS spectrum of Pn(3 ML)/ClAlPc(0.8 ML)/HOPG at 297 K is also plotted for comparison. In main panel, *solid curves*: Convoluted curves with Gaussian functions. *Dotted curves*: The Gaussian function located at the lowest $E_B$ in the convolution. *Dashed lines*: Guide for an exponential-type distribution of the gap states. The *arrows* correspond to boundaries of the distributions of the gap states. *Inset* shows fitting results of the difference spectrum with an exponential function and an arbitrary constant background (*dashed lines*) for 0.8-ML Pn at 60 K and 3-ML Pn. The figure is cited from [8] with permission

magnitude compared with that for the 0.8-ML pentacene, indicating the origin of the DOGS in the 0.8-ML pentacene is not due to impurities and the photoemission from the substrate is negligible for the 3-ML pentacene. Note that there is no sharp DOGS in the gap, demonstrating the large polaron states may be involved in the exponential tail. The DOS values of the exponential tail function $N_0 \exp[-\beta(E_B{}^{th}-E_B]]$ are evaluated, where the UPS intensities are normalized to the most intense point of the pentacene HOMO band and the molecular packing density of the thin-film phase of pentacene is assumed. The values are estimated to be $N_0 = 4 \pm 3 \times 10^{19}$, $5 \pm 2 \times 10^{19}$, and $4 \pm 2 \times 10^{19}$ (at 300, 110, and 60 K), $\beta = 20 \pm 4$, $12 \pm 2$, $15 \pm 3$ (at 300, 110, and 60 K), and $E_B{}^{th} = 0.31$, 0.32, 0.28 (at 300, 110, and 60 K) for the 0.8-ML pentacene, respectively, and $N_0 = 3 \pm 2 \times 10^{19}$, $\beta = 7 \pm 1$, and $E_B{}^{th} = 0.53$ for the 3-ML pentacene [8].

At every temperatures of the 0.8-ML pentacene, the exponential-type DOGS exists at least down to $E_B \sim 0.1$ eV and it extends into the cutoff region of the Fermi-Dirac distribution function. The temperature dependence of the $E_F$ position is small, indicating that the $E_F$ is dominated by the broad energy distribution of DOGS. A temperature dependence of UPS-HOMO shape is observed in the 0.8-ML pentacene spectra, but the origin is not clear so far, although it might be ascribed to changes in the molecular packing structure by thermal agitation. The sharp peak at $E_B \sim 1.24$ eV becomes broader upon cooling probably because of the structural transition of Pn and ClAlPc monolayers.

## 4.4 On the Definition of the Mobility Edge and the Charge Transport Gap

From the ultrahigh sensitivity UPS measurements, it becomes clearer that there are gap states due to significant DOS tailing into the gap even in crystalline films without intentional impurity doping. This is one of the results that come out from the universal features of organic molecular crystals.

The energy distribution of DOGS has profiles of Gaussian and/or exponential functions, which makes it difficult to define the HOMO (LUMO) edge that is necessary in discussing the mobility edge as well as the charge transport gap. For instance, as the gap states are supposed to be spatially localized, the mean distance between the gap states at a same $E_B$ becomes larger with lowering their DOS. No charge transport is expected through electronic states with larger inter-state distance, namely through lower DOGS. As the values of IP and EA are in general obtained by linear extrapolation of the leading slope of UPS HOMO and IPES LUMO features to the baseline, respectively, they are not strictly defined for discussion of the charge transport gap. In experimental fields, therefore, it would be necessary to have a definition of the mobility edge and the charge transport gap, for example, by using a value of DOS and considering the energy required from thermal excitation of carriers for discussing charge transport phenomena in organic semiconductors more quantitatively, especially for hopping-related charge transport.

**Acknowledgements** The authors are very grateful to Dr. M. Ono of Kaiyo Academy for his contribution to constructing the prototype apparatus. They also deeply thank Prof. K. Kudo and Prof. H. Ishii of Chiba University, Prof. M. Nakamura of Nara Institute of Science and Technology, Prof. A. Kahn of Princeton University, Prof. D. Cahen of Weizmann Institute of Science, Prof. N. Koch of Humboldt University, and Prof. A. T. S. Wee and Prof. W. Chen of National University of Singapore for their discussion on origins of the energy level alignment and gap states.

NU is very grateful to the late Prof. K. Seki of Nagoya University for discussion on the organic interface.

The ultrahigh sensitivity UPS project is supported by 21st Century COE program (MEXT) and Global COE program (MEXT).

## References

1. J. Hwang, E.J. Kim, J. Liu, J.L. Bredas, A. Duggal, A. Kahn, J. Phys. Chem. C **111**, 1378 (2007)
2. G. Heimel, S. Duhm, I. Salzmann, A. Gerlach, A. Strozecka, J. Niederhausen, C. Bürker, T. Hosokai, I. Fernandez-Torrente, G. Schulze, S. Winkler, A. Wilke, R. Schlesinger, J. Frisch, B. Bröker, A. Vollmer, B. Detlefs, J. Pflaum, S. Kera, K.J. Franke, N. Ueno, J.I. Pascual, F. Schreiber, N. Koch, Nat. Chem. **5**, 187 (2013)
3. S. Kera, Y. Yabuuchi, H. Yamane, H. Setoyama, K.K. Okudaira, A. Kahn, N. Ueno, Phys. Rev. B **70**, 085304 (2004)
4. H.Y. Mao, F. Bussolotti, D.-C. Qi, R. Wang, S. Kera, N. Ueno, A.T.S. Wee, W. Chen, Org. Electron. **12**, 534 (2011)
5. M. Fahlman, A. Crispin, X. Crispin, S.K.M. Henze, M.P. de Jong, W. Osikowicz, C. Tengstedt, W.R. Salaneck, J. Phys. Condens. Matter **19**, 183202 (2007)
6. H. Fukagawa, S. Kera, T. Kataoka, S. Hosoumi, Y. Watanabe, K. Kudo, N. Ueno, Adv. Mater. **19**, 665 (2007)
7. M. Ono, T. Sueyoshi, Y. Zhang, S. Kera, N. Ueno, Mol. Cryst. Liq. Cryst. **455**, 251 (2006)
8. T. Sueyoshi, H. Fukagawa, M. Ono, S. Kera, N. Ueno, Appl. Phys. Lett. **95**, 183303 (2009)
9. T. Sueyoshi, H. Kakuta, M. Ono, K. Sakamoto, S. Kera, N. Ueno, Appl. Phys. Lett. **96**, 093303 (2010)
10. N. Ueno, S. Kera, K. Kanai, in *Fundamental Electronic Structure of Organic Solids and Their Interfaces by Photoemission Spectroscopy and Related Methods*, ed. by N. Koch, N. Ueno, A.T.S. Wee. The Molecule-Metal Interface, Chap. 7 (Wiley, 2013, Weinheim), pp. 208–212
11. B. Boudaïffa, P. Cloutier, D. Hunting, M.A. Huels, L. Sanche, Science **287**, 1658 (2000)
12. See supplemental material of ref. 16, which is at http://link.aps.org/supplemental/10.1103/PhysRevLett.110.267602
13. H. Fukagawa, H. Yamane, T. Kataoka, S. Kera, M. Nakamura, K. Kudo, N. Ueno, Phys. Rev. B **73**, 245310 (2006)
14. F. Bussolotti, S. Kera, N. Ueno, Phys. Rev. B **86**, 155120 (2012)
15. F. De Angelis, S. Cipolloni, L. Mariucci, G. Fortunato, Appl. Phys. Lett. **88**, 193508 (2006)
16. F. Bussolotti, S. Kera, K. Kudo, A. Kahn, N. Ueno, Phys. Rev. Lett. **110**, 267602 (2013)
17. S. Duhm, G. Heimel, I. Salzmann, H. Glowatzki, R.L. Johnson, A. Vollmer, J.P. Rabe, N. Koch, Nat. Mater. **7**, 326 (2008)
18. W.L. Kalb, S. Haas, C. Krellner, T. Mathis, B. Batlogg, Phys. Rev. B **81**, 155315 (2010)
19. A. Kondo, H. Noguchi, S. Ohnishi, H. Kajiro, A. Tohdo, Y. Hattori, W.-C. Xu, H. Tanaka, H. Kanoh, K. Kaneko, Nano Lett. **6**, 2581 (2006)
20. H. Kajiro, A. Kondo, K. Kaneko, H. Kanoh, Int. J. Mol. Sci. **11**, 3803 (2010)
21. D.V. Lang, X. Chi, T. Siegrist, A.M. Sergent, A.P. Ramirez, Phys. Rev. Lett. **93**, 086802 (2004)
22. I. Salzmann, S. Duhm, R. Opitz, J.P. Rabe, N. Koch, Appl. Phys. Lett. **91**, 051919 (2007)
23. D. Käfer, C. Wöll, G. Witte, Appl. Phys. A **95**, 273 (2009)
24. D. Käfer, L. Ruppel, G. Witte, Phys. Rev. B **75**, 1 (2007)
25. T. Hosokai et al., Chem. Phys. Lett. **544**, 34 (2012)
26. J.H. Kang, D. Da Silva Filho, J.-L. Bredas, X.Y. Zhu, Appl. Phys. Lett. **86**, 152115 (2005)
27. S. Kwon, K.-R. Wee, J.W. Kim, C. Pac, S.O. Kang, J. Chem. Phys. **136**, 204706 (2012)
28. S. Olthof, S.K. Mohapatra, S. Barlow, S. Mehraeen, V. Coropceanu, J.-L. Brédas, S.R. Marder, A. Kahn, Phys. Rev. Lett. **109**, 176601 (2012)
29. W. Han, H. Yoshida, N. Ueno, S. Kera, Appl. Phys. Lett. **103**, 123303 (2013)
30. F. Bussolotti, J-P. Yang, A. Hinderhofer, Y. Huang, W. Chen, S. Kera, A. T. S. Wee, N. Ueno, Phys. Rev. B **89**, 115319 (2014)

31. N. Ueno, S. Kera, K. Kanai, in *Fundamental Electronic Structure of Organic Solids and Their Interfaces by Photoemission Spectroscopy and Related Methods*, ed. by N. Koch, N. Ueno, A.T.S. Wee. The Molecule-Metal Interface, Chap. 7 (Wiley, 2013, Weinheim), p. 179
32. Y. Huang, R. Wang, T.C. Niu, S. Kera, N. Ueno, J. Pflaum, A.T.S. Wee, W. Chen, Chem. Commun. **46**, 9040 (2010)
33. H. Fukagawa, S. Hosoumi, H. Yamane, S. Kera, N. Ueno, Phys. Rev. B **83**, 085304 (2011)

# Chapter 5
# Pentacene Becomes Mott–Hubbard Insulator by Potassium Doping

**Fabio Bussolotti, Satoshi Kera, and Nobuo Ueno**

## 5.1 Introduction

In recent years, organic semiconductors (OSCs) have attracted considerable interest as promising base materials for optoelectronic devices [1–3]. In view of the optimization of the device performance, however, a complete control on the charge transport properties of organic films represents a mandatory step. In this context, $n(p)$-type doping via intercalation of alkali metal atoms was reported to be an efficient method for improving the electrical transport in organic materials [4–7]. Therefore, the electronic properties of alkali-doped organic systems represented the target of a large number of theoretical and experimental investigations [8–14]. Among organic systems, alkali-doped conjugated polymer and fullerene films have been extensively studied for many years [15–21]. In contrast, the impact of the alkali doping on smaller $\pi$-conjugated molecular systems with weak intermolecular interaction remained relatively unclear and less explored. This is despite the possibility, even in such molecular systems, of strong electronic correlation effects [22], as favored by the strong wave-function localization at each molecular site, i.e. by the large on site Coulomb repulsive energy ($U$) with respect to the width ($W$) of the intermolecular bands (see Fig. 5.1)

An insulator-metal-insulator transition as a function of doping was recently observed in electrical transport measurements on alkali $n$-doped phthalocyanine and pentacene thin films [6, 7] (Fig. 5.2). The initial increase of electrical conductivity was ascribed to the filling of the lowest unoccupied molecular orbital (LUMO) by electrons donated from the alkaline atoms. At higher doping level, the organic films become insulator via electronic correlation effects [6, 7, 13].

F. Bussolotti (✉) • S. Kera • N. Ueno
Department of Nanomaterial Science, Graduate School of Advanced Integration Science, Chiba University, Inage-ku, Chiba 263-8522, Japan
e-mail: fabio@restaff.chiba-u.jp

 
H. Ishii et al. (eds.), *Electronic Processes in Organic Electronics*,
Springer Series in Materials Science 209, DOI 10.1007/978-4-431-55206-2_5

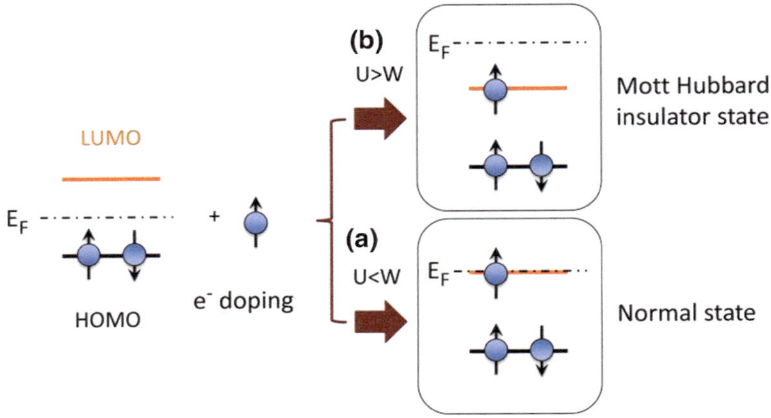


**Fig. 5.1** Schematic mechanism of Mott–Hubbard metal-insulator transition. A single occupied LUMO lies at the Fermi level (normal state (*a*)) if the onsite Coulomb repulsion energy ($U$) is lower than LUMO bandwidth ($W$). When $U > W$ the onsite Coulomb repulsion favors the electron localization thus pushing the single occupied level below the Fermi level position and resulting in a Mott–Hubbard insulator state (*b*)

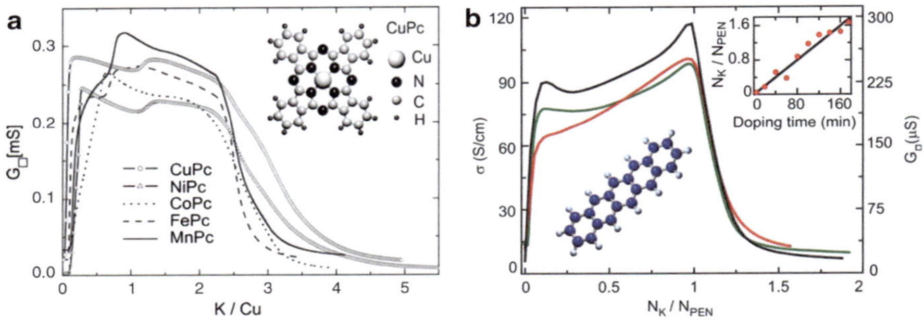


**Fig. 5.2** (**a**) The square conductance G measured at room temperature as a function of potassium concentration (K/Cu) on films of different metal phthalocyanines. *Inset*: structure of a CuPc molecule (Figure adapted with permission from [6], Copyright 2005, Wiley). (**b**) Conductivity (σ) and square conductance G of three different K-doped PEN films as a function of the K doping ratio ($N_K/N_{PEN}$); under the curves the structure of a Pentacene molecule is shown. *Inset*: $N_K/N_{PEN}$ as a function of the doping time (Figure adapted with permission from [7], Copyright 2012, APS)

These results are in contrast with ultraviolet photoelectron spectroscopy (UPS) investigations conducted on similar alkali doped films [10, 11, 14] (Fig. 5.3). In particular, no detectable density of states was observed at the Fermi level ($E_F$), thus indicating that the organic films remained in a stable insulating state [10, 11, 14]. The inconsistency between electrical transport and photoemission measurements was tentatively attributed to structural differences among doped organic films [23]. The electronic properties of organic semiconductors, in fact, are strongly

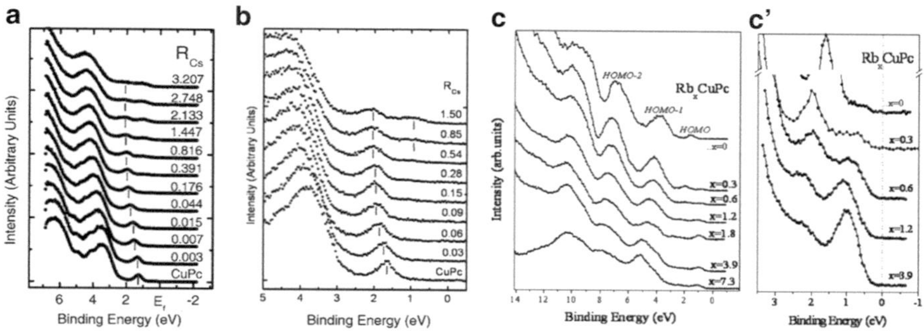

**Fig. 5.3** (**a**) Evolution of the copper phthalocyanine (CuPc) HOMO region in the UPS spectra ($h\nu = 21.128$ eV) as a function of Cs doping ratio $R_{Cs}$. The 15 nm-thick CuPc film was deposited on polycrystalline gold substrate (figure adapted with permission from [10]. Copyright 2008, AIP publishing LLC). (**b**) UPS spectra ($h\nu = 21.128$ eV) of the HOMO region of Cs-doped CuPc film (20 nm) prepared on Ag substrate as a function of Cs doping ratio $R_{Cs}$ (figure adapted with permission from [11]. Copyright 2010, Elsevier). (**c**) Valence-band UPS spectra ($h\nu = 20$ eV) of a thin CuPc film (5 nm) intercalated with rubidium (Rb) as a function of Rb concentration ($x$). Panel (**c'**) shows the enlargement of the energy region close to $E_F$ (*dashed line*) (figure adapted with permission from [14]. Copyright 2008, American Chemical Society). Note that: (i) no density of states near $E_F$ is detected in all the doped CuPc films [panel (**a**)–(**c**)], to indicate a stable insulating state and (ii) progressive broadening of the UPS features is observed as a function of the Cs alkali doping ratio

dependent on the molecular packing due to the planar molecular structure [24]. Moreover, UPS investigations were mainly conducted on polycrystalline or amorphous alkali doped organic thin films [5, 9–11, 14]. A large concentration of structural defects (i.e. grain boundaries, molecular vacancies) limits the diffusion of dopants within the organic layer [25] and leads to an inhomogeneous distribution of dopants in the organic films. Imperfect molecular packing structure originating, for example, by inhomogeneity of the dopant distribution may cause, in principle, site dependence of the intermolecular electronic coupling and the relaxation/polarization energy. These effects may lead to (i) a significative broadening of UPS bandwidth [26, 27] and (ii) additional electronic states in the band gap that can affects, in turn, the energy level alignment (ELA) at the organic/substrate interfaces [28–31]. Despite a careful minimization of the structural defects is essential to elucidate the impact of dopants on the electronic properties, UPS studies on alkali-doped single crystalline (SC) multilayers of small π-conjugated molecules remained, up to now, largely marginal [32].

Here we report on the electronic properties of potassium (K)-doped SC thin film of pentacene (Pn) prepared on Cu(110) as investigated by angle-integrated and resolved UPS techniques. We found that the band dispersion of the highest occupied molecular orbital (HOMO) states of the pristine Pn SC thin film remained unchanged in the $K_1Pn_1$ SC thin film at room temperature (293 K), while a K-doping induced gap state appeared. At 50 K, a band dispersion (~0.08 eV) of the gap state was clearly observed for the $K_1Pn_1$, which is in fairly good agreement with theoretical prediction for $K_1Pn_1$ Mott–Hubbard insulator single crystal. The K-induced gap state was finally identified as the occupied Mott–Hubbard band of the $K_1Pn_1$ Mott–Hubbard insulator. The role of the Mott–Hubbard insulator is discussed in relation to the energy level alignment (ELA).

## 5.2 Preparation of Single Crystalline Films of Pentacene

High-purity (99.999 %) Cu(110) single crystal was cleaned by a series of sputtering-annealing cycles. The quality of the surface was checked by preliminary angle-resolved UPS (ARUPS) investigations. Pn molecules ($C_{22}H_{14}$, Sigma Aldrich), purified by three cycles of vacuum sublimation, were vacuum-deposited onto the Cu(110) clean substrate according to the two step procedure described in [33]: at the first step, a saturated flat lying Pn monolayer was deposited on Cu(110) at 450 K; in the second step, the Pn monolayer was used as a template to grow an upright-standing SC Pn multilayer at 320 K [Fig. 5.4a]. According to Lukas et al. [33] the SC structure of the Pn film is characterized by a surface rectangular unit cell of 5.8 and 7.4 Å along the [1–10] and [001] substrate directions, respectively. These lattice values were confirmed for the present Pn films by the ARUPS data reported in the present contribution. The surface Brillouin zone (SBZ) of the Pn SC multilayer is schematically illustrated in Fig. 5.4b. The film thickness and deposition rate (0.1 nm/min) were monitored by using a quartz microbalance. During the deposition, the pressure remained stably lower than $4 \times 10^{-8}$ Pa.

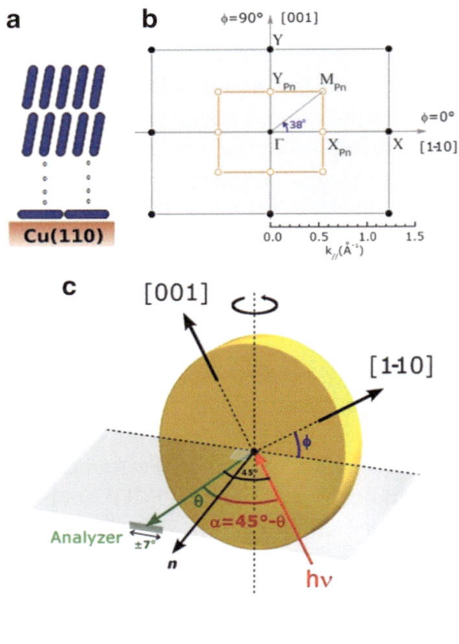


**Fig. 5.4** (**a**) Schematic representation of the upright standing Pn multilayer film structure. (**b**) Sketch of the surface Brillouin zone of upright standing Pn multilayer film on Cu(110). (**c**) Experimental parameters of photoelectron experiment: the photon incidence angle α, electron take-off angle θ, and sample azimuthal angle ϕ (Figure reproduced with permission from [34], Copyright 2012, APS)

After the growth of the Pn multilayer, potassium (K) metal atoms were deposited at 293 K from carefully outgassed SAES S.p.A. dispensers. During K deposition the pressure in the preparation chamber remained below $5 \times 10^{-8}$ Pa. The total number of K atoms per Pn molecule [i.e. the doping level ($x$)] was evaluated from K2p and C1s core level intensity ratio measured by X-ray photoemission spectroscopy (XPS) with Al-$K_{\alpha}$ ($h\nu = 1{,}486.6$ eV), where the ratio was calibrated with respect to (i) the number of C atoms per Pn molecule and (ii) C1s and K2p photoionization cross sections at a photon energy of 1,486.6 eV ($\sigma_{C1s}/\sigma_{K2p} \sim 0.25$) [35].

UPS experiments were performed by using high sensitivity, ultralow background apparatus with a hemispherical electron energy analyzer (MBS A-1) and a monochromatic $HeI_{\alpha}$ ($h\nu = 21.218$ eV) radiation source. Additional details on the experimental apparatus are included in [30, 31]. A complete definition of the UPS experimental geometry including photon incidence angle ($\alpha$), photoelectron emission angle ($\theta$) and the sample azimuthal angle ($\phi$) is shown in Fig. 5.4c. The binding energy scale was referred to the $E_F$ of the substrate. For both angle-integrated and resolved UPS the total energy resolution was set to 30 meV. Angle-integrated spectra were acquired at normal emission ($\theta = 0°$) with an integration angle of $\pm 7°$ where a bias of $-5$ V was applied to the sample to measure the vacuum level (VL). In the ARUPS the angle-resolving range spanned $\pm 7°$ with respect to the lens axis and angular resolution was set to 1°. Wider angular dependence was accessed by rotating the sample with respect to the lens axis. No bias was applied to the sample during ARUPS measurements.

K was deposited until the saturation of VL shift, where the doping level corresponded to $x = 1$. K deposition was then stopped in order to avoid the formation of K metallic clusters. Spectral broadening due to charging effects of the films were not observed during the UPS measurements for both of pristine and K-doped Pn films as judged from observation of a very sharp cut-off of the secondary electrons as well as small band dispersions.

## 5.3 Impact of K Doping on the Film Structure and Energy Levels

Figure 5.5 shows the 293 K-He $I_{\alpha}$ UPS (angle integrated) spectra of the Pn/Cu(110) multilayer (8 nm) as a function of $x$. All the Pn related UPS features are marked by vertical bars. In the pristine Pn film, an intense peak (A) is visible in the secondary electron region [Fig. 5.5a] [36]. Such features in photoemission spectra are generally related to the presence of electronic states above the VL [26]. For pentacene, feature A is associated to a high density of states (DOS) in the conduction band of ordered films presenting an upright standing molecular orientation [26]. The upright standing configuration is generally accompanied by a herringbone type molecular packing in the layers parallel to the substrate surface [37, 38]. The same molecular packing geometry was observed in the (001) plane of Pn single crystals [39].

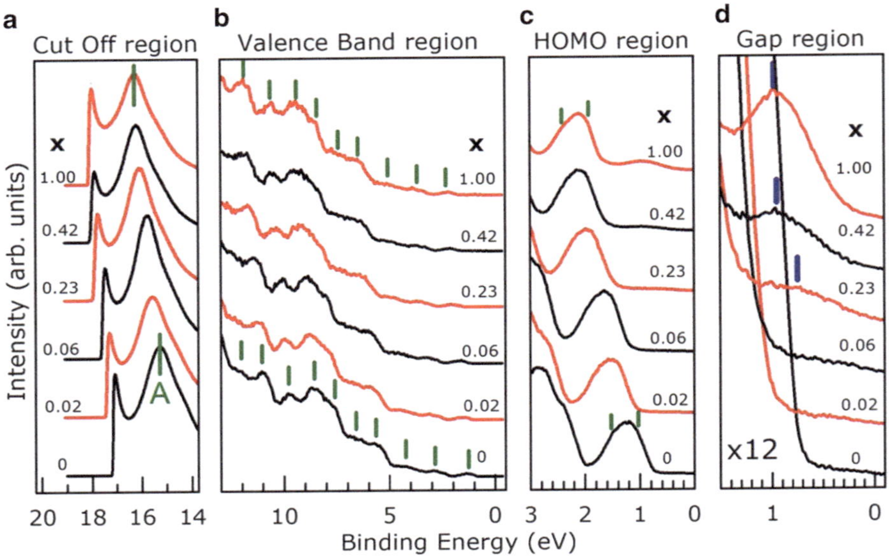


**Fig. 5.5** $HeI_{\alpha}$ UPS spectra (angle integrated) of Pn/Cu(110) multilayer (8 nm) as a function of K-doping level ($x$) in the secondary electron cutoff (**a**), valence band (**b**), HOMO (**c**) and band gap (**d**) regions. All of the spectra were measured at 293 K. The binding energy scale refers to $E_F$ of the substrate. Pn related features are indicated by *vertical bars* (Figure reproduced with permission from [34], Copyright 2012, APS)

A clear energy splitting of the HOMO level was observed [Fig. 5.5c]. The two HOMO features results from the intermolecular HOMO band dispersions with the presence of two nonequivalent molecules in the unit cell [26, 37, 38, 40–42]

The K deposition on the Pn SC film at 293 K induced a progressive shift of the VL and Pn related UPS features towards higher binding energies values [Fig. 5.5a–c]. At the same time, the appearance of a new spectral feature was observed in the gap region [Fig. 5.5d]. For both the HOMO and the gap state, the full width at half maximum (FWHM) was $(0.60 \pm 0.05)$ eV.

The K doping did not introduce any significant broadening of the UPS spectral features of the Pn SC thin film. This is rarely reported in UPS measurements of alkali doped polycrystalline and amorphous organic thin film [10, 11, 14, 32]. Structural defects in the films, in fact, may limit uniform dopant diffusion. Therefore, a larger degree of structural disorder is introduced into the organic film as a consequence of doping, thus resulting in serious broadening of the UPS features (see also Fig. 5.3). Note that, for the present SC Pn thin film, no K-induced changes were detected for the HOMO, conduction band feature A and secondary electron cut-off [Fig. 5.5a, c], all of which are sensitive to the film structure [26, 42]. It is therefore concluded that the K doping did not significantly affect the molecular orientation and the molecular packing geometry, i.e. the present Pn multilayer maintained its SC structure upon K doping. This result also suggests that K atoms are intercalated *between* the Pn stacking layers, in agreement with theoretical studies on $K_1Pn_1$ SC structure [12].

The appearance of gap state in alkali doped organic system is generally associated to polaron formation, as due to the electron transfer from the alkali atoms to the molecules [15, 43, 44]. In the present K-doped Pn SC multilayer, however, the origin of the gap state must be differently explained. In alkali doped organic systems, in fact, polaron formation is generally accompanied by a significant change in the binding energy of the frontier orbitals, where an increase (~0.2–0.3 eV) in the HOMO binding energy was observed [15, 43, 44]. If this is the case also for the present K-doped Pn SC multilayer, the presence of Pn molecules with different charging state (i.e. with different HOMO binding energy) should result in a large broadening of the HOMO lineshape at low and intermediate doping level ($0 < x < 1$). The stability of the HOMO lineshape of Pn SC multilayer after K doping is clearly inconsistent with the polaron formation. As shown in Fig. 5.6, K-induced binding energy shift is the same for the HOMO and the VL (Fig. 5.6). Such a "rigid" energy shift also confirms the absence of polaron formation [45].

For $K_1Pn_1$ SC Mott–Hubbard metal-insulator transition was theoretically predicted by Craciun et al. [7]. In $K_1Pn_1$ SC, Pn LUMO band is half filled.

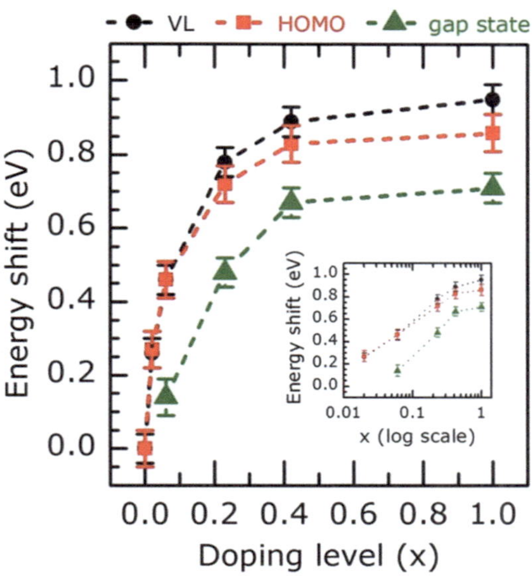


**Fig. 5.6** Summary of the doping-induced binding energy shift of VL (*black circle*), HOMO peak position (*red square*) and gap state peak position (*green triangle*). For the gap state the energy shift was evaluated with respect to the position at $x = 0.02$. The VL and HOMO energy shifts are coincident within experimental error. The *inset* displays semi-logarithmic plot in order to show the linear dependence of the energy shift on $\ln(x)$ (see text for discussion) (Figure reproduced with permission from [34], Copyright 2012, APS)

This corresponds, in principle, to a metallic phase. However, the strong electron–electron correlations determines the splitting of the half filled LUMO band into unoccupied (above $E_F$) and occupied (below $E_F$) bands. $K_1Pn_1$ is therefore driven into an insulating state to become a Mott–Hubbard insulator [7]. The electronic correlation effects are favoured by the low value of LUMO bandwidth ($W$) with respect to the onsite Coulomb repulsion energy $U$ ($U/W \sim 2.1$ [7]). For the present K-doped Pn SC multilayer, Mott–Hubbard transition can explain the origin of the gap state, which can be identified as the occupied Mott–Hubbard band as discussed in the next section. It is also worth to note, that (i) according to theoretical study of $K_1Pn_1$ Mott–Hubbard insulator, the K states are located only at much higher energy above $E_F$, thus suggesting that the role of the K atoms is limited to transferring its electrons to the pentacene molecules [7], and (ii) the stability of binding energy of valence band states upon Mott–Hubbard transition is commonly observed for inorganic compounds [46, 47].

## 5.4 Band Dispersion and Formation of Mott–Hubbard Insulator

For evaluating the impact of the K doping on the intermolecular interaction, the degree of charge delocalization, and electronic correlation, the band structure of pristine Pn and $K_1Pn_1$ SC films were investigated by ARUPS. The results at 293 K are summarized in Figs. 5.7 and 5.8.

Figure 5.7a shows the θ-dependence of ARUPS in the HOMO region of the pristine Pn and $K_1Pn_1$ films. The ARUPS data were acquired along $\Gamma - X_{Pn}$, $\Gamma - Y_{Pn}$, and $\Gamma - M_{Pn}$ directions of the Pn SBZ [see Fig. 5.4b]. The ARUPS spectra remained clearly unaffected by the K doping, unless the rigid binding energy shifts. In particular, for both films, the $H_1$ and $H_2$ features are clearly visible at $\theta = 0°$. These peaks gradually disperse with θ, as highlighted by dashed curves superimposed to the ARUPS data. The positions of $H_1$ and $H_2$ were determined by the least square fitting of the ARUPS data by using Gaussian functions combined with a linear background. Along the $\Gamma - X_{Pn}$ ($\Gamma - Y_{Pn}$) the energy separation is progressively reduced to reach minimum value at $\theta = 16°$ ($\theta = 14°$). For higher θ, the splitting increases again until $\theta = 32°$ ($\Gamma - X_{Pn}$) and $\theta = 28°$ ($\Gamma - Y_{Pn}$), where the normal emission HOMO lineshape is basically recovered. Along the $\Gamma - M_{Pn}$, similar behavior was observed with the dispersion turning points located at around $\theta \sim 22°$ and $\theta \sim 44°$. The band dispersions, $E_B(k_{//})$ of $H_1$ and $H_2$ peaks for pristine and doped films are plotted in Fig. 5.7b, where $k_{//}$ represents the component of the photoelectron wave vector parallel to the surface. The observed $H_1$ and $H_2$ band dispersions of the undoped Pn are in good agreement with previous experimental results [40] and ab initio calculations for Pn bulk phase [24]. Note that, along all the symmetry directions, the $H_1$ and $H_2$ band dispersions of the pristine and K-doped films are very similar in terms of bandwidth and dispersion phase [Fig. 5.5b]. The

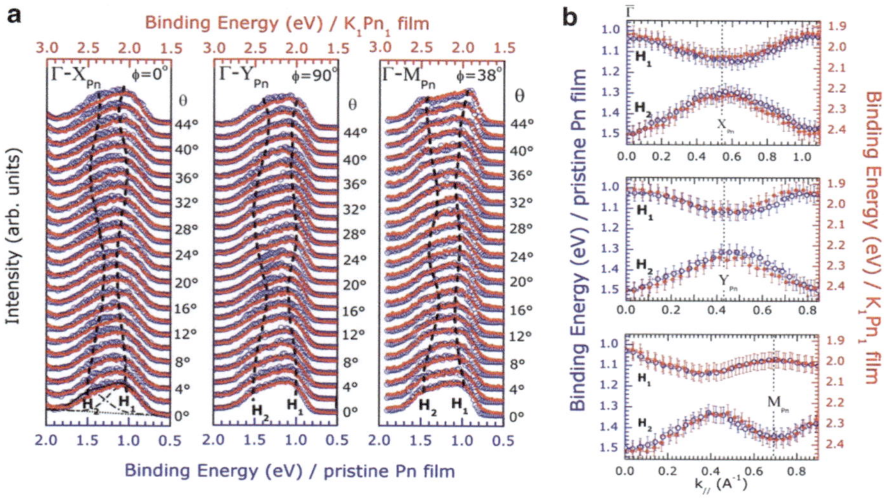


**Fig. 5.7** (**a**) θ-dependence of ARUPS spectra for pristine Pn [*open* (*blue*) *circles*] and $K_1Pn_1$ [*closed* (*red*) *circles*] single crystalline films, as acquired at $\phi = 0°$ ($\Gamma - X_{Pn}$ direction), $\phi = 90°$ ($\Gamma - Y_{Pn}$ direction) and $\phi = 38°$ ($\Gamma - M_{Pn}$ direction). All data were acquired at 293 K. The binding energy positions of the HOMO-derived states ($H_1$ and $H_2$) are indicated by *dashed lines*. $H_1$ and $H_2$ positions were determined by the least square fitting of the ARUPS data by using Gaussian functions on a linear background. An example of ARUPS data fitting is shown for pristine Pn film at $\theta = 0°$ and $\phi = 0°$ (*Dash dotted lines* = Gaussian components, *dotted line* = linear background, *continuous line* = cumulative fitting curve). *Lower* (*blue*) and *upper* (*red*) binding energy scale are for pristine Pn and $K_1Pn$ film, respectively. (**b**) HOMO-band dispersion, $E_B(k_{//})$, for pristine Pn [*open* (*blue*) *circles*] and $K_1Pn$ [*closed* (*red*) *circles*] along the $\Gamma - X_{Pn}$ direction (*upper panel*), $\Gamma - Y_{Pn}$ direction (*middle panel*) and $\Gamma - M_{Pn}$ direction (*lower panel*). *Left* (*blue*) and *right* (*red*) binding energy scales are for pristine Pn and $K_1Pn_1$ multilayer, respectively (Figure reproduced with permission from [34], Copyright 2012, APS)

turning points of $E_B(k_{//})$ curves exist at $k_{//}$=(0.54 ± 0.02) $Å^{-1}$, $k_{//}$=(0.42 ± 0.02) $Å^{-1}$ and $k_{//}$=(0.69 ± 0.02) $Å^{-1}$ along the $\Gamma - X_{Pn}$, $\Gamma - Y_{Pn}$, and $\Gamma - M_{Pn}$ directions, respectively. These values are in excellent agreement with those evaluated from the Pn unit cell lattice parameters [33], i.e. $k^{th}_{//}$=0.542 $A^{-1}$ (along $\Gamma - X_{Pn}$), $k^{th}_{//}$=0.425 $A^{-1}$ (along $\Gamma - Y_{Pn}$), and $k^{th}_{//}$=0.688 $A^{-1}$ (along $\Gamma - M_{Pn}$). Such agreement confirms that both the pristine and K-doped Pn multilayers keeps SC structure. The stability of the band dispersion upon K doping further demonstrates that K atoms are not intercalated among the standing Pn molecules within each stacking layer but between the Pn stacking planes. This conclusion is reasonable, since it is expected that the theoretically predicted increase of the plane separation (~8 %) upon K doping [7] does not significantly affect the film electronic structure as the intermolecular interaction between adjacent Pn layers is much weaker than that within each Pn layer.

Figure 5.8 shows the results of the ARUPS for the gap state in the $K_1Pn_1$ multilayer. No clear band dispersion was detected along the main symmetry

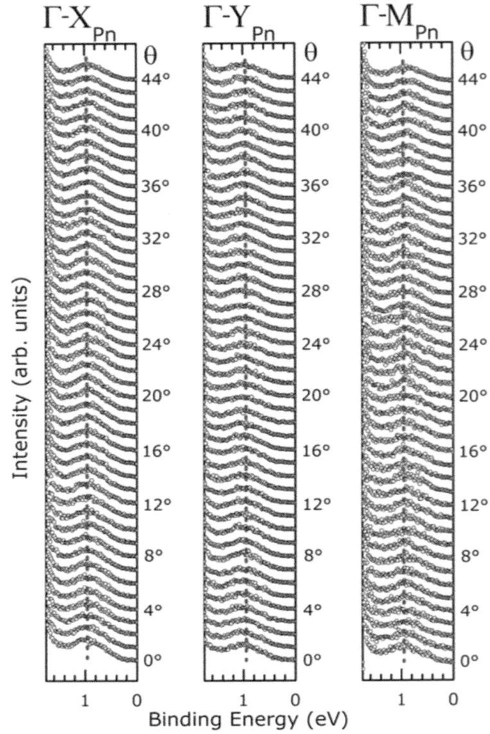


**Fig. 5.8** θ-dependence of gap state ARUPS spectra in $K_1Pn_1$ single crystalline film at 293 K, as acquired along $\Gamma - X_{Pn}$, $\Gamma - Y_{Pn}$, and $\Gamma - M_{Pn}$ directions. *Dashed* (*red*) *lines* indicate the gap state peak position. No energy dispersion was detected (Figure reproduced with permission from [34], Copyright 2012, APS)

directions of the surface unit cell, i.e. the gap state is well localized at 293 K. The differences between the dispersion widths of the gap state and the HOMO can be ascribed to the different spatial distribution of wave functions of the corresponding wave functions. For Pn SC film intercalated with Rubidium (Rb) layers ($Rb_1Pn_1$), for example, the wave function of the gap state was predicted to be localized at the edges of Pn molecules and near to the alkaline atom plane [48], thus limiting the intermolecular π–π interaction.

ARUPS investigation was also conducted on the $K_1Pn_1$ film at 50 K. Figure 5.9a shows the ARUPS results for the HOMO region, as acquired along the $\Gamma - M_{Pn}$ direction. At 50 K, a clear HOMO-band splitting as well as HOMO band dispersion was not observed, in contrast to the 293 K results, even if a small θ-dependence was found for the HOMO band shape. The impact of the temperature on the HOMO spectral lineshape can be ascribed to a temperature-induced change in intermolecular packing structure. It was pointed out theoretically [24] and experimentally [42], in fact, that a slight change in the film structure can deeply affect the electronic band structure of organic films. For the gap state, on the other hand, a small but clear band dispersion (~0.08 eV) was observed along the $\Gamma - M_{Pn}$ direction as shown in Fig. 5.9b. The existence of such dispersive behavior the ARUPS data comparison shown in Fig. 5.9c, where the energy difference of the gap

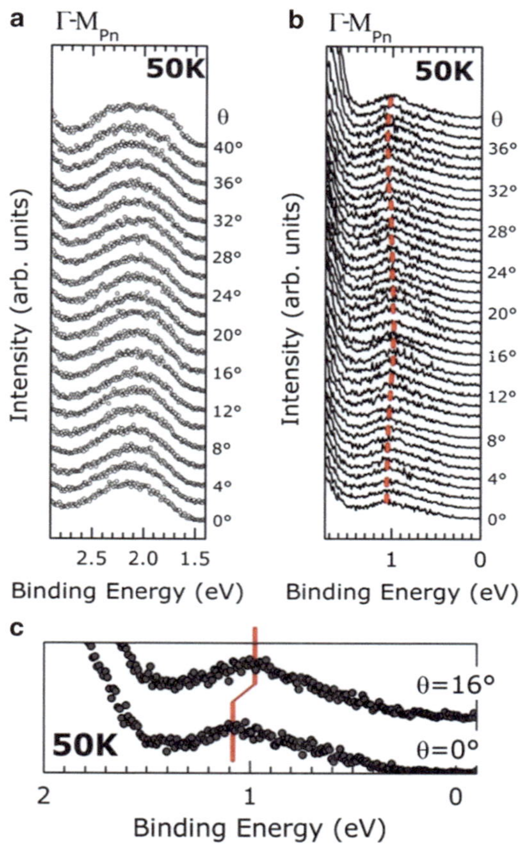


**Fig. 5.9** θ dependence of ARUPS spectra for HOMO (**a**) and gap state (**b**) in $K_1Pn_1$ single crystalline film at 50 K, as acquired along $\Gamma - M_{Pn}$ direction. (**c**) Comparison between the spectra at $\theta = 0°$ and $\theta = 16°$ to clearly show the binding energy change due to the dispersion (Figure reproduced with permission from [34], Copyright 2012, APS)

state positions at $\theta = 0°$ and $\theta = 16°$ (i.e. dispersion turning points) is highlighted. The band dispersion of the gap state also demonstrates that the $K_1Pn_1$ film keeps SC structure. Figure 5.10a displays the gap state band dispersion, at 50 K, of the $K_1Pn_1$ film along the $\Gamma - M_{Pn}$. The dispersion can be quantitatively discussed within the framework of a simple one-dimensional tight binding model: $E_B(k_{//}) = E_c - 2t\cos(\pi k_{//}/G)$ [42] where $E_c$ is the energy of the band center, $t$ is the transfer integral and $G$ is the size of the SBZ. The best fit of the experimental data with a cosine curve is shown in Fig. 5.10c and gives $E_c = (1.021 \pm 0.003)$ eV, $t = -(0.020 \pm 0.003)$ eV and $G = (0.62 \pm 0.01)$ Å$^{-1}$. In particular, at 50 K, the symmetry turning point of the gap-state band ($M^{50K}{}_{Pn,}$) was experimentally found at 0.62 Å$^{-1}$ [Fig. 5.10a], which is located closer to the Γ point than that observed at 293 K ($\Gamma - M_{Pn} = 0.69$ Å$^{-1}$, as evaluated from the HOMO band dispersion at 293 K). The observed contraction of the SBZ at 50 K (see inset in Fig. 5.7a) corresponds, in the real space, to an increase of ~10 % and ~12 % in the size of the unit cell along the [1–10] and [001] directions, respectively. This result is consistent with the possible occurrence of

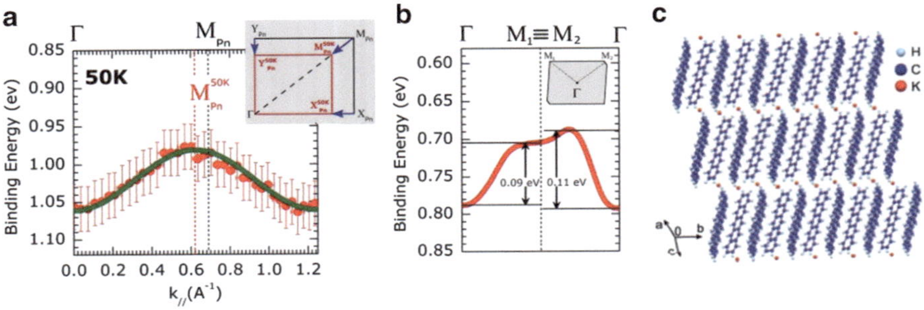


**Fig. 5.10** (**a**) Band dispersion, $E_B(k_{//})$, of the gap state along the $\Gamma - M_{Pn}$ direction at 50 K. *Solid (green) line* represents tight-binding fit of the experimental data [*filled* (*red*) *circles*]. The SBZ edges at 50 K ($M^{50K}{}_{Pn}$) and at 293 K ($M_{Pn}$) are indicated by *dotted lines*. The rectangular SBZ at RT and its shrinkage at 50 K are schematically illustrated in the *inset* (see text for detailed discussion). (**b**) Theoretical band dispersion of lower Mott–Hubbard band along $\Gamma - M_1$ and $\Gamma - M_2$ directions of the SBZ of $K_1Pn_1$ Mott–Hubbard insulator crystal (data adapted from [7]). Sketch of SBZ of the $K_1Pn_1$ Mott–Hubbard insulator crystal used in calculations in [7] is shown in the *inset* of panel (**b**). $M_1$ and $M_2$ come from slightly different crystal structure used in the calculation and $M_1 = M_2$ in the present crystal structure [rectangular SBZ, see Fig. 5.4b]. Experimental dispersion of the gap state [panel (**a**)] agrees well with the theoretical result (after averaging the curves along $\Gamma - M_1$ and $\Gamma - M_2$ directions) [panel (**b**)]. (**c**) *Side view* of the stacked layers of Pn and K in $K_1Pn_1$ SC film, illustrating the potassium intercalation in between the molecular planes (figure adapted from [7]). Note that panel (**c**) represents the structure in the bulk of $K_1Pn_1$ SC film. In the present case of $K_1Pn_1$ multilayer, no K atoms are present on top of the surface Pn layer (Figure reproduced with permission from [34], Copyright 2012, APS)

temperature-induced change in the intermolecular Pn packing structure. Interestingly, a similar lattice expansion after alkaline doping was recently reported for Cesium doped Pn thin film by Annese et al. [32]. In order to determine the lattice parameters of $K_1Pn_1$ SC unit cell at 50 K more precisely, however, detailed structural investigations are required.

Figure 5.10b shows the theoretical band dispersion for the low Mott–Hubbard band of $K_1Pn_1$ Mott–Hubbard insulator crystal, as evaluated along $\Gamma - M_1$ and $\Gamma - M_2$ directions (adapted from [7]). The two directions, $\Gamma - M_1$ and $\Gamma - M_2$, are defined for the SBZ of the $K_1Pn_1$ crystal used in calculations [7] [see inset in Fig. 5.10b]. $M_1$ and $M_2$ symmetry points originate from the slightly different crystal structure used in the calculation with respect to the present Pn SC structure where $M_1 = M_2$, due to the rectangular SBZ, [see Fig. 5.4b]. Interestingly, at 50 K, the experimental band dispersion [Fig. 5.10a] along $\Gamma - M_{Pn}$ direction agrees well with the theoretical prediction [Fig. 5.10b] if we consider $M_1 = M_2$, both in term of bandwidths and dispersion phase. This agreement further indicates that the Mott–Hubbard insulator can be the origin of the gap state in the K-doped Pn SC thin film.

Here we compare the present results with those reported recently for cesium (Cs)-doped *flat lying* Pn SC film on Cu(119) by Annese et al. [32] (Fig. 5.11). They found that the Cs deposition on the flat-lying Pn film determined variation of the overall UPS spectral lineshape. At room temperature, ARUPS indicated the

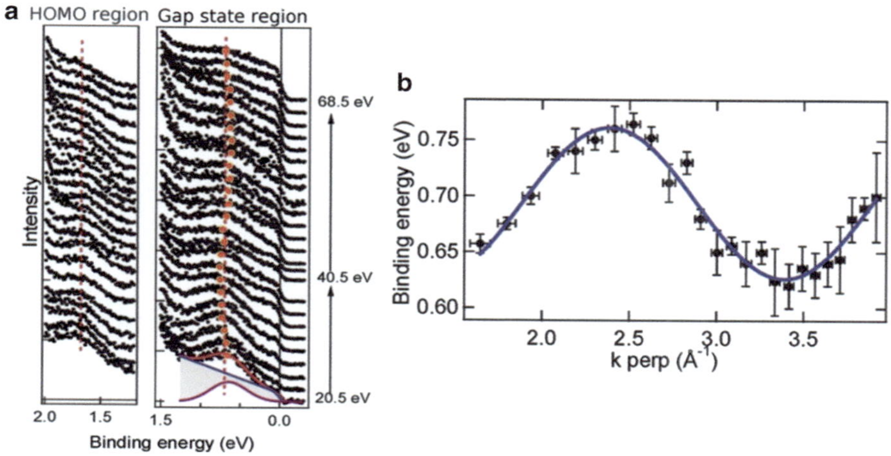


**Fig. 5.11** (**a**) hν-dependent normal emission ARUPS spectra (as measured at room temperature, without data averaging/smoothing) in the HOMO and gap state energy region of Cs-doped pentacene film (doping ratio ~2). Gap state peak positions were determined by least squares fitting method by using a gaussian function (*purple curve*) on polynomial background (*blue curve*). (**b**) Band dispersion (along the substrate normal) of the gap state in panel (**a**). The *solid line* represents the tight binding model fit of the data (figure adapted from [32]. Copyright 2012. American Chemical Society)

formation of a new LUMO derived state at ~0.6–0.7 eV below $E_F$ [Fig. 5.11a]. For their $Cs_2Pn_1$ film a band dispersion (0.13 eV) of the gap state was also observed at room temperature, while no dispersion was detected for the HOMO state, along the direction perpendicular to the substrate surface [Fig. 5.11a]. They also found that the periodicity of the gap state dispersion was related to an expansion of the lattice constant of 0.6 Å (i.e. ~20 %) along the molecular stacking direction (normal to the surface) [Fig. 5.11b]. The present results on $K_1Pn_1$/Cu(110) film agrees *qualitatively* with those on $Cs_2Pn_1$/Cu(119) film [32]. In particular, a large lattice expansion was observed after alkaline doping for both Pn/Cu(110) and Pn/Cu(119) SC thin film. The difference between $Cs_2Pn_1$ and the present $K_1Pn_1$ may be originated from different nature of dopant species and structure of the pristine Pn film (flat lying vs. upright standing).

*Intensity of the gap state and its impact on ELA*. In order to clarify the origin of the gap state and evaluate its impact on the ELA at the organic/metal interface, the gap state angle-integrated UPS spectra are shown in Fig. 5.9 as a function of K doping level. The UPS intensity was plotted on a log scale in order to highlight the K-induced spectral changes. The background contribution was removed and the intensity was normalized as described in [30, 31]. In the pristine Pn SC film a very tiny exponential tail (linear on the log scale of Fig. 5.12) is visible between the HOMO edge (at ~0.7 eV) and $E_F$ (see shaded grey area in Fig. 5.12). Such a density of gap states was already reported for Pn in Pn(monolayer)/ClAl-phthalocyanene/graphite as well a Pn(15 nm)/$SiO_2$ upright standing thin film.

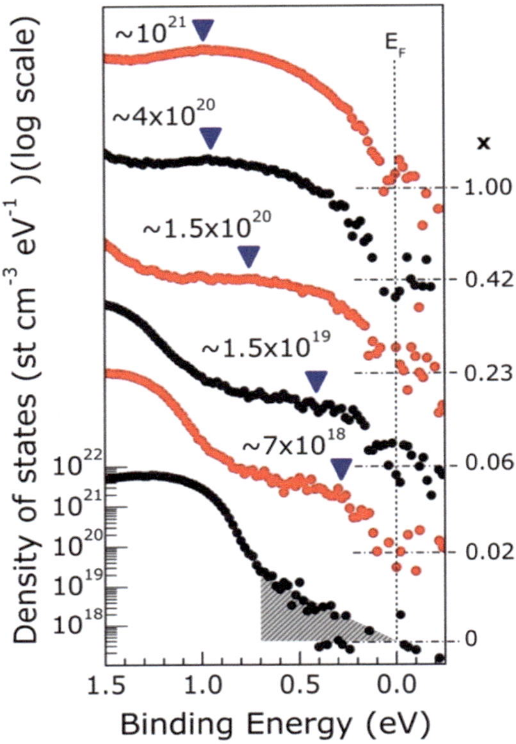


**Fig. 5.12** The density-of-states (DOS) deduced from HeI$_\alpha$ UPS spectra (angle-integrated) of Pn/Cu(110) multilayer as function of K-doping level ($x$) in the HOMO-LUMO gap region. The spectra were acquired at 293 K. Background contribution was subtracted according to same procedure described in [30, 31]. The DOS is plotted on log scale to show tiny K-induced DOS. The molecular packing density of the bulk phase of Pn ($2.9 \times 10^{21}$ cm$^{-3}$) was used to obtain DOS (states cm$^{-3}$ eV$^{-1}$), according to the same procedure described in [28, 30, 31]. The molecular packing density was obtained from the lattice parameters reported in [24] for the Pn bulk phase. Peak position of the gap state is indicated by *blue triangular* markers (see also Fig. 5.5d). Note that clear increase in the DOS of gap states was observed even at the very beginning of the K-doping process ($x = 0.02$). DOS values at peak positions of gap states are shown in the figure. As the DOS values involve photoelectron attenuation effect, ten times larger values must be considered for actual DOS (see main text for details) Figure reproduced with permission from [34], Copyright 2012, APS)

This density of gap state was attributed to the structural imperfections in the Pn molecular packing [30, 31].

Even for the very low doping case ($K_{0.02}Pn_1$ film) the K-induced gap states (indicated by triangle marker in Fig. 5.12) were clearly detected. The intensity of the gap states increases with K doping and the peak position progressively shifts toward higher binding energy side (see Figs. 5.4d, 5.5 and 5.11). No significant photoemission signal was detected at $E_F$ even for $x = 1$, thus suggesting the insulating nature of the K-doped Pn film. In contrast, transport measurements on highly

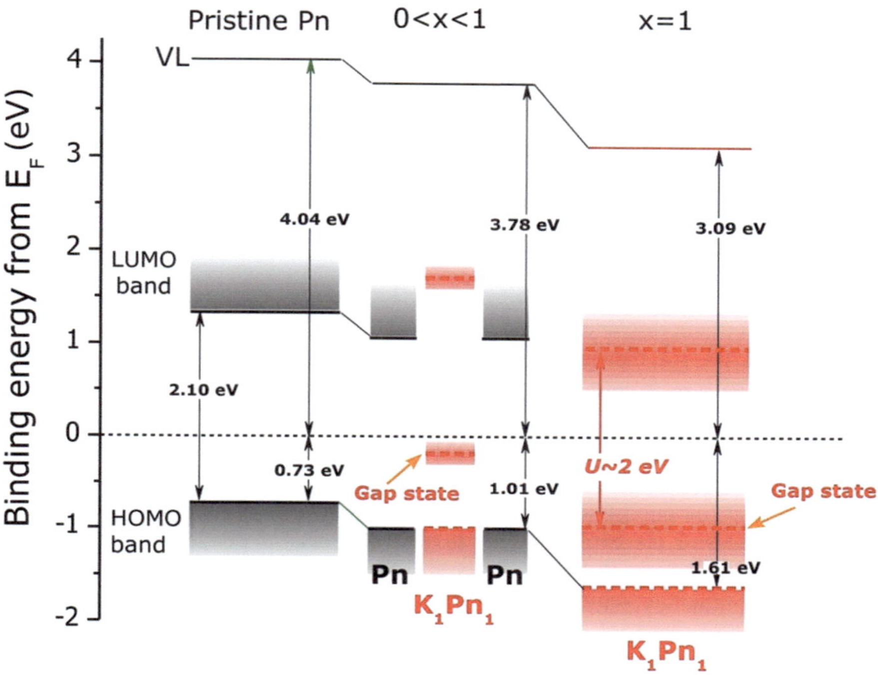


**Fig. 5.13** Summary of energy level alignment in thin films with Pn and $K_1Pn_1$ phases at 293 K. Pn ($K_1Pn_1$) energy level and bandwidth are indicated by *black* (*red*) *line* and by *black* (*red*) *shadow*, respectively. The LUMO position for Pn SC film was taken from [29]. At low and intermediate doping level ($0 < x < 1$), Pn LUMO starts filling to give localized gap state below $E_F$, which originates from $K_1Pn_1$ Mott–Hubbard insulator. Note that the HOMO binding energies of pristine Pn and $K_1Pn_1$ crystals are the same for $0 < x < 1$ according to the UPS results (see text). The energies in the *middle panel* are shown for $x = 0.02$. At $x = 1$, a SC $K_1Pn_1$ Mott insulator is formed and $E_F$ lies in the middle of the Hubbard energy gap $U$ (figure reproduced with permission from [34], Copyright 2012, APS)

ordered $K_xPn_1$ film by Craciun et al. [7] suggested that an insulator-metal-insulator transition appeared depending on $x$. At low and intermediate doping level ($0 < x < 1$), they observed a sharp increase in the Pn film conductivity, resulting in a metallic behavior [7]. The metallic behavior was ascribed to the progressive filling of the LUMO with electrons donated from K atoms [7]. At $x = 1$, electron–electron interactions cause the splitting of the half filled LUMO band and the opening of a Mott–Hubbard gap. The gap opening explained the observed re-entrance of $K_1Pn_1$ film into an insulating state [7]. The evolution of UPS spectra for the present Pn/Cu (110) SC multilayer upon K doping can be rationalized in terms of coexistence of pristine Pn SC and $K_1Pn_1$ Mott insulator at any level of K doping. A similar phase separation was reported in K-doped fullerene films, where they were attributed to the limited solubility of K atoms within the organic matrix host [18]. The impact of such phase coexistence on the ELA of the K doped Pn film is schematically illustrated in Fig. 5.13. For low and intermediate doping level ($0 < x < 1$) the LUMO band is progressively filled with $x$ and a Mott–Hubbard transition occurs to form small $K_1Pn_1$ Mott–Hubbard insulator SCs, which yield gap state below $E_F$ (middle panel in Fig. 5.13). The Mott–Hubbard derived gap states act as electron "donor" levels, which can populate, via thermal excitation, the empty Pn LUMOs above $E_F$. As K-doping level increases, a larger number of donor gap states become available to determine the gradual upward shift of $E_F$ (middle panel in Fig. 5.13). Finally, the $x$-dependent shift of Pn energy levels saturates when a continuous film of $K_1Pn_1$ is formed ($x = 1$) and $E_F$ lies near the middle of the Mott–Hubbard gap (right panel in Fig. 5.13).

As shown in the inset of Fig. 5.6, energy level shifts are linearly dependent on the logarithm of doping level i.e. $\Delta E \propto \alpha\, k_B T\, ln(x)$ for $x < 1$, where $\Delta E$ is the energy level shift, $k_B T$ is the thermal energy (~25 meV at 293 K), and $\alpha$ is a proportional coefficient. This is in agreement with the prediction of the classical theory of doping in semiconductor [10, 49, 50]. In particular a value of $\alpha \sim 7$, $\alpha \sim 6$ and $\alpha \sim 8$ was determined for VL shift, HOMO level shift and gap state energy shift, respectively. The $\alpha$ values of the VL and HOMO shifts are nearly the same, indicating that the energy shifts are mainly related to $E_F$ shift in the energy gap, while a larger $\alpha$ of the gap state may involve also size dependence of the $K_1Pn_1$ binding energy. Values of $\alpha > 1$ were similarly reported by Ding et al. [10, 49] for Cs-*n*-doped phthalocyanine thin film. They suggested that the large value of $\alpha$ originates in the population of the LUMO tailing with electrons from the ionized donor atoms. The LUMO tailing in organic semiconductors can be related to the gap states distribution sensitively originating from structural imperfection [30, 31].

According to the binding energy of the gap state peak in the $K_1Pn_1$ SC film (0.98 eV at 293 K and 1.02 eV at 50 K), Hubbard energy $U$ can be estimated to be ~2.0 eV at 293 K and 50 K. This estimation is in fairly good agreement with the theoretical value of $U = 1.5$ eV [7]. Within the framework of the ELA model illustrated in Fig. 5.13, the intensity of the gap state in UPS is expected to increase with $K_1Pn_1$ crystals size, as effectively observed in Figs. 5.4d and 5.11. As previously pointed out at the end of the Sect. 5.2, the opening of the Mott–Hubbard gap is not expected to modify the binding energy of the $K_1Pn_1$ HOMO level with respect to that of pristine Pn. This is illustrated in middle panel of Fig. 5.13. At low and intermediate doping level ($0 < x < 1$), in fact, a large difference between the HOMO positions of pristine Pn SC film and $K_1Pn_1$ should results in HOMO-lineshape broadening, in contrast with what observed in the photoelectron spectroscopy data [Fig. 5.4c].

Finally we comment on much lower UPS intensity of the gap state than that of the HOMO for the $K_1Pn_1$ SC film. If we assume that (i) the wave function of the occupied Mott–Hubbard insulator state is localized near the interlayer boundary as predicted by theory [7, 48] (see also Fig. 5.8c), i.e. near the bottom edges of upright standing Pn molecules in the surface layer and (ii) K atoms do not exist at the surface, photoelectrons from the gap state are attenuated more than those from the HOMO distributed over each Pn molecule before escape to vacuum. The UPS intensity ratio between the HOMO (distributed over Pn molecule) and gap state (localized near the bottom edge of the first layer Pn), $I_{gap\ state}/I_{HOMO}$, was estimated to be ~0.1 for the $K_1Pn_1$ SC film by using the electron attenuation length (~1 nm), the length of Pn molecule (~1.5 nm), substantial wave-function spread of the Mott–Hubbard state (~0.5 nm; localized on two aromatic rings near the bottom edge of Pn) and the number of electrons in the HOMO (2 electrons) and the Mott–Hubbard insulator state (1 electron). This ratio is in good agreement with the experimental findings [~0.12, see Fig. 5.3d].

## 5.5 Conclusions

The electronic properties of K doped SC Pn thin film was investigated by UPS. The impact of K doping on (i) the electronic band structure of the Pn thin film and (ii) ELA at the organic/substrate interface were carefully evaluated. At 293 K, a progressive shift of the Pn molecular states toward high binding energy side was observed with increase in the K-doping level. At the same time, a gap state appeared below $E_F$. No K-induced broadening in the UPS features was observed, indicating that the Pn SC thin films kept single crystalline structure upon doping. The gap state was attributed to the occupied Mott–Hubbard band of the $K_1Pn_1$ Mott–Hubbard insulator. Such band originated from the splitting of the half filled LUMO band of $K_1Pn_1$ SC film, as induced by electron correlation effects. At 293 K, no impact of the K doping on the HOMO band dispersion was observed and no dispersion for the gap state of the $K_1Pn_1$ SC multilayer was found. In contrast, at 50 K, a small but clear dispersion of the gap state (~0.08 eV) of the $K_1Pn_1$ film was observed, which is in good agreement with theoretical prediction for $K_1Pn_1$ Mott–Hubbard insulator crystal. Moreover significant changes in the HOMO lineshape in K-doped film were observed at 50 K, suggesting also the occurrence of a temperature-induced change in the molecular packing structure. The impact of K doping on the energy level alignment was discussed in terms of coexistence of $K_1Pn_1$ Mott–Hubbard insulator and the Pn SC film. Independent of the doping level no significant density of states was detected at $E_F$, indicating stability of the Mott–Hubbard insulator ground state at least near the surface.

**Acknowledgments** This work was financially supported by the Global-COE Program of MEXT (G03: Advanced School for Organic Electronics, Chiba University) and Grant-in-Aid for Scientific Research (A; Grant No. 20245039).

## References

1. C.W. Tang, S.A. VanSlyke, Appl. Phys. Lett. **51**, 913 (1987)
2. J.H. Burroughes, D.D.C. Bradley, A.R. Brown, R.N. Marks, K. Mackay, R.H. Friend, P.L. Burns, A.B. Holmes, Nature **347**, 539 (1990)
3. C.D. Dimitrakopoulos, P.R.L. Malenfant, Adv. Mat. **14**, 99 (2002)
4. T. Minakata, M. Ozaki, H. Imai, J. Appl. Phys. **74**, 1079 (1993)

5. A.B. Kaiser, Rep. Progr. Phys. **64**, 1 (2001)
6. M.F. Craciun, S. Rogge, M.J.L. den Boer, S. Margadonna, K. Prassides, Y. Iwasa, A.F. Morpurgo, Adv. Mat. **18**, 320 (2006)
7. M.F. Craciun, G. Giovannetti, S. Rogge, G. Brocks, A.F. Morpurgo, J. van den Brink, Phys. Rev. B **79**, 125116 (2009)
8. P.J. Benning, J.L. Martins, J.H. Weaver, L.P.F. Chibante, R.E. Smalley, Science **252**, 1417 (1991)
9. M. Fahlman, D. Beljonne, M. Logdlund, R.H. Friend, A.B. Holmes, J.L. Brédas, W.R. Salaneck, Chem. Phys. Lett. **214**, 327 (1993)
10. H. Ding, Y. Gao, Appl. Phys. Lett. **92**, 053309 (2008)
11. H. Ding, Y. Gao, Org. Electron. **11**, 1786 (2010)
12. A. Hansson, J. Böhlin, S. Stafström, Phys. Rev. B **73**, 184114 (2006)
13. M.G. Betti, F. Crispoldi, A. Ruocco, C. Mariani, Phys. Rev. B **76**, 125407 (2007)
14. F. Evangelista, R. Gotter, N. Mahne, S. Nannarone, A. Ruocco, P. Rudolf, J. Phys. Chem. C **112**, 6509 (2008)
15. G. Iucci, K. Xing, M. Lögdlund, M. Fahlman, W.R. Salaneck, Chem. Phys. Lett. **244**, 139 (1995)
16. K.Z. Xing, M. Fahlman, M. Lögdlund, M. Berggren, O. Inganäs, M.R. Andersson, M. Boman, S. Stafström, G. Iucci, P. Broms et al., Synth. Met. **80**, 59 (1996)
17. G. Greczynski, M. Fahlman, W.R. Salaneck, N. Johansson, D.A. dos Santos, J.L. Brédas, Thin Solid Films **363**, 322 (2000)
18. P.J. Benning, D.M. Poirier, T.R. Ohno, Y. Chen, M.B. Jost, F. Stepniak, G.H. Kroll, J.H. Weaver, J. Fure, R.E. Smalley, Phys. Rev. B **45**, 6899 (1992)
19. F. Sunqi, Z. Xing, W. En, F. Jishi, M. Jinchang, G. Zhennan, Q. Jiuxin, Z. Xihuang, J. Zhaoxia, Y. Bo, Solid State Commun. **80**, 639 (1991)
20. Y.W. Park, Synth. Met. **56**, 3220 (1993)
21. L. DeGiorgi, J. Low Temp. Phys. **105**, 1671 (1996)
22. R. Mitsuhashi, Y. Suzuki, Y. Yamanari, H. Mitamura, T. Kambe, N. Ikeda, H. Okamoto, A. Fujiwara, M. Yamaji, N. Kawasaki et al., Nature **464**, 76 (2010)
23. O.V. Molodtsova, V.M. Zhilin, D.V. Vyalikh, V.Y. Aristov, M. Knupfer, J. Appl. Phys. **98**, 093702 (2005)
24. H. Yoshida, N. Sato, Phys. Rev. B **77**, 235205 (2008)
25. S.D. Ha, A. Kahn, Phys. Rev. B **80**, 195410 (2009)
26. H. Fukagawa, H. Yamane, T. Kataoka, S. Kera, M. Nakamura, K. Kudo, N. Ueno, Phys. Rev. B **73**, 245310 (2006)
27. S. Kera, H. Yamane, N. Ueno, Progr. Surf. Sci. **84**, 135 (2009)
28. D.V. Lang, X. Chi, T. Siegrist, A.M. Sergent, A.P. Ramirez, Phys. Rev. Lett. **8**, 086802 (2004)
29. J. Hwang, A. Wan, A. Kahn, Mat. Sci. Eng. R **64**, 1 (2009)
30. T. Sueyoshi, H. Fukagawa, M. Ono, S. Kera, N. Ueno, Appl. Phys. Lett. **95**, 183303 (2009)
31. F. Bussolotti, S. Kera, K. Kudo, A. Kahn, N. Ueno, Phys. Rev. Lett. **110**, 267602 (2013)
32. E. Annese, J. Fujii, I. Vobornik, G. Rossi, J. Phys. Chem. C **116**, 2382 (2012)
33. S. Lukas, S. Söhnchen, G. Witte, C. Wöll, Chem. Phys. Chem. **5**, 266 (2004)
34. F. Bussolotti, S. Kera, N. Ueno, Phys. Rev. B **86**, 155120 (2012)
35. J.J. Yeh, L. Lindau, Atom. Data Nucl. Data **32**, 1 (1985)
36. N. Ueno, S. Kiyono, T. Watanabe, Chem. Phys. Lett. **46**, 89 (1977)
37. J. Cornil, D. Beljonne, J.P. Calbert, J.L. Brédas, Adv. Mat. **13**, 1053 (2001)
38. Y.C. Cheng, R.J. Silbey, D.A. da Silva Filho, J.P. Calbert, J. Cornil, J.L. Brédas, J. Chem. Phys. **118**, 3764 (2003)
39. C.C. Mattheus, G.A. de Wijs, R.A. de Groot, T.T.M. Palstra, J. Am. Chem. Soc. **125**, 6323 (2003)
40. H. Yamane, E. Kawabe, D. Yoshimura, R. Sumii, Y. Ouchi, N. Ueno, K. Seki, Phys. Status Solidi **245**, 793 (2008)
41. N. Ueno, S. Kera, K. Sakamoto, K. Okudaira, Appl. Phys. A **92**, 495 (2008)

42. N. Ueno, S. Kera, Progr. Surf. Sci. **83**, 490 (2008)
43. N. Koch, A. Rajagopal, J. Ghijsen, R.L. Johnson, G. Leising, J.J. Pireaux, J. Phys. Chem. B **104**, 1434 (2000)
44. N. Koch, H. Oji, E. Ito, E. Zojer, H. Ishii, G. Leising, K. Seki, Appl. Surf. Sci. **175**, 764 (2001)
45. T. Schwieger, H. Peisert, M. Golden, M. Knupfer, J. Fink, Phys. Rev. B **66**, 155607 (2002)
46. M. Imada, A. Fujimori, Y. Tokura, Rev. Mod. Phys. **70**, 1039 (1998)
47. R. Eguchi, M. Taguchi, M. Matsunami, K. Horiba, K. Yamamoto, Y. Ishida, A. Chainani, Y. Takata, M. Yabashi, D. Miwa et al., Phys. Rev. B **78**, 075115 (2008)
48. Y. Shichibu, K. Watanabe, Jpn. J. Appl. Phys. **42**, 5472 (2003)
49. Y. Gao, L. Yan, Chem. Phys. Lett. **380**, 451 (2003)
50. S.M. Sze, *Physics of Semiconductor Devices, Section 1.4.2* (Wiley, New York, 1981)

# Chapter 6
# Vertical Bonding Distances Impact Organic-Metal Interface Energetics

**Steffen Duhm, Christoph Bürker, Takuya Hosokai, and Alexander Gerlach**

## 6.1 Introduction

Understanding contact formation is one of the key issues in *organic electronics*. Especially the energy-level alignment (ELA) at organic-metal interfaces is of pivotal importance for device performance as it controls the charge injection into active layers of organic devices [1]. The mechanism of ELA is subject of intense research and discussions [2–4] and is still not understood in sufficient detail. However, it is well accepted that the ELA is not simply controlled by the work function ($\phi$) of the metal substrate and the electron affinity (EA) and ionization energy (IE) of the conjugated organic molecule (COM), which forms the adsorbate, see Fig. 6.1. In addition to solid state effects, such as polarization or the impact of the packing structure, also the chemical structure of the COM plays an important role in formation of the contact between the metal and the first molecular layer. Here, in turn, the vertical bonding distance ($d_H$) and adsorption induced conformation changes are crucial for the formation of interface dipoles. Some of the manifold reasons for them will be discussed in this chapter. These interface dipoles lead to vacuum-level shifts ($\Delta VL$) at the interface between the metal and the organic

S. Duhm (✉)
Institute of Functional Nano & Soft Materials (FUNSOM), Soochow University, 199 Ren-Ai Road, Suzhou 215123, P. R. China
e-mail: duhm@suda.edu.cn

C. Bürker • A. Gerlach (✉)
Institut für Angewandte Physik, Universität Tübingen, Auf der Morgenstelle 10, 72076 Tübingen, Germany
e-mail: alexander.gerlach@uni-tuebingen.de

T. Hosokai
Research Institute of Instrumentation Frontier (RIIF), National Institute of Advanced Industrial Science and Technology (AIST), Umezono 1-1-1, Tsukuba, Ibaraki, 305-8568, Japan



H. Ishii et al. (eds.), *Electronic Processes in Organic Electronics*, Springer Series in Materials Science 209, DOI 10.1007/978-4-431-55206-2_6

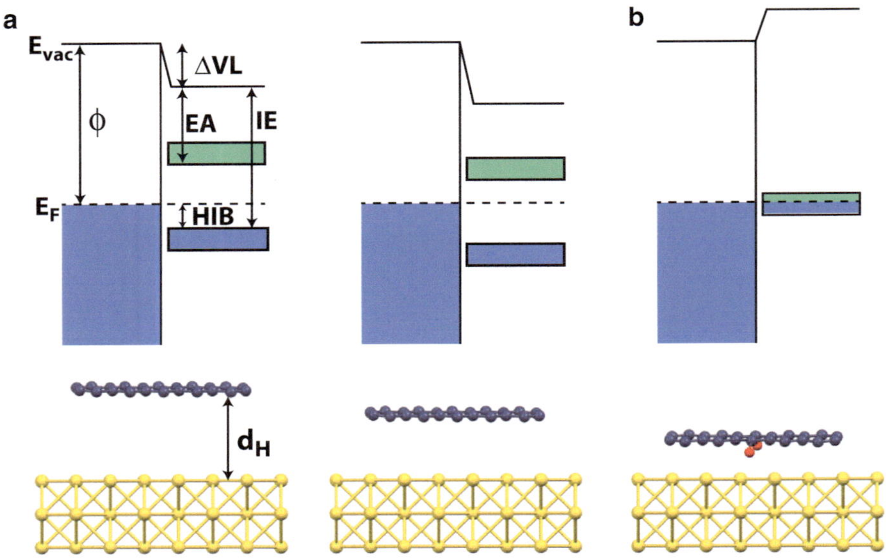


**Fig. 6.1** Schematics of the relationship of bonding distance ($d_H$) and adsorption induced molecular distortions and the energy-level alignment at organic-metal interfaces in case of (**a**) a physisorbed and (**b**) a chemisorbed molecular monolayer. The vacuum-level ($E_{vac}$), the vacuum-level shift ($\Delta VL$), the Fermi-level ($E_F$), the work function ($\phi$), the electron affinity (EA), the ionization energy (IE) and the hole injection barrier (HIB) are plotted for a clean metal surface and a monolayer of a COM. The schematics are simplified and do not apply for all metal/COM combinations

monolayer. Notably, further vacuum-level shifts can occur for higher (multilayer) coverages, which are related to, e.g., band bending [5, 6] or gap-states [7, 8]. It is important to realize that for multilayer coverages $\Delta VL$ might be caused by different effects than in the monolayer regime. It can be even helpful to define an "effective substrate work function", which is the work function of the metal substrate corrected by $\Delta VL$ at the interface. This effective work function is then responsible for the ELA of organic multilayers. Thus, rationalizing the origins of interface dipoles is crucial to understand (and finally to control) the ELA of organic layers on metal substrates.

The ELA is often measured by ultraviolet photoelectron spectroscopy (UPS) [9–12], and X-ray photoelectron spectroscopy (XPS) can give insight in chemical interactions at the interface. However, to fully understand the mechanisms of ELA also the monolayer structure has to be known. The lateral structure can be accessed with scanning probe and/or diffraction techniques [13, 14]. For vertical adsorption heights the X-ray standing wave (XSW) technique [15, 16], which will be shortly reviewed in Sect. 6.2, is a powerful tool. The combination of diffraction and spectroscopy allows to precisely determine element-specific bonding distances of molecular (sub)monolayers on single crystal substrates [17]–see the selected results for different COMs on noble metal surfaces in Table 6.1. A comprehensive electronic and structural characterization of organic-metal interfaces bridges the gap to theory, which nowadays is able to model these systems and, in some cases, to explain the different electronic properties and bonding distances [18–20].

While many effects can contribute to interface dipoles [2, 21, 22], the focus within this chapter is on those which are related to (element-specific) bonding distances. In general, dipoles at an organic-metal interface will lead to a change in the vacuum level $\Delta VL$ according to the Helmholtz equation:

$$\Delta VL = \frac{e p_z N}{\varepsilon_0 \varepsilon} \tag{6.1}$$

with $e$ the elementary charge, $p_z$ the dipole moment perpendicular to the substrate surface, $N$ the dipole density, $\varepsilon_0$ the vacuum permittivity and $\varepsilon$ the dielectric constant of the monolayer [23]. For physisorbed and weakly interacting systems the main reason for an interface dipole at the organic-metal interface is the so-called push-back effect [2, 3, 24], which is the result of Pauli repulsion of metal electron density spilling out in the vacuum by the sheer presence of the COM overlayer,

**Table 6.1** Summary of element-specific bonding distances ($d_H$) of some COMs on metal single crystal substrates measured with the XSW technique

| Molecule | Substrate | Carbon (Å) | Other atoms (Å) | Reference |
|---|---|---|---|---|
| DIP | Au(111)<br>Ag(111)<br>Cu(111) | 3.17<br>3.01<br>2.51 | | [20]<br>[20]<br>[20] |
| PEN | Ag(111)<br>Cu(111) | 3.12<br>2.34 | | [27]<br>[28] |
| PFP | Ag(111)<br>Cu(111) | 3.16<br>2.98 | F: 3.16<br>F: 3.08 | [29]<br>[28] |
| P2O | Ag(111)<br>Cu(111) | 3.32<br>2.34 | O: 3.35<br>O: 2.02 | [30]<br>[30] |
| P4O | Au(111)<br>Ag(111)<br>Cu(111) | 3.35<br>2.69<br>2.25 | N/A<br>O: 2.43<br>O: 1.98 | [30]<br>[30]<br>[30] |
| ZnPc | Cu(111) | 2.49 | Zn: 2.25<br>N: 2.55 | [31] |
| $F_{16}$ZnPc | Cu(111) | 2.66 | Zn: 2.58<br>N: 2.85<br>F: 3.15 | [31] |
| GaClPc | Cu(111) | 4.44 | Ga: 4.21<br>N: 4.71<br>Cl: 1.88 | [32] |

In most cases the nominal coverage is below one monolayer. The actual bonding distance can depend critically on the coverage, which is not reflected in this table. Note that the bonding distances on Au(111) substrates listed here are slightly decreasing if the surface reconstruction is taken into account [20, 33]. A more detailed summary of COM bonding distances can be found in [17]. The acronyms of the COMs stand for: DIP: diindenoperylene; PEN: pentacene; PFP: perfluoropentacene; P2O: 6,13-pentacenequinone; P4O: 5,7,12,14-pentacenetetrone; Pc: phthalocyanine which leads to a decrease in the effective work function.[1] The vacuum-level shift caused by the push-back effect is in the range of up to 1 eV [2, 3, 24] and can thus substantially increase hole injection barriers (HIBs, defined as the energetic distance of the substrate Fermi-level to the onset of the highest occupied molecular orbital [HOMO] of the adsorbate) at organic-metal interfaces (Fig. 6.1a). This $\Delta VL$ is naturally increasing with decreasing bonding distance [25, 26] until the organic adsorbate forms chemical bonds to the substrate. In Sect. 6.3 we will discuss how the adsorption height of large hydrocarbon molecules is determined by the interplay of Pauli repulsion and van der Waals (vdW) attraction and how functional side-groups can change the bonding distance.

In Sect. 6.4 we will discuss other origins of interface dipoles of weakly interacting systems. This can be, of course, the adsorption of polar molecules [21, 23]. Here, in the first place, the orientation of the molecule on the surface determines the $p_z$-component of the interfacial dipole moment and thus the sign and the magnitude of $\Delta VL$ according to Eq. (6.1). But even for weakly interacting systems also non-polar molecules can become distorted upon adsorption on a metal surface by site-specific interactions and then comprise a dipole moment [28, 31].

For strongly interacting systems the adsorption process commonly induces such conformation changes for the molecules in the monolayer [30, 34, 35], which are often closely related to the formation of an organic-metal charge transfer complex (CTC) and the (partial) filling of the former lowest unoccupied molecular orbital (LUMO) of the organic adsorbate. Such a charge transfer leads to an additional interface dipole which affects the energy-level alignment (Fig. 6.1b). Thus, CTC formation can be used to tailor effective work functions by pre-covering metal electrodes with strong electron accepting molecules [36]. Recently, we could identify *surface-induced aromatic stabilization* as one possible driving force for molecules to undergo a charge transfer with inert metal substrates [30]. In Sect. 6.5 we will review this concept with special attention to the role of bonding distances in the charge transfer and the energy-level alignment process. In Sect. 6.6 we will summarize and give a short outlook how the ELA at organic-metal interfaces could be potentially tuned by achieving control over bonding distances at these interfaces by rational design of molecular structures.

## 6.2 The X-Ray Standing Wave Technique

In short, the bonding distances $d_H$ of molecules adsorbed on single crystal surfaces are determined by measuring the element-specific absorption in an X-ray standing wave field. For reasons to be explained below, substrate Bragg reflections close to

[1] In the literature usually this decrease in $\phi$ by the push-back effect is attributed to an interface dipole, although strictly speaking an existing interface dipole (at the vacuum-metal interface) is reduced by the organic-metal interface formation.

$\theta_B = 90°$ are used to create the standing wave field with a certain spatial modulation. By scanning the photon energy of the incident wave through the Bragg condition one may vary the amplitude and phase of the XSW field. Depending on the bonding distance of the molecules, the X-ray absorption—and correspondingly the measured photoelectron yield for a given element—changes in a characteristic form with the photon energy.

### 6.2.1 The X-Ray Standing Wave Field

The XSW field is produced by the coherent superposition of the incoming plane wave with the electrical field vector $\mathbf{E}_0$ and the Bragg-reflected wave $\mathbf{E}_H$ as depicted in Fig. 6.2. The (normalized) spatial intensity variation $I^{SW}(\mathbf{r})$ above the substrate is then given as

$$I^{SW}(\mathbf{r}) = \frac{|\mathbf{E}_0 + \mathbf{E}_H|^2}{|E_0|^2} = 1 + \left|\frac{E_H}{E_0}\right|^2 + 2C\left|\frac{E_H}{E_0}\right| \cos(\nu - 2\pi\mathbf{H}\cdot\mathbf{r}), \quad (6.2)$$

where $\nu$ is the phase of the complex amplitude ratio, $E_H/E_0 = |E_H/E_0|\exp(i\nu)$, $\mathbf{H}$ the reciprocal lattice vector, $C$ the polarization factor ($C = 1$ for $\sigma$-polarization; $C = \cos 2\theta_B$ for $\pi$-polarization), and $\theta_B$ the Bragg angle.

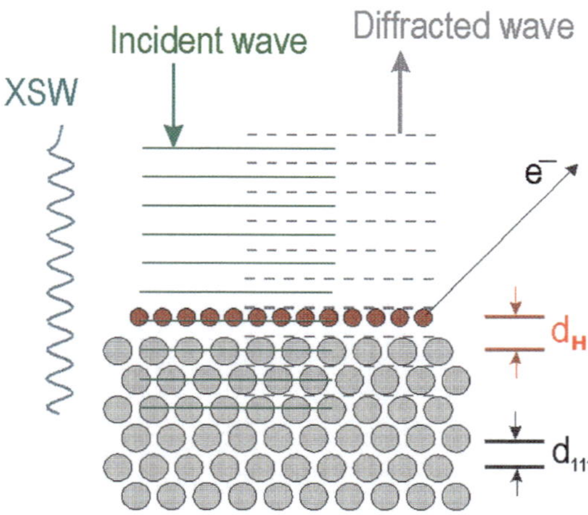


**Fig. 6.2** A standing wave with the periodicity of the substrate lattice is generated by the reflection of the incident wave from the single crystal substrate. By scanning the photon energy of the incident wave through the Bragg condition the XSW field above the surface changes in a characteristic way. The emitted photoelectrons that are generated by inelastic scattering processes can be used as element-specific signals that reflect the average position $d_H$ of the atoms within the adsorbed molecule

The photon energy dependence both of $\nu$ and the reflectivity $R = |E_H/E_0|^2$ are well-known within the framework of dynamical diffraction theory: The ratio of the complex amplitudes $E_0$ and $E_H$ of $\sigma$-polarized X-rays is given by [16]

$$\frac{E_H}{E_0} = \sqrt{\chi_H/\chi_{\bar{H}}}\left(\eta \pm \left(\eta^2 - 1\right)^{1/2}\right) \tag{6.3}$$

for a symmetric Bragg reflection. Here, $\chi_H$ is the Hth-order Fourier component of the complex crystal susceptibility $\chi(\mathbf{r})$, which is related to the dielectric function $\varepsilon(\mathbf{r})$ via $\chi(\mathbf{r}) = \varepsilon(\mathbf{r}) - 1$. The susceptibility components[2] $\chi_H$ are calculated from the structure factor $F_H$ via

$$\chi_H = \frac{r_0\lambda^2}{\pi V_c} F_H, \tag{6.4}$$

where $r_0$ is the classical electron radius, $\lambda$ the wavelength, and $V_c$ the volume of the unit cell for which the structure factor has been derived. The complex variable $\eta$ in Eq. (6.3), which for Bragg reflections close to $\theta_B = 90°$ is given by [15]

$$\eta = \frac{2(\Delta E/E_B)\sin^2\theta_B - \chi_0}{\left(\chi_H \chi_{\bar{H}}\right)^{1/2}}, \tag{6.5}$$

is a function of the energy difference $\Delta E = E_B - E$. On the basis of these equations the reflectivity $R$ and the phase $\nu$ can be calculated, see Fig. 6.3.

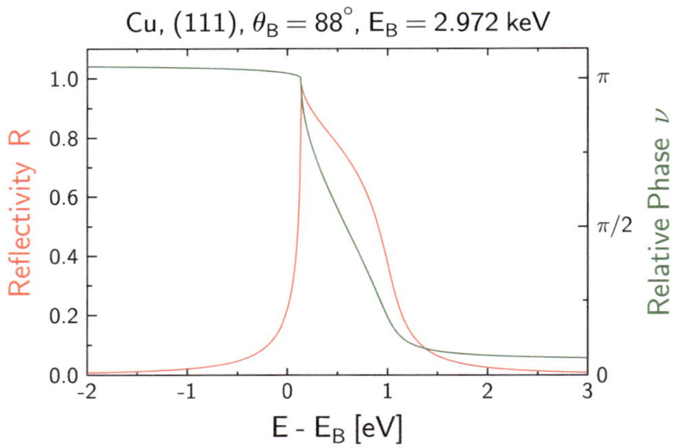


**Fig. 6.3** Intrinsic energy dependence of the reflectivity $R$ and phase $\nu$ for $\theta_B = 88°$ calculated with Eq. (6.3)–(6.5) for the 111-reflection of copper

[2] Note, that for a centrosymmetric crystal, $\chi_H = \chi_{\bar{H}}^*$ and therefore $\left(\chi_H \chi_{\bar{H}}\right)^{1/2} = |\chi_H|$.

Since the phase $\nu$ generally varies asymptotically by $\pi$ when going through the Bragg condition, the X-ray standing wave field moves by half a diffraction plane spacing within this energy range.

It can be shown that the angular width $W_\theta$ of the Bragg reflection varies as

$$W_\theta \propto d_s^2 F_H \tan\theta_B, \tag{6.6}$$

where $d_s$ is diffraction plane spacing. Only because for Bragg angles close to 90° the width increases strongly, typical crystal imperfections (mosaicity) can be tolerated.

## 6.2.2 The Photoelectron Yield

As indicated in Fig. 6.2 the XSW technique requires measurements of the photo yield $Y_p$ [37] which characterizes the position of the atoms in the interference field. For this purpose the intensities of core-level photoelectrons are commonly used. If for simplicity we only consider the so-called dipole approximation, $Y_p$ is proportional to the square of the electric field at the center $\mathbf{r}$ of the atoms and thus proportional to $I^{SW}$.[3] For systems with all atoms on equivalent sites one obtains

$$Y_p(\mathbf{r}) = 1 + R + 2C\sqrt{R}\cos(\nu - 2\pi\mathbf{H}\cdot\mathbf{r}), \tag{6.7}$$

with the energy-dependent reflectivity $R$ and phase $\nu$. Figure 6.4 illustrates the effect that different positions $\mathbf{r}$ have on the variation of the photo yield.

For an (partly) incoherent system with atoms on non-equivalent sites a normalized distribution function $n(\mathbf{r})$ that describes the spread of atoms around their average lattice position has to be introduced. Equation (6.7) is then generalized to

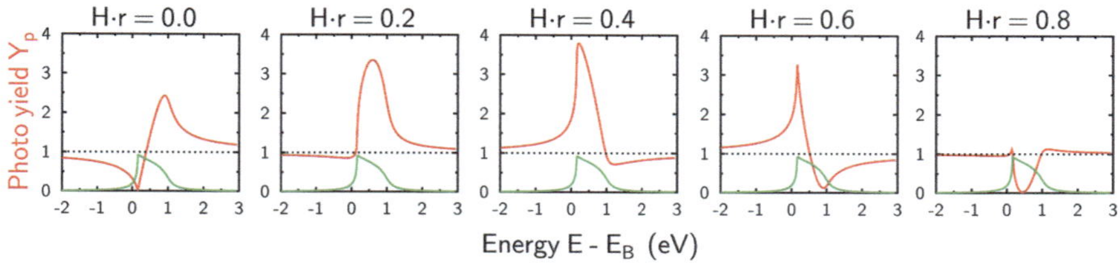


**Fig. 6.4** Theoretical normalized photo yields for different values of $\mathbf{H}\cdot\mathbf{r}$ in Eq. (6.7). The high sensitivity of $Y_p$ to the position of the atom $\mathbf{r}$ is demonstrated. Taken from Ref. [17] with copyright permission

[3] Depending on the experimental setup, i.e. particularly the emission angle, non-dipole corrections [15] have to be included.

$$Y_p = \int Y_p(\mathbf{r})\; n(\mathbf{r})d\mathbf{r} \tag{6.8}$$

$$= 1 + R + 2C\sqrt{R}\int n(\mathbf{r})\cos(\nu - 2\pi\mathbf{H}\cdot\mathbf{r})d\mathbf{r}. \tag{6.9}$$

A more manageable form of this averaging equation is obtained by introducing two new parameters: the *coherent position* $P_H = \mathbf{H}\cdot\mathbf{r}$ and the *coherent fraction* $f_H$.

$$Y_p = 1 + R + 2C\sqrt{R}f_H\cos(\nu - 2\pi P_H). \tag{6.10}$$

This is the working equation for most XSW experiments since $P_H$ and $f_h$ contain all structural information on the specific system. The coherent position $0 \leq P_H \leq 1$ now gives the position relative to the diffraction planes associated with the vector **H**: The average bonding distance of the atoms is simply given by

$$d_H = d_{hkl}(n + P_H), \tag{6.11}$$

where $n$ is an integer and $d_{hkl}$ the substrate lattice plane spacing corresponding to the Bragg reflection. The coherent fraction $0 \leq f_H \leq 1$, on the other hand, describes the degree of coherent order in the system. This is best illustrated by considering two extreme cases: A random distribution of atoms on the surface, e.g. a thick incoherent film of molecules, with $f_H = 0$ results in $Y_p = 1 + R$, i.e. depending on the absorption $Y_p \leq 2$. A perfectly ordered ad-layer with $f_H = 1$, in contrast, amplifies the photo yield to $Y_p \leq 4$.

As shown in Fig. 6.5 for three different XSW signals, i.e. carbon, nitrogen, and zinc atoms in ZnPc (molecular structure in the inset of Fig. 6.5a) on Cu(111), one can determine the coherent fraction and the coherent position for each atomic species and thereby derive the average bonding distances (Fig. 6.5c) for these atoms with high precision.

## 6.3 Physisorption: Interplay of Pauli Repulsion and van der Waals Attraction

The simplest case of adsorption of a COM on a metal substrate is the case of physisorption, where the bonding distance itself is the result of the interplay of van der Waals attraction and Pauli repulsion between adsorbate and substrate. Just recently it has become possible to reliably calculate bonding distances by DFT including vdW interactions [18–20] and to quantitatively access the contributions of these competing effect to the equilibrium adsorption height.

For the prototypical large hydrocarbon molecule diindenoperylene (DIP) the measured averaged bonding distances on the (111) surfaces of coinage metals are decreasing in the order Au, Ag, Cu (Table 6.1). Model calculations show that the vdW interaction energy is increasing with decreasing bonding distance [20].

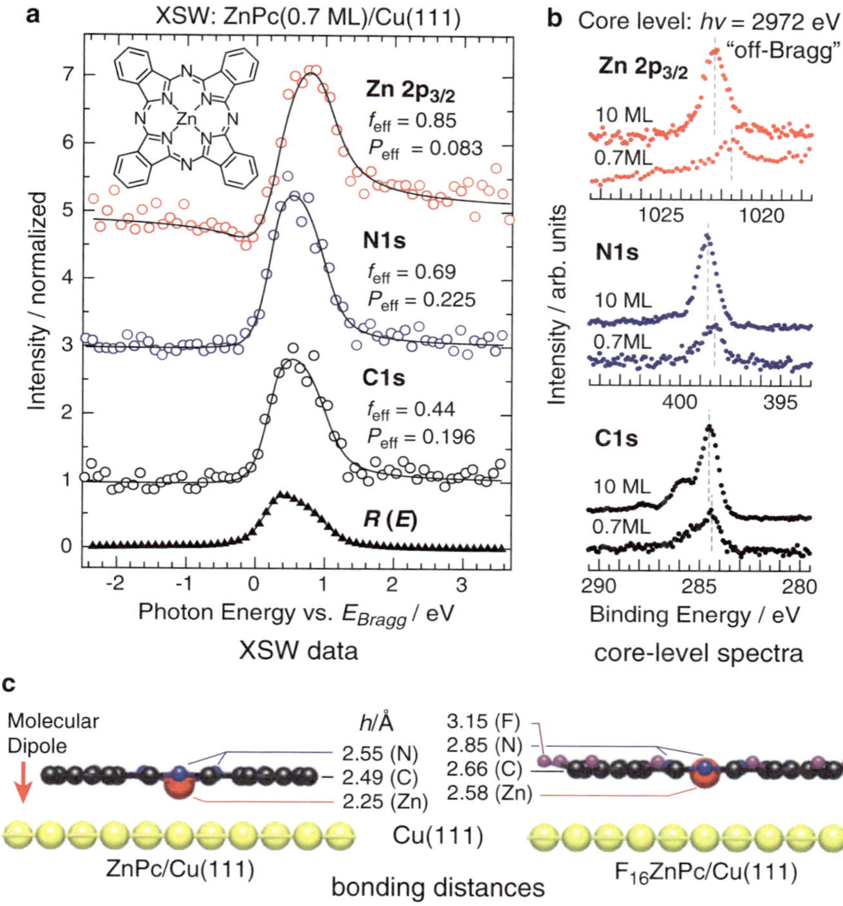


**Fig. 6.5** (**a**) XSW data analysis for ZnPc on Cu(111). The coherent fraction and the coherent position are determined by least-square fits (*solid lines*) to the experimental data. (**b**) Core-level spectra for one particular photon energy. (**c**) Element-specific bonding distances for ZnPc and $F_{16}$ZnPc/Cu(111) on Cu(111). Taken from Ref. [31] with copyright permission

For DIP on Cu(111), the bonding distance is small and the atomic density relatively high, i.e. the main part of the vdW energy comes from interaction of DIP with the first and second substrate layer. On Au(111) and Au(111), the bonding distances are larger and the atomic densities lower, and the main contribution to the overall vdW energy emerges from interaction between the molecule and the topmost substrate layer (Fig. 6.6). This can be rationalized by the vdW radii $r_{vdW}$ of the metal atoms which are decreasing in the order Au, Ag, Cu. With decreasing $r_{vdW}$, DIP can come closer to the metal and Pauli repulsion sets in at a shorter distance to the surface. However, this simple picture only holds for physisorbed systems and the border to chemisorption is fluent and particularly cannot be drawn at a bonding distance

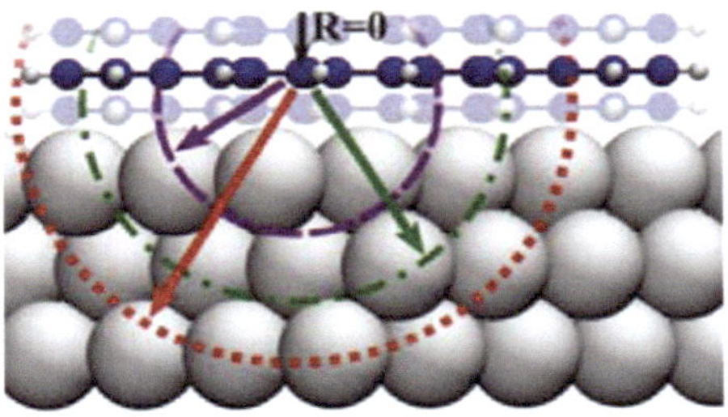


**Fig. 6.6** Schematic of the vdW interaction of the DIP molecule with the substrate. A carbon atom interacts with a certain number of substrate atoms, which lie within a sphere with certain radius, marked by an arrow. For substrates with high atomic density and at a smaller bonding distance, more substrate atoms are lying within that sphere, and the vdW interaction becomes stronger. Taken from Ref. [20] with copyright permission

where $r_{vdW}$ of adsorbate and substrate start to overlap [33]. In the case of chemisorption also other factors like covalent bonds or charge transfer reactions have to be taken into account in addition to Pauli repulsion and vdW interaction.

This can be demonstrated by comparing DIP with the smaller hydrocarbon pentacene (PEN). For PEN, the bonding distance on Ag(111) is increased compared to DIP, whereas it is decreased on Cu(111) substrates (Table 6.1). On Ag(111) the bonding type of PEN can be described as soft chemisorption [27] leading to an unique adsorption behavior with highly mobile molecules below monolayer coverage [38, 39]. As a consequence, the bonding distance depends critically on the coverage [27], which makes a straightforward comparison with the virtually not coverage dependent bonding distance of DIP on Ag(111) difficult. PEN on Cu(111) is clearly chemisorbed [28], which explains the small bonding distance.

So far, we only looked on pure hydrocarbon systems. However, functional side-groups can dramatically impact adsorption heights and bonding types. Perfluorination of PEN leads to an increase of 0.64 Å in the averaged bonding distance of the carbon atoms compared to PEN/Cu(111) and to an adsorption induced distortion of perfluoropentacene (PFP) with the fluorine atoms bent 0.10 Å away from the carbon skeleton (Table 6.1). In contrast to PEN, PFP on Cu (111) can be regarded as physisorbed [28]. The molecular distortion adds an additional dipole moment to the interface, which counteracts the decrease in the vacuum level by the push-back effect. The total $\Delta VL$ upon monolayer formation (measured by UPS) is thus only $-0.35$ eV for PFP on Cu(111), but $-0.90$ eV for PEN on the same substrate [28]. For also physisorbed PFP on Ag(111) [29], the carbon bonding distance is larger than on Cu(111) and allows PFP to retain its planar gas phase conformation also in the monolayer on the metal substrate (Table 6.1). In such a way, adding functional side-groups emerges as suitable method to actively tune interface energetics. Before we discuss the dramatic effects of oxygen side-groups on bonding distances, interface energetics and interaction strength of PEN on metal substrates, we consider other weakly interacting systems in the next section. These are defined here that—despite a possible charge rearrangement upon interface formation and possible signs of chemisorptive behavior—the organic monolayer still remains semiconducting at the metal substrate.

## 6.4 Charge Rearrangements of Weakly Interacting Systems

Pure (fluorinated) hydrocarbons are an excellent testing ground for the concepts which have been developed to rationalize the different bonding distances of COMs. However, these systems offer few degrees of freedom to actively control bonding distances and thus interaction strength by adding functional side-groups. For this purpose phthalocyanines (Pcs) evolved as model molecules to study organic/metal interface formation and various XSW studies are available [31, 32, 40–45]. The size of the central metal atom determines whether in the gas phase the molecules are planar or not and are thus non-polar or polar. Additional intramolecular dipoles can be introduced by adding moieties like Cl or O to the central atom.

Depending on the central atom site-specific interactions can appear, which can lead to additional molecular dipole moments at the organic-metal interface and might change the overall interaction strength between adsorbate and substrate. For example, ZnPc is planar in the gas phase, however, adsorbed on Cu(111) the Zn atom is protruding towards the substrate (Fig. 6.5c), which leads to the formation of such an intramolecular dipole moment between the Zn atom and the Pc-skeleton. Like in the case of PEN, also for ZnPc fluorination of the molecule leads to a repulsive interaction with the substrate, the molecular backbone is pushed away from the surface and the fluorine atoms are bent upwards (Fig. 6.5c). The adsorption height difference between Zn and the Pc-skeleton is smaller for $F_{16}$ZnPc/Cu(111) than for ZnPc/Cu(111) and thus the intramolecular dipole moment is decreasing. In addition, another intramolecular dipole moment is created by the adsorption height difference of the fluorine atoms and the Pc-skeleton. All these site-specific interactions lead to electronic relaxations and results in interface states, which have been measured by angle-resolved UPS [31].

For molecules with permanent molecular dipoles, the adsorption orientation in the monolayer is obviously crucial for the ELA. For substituted Pcs and related molecules, a single hetero-atom (like O or Cl) is pointing out of the large Pc-skeleton and one might naively expect that in a monolayer of such COMs on a metal the "hetero-atom up" orientation would be the most stable one. However, experiments show that both orientations are possible [13, 46, 47]. Moreover, the adsorption can lead to changes in the bond lengths and thus to a change in the intramolecular dipole moment.

For submonolayer coverages of GaClPc on Cu(111) XSW reveals that Cl-down is the preferred adsorption geometry (Table 6.1). The Cl-down configuration leads to a libration of GaClPc on the surface and by a precise analysis of coherent fractions also the libration amplitude could be determined [32]. Based on this experimental data the charge rearrangements across the organic-metal interface could be calculated (Fig. 6.7) and the vacuum-level shift can be disentangled into a contribution which is related to the dipole moment of GaClPc ($\Delta VL_{dip}$) and a contribution related to the charge rearrangements upon adsorption ($\Delta VL_{bond}$):

$$\Delta VL = \Delta VL_{dip} + \Delta VL_{bond}. \tag{6.12}$$

The charge rearrangements, which result in $\Delta VL_{bond}$, take mainly place around the Cl atom (Fig. 6.7), which is suggested to be covalently bound to Cu(111) [32]. Interestingly, the large bonding distances of the main part of the molecule (for carbon $d_H$ is almost twice as for the related ZnPc, Table 6.1), lead to the absence of a push-back effect at this particular organic-metal interface [32]. The Ga-Cl bond length is slightly increased by the adsorption compared to the gas phase and for the calculated $\Delta VL_{dip}$ the experimentally determined bond length can be used. However, as it becomes apparent from Fig. 6.7, in addition to the dipole moments originated in a molecular distortion/change of bond lengths (which could be measured by XSW), also charge rearrangements which can be regarded as a kind of polarization can take place at organic-metal interfaces [22, 32, 48]. Thus, in order to precisely assess interface dipoles, calculations at an advanced level are necessary.

## 6.5 Oxygen Side-Groups Lead to Surface-Induced Aromatic Stabilization

If a charge transfer complex with net electron transfer into the COM is formed at the organic-metal interface, the former LUMO of the molecule becomes (partially) filled (Fig. 6.1b) and can thus be easily measured with conventional UPS [30, 36] [4]. Such charge transfer complexes can then be used to tune effective metal work functions and thus the HIB of subsequently deposited COM layers [36, 50, 51], and offer the possibility of "soft" metallic contacts into COM layers [30, 52].

The CTC formation at organic-metal interfaces often involves adsorption induced conformation changes, as was shown experimentally e.g. for 2,3,5,6-tetrafluoro-7,7,8,8-tetracyanoquinodimethane (F4-TCNQ) [35], 3,4,9,10-perylene-tetracarboxylic-dianhydride (PTCDA) [33, 34, 53, 54], 1,4,5,8-naphtalene-tetracarboxylic-dianhydride (NTCDA) [55] and the pentacene oxo-derivatives 6,13-pentacenequinone (P2O) and 5,7,12,14-pentacenetetrone (P4O) [30]. Moreover, as CTC formation involves donation and backdonation of charges, the molecules rehybridize on the metal surface and changing bond length—including vertical distortions of molecules on a surface—play a pivotal role in the formation of charged organic monolayers on metal substrates.

[4] "Conventional UPS" is used here in contrast to "ultrahigh-sensitivity UPS" [8, 49], which can also detect tiny density of states which are below the detection limit of most current UPS set-ups.

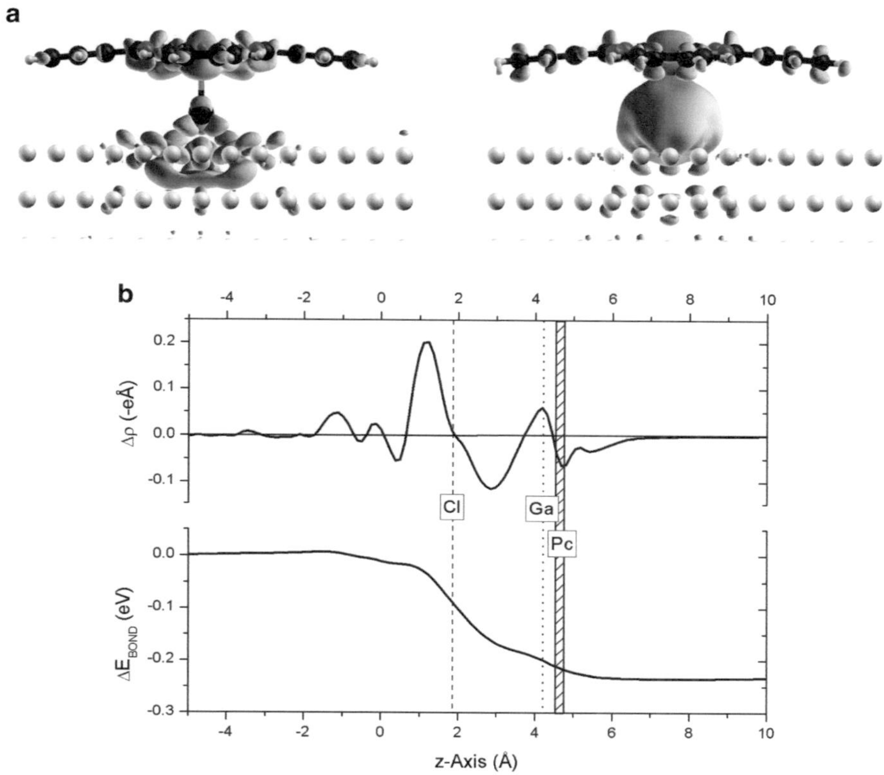


**Fig. 6.7** Calculated charge rearrangement upon adsorption of GaClPc on Cu(111). (**a**) Electron density increase (*left*) and electron density decrease (*right*). (**b**) Plane integrated charge rearrangements $\Delta\rho$ as function of vertical position and impact on the bond dipole ($\Delta E_{bond}$). Zero is set to the topmost metal plane. Taken from Ref. [32] with copyright permission

Whether organic semiconductors on a metal surface undergo a charge transfer beyond Fermi-level pinning [4, 9, 56] with the substrate, i.e. form a *metallic organic monolayer*, or maintain their finite fundamental gap does not simply depend on the EA and IE of the molecule and $\phi$ of the substrate. Also the bonding distance alone cannot be taken as explicit indicator for CTC formation. For example, PEN and P2O have almost identical DFT calculated gas phase electron affinities ($\approx 1.50$ eV, [30]) and the same carbon bonding distance on Cu(111) (2.34 Å, Table 6.1). For PEN on Cu(111) no distinct filling of the LUMO could be found by UPS [28], whereas the LUMO of P2O becomes partially filled in the monolayer on the same substrate [30]. The related P4O even exhibits an completely filled former LUMO in the monolayer on Cu(111) [30]. The CTC formation of P2O and P4O is accompanied by a bending of the oxygen atoms towards the Cu(111) surface (Table 6.1 and Fig. 6.8a) and large chemical shifts in the C1s and O1s core-levels as measured by XPS [30].

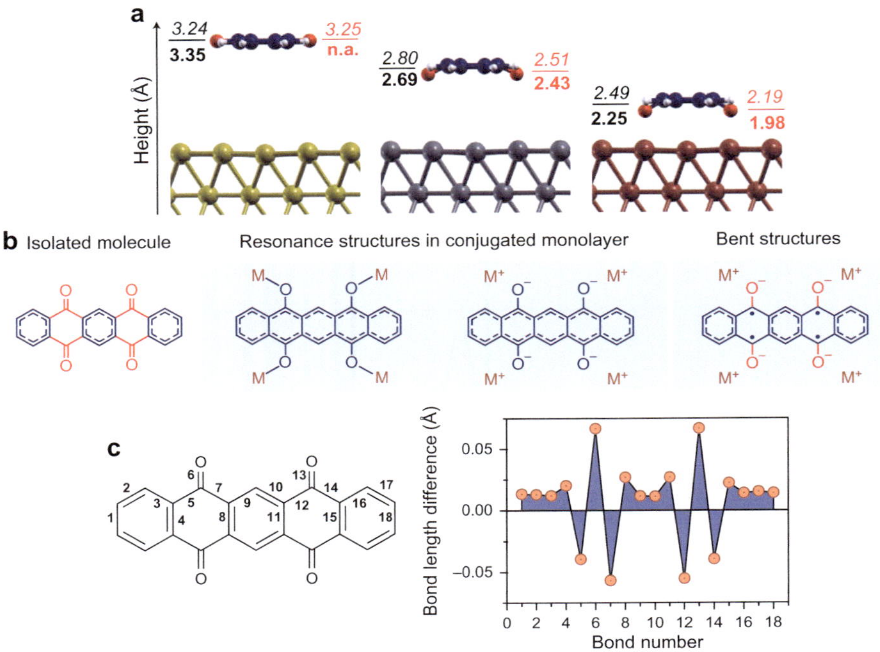


**Fig. 6.8** (**a**) Experimentally determined (*bold*) and DFT-calculated averaged bonding distance (*italic*) of the carbon atoms (*black*) and the oxygen atoms (*red*) of P4O on (from *left* to *right*) Au (111), Ag(111) and Cu(111), respectively. (**b**) Resonance structures of P4O. (**c**) Calculated bond length changes of P4O upon adsorption on Cu(111). Taken from Ref. [30] with copyright permission

The ability of P2O and P4O to form a CTC on Cu(111) (in contrast to PEN) is related to the keto-groups of the molecules, which break the conjugation along the molecular backbone (Fig. 6.8b). In fact, an isolated P4O molecule could be also regarded as three benzene molecules, which are fused by the keto-groups, rather than a PEN derivative (and similar for P2O). Upon CTC formation the molecules rehybridize and can extend their conjugated $\pi$-system across the entire molecule (Fig. 6.8b). In such a way, P2O and P4O become more PEN-like by the process of *surface-induced aromatic stabilization* [30]. This rehybridization involves naturally also bond length changes (Fig. 6.8a) and leads to the out-of-plane bending of the C-O bonds, which could be experimentally determined with XSW (Table 6.1 and Fig. 6.8a).

Similar motifs of surface-induced aromatic stabilization could be also found for other molecules forming CTC on metal surfaces [34, 35, 53–55]: in all cases the conjugation is broken by functional side-groups and can be (partly) re-established by a charge transfer from a metal substrate. However, not all of these COMs form a CTC on *each* metal surface. For example, P4O on Au(111) is physisorbed with a planar adsorption geometry and large bonding distances, whereas P4O on Ag(111) and Cu(111) forms a CTC with a bent adsorption geometry (Fig. 6.8a). In a figurative description, in the initial adsorption process the bonding distance might be determined by the interplay of Pauli repulsion and vdW interaction (as in the case of physisorption) and only if the molecule comes close enough to the surface CTC formation can set in. It is, however, important to bear in mind that coming close enough to the surface is essential but not sufficient to form a CTC as was shown by comparing PEN and P2O on Cu(111).

Noteworthy, for *metallic* organic monolayers—in contrast to semiconducting organic layers—also a work function is defined (Fig. 6.1b), which consequently determines the position of the vacuum-level. In such a way, the ELA of COMs which undergo surface-induced aromatic stabilization on a given substrate becomes independent from the metal substrate work function [57]. This means that interface phenomena like push-back effect or adsorption induced dipole moments, which have been discussed above for weakly interacting systems, do not translate directly into vacuum-level shifts and in addition the (unknown) chemical potential of the metallic organic monolayer plays a crucial role in the ELA.

## 6.6 Conclusions

The first layer of an organic thin film on a metal substrate is crucial for the energy-level alignment of further deposited organic layers as it determines the effective substrate work function. Element-specific bonding distances can impact interface dipoles in various ways. For weakly interacting system these include:

- the push-back effect,
- permanent molecular dipoles,
- adsorption induced molecular distortions,
- site-specific structural relaxations.

Measuring bonding distances is thus essential for the understanding of contact formation and the electronic structure at these interfaces. With this knowledge one can benchmark DFT calculations [19, 20] and predict interface dipoles [25, 26]. Moreover, one can now try to tune the bonding distances by adding certain functional groups to selected COMs. Note, however, that because the metal/COM interaction mechanism can be rather complex, tailoring the properties is a challenging task. Especially, we almost neglected *inter*molecular interaction in the (sub)monolayer, which can also have a considerable impact on bonding distances as shown in the pioneering work of C. Kumpf et al. [40, 41, 43, 44].

The magnitude of the vacuum-level shift caused by the push-back effect depends critically on the bonding distance [25, 26], which for physisorbed systems is determined by the interplay of Pauli repulsion and vdW interaction, which cannot be actively tuned. However, bulky spacer groups can hinder the conjugated core of a COM to come closer to a surface [58]. Also, the Cl-atom in GaClPc hinders the

Pc-skeleton to come close to the Cu(111) surface, which even leads to the absence of a push-back effect [32]. In this case, however the molecule itself has a dipole moment and manipulating such permanent molecular dipoles might be the most straightforward way to control vacuum-level shifts at organic-metal interfaces.

Intramolecular dipoles caused by adsorption induced distortion are more surprising and can even occur for weakly interacting systems like PFP on Cu(111) [28] by the repulsive interaction of saturated fluorine with the substrate. The same holds for fluorination of ZnPc [31] and CuPc [41, 42]. As the interaction of the unfluorinated parent molecules with Cu(111) is very different, (per)fluorination might be regarded as a general method to weaken the interaction at organic-metal interfaces and to introduce intramolecular dipole moments to non-polar molecules. The latter only holds if the bonding distance of the unfluorinated parent molecule is not already too large [like in the case of PEN/Ag(111), which leads to a planar adsorption geometry also for PFP]. Note that this rule does not necessarily hold for strongly chemisorbed systems like F4-TCNQ on Cu(111). Also here the fluorine atoms are bent upwards with respect to carbon and nitrogen atoms [35], however the interaction with the substrate is believed to *in*crease upon fluorination [35, 59].

Site-specific electronic relaxations like for ZnPc and $F_{16}$ZnPc on Cu(111) can lead to charge transfer across the organic-metal interface, which does not involve a (partial) filling of the COM's LUMO [31]. However, these electronic relaxations are related to site-specific geometric relaxations, which also lead to additional interface dipoles.

As mentioned before, for strongly interacting systems the bonding distances cannot very easily be connected to interaction strength and also the assignment of interface dipoles becomes more complicated as charged organic monolayers cannot be treated *uncoupled* from the metal substrate. Moreover, the ELA for multilayers is in this case also controlled by the chemical potential of the metallic monolayer. The main driving force for surface-induced aromatic stabilization is the chemical structure of the adsorbate and affinity-levels play a minor role. However, whether a COM can form a CTC on a specific metal surface depends on whether it can come close enough to the surface in the initial adsorption process (Fig. 6.8a), so bonding distances are still important in this context.

It is important to realize that a bonding distance on its own provides only limited insight and that one has to be careful when interpreting those numbers. For the same COM on different metals the vdW radii of the substrate atoms play a crucial role particularly for physisorbed systems. For different COMs on the same substrate the situation is even more complex. At the same time, the precise bonding distances obtained with the XSW technique are indispensable to unravel the organic-metal interface energetics. Together with other experimental techniques like UPS and with theoretical support we are coming closer to fully understand at least certain model systems.

**Acknowledgements** All measurements of bonding distances discussed in this chapter have been performed at beamline ID32 [60] at the European Synchrotron Radiation Facility (ESRF) in Grenoble (France). Analysis of experimental data was done using the software package `dare` (developed at the ESRF). The beamtimes have been supported in an outstanding way by the staff of ID32, namely T.-L. Lee, Y.Y. Mi, J. Roy, P. Rajput, J. Duvernay, B. Detlefs and J. Zegenhagen. We thank J. Niederhausen S. Kera, N. Ueno, G. Heimel, I. Salzmann, N. Koch, H. Yamane, E. Zojer and F. Schreiber for many fruitful discussions and most of them also for help during the beamtimes. The work which resulted in the publications [20, 27–32] has been supported by the G-COE of Chiba University. In addition, TH and SD gratefully acknowledge financial support by scholarships from the Alexander von Humboldt-Foundation and the Japan Society for the Promotion of Science, respectively. SD is also affiliated with the Soochow University-Western University Centre for Synchrotron Radiation Research.

## References

1. N. Koch, J. Phys.: Condens. Matter **20**, 184008 (2008)
2. H. Ishii, K. Sugiyama, E. Ito, K. Seki, Adv. Mater. **11**, 605 (1999)
3. A. Kahn, N. Koch, W.Y. Gao, J. Poly. Sci. B **41**, 2529 (2003)
4. S. Braun, W.R. Salaneck, M. Fahlman, Adv. Mater. **21**, 1450 (2009)
5. H. Ishii, N. Hayashi, E. Ito, Y. Washizu, K. Sugi, Y. Kimura, M. Niwano, Y. Ouchi, K. Seki, Phys. Stat. Sol. A **201**, 1075 (2004)
6. I. Lange, J.C. Blakesley, J. Frisch, A. Vollmer, N. Koch, D. Neher, Phys. Rev. Lett. **106**, 216402 (2011)
7. H.Y. Mao, F. Bussolotti, D.C. Qi, R. Wang, S. Kera, N. Ueno, A.T.S. Wee, W. Chen, Org. Electron. **12**, 534 (2011)
8. F. Bussolotti, S. Kera, K. Kudo, A. Kahn, N. Ueno, Phys. Rev. Lett. **110**, 267602 (2013)
9. J. Hwang, A. Wan, A. Kahn, Mater. Sci. Eng. R **64**, 1 (2009)
10. N. Ueno, S. Kera, Prog. Surf. Sci. **83**, 490 (2008)
11. N. Ueno, in *Fundamental aspects on organic semiconductors and their interfaces: Experimental electronic structure of organic-device related systems*, ed. by H. Ishii, K. Kudo, T. Nakayama, N. Ueno (Springer, NewYork, 2014)
12. Y. Nakayama, S. Duhm, Q. Xin, S. Kera, H. Ishii, N. Ueno, in *Ultraviolet photoelectron spectroscopy (UPS) I: Band dispersion measurements of "Insulating" organic single crystals*, ed. by H. Ishii, K. Kudo, T. Nakayama, N. Ueno (Springer, NewYork, 2014),
13. R. Otero, J.M. Gallego, A.L. Vázquez de Parga, N. Martín, R. Miranda, Adv. Mater. **23**, 5148 (2011)
14. G. Witte, C. Wöll, Phys. Stat. Sol. A **205**, 497 (2008)
15. D.P. Woodruff, Rep. Prog. Phys. **68**, 743 (2005)
16. J. Zegenhagen, Surf. Sci. Rep. **18**, 199 (1993)
17. A. Gerlach, C. Bürker, T. Hosokai, F. Schreiber, in *The Molecule-Metal Interface*, ed. by N. Koch, N. Ueno, A.T.S. Wee (Wiley-VCH, 2013), pp. 153–172
18. G. Mercurio, E.R. McNellis, I. Martin, S. Hagen, F. Leyssner, S. Soubatch, J. Meyer, M. Wolf, P. Tegeder, F.S. Tautz, K. Reuter, Phys. Rev. Lett. **104**, 036102 (2010)
19. V.G. Ruiz, W. Liu, E. Zojer, M. Scheffler, A. Tkatchenko, Phys. Rev. Lett. **108**, 146103 (2012)
20. C. Bürker, N. Ferri, A. Tkatchenko, A. Gerlach, J. Niederhausen, T. Hosokai, S. Duhm, J. Zegenhagen, N. Koch, F. Schreiber, Phys. Rev. B **87**, 165443 (2013)
21. A. Natan, L. Kronik, H. Haick, R. Tung, Adv. Mater. **19**, 4103 (2007)
22. F. Rissner, G.M. Rangger, O.T. Hofmann, A.M. Track, G. Heimel, E. Zojer, ACS Nano **3**, 3513 (2009)
23. H. Fukagawa, H. Yamane, S. Kera, K.K. Okudaira, N. Ueno, Phys. Rev. B **73**, 041302 (R) (2006)
24. P.S. Bagus, V. Staemmler, C. Wöll, Phys. Rev. Lett. **89**, 096104 (2002)

25. K. Toyoda, I. Hamada, K. Lee, S. Yanagisawa, Y. Morikawa, J. Chem. Phys. **132**, 134703 (2010)
26. P.C. Rusu, G. Giovannetti, C. Weijtens, R. Coehoorn, G. Brocks, Phys. Rev. B **81**, 125403 (2010)
27. S. Duhm, C. Bürker, J. Niederhausen, I. Salzmann, T. Hosokai, J. Duvernay, S. Kera, F. Schreiber, N. Koch, N. Ueno, A. Gerlach, ACS Appl. Mater. Interfaces **5**, 9377 (2013)
28. N. Koch, A. Gerlach, S. Duhm, H. Glowatzki, G. Heimel, A. Vollmer, Y. Sakamoto, T. Suzuki, J. Zegenhagen, J.P. Rabe, F. Schreiber, J. Am. Chem. Soc. **130**, 7300 (2008)
29. S. Duhm, S. Hosoumi, I. Salzmann, A. Gerlach, M. Oehzelt, B. Wedl, T.L. Lee, F. Schreiber, N. Koch, N. Ueno, S. Kera, Phys. Rev. B **81**, 045418 (2010)
30. G. Heimel, S. Duhm, I. Salzmann, A. Gerlach, A. Strozecka, J. Niederhausen, C. Bürker, T. Hosokai, I. Fernandez-Torrente, G. Schulze, S. Winkler, A. Wilke, R. Schlesinger, J. Frisch, B. Bröker, A. Vollmer, B. Detlefs, J. Pflaum, S. Kera, K.J. Franke, N. Ueno, J.I. Pascual, F. Schreiber, N. Koch, Nat. Chem. **5**, 187 (2013)
31. H. Yamane, A. Gerlach, S. Duhm, Y. Tanaka, T. Hosokai, Y.Y. Mi, J. Zegenhagen, N. Koch, K. Seki, F. Schreiber, Phys. Rev. Lett. **105**, 046103 (2010)
32. A. Gerlach, T. Hosokai, S. Duhm, S. Kera, O.T. Hofmann, E. Zojer, J. Zegenhagen, F. Schreiber, Phys. Rev. Lett. **106**, 156102 (2011)
33. S.K.M. Henze, O. Bauer, T.L. Lee, M. Sokolowski, F.S. Tautz, Surf. Sci. **601**, 1566 (2007)
34. A. Hauschild, K. Karki, B.C.C. Cowie, M. Rohlfing, F.S. Tautz, M. Sokolowski, Phys. Rev. Lett. **94**, 036106 (2005)
35. L. Romaner, G. Heimel, J.L. Brédas, A. Gerlach, F. Schreiber, R.L. Johnson, J. Zegenhagen, S. Duhm, N. Koch, E. Zojer, Phys. Rev. Lett. **99**, 256801 (2007)
36. N. Koch, S. Duhm, J.P. Rabe, A. Vollmer, R.L. Johnson, Phys. Rev. Lett. **95**, 237601 (2005)
37. T. Fujikawa, in *Theory of photoelectron spectroscopy*, ed. by H. Ishii, K. Kudo, T. Nakayama, N. Ueno (Springer, NewYork, 2014)
38. D.B. Dougherty, W. Jin, W.G. Cullen, J.E. Reutt-Robey, S.W. Robey, J. Phys. Chem. C **112**, 20334 (2008)
39. M. Eremtchenko, R. Temirov, D. Bauer, J.A. Schaefer, F.S. Tautz, Phys. Rev. B **72**, 115430 (2005)
40. C. Stadler, S. Hansen, I. Kröger, C. Kumpf, E. Umbach, Nat. Phys. **5**, 153 (2009)
41. I. Kröger, B. Stadtmüller, C. Kleimann, P. Rajput, C. Kumpf, Phys. Rev. B **83**, 195414 (2011)
42. A. Gerlach, F. Schreiber, S. Sellner, H. Dosch, I.A. Vartanyants, B.C.C. Cowie, T.L. Lee, J. Zegenhagen, Phys. Rev. B **71**, 205425 (2005)
43. I. Kröger, B. Stadtmüller, C. Stadler, J. Ziroff, M. Kochler, A. Stahl, F. Pollinger, T.L. Lee, J. Zegenhagen, F. Reinert, C. Kumpf, New J. Phys. **12**, 083038 (2010)
44. C. Stadler, S. Hansen, F. Pollinger, C. Kumpf, E. Umbach, T.L. Lee, J. Zegenhagen, Phys. Rev. B **74**, 035404 (2006)
45. I. Kröger, P. Bayersdorfer, B. Stadtmüller, C. Kleimann, G. Mercurio, F. Reinert, C. Kumpf, Phys. Rev. B **86**, 195412 (2012)
46. J. Gantz, D. Placencia, A. Giordano, S.R. Marder, N.R. Armstrong, J. Phys. Chem. C **117**, 1205 (2013)
47. Y.L. Huang, W. Chen, F. Bussolotti, T.C. Niu, A.T.S. Wee, N. Ueno, S. Kera, Phys. Rev. B **87**, 085205 (2013)
48. S. Duhm, Q. Xin, N. Koch, N. Ueno, S. Kera, Org. Electron. **12**, 903 (2011)
49. N. Ueno, T. Sueyoshi, F. Bussolotti, S. Kera, in *Ultraviolet photoelectron spectroscopy (UPS) III: Direct study of "invisible" band gap states by ultrahigh-sensitivity UPS*, ed. by H. Ishii, K. Kudo, T. Nakayama, N. Ueno (Springer, NewYork, 2014)
50. J. Niederhausen, P. Amsalem, J. Frisch, A. Wilke, A. Vollmer, R. Rieger, K. Müllen, J.P. Rabe, N. Koch, Phys. Rev. B **84**, 165302 (2011)
51. S. Duhm, H. Glowatzki, V. Cimpeanu, J. Klankermayer, J.P. Rabe, R.L. Johnson, N. Koch, J. Phys. Chem. B **110**, 21069 (2006)

52. H. Glowatzki, B. Bröker, R.P. Blum, O.T. Hofmann, A. Vollmer, R. Rieger, K. Müllen, E. Zojer, J.P. Rabe, N. Koch, Nano Lett. **8**, 3825 (2008)
53. A. Gerlach, S. Sellner, F. Schreiber, N. Koch, J. Zegenhagen. Phys. Rev. B **75**, 045401 (2007)
54. O. Bauer, G. Mercurio, M. Willenbockel, W. Reckien, C.H. Schmitz, B. Fiedler, S. Soubatch, T. Bredow, F.S. Tautz, M. Sokolowski, Phys. Rev. B **86**, 235431 (2012)
55. C. Stadler, S. Hansen, A. Schöll, T.L. Lee, J. Zegenhagen, C. Kumpf, E. Umbach, New J. Phys. **9**, 50 (2007)
56. N. Koch, A. Vollmer, Appl. Phys. Lett. **89**, 162107 (2006)
57. S. Duhm, A. Gerlach, I. Salzmann, B. Bröker, R.L. Johnson, F. Schreiber, N. Koch, Org. Electron. **9**, 111 (2008)
58. E.R. McNellis, G. Mercurio, S. Hagen, F. Leyssner, J. Meyer, S. Soubatch, M. Wolf, K. Reuter, P. Tegeder, F.S. Tautz, Chem. Phys. Lett. **499**, 247 (2010)
59. T.C. Tseng, C. Urban, Y. Wang, R. Otero, S.L. Tait, M. Alcamí, D. Écija, M. Trelka, J.M. Gallego, N. Lin, M. Konuma, U. Starke, A. Nefedov, A. Langner, C. Wöll, M.A. Herranz, F. Martín, N. Martín, K. Kern, R. Miranda, Nat. Chem. **2**, 374 (2010)
60. J. Zegenhagen, B. Detlefs, T.L. Lee, S. Thiess, H. Isern, L. Petit, L. André, J. Roy, Y.Y. Mi, I. Joumard, J. Electron Spectrosc. Relat. Phenom. **178–179**, 258 (2010)

# Chapter 7
# Structure Matters: Combining X-Ray Scattering and Ultraviolet Photoelectron Spectroscopy for Studying Organic Thin Films

**Alexander Hinderhofer, Keiichirou Yonezawa, Kengo Kato, and Frank Schreiber**

## 7.1 Introduction

In this chapter the relationship between organic film structure and ultraviolet photoelectron spectroscopy (UPS) data is discussed. As a useful method for obtaining detailed structural data we first summarize shortly the advantages of X-ray scattering. Here, of course we cannot include the full body of literature on interface-sensitive X-ray scattering but rather refer to general references. By combining such structural data and electronic information from UPS new insights in the fundamental principles of organic electronics can be obtained. On the basis of single layer and heterostructures we discuss the dependence of the electronic level alignment and the spectral shape of the HOMO band on the structural properties of organic thin films. The crystallinity and therefore also the electronic properties of an organic thin film can be tuned by controlling growth parameters such as the substrate temperature. The examples are drawn from our own work in order to specifically relate to other chapters in this book, which is not intended to imply that there are no others [1–3].

A. Hinderhofer (✉) • K. Yonezawa • K. Kato
Department of Nanomaterial Science, Graduate School of Advanced Integration Science, Chiba University, Inage-ku, Chiba 263-8522, Japan
e-mail: alexander.hinderhofer@chiba-u.jp; k.yonezawa@chiba-u.jp; acwa3955@chiba-u.jp

F. Schreiber
Institute for Applied Physics, University of Tübingen, Auf der Morgenstelle 10, 2076 Tübingen, Germany
e-mail: frank.schreiber@uni-tuebingen.de



H. Ishii et al. (eds.), *Electronic Processes in Organic Electronics*,
Springer Series in Materials Science 209, DOI 10.1007/978-4-431-55206-2_7

## 7.2 X-Ray Scattering

### 7.2.1 X-Ray Reflectivity

X-ray reflectivity (XRR) is a powerful tool to probe structure and morphology of organic thin films. Here only the basic concepts of XRR are described. For the details on this method we refer to Refs. [4, 5]. Figure 7.1 depicts the simplified scattering geometry of XRR, for which the angle of incidence $\Theta$ and the detector angle $\Omega$ are kept equal: $\Theta = \Omega$. Then the complete momentum transfer $q$ has only a non-zero component perpendicular to the substrate ($q_z$) and can be written as:

$$q_z = \frac{4\pi}{\lambda} \sin \Theta \tag{7.1}$$

Using the Fresnel coefficients the reflectivity $R$ of a multilayer can be modeled within the framework of dynamical scattering theory with a recursive formalism described by Parratt [6].

The electron density profile of a sample can be extracted along the surface normal by fitting the experimental data. That is, only information about the out-of-plane sample structure is obtained. In Fig. 7.2a a typical XRR dataset from an organic thin film diindenoperylene (DIP) is shown. From fitting such a dataset the following physical parameters, which are actually the free parameters of the fit, can be obtained:

- **Average electron density** The average electron density $\rho_e$ of a sample is directly connected to the total reflection edge in XRR data.
- **Film thickness** From the periodicity of the Kiessig or thickness oscillations the average thickness $d$ of a thin film can be determined.
- **Roughness** From the damping of the Kiessig oscillations the roughness $\sigma_{\text{rms}}$ of a thin film can be determined.
- **Out-of-plane lattice spacing** In organic thin films molecules are often ordered in a crystal, which results in a periodic variation in the electron density of a thin film, from which the out-of-plane lattice spacing can be determined (Fig. 7.2b).

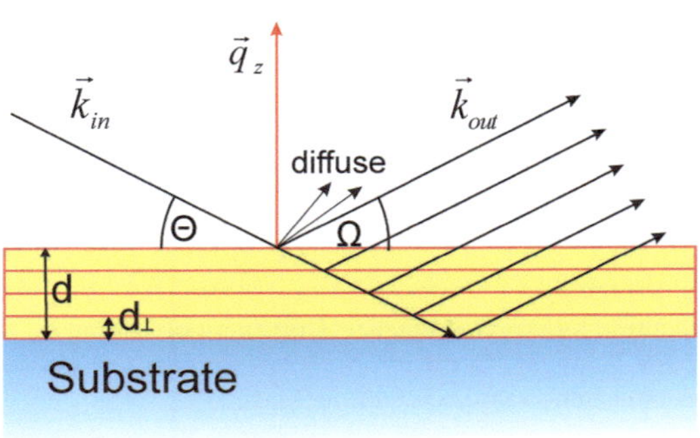


**Fig. 7.1** Scattering geometry for X-ray reflectivity (XRR) on a thin film. $d$ corresponds to the film thickness and $d_\perp$ is the out-of-plane layer spacing. For XRR the momentum transfer $q_z$ is perpendicular to the substrate

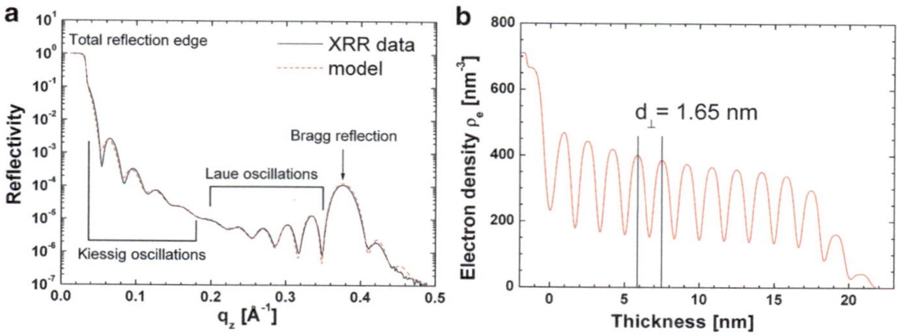


**Fig. 7.2** (**a**) XRR data from a 20 nm DIP film grown on $SiO_2$ fitted with a Parratt-model. The fitted electron density versus film thickness is shown in (**b**)

Since the molecules are often oriented with their long axis along the growth direction the periodicity is on a length scale larger than 1 nm. Constructive interference from waves scattered at the crystal planes gives rise to Bragg reflections at certain angles. The position of Bragg reflections in XRR may be approximated by Bragg's law:

$$n\lambda = 2d_{\perp} \sin \Theta \tag{7.2}$$

However, in contrast to scattering from crystal powders, for XRR, the Bragg peak positions may in general be shifted due to multiple scattering.

- **Coherently scattering island size** From the periodicity of the side fringes of the Bragg reflection (Laue oscillations) the coherently scattering crystal size can be determined. If the periodicity of the Laue oscillations and Kiessig oscillations is equal, the film is coherently ordered for the complete film thickness.

Modeling XRR data, such as presented in Fig. 7.2a, with the Parratt-formalism results in a complete electron density profile of a thin film as shown in Fig. 7.2b, from which the thin film parameters described above can be extracted. For organics, the absorption $\beta$ is in the order of $1 \times 10^{-9}$. Thus, for the derivation of the intensity of specularly reflected X-rays from organic thin films $\beta$ is neglected.

### 7.2.2 *Grazing Incidence X-Ray Diffraction*

To gain knowledge about lateral structures, grazing incidence x-ray diffraction (GIXD) can be performed. Here, only the main concepts of GIXD are presented, for details it is referred to Ref. [7].

In GIXD, the angle of incidence $\Theta$ is near the total reflection edge with an angle of $\alpha_c$. In this case we observe total external reflection and the transmitted wave is

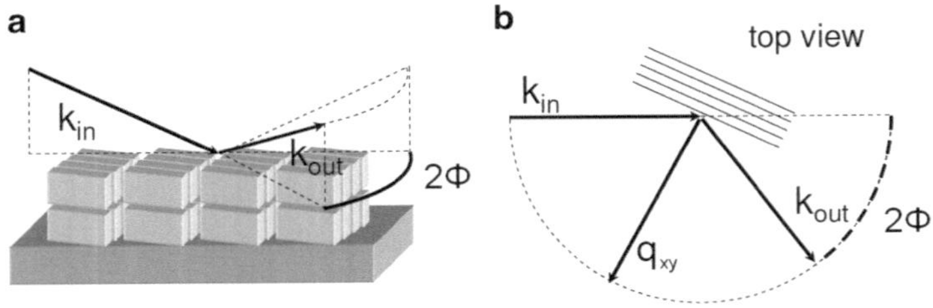


**Fig. 7.3** (**a**) In GIXD the X-ray beam has an angle of incidence near the critical angle and the detector angle is varied parallel to the substrate by the angle $2\Phi$. (**b**) X-rays are diffracted by the crystalline in-plane lattice by an angle $2\Phi$, yielding information about the lattice spacing within the plane. Picture taken and modified from Ref. [8]

very weak. It propagates along the surface with a penetration depth of $\Lambda = 12k\alpha_c$, which is the distance at which the intensity falls of by a factor of $1/e$. Due to the finite penetration depth, it is called an evanescent wave. For GIXD the out-of-plane detector angle is kept equal to the angle of incidence $\Omega = \Theta$. The in-plane crystal structure is probed by varying the in-plane angle $2\Phi$ related to the in-plane momentum transfer $q_{xy} \approx 4\pi/\lambda \sin \Phi$ (Fig. 7.3).

Samples studied in the following correspond to an in-plane powder. Thus, in the in-plane direction crystalline domains do not have a preferred orientation, However, perpendicular to the sample surface (i.e. out-of-plane), the lattice planes are all oriented parallel to the sample surface. Therefore probing the in-plane structure by GIXD can be understood in terms of powder diffraction. For acquisition of this powder diffraction pattern it is possible to use a point detector and perform a $2\Phi$ scan. Alternatively, it is possible to use an area detector, thereby acquiring scattering data for a whole range of exit angles in the in-plane direction. With an area detector additional information in the out-of-plane direction $q_z$ is also resolved. The momentum transfer in each direction is calculated from the following equations:

$$q_{xy} = \frac{2\pi}{\lambda}\sqrt{(\sin \Phi \cos \Omega)^2 + (\cos \Omega \cos \Phi - \cos \Theta)^2} \tag{7.3}$$

$$q_z = \frac{2\pi}{\lambda}(\sin \Theta + \sin \Omega) \tag{7.4}$$

### 7.2.3 *Coherent Island Size and Scherrer Formula*

The peak width of a Bragg reflection depends on the number lattice planes, which are scattering the incoming X-rays coherently. Lower limits of the coherent in-plane island size $l_s$ can be determined by the Scherrer formula [5]:

$$l_s = 2\pi/\mathrm{FWHM} \cdot 0.9394 \cdot K_s \tag{7.5}$$

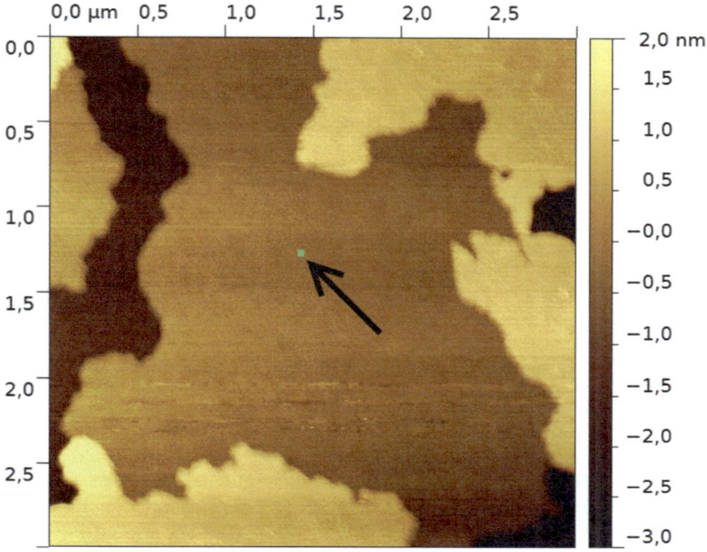


**Fig. 7.4** AFM image of a thin DIP film ($d \sim 2$ nm) on $SiO_2$ showing crystalline islands on the order of several $\mu$m. The *green dot* pointed at by the arrow depicts the average defect free coherent island size $l_s \sim 30$ nm as determined by GIXD

Here $K_s = 1.0747$ is the Scherrer constant for spherical grains and FWHM is the full width half maximum of the peak. Here, Eq. 7.5 is mainly used to determine the coherent in-plane island size $l_s$. However the same formula could be used to estimate the coherent out-of-plane island size $l_z$. When the instrumental broadening of the diffractometer is not included in the calculations, only lower limits of $l_s$ and $l_z$ can be obtained.

For organic thin films grown on amorphous substrates like $SiO_2$, $l_s$ does often not exceed $\sim 50$ nm, which is in contrast to the much larger domain sizes visualized by atomic force microscopy (AFM) or X-ray spectromicroscopy [9, 10]. For example Fig. 7.4 shows an AFM image of thin DIP film on $SiO_2$ with domain sizes of several $\mu$m. The green dot in this image represents the average defect free coherent island size $l_s \sim 30$ nm of this sample as determined by GIXD.

## 7.3 Probing Depths of X-Ray Scattering and UPS

When combining data from different methods always care have to be taken. Both X-ray scattering and UPS have usually a rather large probing area in the range of $mm^2$. However, the probing depths of both methods depend on different parameters and can be very different.

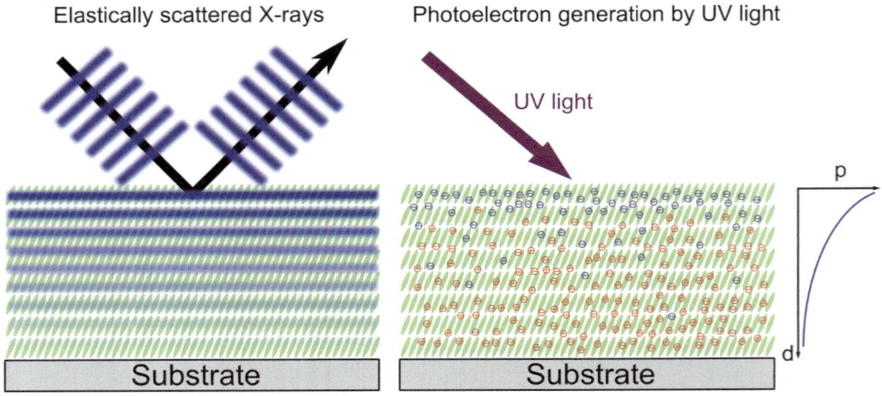


**Fig. 7.5** Illustration of probing depth for GIXD (*left*) and UPS (*right*). For GIXD, the intensity of the evanescent wave decays exponentially from the surface dependent on photon energy and angle of incidence. For UPS the probability $p$ of generated photoelectrons to leave the sample unscattered decreases exponentially for larger film thicknesses $d$. For example the *"blue"* electrons may leave the film unscattered, while the *"red"* electrons will be scatters inelastically before reaching the film surface

For UPS the probing depth is limited by inelastic mean free path IMFP of the generated photoelectrons, i.e. the average length a free electron can travel in the film before it is scattered inelastically. The IMFP depends mainly on the kinetic energy of the electron transferred from an incident photon and does only weakly depend on the material. The IMFP for different kinetic energies is therefore often presented as a "universal curve" valid for all materials. Since the penetration depth of the incident UV-light is usually much larger than the IMFP of the excited photoelectrons we assume that the amount of electrons generated is nearly equal at all thicknesses through the film. Then, as illustrated in Fig. 7.5, the probability $p$ for an electron to reach the surface without being inelastically scattered decreases exponentially with larger distances $d$ between surface and place of generation.

In contrast to UPS, the probing depth of X-ray scattering techniques depend on photon energy, angle of incidence and the dielectric function of the thin film. For most scattering geometries, for example XRR, the angle of incidence relative to the sample surface is above the total reflection edge. Thus the X-ray beam is penetrating completely into the sample until it is absorbed or reflected, resulting in a probing depth which is usually much larger than the film thickness. For GIXD, the angle of incidence is below the total reflection edge, which means that the X-ray beam is totally reflected at the surface and only an evanescent wave, with exponentially decreasing intensity (Fig. 7.5) is penetrating into the film. By adjusting the angle of incidence (or photon energy) for a given sample the probing depth can be varied, but due to the intensity decay versus thickness the signal-weight is always biased to the surface area, similar as is the case for UPS.

## 7.4 Example 1: DIP at Different Substrate Temperatures

The structure of organic semiconductor thin films is frequently modified by varying the substrate temperature during deposition to tune the electrical and optical characteristics of the films. In general, at high substrate temperatures ($T$) molecular mobilities are high leading to enhanced crystallinity and grain size in the grown films [2, 11]. In contrast, deposition at low temperatures is applied to obtain less crystalline or amorphous films [9, 12–17]. For low $T$ growth, it is often not clear, if the film undergoes a structural or morphological transition upon heating to room temperature, because characterization is mostly done after growth at room temperature. In particular, properties like crystallinity, molecular orientation and roughness of the film surface are important for device applications with organic heterostructures, where the top surface of the first layer serves as a template for subsequent layers [3, 18, 19].

Here, we study the morphological and electronic impact of post-growth heating on low $T$ deposited organic thin films by X-ray reflectivity (XRR) and ultraviolet photoelectron spectroscopy (UPS). For the experiments we choose diindenoperylene (DIP, $C_{32}H_{16}$, inset Fig. 7.6b) as an organic material with high relevance for applications [20–23]. For example, organic solar cells with DIP as electron donor reached high fill factors and power conversion efficiencies of more than 4 %. DIP is deposited on two different substrates (indium-tin-oxide (ITO) and silicon dioxide ($SiO_2$)), because the film characteristics on both substrates differ in crystallinity [15, 16]. X-ray diffraction techniques [5] were applied to determine the bulk crystal structure and surface roughness ($\sigma_{RMS}$). UPS was used to determine the surface electronic structure, which depends on the crystallinity, orientation and

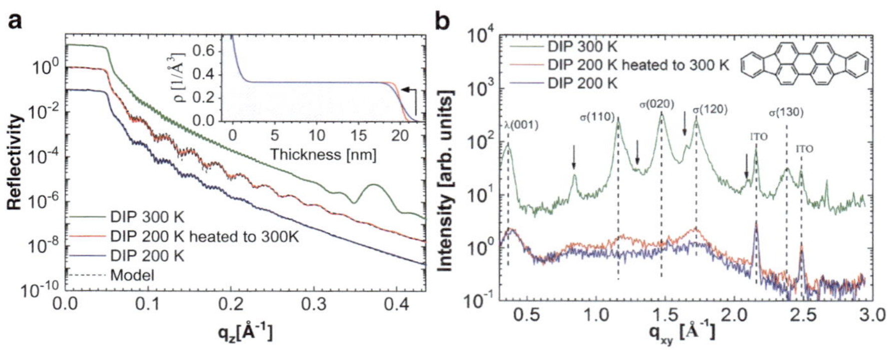


**Fig. 7.6** (**a**) XRR and (**b**) GIXD data of a DIP film ($d = 20$ nm) grown on ITO at 200 K measured directly after growth at 200 K and after slow heating ($\sim 1$ h) to 300 K. For comparison, data from a film grown and measured at 300 K is also shown. The inset in (**a**) shows the modeled electron density of the 200 K film directly after growth (*blue*) and after heating (*red*). The inset in (**b**) shows the molecular structure of DIP. Reprinted with permission from Ref. [31]

uniformity of domains at the surface [24–26]. The combination of both techniques, therefore, provides information on the change of structure and morphology of the films for the entire thickness region.

### 7.4.1 Experimental

Organic thin films of DIP were grown on silicon wafers with native $SiO_2$ (surface roughness $\sigma_{RMS} = 0.3$ nm) or on ITO-coated glass substrates (ITO thickness: 130 nm, $\sigma_{RMS} = 0.95$ nm) under ultra high vacuum (UHV) conditions (base pressure $< 6 \cdot 10^{-9}$ mbar) by thermal evaporation. Before deposition, substrates were cleaned ultrasonically with acetone, isopropanol, and ultra pure water, followed by heating to 700 K in the UHV growth chamber. The growth rate was between 0.1 and 0.3 nm/min monitored by XRR and a quartz crystal microbalance. Substrate temperatures during growth and measurements were controlled with liquid nitrogen cooling in a range of $T = 200 - 300$ K.

In situ XRR and grazing incidence X-ray diffraction (GIXD) for the ITO samples were performed at beamline ID10B ($\lambda = 0.092$ nm) of the ESRF in Grenoble, France. XRR for $SiO_2$ samples were performed at the X04SA beamline of the Swiss Light Source, Paul Scherrer Institut, Villigen, Switzerland ($\lambda = 0.10$ nm). Peak indexing of DIP is based on the crystal structure (P21/a polymorph) reported in Ref. [27]. Modeling of XRR data was done with Motofit [28].

He I UPS experiments were performed with a home-built UHV system equipped with a PHOIBOS-HSA100 analyzer with an energy resolution of 60 meV [29]. UPS were measured at a light incident angle of 45° and electron emission angles of 0° (normal emission). After growth in a UHV preparation chamber the samples were directly transferred to the measurement chamber without breaking the vacuum and with keeping the substrate temperature at 200 K. The vacuum level (VL) was obtained by applying a sample bias of − 5 V during the UPS measurements.

### 7.4.2 Results

Figure 7.6a shows XRR data from a DIP film with a thickness of $d = 20$ nm grown on ITO at 200 K and measured at 200 K directly after growth and after slow heating ($\sim 1$ h) to 300 K. Both films show no out-of-plane Bragg reflections, indicating weak order in this direction. We applied a three layer model (glass-ITO-DIP) to fit the electron densities $\rho$ of the as-grown and the annealed films (inset Fig. 7.6a). From the electron densities of both films it is evident that the roughness is reduced by 50 % during annealing from initially $\sigma_{RMS} = 0.90$ nm to $\sigma_{RMS} = 0.45$ nm at 300 K. For comparison, XRR from a film grown and measured at 300 K is also shown. This film is crystalline and exhibits an out-of-plane lattice spacing of 1.69 nm corresponding to textured growth of the strained high temperature

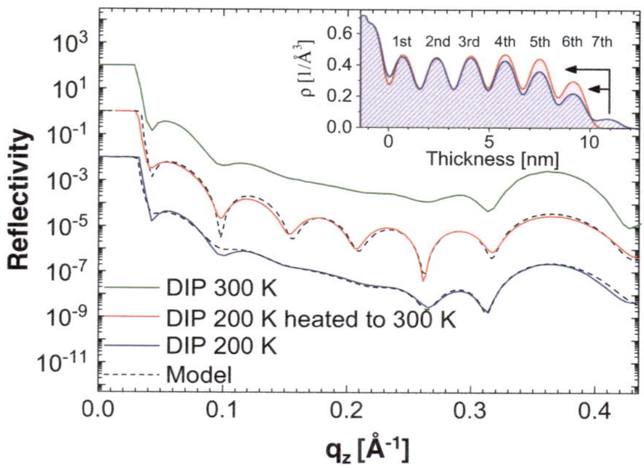


**Fig. 7.7** XRR data of DIP films ($d = 10$ nm) grown on $SiO_2$ at two different substrate temperatures (200 K and 300 K). For the 200 K DIP film, XRR data directly after growth at 200 K are shown together with data after slow heating (∼ 1 h) to 300 K. The inset shows the modeled electron density of the 200 K film directly after growth (*blue*) and the heated 200 K film (*red*). Reprinted with permission from Ref. [31]

phase of DIP (HT-phase) [27, 30] with the (001) plane parallel to the substrate ($\sigma$-structure). The roughness of this film is significantly higher compared to low $T$ deposited films ($\sigma_{RMS} = 2.6$ nm).

Figure 7.6b shows GIXD data of the 200 K DIP film on ITO before and after heating. Before heating the film shows only very broad Bragg reflections with the most intense feature stemming from domains with nearly lying DIP molecules ($\lambda$(001) of the HT-phase). During heating the film crystallizes partly in the DIP HT-phase as seen by the slight intensity increase of the $\sigma$(110) and $\sigma$(120) reflections. However, all reflections both from the $\lambda$- and $\sigma$-structure remain broad with a coherent island size of less than 5 nm estimated with the Scherrer formula [5]. GIXD data from a crystalline film grown and measured at 300 K (Fig. 7.6b) exhibits mainly Bragg reflections corresponding to textured growth of the DIP HT-phase ($\sigma$-structure) [27]. Bragg reflections marked with stars stem presumably from a DIP low temperature phase as suggested in Refs. [15, 16, 27]. XRR and GIXD data show that during heating to room temperature the bulk DIP film on ITO crystallizes only partly, however, the reorganization of the surface molecules yield a very low roughness.

In Ref. [15] it was shown that low $T$ deposition of DIP on $SiO_2$ yield films, which are more crystalline than films deposited on rough substrates like ITO. In the following, we report on an in situ study of such a film, in order to test if the surface smoothing observed for amorphous DIP films on ITO is also present for crystalline films upon heating to room temperature.

Figure 7.7 shows XRR data from a DIP film ($d = 10$ nm) grown on $SiO_2$ at 200 K measured at 200 K directly after growth and after slow heating to 300 K. From the modeled electron densities of the heated and the as-grown 200 K DIP film (inset Fig. 7.7) we find that both films are crystalline and exhibit an out-of-plane lattice spacing of 1.69 nm corresponding again to textured growth of the strained DIP HT-phase ($\sigma$-structure) [27, 30]. However, the layer fillings of these films show

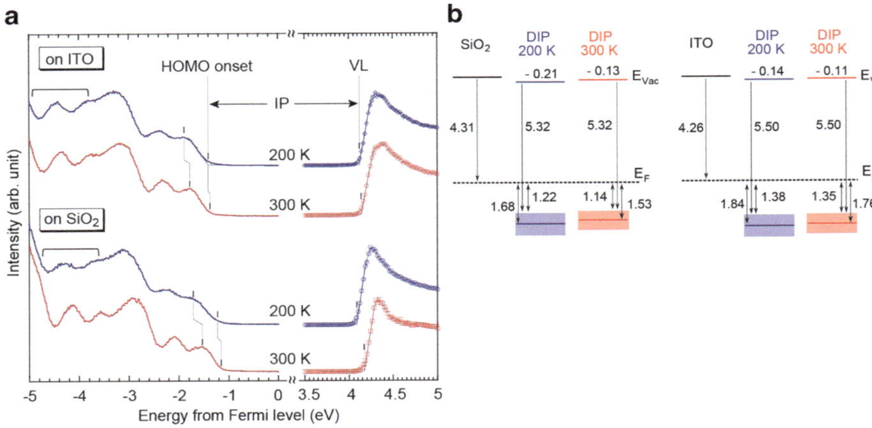


**Fig. 7.8** (**a**) UPS data of DIP films grown on $SiO_2$ ($d = 10$ nm) and ITO ($d = 20$ nm) at 200 K and after slow heating (∼4 h) to 300 K. (**b**) Schematic energy level diagram. The HOMO width is given by the difference of the HOMO peak position and the HOMO onset. Reprinted with permission from Ref. [31]

significant differences (inset Fig. 7.11), resulting in roughnesses of $\sigma_{RMS} = 1.5$ nm for the as-grown film and $\sigma_{RMS} = 0.75$ nm for the heated film. This observation is rationalized by a molecular 'downhill' current from the top layer (7th) to the partly filled lower lying layers (6th, 5th, 4th), thereby leaving the out-of-plane crystal structure and the intermediate electron density unchanged. For comparison, XRR from a film grown and measured at 300 K is also shown. This film has the same out-of-plane lattice spacing (1.69 nm) and a similar roughness as the 200 K film without annealing ($\sigma_{RMS} = 1.6$ nm).

The observation of the roughness reduction by ∼50 % shows that the reorganization of surface molecules is not exclusively a feature of amorphous films. Surprisingly, also for a bulk crystalline film deposited at low $T$ the surface is smoothed upon heating.

### 7.4.3 UPS

We employed UPS to detect the correlation between structure and the molecular electronic states upon heating, since the electronic states near the surface play a significant role in the energy level alignment of organic heterostructures. The probing depth of UPS is ∼1 nm, which means that the bulk of the film does not contribute to the measured data. Figure 7.8 shows UPS data of DIP films prepared under similar conditions as those as presented above. The ionization potential (IP) of DIP (200 K) on $SiO_2$ (5.32 eV) and on ITO (5.5 eV) was determined from the onset of the highest occupied molecular orbital (HOMO) and the vacuum level (VL). The IP of DIP depends on the orientation of the molecules at the surface [32]. An IP of 5.32 eV is consistent with a nearly upright oriented DIP film [20, 33]. The IP of lying DIP is larger by $\sim 0.4$ eV compared to nearly upright standing DIP as was shown in Refs. [34–36]. The DIP film on ITO is nearly amorphous and exhibits therefore no preferred molecular orientation. In addition, the size of orientational domains is very small, which results in a common vacuum level. Consistently, the IP of 5.5 eV of this film corresponds to an average of standing and lying DIP.

After heating the IP of both films did not change significantly, indicating that the DIP molecules did not reorient on average. However we detect a parallel shift of all valence features towards the Fermi level, which is attributed partly to an interfacial dipole effect and partly to the reduction of gap states (Fig. 7.8b). Dipole effects are associated with a parallel shift of the HOMO and the vacuum level. A HOMO shift stemming from a reduction of gap states is recognized by a reduced peak width of the valence states (gap state effect). For DIP on $SiO_2$ (ITO) the HOMO shift is $\Delta E = 0.15$ eV ($\Delta E = 0.08$ eV) with a contribution of 0.08 eV (0.03 eV) from the dipole effect and 0.07 eV (0.05 eV) from the gap state effect. The dipole effect may be attributed either to an interfacial dipole between the surface layer and its underlayer induced by a slightly different electron density at the interface [37], and/or a temperature dependent level alignment between the substrates and the DIP. The spectral broadening at low $T$ is caused by different polarization environments associated with disorder of the molecules at the nearest-neighbor level. The peak narrowing upon heating indicates therefore that the surface molecules are more uniformly ordered for both systems which results in more similar polarization environments for them. Since the density of gap states is reduced, the HOMO level shifts closer to the Fermi level. This effect is explained in detail in Refs. [38–40].

When comparing the DIP-on-$SiO_2$ and the DIP-on-ITO systems several differences can be found as shown in Fig. 7.9: First, at 200 K molecules at the film surface on $SiO_2$ orient basically nearly upright ($\sigma$-orientation), which is consistent with the data in Fig. 7.6. Nevertheless, at this temperature the spectral features of DIP-on-$SiO_2$ exhibit a similar or even stronger broadening compared to DIP-on-ITO, which can be observed most easily from the energy region around $-4$ eV marked with brackets in Fig. 7.8. The broad spectrum of DIP-on-$SiO_2$ implies that, in spite of better bulk crystallinity, the surface material in films on $SiO_2$ does not show a better in-plane order than DIP-on-ITO. Second, upon heating peak narrowing and the spectral shift are both significantly more pronounced for the DIP-on-$SiO_2$ system compared to the DIP-on-ITO system. This implies that the ordering effect of the surface material is more pronounced if the crystallinity of the initial material is better, which is the case for DIP-on-$SiO_2$ (Fig. 7.9).

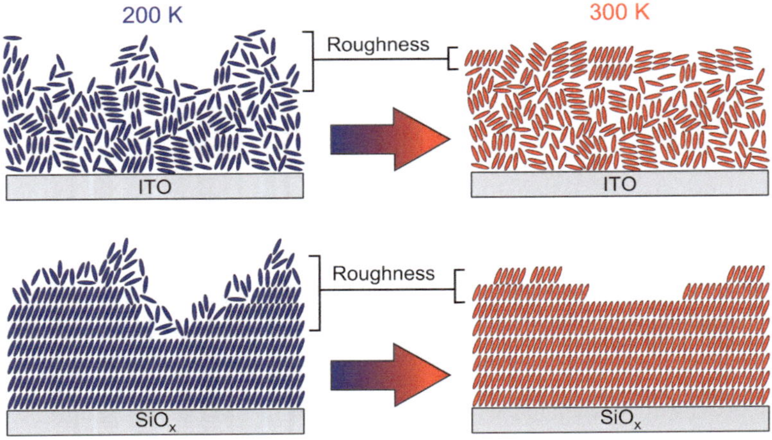


**Fig. 7.9** A sketch of DIP thin films on ITO and $SiO_2$ illustrating the surface smoothing due to crystallization upon heating. Reprinted with permission from Ref. [31]

## 7.5 Example 2: $C_{60}$ on DIP

For growth of organic thin films the structure and morphology depends strongly on the substrate, onto which they are deposited [2, 41]. This was demonstrated for example by the surface modification of an inorganic substrate with an organic self assembled monolayer (SAM), which influenced the resulting growth behavior [42–49]. For such heterostructures the growth behavior of the top layer is mostly discussed in terms of surface energies, although also some degree of azimuthal alignment has been observed [49]. In contrast, a close relationship between two organic layers can be observed in organic-organic heteroepitaxy [3, 18, 50–52]. There, the growth behavior of the deposited compound depends on the, usually anisotropic, potential surface of both involved materials. The control of the top layer morphology by tuning of the bottom layer was also discussed as templating, particularly for changing the molecular orientation relative to the surface (standing orientation vs. lying orientation) [16, 53].

The structural relationship at an organic-organic hetero-interface resulting from the non-equilibrium growth process has a large impact on electrical properties, *inter alia* charge carrier generation and transport [1, 54]. For example for organic field effect transistors it was shown that an organic templating layer may improve the electronic mobility of the active material substantially [18, 55–57]. In this regard, an important point for small-molecule organic semiconductors can be the orientation of the molecules. Frequently, there is at least a competition between lying down and standing up orientation [35, 45], which depends strongly on the underlying substrate. Other systems, such as PTCDA exhibit a very strong tendency to form lying-down structures, almost independent of the substrate [17, 49, 58–60]. In this context, $C_{60}$ is a rather unique case in the area of small-molecule organic

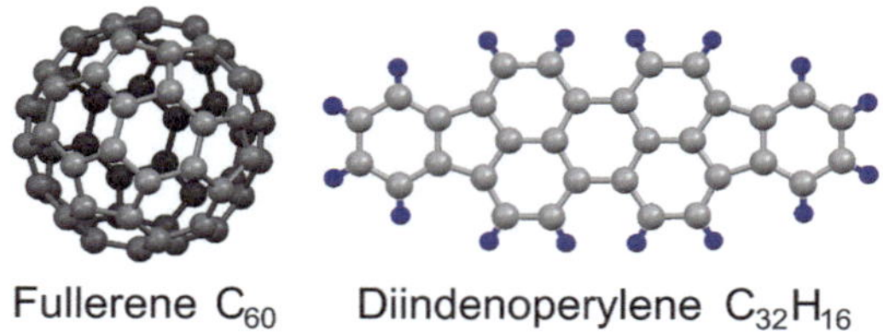


**Fig. 7.10** Sketch of fullerene $C_{60}$ and diindenoperylene (DIP, $C_{32}H_{16}$). Reprinted with permission from Ref. [71]

semiconductors, since it exhibits essentially rotational symmetry. The issue of lying-down vs. standing-up orientation does not complicate matters, and the orientational degrees of freedom in structure formation enter basically only via the orientation of the lattice planes and their distribution or alignment. $C_{60}$ was shown to grow with low structural order on several inorganic substrates like $SiO_2$ [61], quartz glass [62] and sapphire [55] but crystallizes well on organics like pentacene [55, 63] or sexiphenyl [64, 65].

Here, we study the influence of a diindenoperylene (DIP, Fig. 7.10) templating layer [27, 30, 66–68] on the growth and electronic structure of $C_{60}$. The combination of $C_{60}$ and DIP in a heterostructure was shown to exhibit excellent photovoltaic performance [20, 69], which is related to the high exciton diffusion length in DIP [23, 70] and the favorable energy level alignment of both materials [20, 21, 33]. In addition, we investigate to which extent structural properties like roughness, domain size and crystallinity of the DIP templating layer influence the growth of $C_{60}$.

To study the structure of $C_{60}$ thin films we compare first a reciprocal space map of a $C_{60}$ film grown on $SiO_2$ ($C_{60}/SiO_2$; Fig. 7.11a) with data from $C_{60}$ grown on DIP ($C_{60}$/DIP; Fig. 7.11b). The $C_{60}/SiO_2$ film exhibits broad diffraction rings indicating crystalline domains without preferred orientation. Indexing is done according to the $C_{60}$ *fcc*-structure reported in Ref. [72]. One reflection, indexed as $C_{60}$# in the bottom GIXD data, does not stem from the $C_{60}$ *fcc*-structure. Its $q$-value ($q = 0.725\,\text{Å}^{-1}$) coincides with the (100) reflection from the $C_{60}$ *hcp* structure. This observation is in agreement with single crystal growth, where a small fraction of crystals adopt *hcp* packing [72]. Note that in the indexing of GIXD data at the bottom of Fig. 7.11a only one index for each reflection is given, since other reflections with the same $|q|$ cannot be distinguished. The scattering data of the $C_{60}/SiO_2$ film shown here is in agreement with data presented in Ref. [61].

Figure 7.11b displays 2-dimensional GIXD data from a $C_{60}$/DIP heterostructure. At the bottom the integrated GIXD intensity is shown. Indexing is done again according to the $C_{60}$ *fcc*-structure. Compared to $C_{60}$ grown on $SiO_2$ (Fig. 7.11a) the diffraction pattern of $C_{60}$ grown on DIP shows significant differences. The distribution of Bragg reflections reveals the alignment of the *fcc*-(111) crystal plane parallel to the substrate. Only a small fraction of crystallites nucleates with a random orientation as indicated by the weak $C_{60}$ (111)* index. Note that the Bragg reflection at $q_{xy} = 0.725\,\text{Å}^{-1}$ stems not from the *hcp*-structure as in Fig. 7.11a. InZstead, this peak is the projection of the *fcc*-(11-1) Bragg reflection onto the $q_{xy}$ plane. The width of the *fcc* Bragg reflections in $q_z$ (out-of-plane) is

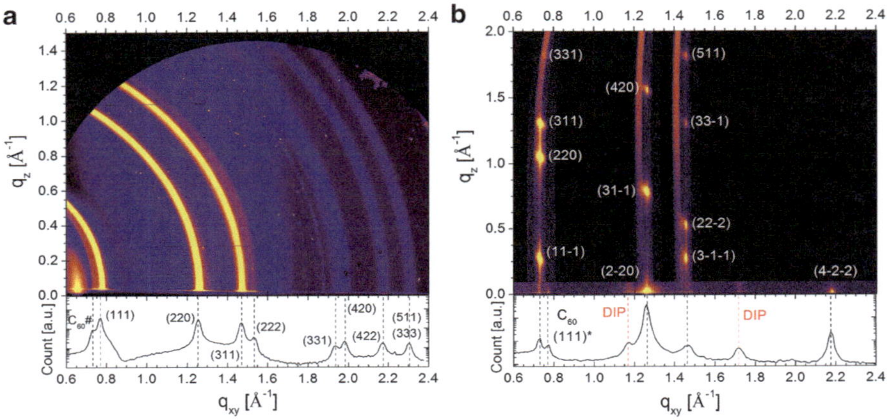


**Fig. 7.11** (**a**) Reciprocal space map from a 60 nm $C_{60}$ film recorded with a MARCCD area detector. At the bottom additional GIXD data measured with a point detector at $q_z = 0.02\,\text{Å}^{-1}$ is shown. (**b**) 2-dimensional GIXD data from a 15 nm $C_{60}$ film grown on a DIP templating layer ($d = 4$ nm) indexed according to the $C_{60}$ *fcc*-structure. The data consist of four detector scans at a fixed angle of incidence at 0.1 ° performed with a PILATUS II area detector. Images from each data point were transformed into $q$ coordinates and then assembled into one image. One scan was performed parallel to the substrate plane, for which at the bottom the integrated GIXD data is shown. Three scans were performed along the $C_{60}$ crystal truncation rods. Reprinted with permission from Ref. [71]

relatively large because of the small crystal size in $q_z$ direction limited by the film thickness of 15 nm. The in-plane coherent crystal sizes $l_s$ of both $C_{60}$ films were determined with the Scherrer formula to be $l_s = 7$ nm for the $C_{60}$/$SiO_2$ film and $l_s = 28$ nm for the $C_{60}$/DIP film. For the determination of $l_s$ of both $C_{60}$ films we choose the (2-20) and (4-2-2) reflections and averaged over the obtained values. The difference in $l_s$ is a clear evidence for the improved crystal quality and reduction of crystal defect density in the $C_{60}$/DIP film compared to the $C_{60}$/$SiO_2$ film.

### 7.5.1 UPS

Since $C_{60}$/DIP films are highly relevant for photovoltaic applications [20], we study whether the templating effect demonstrated above influences the electronic structure of $C_{60}$.

Figure 7.12a shows UPS data of the highest occupied molecular orbital (HOMO) region of $C_{60}$ grown on DIP and $SiO_2$. The overall shape of the HOMO and HOMO-1 states in Fig. 7.12a correspond to spectra measured for $C_{60}$ thin films on various inorganic and organic substrates/films [63, 75–77]. The similarity to the gas phase spectra of $C_{60}$ [78] indicates only weak interaction between molecules in the thin film. Apart from a spectral shift of 150 meV, resulting from the different energy level

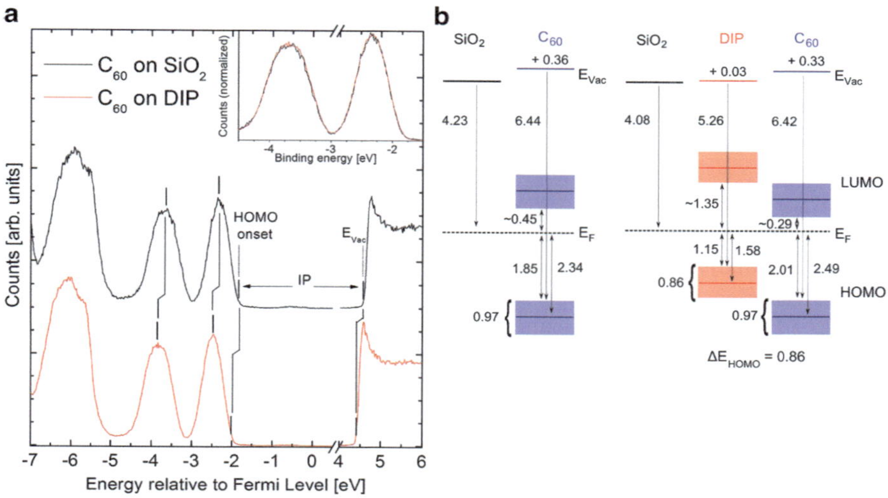


**Fig. 7.12** (**a**) UPS data from $C_{60}$ ($d = 13$ nm) grown on $SiO_2$ and grown on DIP ($d = 4$ nm). The inset shows the normalized HOMO regions of both datasets with the $C_{60}$/DIP data shifted by 150 meV. (**b**) Sketch of the electronic level alignment from the UPS data in (**a**). LUMO levels were taken from Refs. [73, 74]. All values are given in eV and have an error bar of $\pm$ 0.05 eV. The width of the HOMO state is determined by the respective onsets. Reprinted with permission from Ref. [71]

alignment (ELA) of the $C_{60}$ film to the DIP layer and the $SiO_2$ substrate, both spectra from the $C_{60}$ films are essentially identical. This can be seen more clearly in the inset in Fig. 7.12a, where the data are overlayed and the $C_{60}$/DIP data were shifted by 150 meV. In addition, also the ionization potentials (IP) determined by the secondary electron cutoff and the HOMO onset are equal (6.4 eV) for both $C_{60}$ films. Strong changes in structure and domain orientation for organic thin films lead to a significant change in spectral width of the HOMO or the IP depending on the molecular system [31, 32, 79]. However, our data demonstrates that at room temperature thin films of $C_{60}$ do not show these effects because of the reasons described below.

In general several mechanisms may influence the HOMO-band width of a molecular material [24, 80]. For instance in a non-uniform or disordered film the polarization energy (final state effect) and the intermolecular interaction (initial state effect) is locally different, resulting in slightly different ionization energies at different positions and therefore in spectral broadening. Another broadening effect is associated with band dispersion due to delocalization of the electronic states, which can be observed only for large single crystals of $C_{60}$ [81] and can therefore be omitted in the following discussion. In comparison to the $C_{60}$/$SiO_2$ film we may expect spectral sharpening for the $C_{60}$/DIP film, because of the improved crystal quality. The island size of the $C_{60}$/$SiO_2$ film is much smaller compared to the $C_{60}$/DIP film, implying that the density of crystal defects is higher in the $C_{60}$/$SiO_2$ film compared to the $C_{60}$-on-DIP film. However, the absence of any difference in the width of the $C_{60}$ states implies that a small coherent island size in a $C_{60}$ film has no significant impact on the polarization energy/intermolecular interaction. This observation may be rationalized by the high-symmetry shape of the single $C_{60}$ molecule and its rotation at room temperature [76, 82], which results in a much smaller local polarization/interaction variation due to crystal defects in comparison to anisotropic rod-like molecules.

For the mechanisms of energy level alignment at the organic-inorganic and organic-organic interfaces several different models have been discussed [39, 83–88] without a definitive conclusion in the literature. Figure 7.12b summarizes the ELA of two measured samples. For a low work function substrate like $SiO_2$ (WF = 4.23 eV) the $C_{60}$ LUMO (lowest unoccupied molecular orbital) level is presumably located at the substrate Fermi level leading to an interface dipole of 0.36 eV. For organic-organic heterostructures the ELA of the top layer is typically governed by the energetic position of the bottom layer. The difference of the HOMO onsets of DIP and $C_{60}$ as determined here is $\Delta E_{HOMO} = 0.86$ eV, similar to values reported for a DIP/$C_{60}$ heterostructure on PEDOT:PSS [20] and for the vice versa heterostructure (DIP-on-$C_{60}$) [33].

## 7.6 Summary

In summary, we presented two studies on how X-ray scattering can be combined with UPS to obtain detailed insights in the properties of organic thin films:

In the first example we observed the surface smoothing of low $T$ deposited organic thin films upon heating to room temperature. With the combined results from X-ray scattering and UPS we illustrated the process of surface smoothing by crystallization as in Fig. 7.9. After low $T$ deposition on ITO, the DIP film is nearly amorphous and also rough. Upon heating the surface material is crystallizing in domains without any predominant texture, whereas the bulk material does not strongly reorganize, because of lower mobility of these molecules. This crystallization process is associated with a molecular "downhill" current, which smooths the surface. In contrast to deposition on ITO, the DIP film deposited on $SiO_2$ is, except for the top surface material, already crystalline with a preferred orientation of the domains ($\sigma$-orientation). Upon heating to room temperature, the surface material is also crystallizing and exhibits thereby a similar smoothing as the film on ITO. Therefore, we conclude that the smoothing effect observed is qualitatively irrespective of the degree of bulk crystallinity. In addition, due to the crystallization the density of gap states at the surface is changed leading to a shift of the valence band features towards the Fermi level. The observed post-growth smoothing and crystallization is of importance for the growth of organic heterostructures, where the top surface of the first layer serves as a template for subsequent layers.

In the second example we demonstrated that the structural order of $C_{60}$ is significantly improved by inserting a DIP templating layer between the $SiO_2$ substrate and $C_{60}$ film. In contrast to growth on an amorphous substrate like $SiO_2$,

$C_{60}$ grown on a DIP film exhibits alignment of *fcc*-domains with the (111) plane parallel to the substrate and a significant increase of the coherent in-plane island size $l_s$ by a factor of $\sim 4$. UPS measurements revealed that the spectral broadening of the $C_{60}$ HOMO region interestingly do not depend significantly on the structural order in the $C_{60}$ film. This observation is in strong contrast to the data presented in the first example study where an increased of structural order in DIP lead to a significant sharpening of the HOMO band. This unusual behavior can be rationalized by the highly symmetric shape of the $C_{60}$ molecule.

**Acknowledgements** The authors thank T. Hosokai, F. Bussolotti and S. Duhm for helpful discussions and P. Willmott and S. Leake at the Swiss Light Source whose great efforts have made these experiments possible. This research project has been supported by the DFG, the G-COE program at Chiba University. A.H. acknowledges support from the JSPS.

## References

1. W. Brütting, S. Berleb, A.G. Mückl, Organic Electronics **2**(1), 1 (2001). URL http://www.sciencedirect.com/science/article/B6W6J-42SXF0X-1/1/e711a4bc688f027357e9b1b2be2ab1d7
2. G. Witte, C. Wöll, J. Mater. Res. **19**(7), 1889 (2004). DOI 10.1557/JMR.2004.0251
3. A. Hinderhofer, F. Schreiber, ChemPhysChem **13**(3), 628 (2012). DOI 10.1002/cphc.201100737. URL http://dx.doi.org/10.1002/cphc.201100737
4. J. Als-Nielsen, D. McMorrow, *Elements of Modern X-ray Physics* (Wiley, New York, 2001)
5. M. Birkholz, *Thin Film Analysis by X-Ray Scattering* (Wiley-VCH, Weinheim, 2006)
6. L.G. Parratt, Phys. Rev. **95**, 359 (1954)
7. H. Dosch, *Critical Phenomena at Surfaces and Interfaces: Evanescent X-Ray and Neutron Scattering* (Springer, Berlin, 1992)
8. S. Kowarik, Real-time studies of thin film growth of organic semiconductors. Ph.D. thesis, Wadham College, Oxford (2006)
9. R. Matsubara, M. Sakai, K. Kudo, N. Yoshimoto, I. Hirosawa, M. Nakamura, Organic Electronics **12**, 195 (2011)
10. S. Kowarik, K. Broch, A. Hinderhofer, A. Schwartzberg, J.O. Osso, D. Kilcoyne, F. Schreiber, S.R. Leone, J. Phys. Chem. C **114**(30), 13061 (2010). DOI 10.1021/jp103713z. URL http://pubs.acs.org/doi/abs/10.1021/jp103713z
11. W. Brütting (ed.), *Physics of Organic Semiconductors* (Wiley-VCH, Weinheim, 2005)
12. D. Käfer, C. Wöll, G. Witte, Appl. Phys. A **95**(1), 273 (2009). URL http://dx.doi.org/10.1007/s00339-008-5011-3
13. C.D. Dimitrakopoulos, P.R.L. Malenfant, Adv. Mater. **14**, 99 (2002)
14. K. Walzer, B. Maennig, M. Pfeiffer, K. Leo, Chem. Rev. **107**(4), 1233 (2007). DOI 10.1021/cr050156n. URL http://pubs.acs.org/doi/abs/10.1021/cr050156n
15. S. Kowarik, A. Gerlach, S. Sellner, L. Cavalcanti, O. Konovalov, F. Schreiber, Appl. Phys. A **95**(1), 233 (2009). URL http://dx.doi.org/10.1007/s00339-008-5012-2
16. A. Hinderhofer, T. Hosokai, C. Frank, J. Novák, A. Gerlach, F. Schreiber, J. Phys. Chem. C **115**, 16155 (2011)
17. B. Krause, F. Schreiber, H. Dosch, A. Pimpinelli, O.H. Seeck, Europhys. Lett. **65**(3), 372 (2004). URL http://stacks.iop.org/0295-5075/65/i=3/a=372
18. J. Yang, D. Yan, Chem. Soc. Rev. **38**, 2634 (2009). DOI 10.1039/B815723P. URL http://dx.doi.org/10.1039/B815723P

19. A. Hinderhofer, A. Gerlach, S. Kowarik, F. Zontone, J. Krug, F. Schreiber, EPL **91**(5), 56002 (2010). DOI 10.1209/0295-5075/91/56002. URL http://stacks.iop.org/0295-5075/91/i=5/a=56002
20. J. Wagner, M. Gruber, A. Hinderhofer, A. Wilke, B. Bröker, J. Frisch, P. Amsalem, A. Vollmer, A. Opitz, N. Koch, F. Schreiber, W. Brütting, Adv. Funct. Mater. **20**, 4295 (2010). URL http://dx.doi.org/10.1002/adfm.201001028
21. U. Hörmann, J. Wagner, M. Gruber, A. Opitz, W. Brütting, Phys. Stat. Sol. (RRL) **5**(7), 241 (2011). DOI 10.1002/pssr.201105238. URL http://dx.doi.org/10.1002/pssr.201105238
22. M. Horlet, M. Kraus, W. Brütting, A. Opitz, Appl. Phys. Lett. **98**(23), 233304 (2011). DOI 10.1063/1.3598423. URL http://link.aip.org/link/?APL/98/233304/1
23. D. Kurrle, J. Pflaum, Appl. Phys. Lett. **92**(13), 133306 (2008). DOI 10.1063/1.2896654. URL http://link.aip.org/link/?APL/92/133306/1
24. N. Ueno, S. Kera, Prog. Surf. Sci. **83**(1012), 490 (2008). DOI 10.1016/j.progsurf.2008.10.002. URL http://www.sciencedirect.com/science/article/pii/S0079681608000567
25. H. Ishii, K. Sugiyama, E. Ito, K. Seki, Adv. Mater. **11**(8), 605 (1999). DOI 10.1002/(SICI)1521-4095(199906)11:8⟨605::AID-ADMA605⟩3.0.CO;2-Q. URL http://dx.doi.org/10.1002/(SICI)1521-4095(199906)11:8⟨605::AID-ADMA605⟩3.0.CO;2-Q
26. N. Koch, ChemPhysChem **8**(10), 1438 (2007). DOI 10.1002/cphc.200700177. URL http://dx.doi.org/10.1002/cphc.200700177
27. M.A. Heinrich, J. Pflaum, A.K. Tripathi, W. Frey, M.L. Steigerwald, T. Siegrist, J. Phys. Chem. C **111**, 18878 (2007)
28. A. Nelson, J. Appl. Crystallogr. **39**(2), 273 (2006). DOI 10.1107/S0021889806005073. URL http://dx.doi.org/10.1107/S0021889806005073
29. T. Hosokai, M. Horie, T. Aoki, S. Nagamatsu, S. Kera, K.K. Okudaira, N. Ueno, J. Phys. Chem. C **112**(12), 4643 (2008). DOI 10.1021/jp710835b. URL http://dx.doi.org/10.1021/jp710835b
30. S. Kowarik, A. Gerlach, S. Sellner, F. Schreiber, L. Cavalcanti, O. Konovalov, Phys. Rev. Lett. **96**(12), 125504 (2006). DOI 10.1103/PhysRevLett.96.125504. URL http://link.aps.org/abstract/PRL/v96/e125504
31. A. Hinderhofer, T. Hosokai, K. Yonezawa, A. Gerlach, K. Kato, K. Broch, C. Frank, J. Novak, S. Kera, N. Ueno, F. Schreiber, Appl. Phys. Lett. **101**, 033307 (2012)
32. S. Duhm, G. Heimel, I. Salzmann, H. Glowatzki, R.L. Johnson, A. Vollmer, J.P. Rabe, N. Koch, Nat. Mater. **7**(4), 326 (2008). DOI 10.1038/nmat2119. URL http://dx.doi.org/10.1038/nmat2119
33. A. Wilke, P. Amsalem, J. Frisch, B. Brker, A. Vollmer, N. Koch, Appl. Phys. Lett. **98**(12), 123304 (2011). DOI DOI:10.1063/1.3571286. URL http://dx.doi.org/doi/10.1063/1.3571286
34. Y.L. Huang, W. Chen, H. Huang, D.C. Qi, S. Chen, X.Y. Gao, J. Pflaum, A.T.S. Wee, J. Phys. Chem. C **113**(21), 9251 (2009). URL http://dx.doi.org/10.1021/jp810804t
35. A.C. Dürr, N. Koch, M. Kelsch, A. Ruehm, J. Ghijsen, R.L. Johnson, J.J. Pireaux, J. Schwartz, F. Schreiber, H. Dosch, A. Kahn, Phys. Rev. B **68**, 115428 (2003)
36. J.Q. Zhong, H.Y. Mao, R. Wang, D.C. Qi, L. Cao, Y.Z. Wang, W. Chen, J. Phys. Chem. C **115**(48), 23922 (2011). DOI 10.1021/jp208645f. URL http://pubs.acs.org/doi/abs/10.1021/jp208645f
37. H. Yamane, Y. Yabuuchi, H. Fukagawa, S. Kera, K.K. Okudaira, N. Ueno, J. Appl. Phys. **99**(9), 093705 (2006). DOI 10.1063/1.2192978. URL http://link.aip.org/link/?JAP/99/093705/1
38. T. Sueyoshi, H. Kakuta, M. Ono, K. Sakamoto, S. Kera, N. Ueno, Appl. Phys. Lett. **96**(9), 093303 (2010). DOI 10.1063/1.3332577. URL http://link.aip.org/link/?APL/96/093303/1
39. H.Y. Mao, F. Bussolotti, D.C. Qi, R. Wang, S. Kera, N. Ueno, A.T.S. Wee, W. Chen, Organic Electronics **12**(3), 534 (2011). DOI DOI:10.1016/j.orgel.2011.01.003. URL http://www.sciencedirect.com/science/article/pii/S1566119911000127
40. T. Hosokai, H. Machida, A. Gerlach, S. Kera, F. Schreiber, N. Ueno, Phys. Rev. B **83**, 195310 (2011). DOI 10.1103/PhysRevB.83.195310. URL http://link.aps.org/doi/10.1103/PhysRevB.83.195310

41. F. Schreiber, Phys. Stat. Sol. **201**, 1037 (2004)
42. D. Käfer, L. Ruppel, G. Witte, Phys. Rev. B **75**(8), 085309 (2007). DOI 10.1103/PhysRevB.75.085309. URL http://link.aps.org/abstract/PRB/v75/e085309
43. T.V. Desai, A.R. Woll, F. Schreiber, J.R. Engstrom, J. Phys. Chem. C **114**(47), 20120 (2010). DOI 10.1021/jp107518f. URL http://pubs.acs.org/doi/abs/10.1021/jp107518f
44. X.N. Zhang, E. Barrena, D.G. de Oteyza, E.D. Souza, H. Dosch, J. Appl. Phys. **104**(10), 104308 (2008). DOI 10.1063/1.2977726. URL http://link.aip.org/link/?JAP/104/104308/1
45. T.V. Desai, S. Hong, A.R. Woll, K.J. Hughes, A.P. Kaushik, P. Clancy, J.R. Engstrom, J. Chem. Phys. **134**(22), 224702 (2011). DOI 10.1063/1.3591965. URL http://link.aip.org/link/?JCP/134/224702/1
46. R. Hayakawa, A. Turak, X. Zhang, N. Hiroshiba, H. Dosch, T. Chikyow, Y. Wakayama, J. Chem. Phys. **133**(3), 034706 (2010). DOI 10.1063/1.3456733. URL http://link.aip.org/link/?JCP/133/034706/1
47. H. Zhu, Q.L. Li, X.J. She, S.D. Wang, Appl. Phys. Lett. **98**(24), 243304 (2011). DOI 10.1063/1.3599579. URL http://link.aip.org/link/?APL/98/243304/1
48. H. Yang, T.J. Shin, M.M. Ling, K. Cho, C.Y. Ryu, Z. Bao, J. Am. Chem. Soc. **127**(33), 11542 (2005). DOI 10.1021/ja052478e. URL http://pubs.acs.org/doi/abs/10.1021/ja052478e
49. M.C. Gerstenberg, F. Schreiber, T.Y.B. Leung, G. Bracco, S.R. Forrest, G. Scoles, Phys. Rev. B **61**(11), 7678 (2000). DOI 10.1103/PhysRevB.61.7678
50. S.R. Forrest, Chem. Rev. **97**(6), 1793 (1997). DOI 10.1021/cr941014o. URL http://pubs.acs.org/doi/abs/10.1021/cr941014o
51. A. Sassella, M. Campione, A. Borghesi, Rivista del Nuovo Cimento **31**, 457 (2008)
52. L. Raimondo, M. Moret, M. Campione, A. Borghesi, A. Sassella, J. Phys. Chem. C **115**(13), 5880 (2011). DOI 10.1021/jp111754r. URL http://pubs.acs.org/doi/abs/10.1021/jp111754r
53. P. Sullivan, T.S. Jones, A.J. Ferguson, S. Heutz, Appl. Phys. Lett. **91**(23), 233114 (2007). DOI 10.1063/1.2821229. URL http://link.aip.org/link/?APL/91/233114/1
54. P. Peumans, A. Yakimov, S.R. Forrest, J. Appl. Phys. **93**(7), 3693 (2003). DOI 10.1063/1.1534621. URL http://link.aip.org/link/?JAP/93/3693/1
55. K. Itaka, M. Yamashiro, J. Yamaguchi, M. Haemori, S. Yaginuma, Y. Matsumoto, M. Kondo, H. Koinuma, Adv. Mater. **18**(13), 1713 (2006). URL http://dx.doi.org/10.1002/adma.200502752
56. M. Kraus, S. Richler, A. Opitz, W. Brütting, S. Haas, T. Hasegawa, A. Hinderhofer, F. Schreiber, J. Appl. Phys. **107**(9), 094503 (2010)
57. M. Haemori, J. Yamaguchi, S. Yaginuma, K. Itaka, H. Koinuma, Jpn. J. Appl. Phys. **44**(6A), 3740 (2005). DOI 10.1143/JJAP.44.3740. URL http://jjap.jsap.jp/link?JJAP/44/3740/
58. P. Fenter, F. Schreiber, L. Zhou, P. Eisenberger, S.R. Forrest, Phys. Rev. B **56**, 3046 (1997). URL http://dx.doi.org/10.1103/PhysRevB.56.3046
59. D.E. Hooks, T. Fritz, M.D. Ward, Adv. Mater. **13**, 227 (2001). DOI 10.1002/1521-4095(200102)13:4⟨227::AID-ADMA227⟩3.0.CO;2-P. URL http://dx.doi.org/10.1002/1521-4095(200102)13:4⟨227::AID-ADMA227⟩3.0.CO;2-P
60. L. Kilian, A. Hauschild, R. Temirov, S. Soubatch, A. Schöll, A. Bendounan, F. Reinert, T.L. Lee, F.S. Tautz, M. Sokolowski, E. Umbach, Phys. Rev. Lett. **100**, 136103 (2008). DOI 10.1103/PhysRevLett.100.136103. URL http://link.aps.org/doi/10.1103/PhysRevLett.100.136103
61. T.B. Singh, N.S. Sariciftci, H. Yang, L. Yang, B. Plochberger, H. Sitter, Appl. Phys. Lett. **90**(21), 213512 (2007). DOI 10.1063/1.2743386. URL http://link.aip.org/link/?APL/90/213512/1
62. S. Yim, T.S. Jones, Appl. Phys. Lett. **94**(2), 021911 (2009). DOI 10.1063/1.3072805. URL http://link.aip.org/link/?APL/94/021911/1
63. I. Salzmann, S. Duhm, R. Opitz, R.L. Johnson, J.P. Rabe, N. Koch, J. Appl. Phys. **104**(11), 114518 (2008). DOI 10.1063/1.3040003. URL http://link.aip.org/link/?JAP/104/114518/1
64. W. Chen, H. Zhang, H. Huang, L. Chen, A.T.S. Wee, ACS Nano **2**(4), 693 (2008). DOI 10.1021/nn800033z. URL http://pubs.acs.org/doi/abs/10.1021/nn800033z

65. J.Q. Zhong, H. Huang, H.Y. Mao, R. Wang, S. Zhong, W. Chen, J. Chem. Phys. **134**(15), 154706 (2011). DOI 10.1063/1.3582789. URL http://link.aip.org/link/?JCP/134/154706/1
66. U. Heinemeyer, R. Scholz, L. Gisslén, M.I. Alonso, J.O. Ossó, M. Garriga, A. Hinderhofer, M. Kytka, S. Kowarik, A. Gerlach, F. Schreiber, Phys. Rev. B **78**, 085210 (2008). URL http://dx.doi.org/10.1103/PhysRevB.78.085210
67. A.C. Dürr, F. Schreiber, K.A. Ritley, V. Kruppa, J. Krug, H. Dosch, B. Struth, Phys. Rev. Lett. **90**, 016104 (2003)
68. R. Scholz, L. Gisslen, B.E. Schuster, M.B. Casu, T. Chassé, U. Heinemeyer, F. Schreiber, J. Chem. Phys. **134**, 014504 (2011). URL http://dx.doi.org/doi:10.1063/1.3514709
69. J. Wagner, M. Gruber, A. Wilke, Y. Tanaka, K. Topczak, A. Steindamm, U. Hörmann, A. Opitz, Y. Nakayama, H. Ishii, J. Pflaum, N. Koch, W. Brütting, J. Appl. Phys. **111**(5), 054509 (2012). DOI 10.1063/1.3692050. URL http://link.aip.org/link/?JAP/111/054509/1
70. R.R. Lunt, N.C. Giebink, A.A. Belak, J.B. Benziger, S.R. Forrest, J. Appl. Phys. **105**(5), 053711 (2009). DOI 10.1063/1.3079797. URL http://link.aip.org/link/?JAP/105/053711/1
71. A. Hinderhofer, A. Gerlach, K. Broch, T. Hosokai, K. Yonezawa, K. Kato, S. Kera, N. Ueno, F. Schreiber, J. Phys. Chem. C **117**(2), 1053 (2013). DOI 10.1021/jp3106056. URL http://pubs.acs.org/doi/abs/10.1021/jp3106056
72. J.L. de Boer, S. van Smaalen, V. Petricek, M. Dusek, M.A. Verheijen, G. Meijer, Chem. Phys. Lett. **219**(5–6), 469 (1994). DOI DOI:10.1016/0009-2614(94)00110-3. URL http://www.sciencedirect.com/science/article/B6TFN-44J6FC9-M4/2/165963913b89bd08a6214011d478c2c0
73. R.W. Lof, M.A. van Veenendaal, B. Koopmans, H.T. Jonkman, G.A. Sawatzky, Phys. Rev. Lett. **68**, 3924 (1992). DOI 10.1103/PhysRevLett.68.3924. URL http://link.aps.org/doi/10.1103/PhysRevLett.68.3924
74. S. Krause, Determination of the transport levels in thin films of organic semiconductors. Ph.D. thesis, Universität Würzburg (2008)
75. J.H. Weaver, J.L. Martins, T. Komeda, Y. Chen, T.R. Ohno, G.H. Kroll, N. Troullier, R.E. Haufler, R.E. Smalley, Phys. Rev. Lett. **66**, 1741 (1991). DOI 10.1103/PhysRevLett.66.1741. URL http://link.aps.org/doi/10.1103/PhysRevLett.66.1741
76. S. Hasegawa, T. Miyamae, K. Yakushi, H. Inokuchi, K. Seki, N. Ueno, Phys. Rev. B **58**, 4927 (1998). DOI 10.1103/PhysRevB.58.4927. URL http://link.aps.org/doi/10.1103/PhysRevB.58.4927
77. O.V. Molodtsova, M. Knupfer, J. Appl. Phys. **99**(5), 053704 (2006). DOI 10.1063/1.2175468. URL http://link.aip.org/link/?JAP/99/053704/1
78. T. Liebsch, O. Plotzke, F. Heiser, U. Hergenhahn, O. Hemmers, R. Wehlitz, J. Viefhaus, B. Langer, S.B. Whitfield, U. Becker, Phys. Rev. A **52**, 457 (1995). DOI 10.1103/PhysRevA.52.457. URL http://link.aps.org/doi/10.1103/PhysRevA.52.457
79. H. Fukagawa, H. Yamane, T. Kataoka, S. Kera, M. Nakamura, K. Kudo, N. Ueno, Phys. Rev. B **73**, 245310 (2006). DOI 10.1103/PhysRevB.73.245310. URL http://link.aps.org/doi/10.1103/PhysRevB.73.245310
80. S. Kera, H. Yamane, N. Ueno, Prog. Surf. Sci. **84**, 135 (2009). DOI 10.1016/j.progsurf.2009.03.002. URL http://www.sciencedirect.com/science/article/pii/S007968160900029X
81. P. He, S. Bao, C. Yu, Y. Xu, Surf. Sci. **328**(3), 287 (1995). DOI 10.1016/0039-6028(95)00036-4. URL http://www.sciencedirect.com/science/article/pii/0039602895000364
82. R. Tycko, G. Dabbagh, R.M. Fleming, R.C. Haddon, A.V. Makhija, S.M. Zahurak, Phys. Rev. Lett. **67**, 1886 (1991). DOI 10.1103/PhysRevLett.67.1886. URL http://link.aps.org/doi/10.1103/PhysRevLett.67.1886
83. H. Fukagawa, S. Kera, T. Kataoka, S. Hosoumi, Y. Watanabe, K. Kudo, N. Ueno, Adv. Mater. **19**(5), 665 (2007). URL http://dx.doi.org/10.1002/adma.200601678
84. M.T. Greiner, M.G. Helander, W.M. Tang, Z.B. Wang, J. Qiu, Z.H. Lu, Nat. Mater. **11**(1), 76 (2012). DOI 10.1038/nmat3159. URL http://dx.doi.org/10.1038/nmat3159
85. S. Braun, W.R. Salaneck, M. Fahlman, Adv. Mater. **21**(14–15), 1450 (2009). DOI 10.1002/adma.200802893. URL http://dx.doi.org/10.1002/adma.200802893

86. H. Vazquez, W. Gao, F. Flores, A. Kahn, Phys. Rev. B **71**(4), 041306 (2005). DOI 10.1103/PhysRevB.71.041306. URL http://link.aps.org/abstract/PRB/v71/e041306
87. M. Linares, D. Beljonne, J. Cornil, K. Lancaster, J.L. Brédas, S. Verlaak, A. Mityashin, P. Heremans, A. Fuchs, C. Lennartz, J. Idé, R. Méreau, P. Aurel, L. Ducasse, F. Castet, J. Phys. Chem. C **114**(7), 3215 (2010). DOI 10.1021/jp910005g. URL http://pubs.acs.org/doi/abs/10.1021/jp910005g
88. J. Ivanco, Thin Solid Films **520**(11), 3975 (2012). DOI 10.1016/j.tsf.2012.01.035. URL http://www.sciencedirect.com/science/article/pii/S0040609012000739

# Chapter 8
# Photoelectron Yield Spectroscopy for Organic Materials and Interfaces

**Hisao Ishii, Hiroumi Kinjo, Tomoya Sato, Shin-ichi Machida, and Yasuo Nakayama**

## 8.1 Introduction of Photoelectron Yield Spectroscopy

Information on the electronic structures of various organic semiconductors is indispensable for understanding the operation of organic electronics and improve the device performance. As described in other chapters, such information has been widely investigated by using photoemission spectroscopy such as ultraviolet photoelectron spectroscopy (UPS) and X-ray photoelectron spectroscopy (XPS). Actually, the bulk and interfacial valence electronic structures of various organic semiconductors have been investigated by UPS, but there are several drawbacks as follows.

Firstly, the environment for UPS measurement should be restricted to high or ultrahigh vacuum (UHV) condition. This is because the mean free path of photoelectrons should be longer than the dimension of the measurement chamber to exclude the inelastic scattering of photoelectrons by residual gas molecules. On the other hand, organic electronic devices work practically not in vacuum but in atmospheric air environment. It is well known that the air, especially, oxygen and moisture influence the performance and degradation process of organic devices [1]. In addition, very recently the possibility of liquid organic semiconductors was proposed for liquid organic light emitting diodes [2]. Because of finite vapor pressure, UPS is not applicable for liquid system, although some efforts to measure UPS of liquid system in vacuum environment by using micro liquid jet were reported in literature [3]. Thus, in order to investigate practical aspect of organic

H. Ishii (✉)
Center for Frontier Science, Chiba University, Chiba, Japan
e-mail: ishii130@faculty.chiba-u.jp

H. Kinjo • T. Sato • S. Machida • Y. Nakayama
Department of Nanomaterials science, Graduate School of Advanced Integration Science, Chiba University, Chiba, Japan



H. Ishii et al. (eds.), *Electronic Processes in Organic Electronics*,
Springer Series in Materials Science 209, DOI 10.1007/978-4-431-55206-2_8

devices, the observation of the electronic structures in atmospheric condition is strongly desired.

Secondly, *charge-up* problem is serious for UPS measurement of insulating materials such as organic semiconductors. When photoemission occurs in an organic film, holes are left in the film. Such holes are attracted by their image charge in the conducting substrate underneath the film, and ideally can escape to the earth through the substrate making the system neutral. In practical system, there exist many charge carrier traps to capture the traveling holes. If the sample is good conductor with charge carriers, the trapped charge can be neutralized by the redistribution of charge carriers in the sample. Unfortunately this is not a case for conventional organic semiconductors. If the sample is positively charged, photoelectrons are decelerated inducing spectral shift or deformation. At worst, the emission of photoelectron is suppressed. This is called *charge up*. To avoid this, sample neutralization with an electron flood gun and the usage of very thin films are common maneuver [4, 5]. Flood gun is not effective because of possible sample damage by electron irradiation. Usually, thin film of less than several tens nm is used and the excitation light intensity is reduced by inserting a mesh in light path. Such thickness limitation restricts the application of UPS for organic films. For example, the typical thickness of organic devices is several hundreds nm, and the UPS measurement for device-scale film is not possible. In the case of organic single crystals, which has attracted much attention due to their high performance, the typical thickness is in the order of micro-meter, and conventional UPS measurement is not available unless the crystal is illuminated by additional laser light to increase photoconductivity as described in Chap. 2. As a consequence, techniques for probing the electronic structures of insulating material, independently of the charging effect, have been highly anticipated.

To overcome these problems in UPS, photoelectron yield spectroscopy (PYS) can be applied to investigate the occupied electronic states of materials. Figure 8.1 shows the principle of PYS for organic sample. In PYS, the quantum yield of photoelectron ($Y$), which is the number of emitted photoelectrons per photon absorbed, is detected as a function of incident photon energy ($h\nu$). It should be noted that physical process for PYS is the same as photoelectron spectroscopy (PES) where photoelectrons are emitted from the sample surface upon UV irradiation; photoemission event is common, but the photon energy range and detection scheme of photoelectrons are different. If $h\nu$ becomes greater than the threshold ionization energy ($I$) of an organic sample during $h\nu$ scan, the value of $Y$ starts to increase. Thus by determining the threshold of the spectrum, the value of $I$ can be evaluated. In the case of metal sample, the work function of the sample can be deduced in similar way. In contrast to PES, not the kinetic energy but total number of photoelectrons is essential to PYS leading to some advantages to overcome the problems of UPS: (i) PYS is available not only in vacuum but also in air. (ii) Sample charging can be practically avoided as described later.

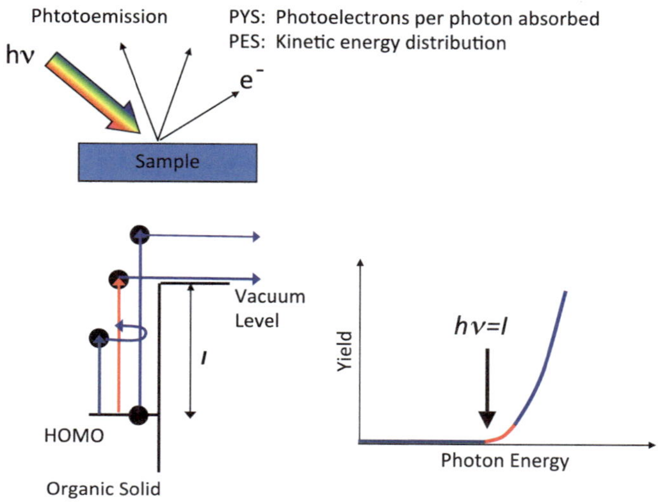


**Fig. 8.1** Principle of photoelectron yield spectroscopy in the case of organic materials

## 8.2 Analysis of the Threshold Region of PYS

Regarding of theory to evaluate photoemission yield, Pope and Swenberg reported a detailed review [6]. Here the simple principle will be described after the literature.

According to the *three-step* model [7], photoemission event consists of three steps; (i) optical excitation of an electron at initial energy $E$, (ii) electron transport towards the surface, and (iii) escape through the surface to vacuum. During the steps (ii) and (iii), primary photoelectron suffers from inelastic scattering and secondary electrons with lower kinetic energy are produced. Thus, the total photoemission current $I(E, h\nu)$ can be expressed by

$$I(E, h\nu) = I_p(E, h\nu) + I_s(E, h\nu). \tag{8.1}$$

Here the $I_p$ and $I_s$ are the current of primary and secondary electrons, respectively. To evaluate $I_s$, scattering processes such as electron–electron and electron–phonon scattering should be considered. But, if we focus on the threshold region of PYS, photoelectrons are mainly primary, and the contribution of $I_s$ will be neglected hereafter. The primary term can be expressed according to the three steps as

$$I(E, h\nu) = P(E, h\nu)X(E)T(E), \tag{8.2}$$

where $P(E, h\nu)$ is the probability of exciting an electron to a state of energy $E$ with photon energy $h\nu$, $X(E)$ is the transmission probability for the electron at the level $E$, and $T(E)$ is a surface escape function. According to Fermi's golden rule, $P(E, h\nu)$ can be described as

$$\begin{aligned} P(E, h\nu) &= A|\langle \Psi_{\mathrm{f}}|r|\Psi_{\mathrm{i}}\rangle|^2 N_{\mathrm{i}}(E - h\nu)N_{\mathrm{f}}(E) \\ &= AN_{\mathrm{i}}^{\mathrm{opt}}(E - h\nu)N_{\mathrm{f}}^{\mathrm{opt}}(E) \end{aligned} \tag{8.3}$$

where $\psi_{\mathrm{i}}$ and $\psi_{\mathrm{f}}$ are the initial and final states wave function, respectively, $N_{\mathrm{i}}$ and $N_{\mathrm{f}}$ are the density of states (DOS) of initial and final states, $A$ is a constant. If the matrix element is effectively included to $N$ term, $P$ can be expressed by the product of initial and final optical density of states ($N_{\mathrm{i}}^{\mathrm{opt}}$ and $N_{\mathrm{f}}^{\mathrm{opt}}$). Transmission probability can be simply assumed as

$$X(E) = \frac{L(E)}{L(E) + l_a(h\nu)}, \tag{8.4}$$

where $L(E)$ is an isotropic scattering length of electrons, and $l_{\mathrm{a}}$ is an absorption depth.

Surface escape function can be simplified as

$$T(E) = \begin{matrix} 0; & E \leq E_{\mathrm{vac}} \\ \frac{1}{2}\left(1 - \sqrt{E_{\mathrm{vac}}/E}\right); & E > E_{\mathrm{vac}} \end{matrix}, \tag{8.5}$$

under assumption of isotropic distribution of photoelectrons. Here $E_{\mathrm{vac}}$ is the energy of vacuum level of the sample. Finally, the photoemission yield for $h\nu$ can be described as

$$Y(h\nu) \propto \frac{\int_{E_{vac}}^{h\nu} P(E, h\nu)X(E)T(E)dE}{\int_0^{h\nu} N_i^{opt}(E - h\nu)N_f^{opt}(E)dE} \tag{8.6}$$

In PYS, the threshold energy for photoemission is essential, but practically the precise determination of the threshold is not easy, because the observed yield shows finite intensity even far below the ionization energy. That is, if we magnify the threshold region in the yield curve, some structure can be seen close to detection limit. Thus, the yield curve in the threshold region is fitted by assuming some function form to determine the threshold energy $E_{\mathrm{T}}$.

In the case of metal sample, the yield curve shows so-called Fowler function curve. Fowler described the yield function of photoemission from metal sample with work function $\Phi$ as the following equations [8].

$$
\begin{aligned}
&Y = bT^2F\left(\frac{h\nu-\Phi}{kT}\right), \quad ln\left(\frac{Y}{T^2}\right) - lnb = ln\left[F\left(\frac{h\nu-\Phi}{kT}\right)\right]\\
&F(\mu) = e^{\mu} - \frac{e^{2\mu}}{2^2} + \frac{e^{3\mu}}{3^2}\cdots \text{ for } \mu \le 0 \\
&= \frac{\pi^2}{6} + \frac{1}{2}\mu^2 - \left[e^{-\mu} - \frac{e^{-2\mu}}{2^2} + \frac{e^{-3\mu}}{3^2}\cdots\right] \text{ for } \mu \le 0.
\end{aligned}
\tag{8.7}
$$

Figure 8.2a shows an example of Fowler plot of a contaminated gold film. The ln $(Y/T^2)$ is plotted as a function of $(h\nu\text{-}\Phi)/kT$. From the best fitting, the work function was determined as 4.26 eV. If the hν is not close to the threshold, the equation can be approximated as quadratic function [9].

$$Y \propto (h\nu - \Phi)^2. \tag{8.8}$$

Thus, the value of Φ can be easily determined as the threshold in $h\nu - Y^{1/2}$ plot as shown in Fig. 8.2b. This value corresponds to the work function at zero temperature. Except the case of Fowler plot, the yield curve is usually analyzed by using power dependence; $Y \propto (h\nu - E_T)^n$. The Eq. (8.8) for metal sample corresponds to the case $n = 2$. In the case of inorganic semiconductors, Kane proposed $n$ values for various cases. $n = 1$ and 2 are indicated to the case of direct and indirect optical excitation, respectively. Under the consideration of surface processes, it can vary among $n = 1$, 3/2, 2, 5/2 depending on the surface properties [10]. For a large number of organic materials, a cubic dependence,

$$Y \propto (h\nu - E_T)^3, \tag{8.9}$$

is observed near threshold [6]. A variety of phenomenological explanations of cubic dependence have been proposed [11, 12]. It should be noted that there are some theory to explain $n = 3$ but this value is still empirical. Pope et al. wrote that caution should be exercised in overstressing the significance of the good fit [6]. They showed an example where all of $Y^{1/3}$, $Y^{1/4}$, and $Y^{1/5}$ plots show good linearity for experimental result of tetracene reported in the [11]. $n = 2$ is also often used for organic semiconductors in analogy to inorganic semiconductors without serious consideration. At this present, the fitting by $n = 2$, 5/2, 3 should be tried to determine the threshold, and the plot with best linearity should be adopted. But if we compare the difference in ionization energy among the different materials, $n$ is better to be fixed, because the threshold energy depends on the value of $n$.

In addition to ionization threshold, PYS can give us rough insight for density-of-states near threshold region. If $X(E)$, $T(E)$, $P(E, h\nu)$, and $N_f^{opt}(E)$ in Eqs. (8.3) and (8.6) can be regarded constant, the first derivative of the yield with respect to energy, $dY/d(h\nu)$, reflects an effective density of states around valence top region [13]. The validity of this assumption is not well established. Practically, the situation strongly depends on material. Caution should be paid upon this kind of analysis of PYS spectrum.

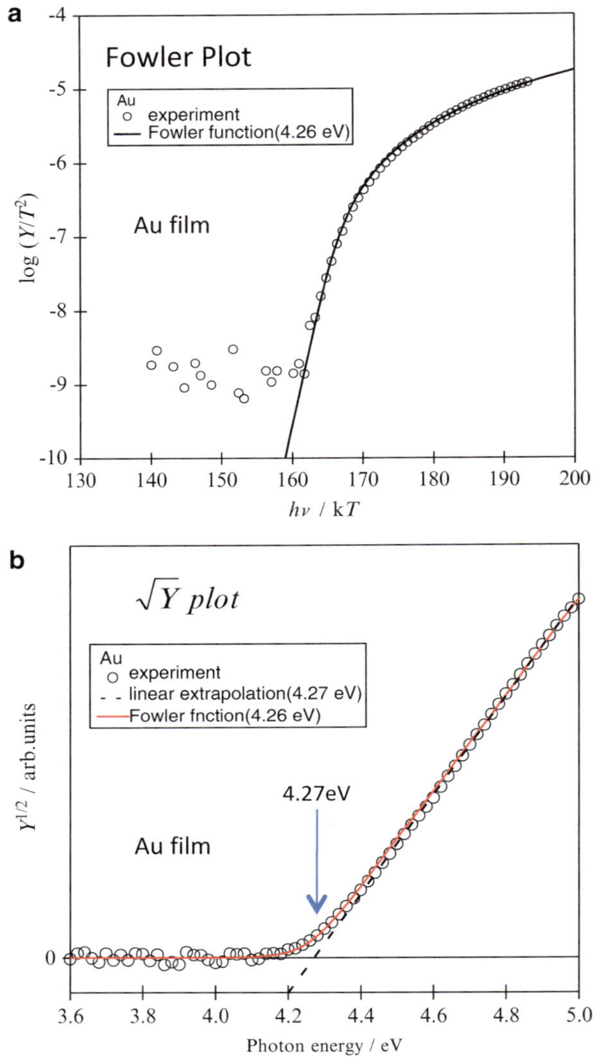


**Fig. 8.2** (**a**) An example of Fowler plot analysis to determine the work function of a gold film from PYS spectrum. (**b**) The plot of the square root of the yield to determine the work function at zero temperature approximation

## 8.3 Detection Methods of Photoemission Yield

Primary advantage of PYS is that it can be applied in both vacuum and non-vacuum atmosphere. For vacuum system, PYS has a long history. Very early works were basic experiments to verify quantum mechanics for some metal surfaces (For example, [14]). Fowler gave a theoretical interpretation to determine the work function from the experiment [8]. Later, it was established as spectroscopic method to investigate the electronic structures of inorganic materials [13, 15, 16]. PYS was also applied to investigate interfacial electronic structures of inorganic semiconductors [17, 18]. Except early stage, electron multiplier was used to sensitively detect photoelectrons. Because the multiplier can work only in high or ultrahigh vacuum, this type of PYS measurements was limited to vacuum environment.

The efforts to perform PYS in non-vacuum environment were made by various research groups. Pope et al. performed atmospheric PYS measurements for small organic single crystals. Their apparatus was based on the Millikan chamber. The charged small (μm-scaled) crystals are suspended like a droplet of Millikan experiment. The photoemission event changes the charge of the crystal inducing the movement of the crystal. The movement can be detected by using optical method with extremely high sensitivity enough to detect two-photon process by conventional light source [6]. Because the sample form is much limited to small particle or crystal to be suspended, this technique cannot be standard technique for PYS in air.

Practically established method to detect photoelectrons in air was proposed by Kirihara and Uda et al. [19]. They developed *low energy electron counter*. Electrons ejected from a sample surface to air attach to oxygen molecules to form oxygen anions. This counter can detect such oxygen anions. This is like a discharge-type counter, and once the oxygen anions enter the discharge region under bias voltage, it initiates continuous discharge to induce electrical signal to be detected. This method enabled PYS measurements in air practically, and the commercial apparatus has been widely utilized in the device field. Because oxygen in air can absorb high energy photon ($\lambda < 240$ nm), the usage of this technique was initially limited to detect ionization energy less than 6 eV. Later, the optimization of optical arrangement to reduce the absorption by oxygen, the limitation is now extended to less than 7 eV. The existence of oxygen molecule is not necessary condition, but it is much preferred for stable measurement with good S/N ratio (Y. Nakajima, personal communication). The saturation limit of the detector is several thousands count per second because the detector is not always open to signal (due to reset process to stop the continuous discharge), but the limitation is not problem for usual samples.

The other approach to detect photoelectron in air is to measure current to compensate the sample after the photoemission. Historically, Smith demonstrated the detection of photoelectron emission from Al and Ni surface in air just by measuring photoemission current, although the $h\nu$ was not scanned [20]. Later several group including us developed apparatus to detect the emission of photoelectron in air as current [21–23]. Current detection has some advantages; wide dynamic range and feasibility in both vacuum and non-vacuum environment. In current detection in air, the application of electric field to extract photoelectrons from the sample is necessary. This is because the mean-free-path of photoelectron in atmospheric condition is less than μm order, leading the formation of negative space charge close to the sample surface under no field. Such space charges suppress the further photoemission. For that reason, finite field to support electron emission is usually applied to the sample intentionally or unintentionally. Here it should be noted that the application of voltage to the sample to extract electrons is necessary not only for atmospheric measurement but also for vacuum measurement. A bias to ensure the vacuum level of the sample higher than that of detector; otherwise the observed ionization energy (or work function) should be smaller by the difference of the vacuum level). This situation is the same as photoemission spectroscopy; a bias is necessary to detect secondary cut-off to correctly determine the position of vacuum level.

## 8.4 Apparatus for PYS measurement

Here, as an example, our apparatus for PYS is described. Figure 8.3 shows the diagram of our apparatus. The apparatus consists of a light source, main chamber, preparation chamber, and atmospheric chamber. Monochromatized photon from Xe

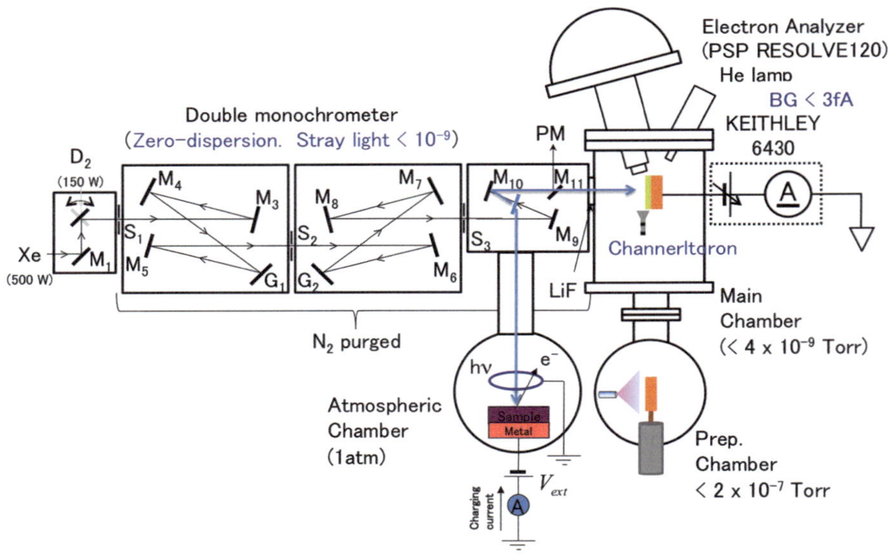


**Fig. 8.3** Our apparatus of PYS and high sensitivity photoemission spectroscopy. The system consists of monochromatic light source and three chambers; main chamber, preparation chamber and atmospheric chamber. See the text for details

(500 W) or $D_2$ (150 W) lamps is used as an excitation light. To suppress the absorption of the light by oxygen, the inside of the monochromator is purged by dry nitrogen gas, and photon energy can be scanned from 1.5 to 8 eV. Aiming at eliminating stray light, a zero-dispersion type of double monochromator (Bunkoukeiki M25GTM-DZ) was adopted. The relative intensity of the stray light was less than $1 \times 10^{-9}$ at a wavelength of 632.8 nm. Such extremely low level of stray light is important to detect weak density of states like gap states above HOMO. Excitation light is introduced to the main chamber. The typical size of the incident light was $5 \times 3.5\ mm^2$. By using electrical feedthrough with guard line, the sample is connected to sub-pico ammeter (Keithley 6430) to measure the photocurrent. The background level was less than 3fA. A Channeltron is also equipped to perform high-sensitivity measurement of PYS. An electron analyzer (PSP Resolve120) for photoemission spectroscopy is attached to the main chamber. This is available to perform high-sensitivity photoemission spectroscopy to detect weak states like gap states [24]. Sample evaporation can be performed in the preparation chamber, and the prepared sample can be transferred to the main chamber. By switching mirror, the excitation light can be introduced into the atmospheric chamber, where PYS is performed in atmospheric condition. The light is incident vertically onto the sample surface. This geometry is essential to measure liquid surface. A ring-shaped electrode to extract electrons is set in front of the sample. The electrode is grounded, and the sample is connected to an ammeter. By negatively biasing the sample, the photoemission can be kept even in air. The sample substrate can be horizontally located, enabling the measurement of also liquid sample.

## 8.5 Application of PYS to Organic Materials

### 8.5.1 *Examples of PYS for Organic Films and air-Exposure Effect on Their Ionization Energy*

Here, examples of PYS measurements for organic films are described. Figure 8.4 shows the PYS spectra of the evaporated amorphous film and single crystal of rubrene [25]. Rubrene is one of the most promising materials for organic field effect transistors due to its high hole mobility in the single-crystalline phase [26]. After subtracting the background level, the cubic root of yield is plotted as a function of photon energy $h\nu$. Both spectra show linear shape at the threshold region. From linear extrapolation, the threshold ionization ($I_s$) energy was determined. $I_s$ of the evaporated film was about 5.3 eV while that of single crystal 4.9 eV. In comparison to the amorphous film, $I_s$ of the rubrene SC is reduced by 0.4 eV. Although the reasons for this $I_s$ reduction have not been completely specified, the upshift of the HOMO edge due to band formation probably plays an essential role in this behavior. This result indicates that the ionization energy of evaporated film is not

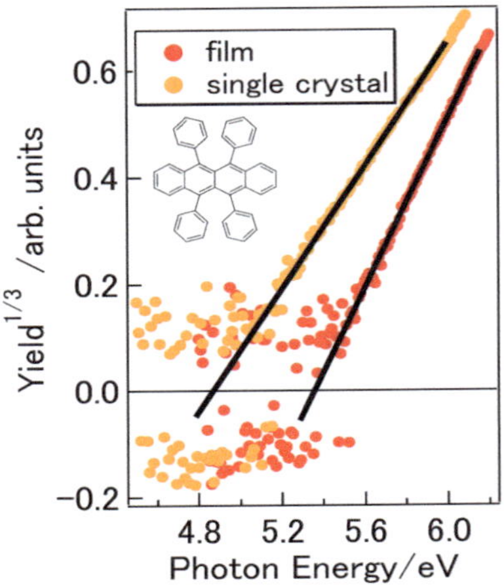


**Fig. 8.4** PYS spectra of a film and single crystal of rubrene

good measure of that of the single crystal: the direct observation of organic single crystals is necessary although conventional photoemission spectroscopy is not easy to perform due to charge-up problem.

It has long been well known that rubrene molecule changes into rubrene-endoperoxide ($RubO_2$) quite sensitively in the presence of oxygen and visible light [27]. Such atmospheric effect to rubrene was monitored in-situ as shown in Fig. 8.5. The spectra of the SC were substantially unchanged by visible and near-ultraviolet light illumination in air (Fig. 8.5a), whereas those of the film show apparent degradation of the spectral amplitude (Fig. 8.5b). This change corresponds to the decrease in unreacted rubrene underneath the $RubO_2$ layers whose contribution to $Y$ occurs only when $h\nu$ is greater than 6 eV. These results prove the reported durability of the rubrene SC to the photo-oxidation compared to its amorphous phase [28].

Next the affect of air-exposure to ionization energy of organic materials is examined. As mentioned in Sect. 8.1, the electric properties of organic semiconductors are often affected by air. Our PYS method by current-mode detection can determine ionization energy of organics both in vacuum and atmospheric conditions. Figure 8.6 shows the comparison of ionization energy of various organic semiconductors between in vacuum and in air. First, $I_s$ of each material was measured in vacuum, and then the sample was purged and $I_s$ was measured in air. In the case of Au film, large difference was observed. This is because the work function of metal surface is sensitive to surface contamination. For organic semiconductors, the difference is not so large. The change of $I_s$ due to air exposure often seems less than 0.1 eV [29]. But in the case of rubrene film and single crystal, the change is almost 0.2 eV. The special case is titanyl phthalocyanine (TiPc), where progressive change of ionization energy with additional tail structure was induced by exposing to water vapor as shown in Fig. 8.7 [30]. Here $I_t$ was the onset energy of the tail structure. As shown in Fig. 8.8, the first exposure to water vapor did not much affect both $I_0$ and $I_t$, while the following evacuation-exposure cycles caused

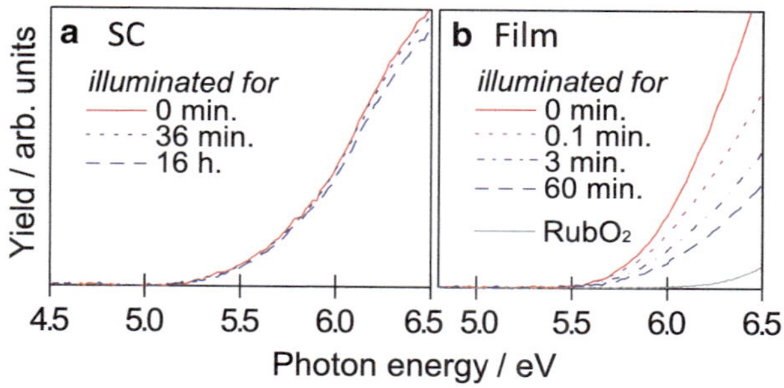


**Fig. 8.5** Time evolution of PYS spectra after illuminating in air. (**a**) PYS of rubrene single crystal, (**b**) PYS of rubrene film. Reproduced from [25] with permission of the American Institute of Physics

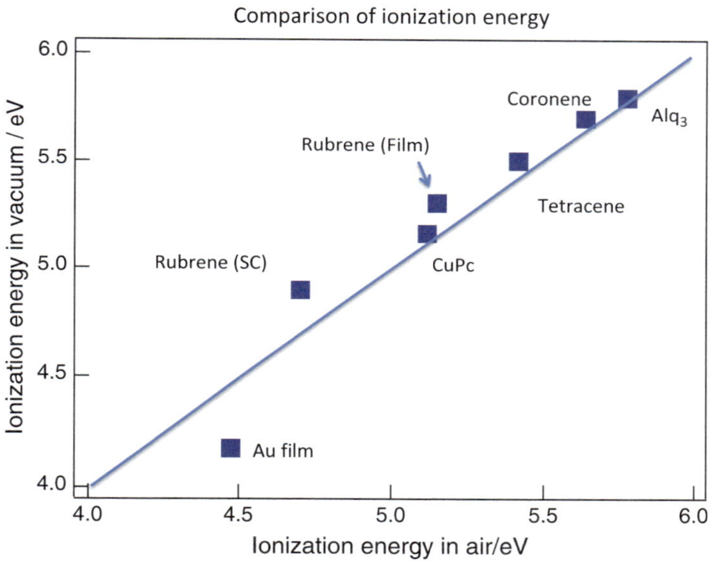


**Fig. 8.6** Comparison of ionization energy between in vacuum and in air. The data points are from literature; Au film, CuPc, Tetracene, coronene, and $Alq_3$ [29], and Rubrene [25]. The *line* corresponds to the case of no difference between vacuum and air

mostly reversible decrease and increase of $I_0$ and $I_t$ by about 0.2 and 0.4 eV, respectively. The accumulation of water molecule in the film is a possible reason for this change. In summary, the atmospheric effect on the ionization energy of organic semiconductors seems to be not so serious for most materials; the possible change may be 0.1–0.2 eV, but there exists exceptional case like TiPC. So caution should be paid to handle the ionization energy reported by various methods in various atmospheric conditions.

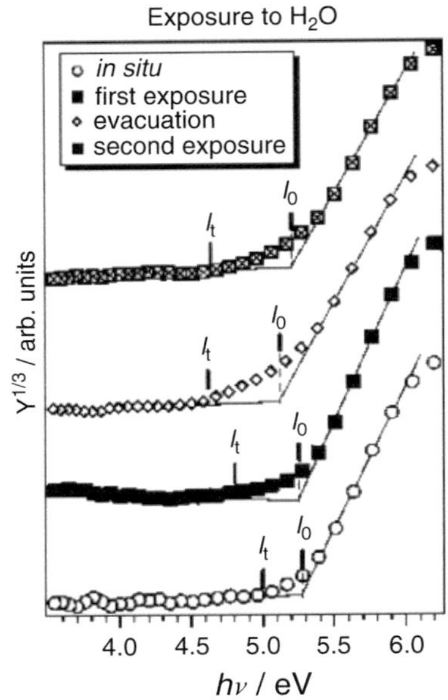


**Fig. 8.7** The observed PYS spectra of the TiOPc film at repeated cycles of $H_2O$ exposure and evacuation. *Vertical broken bars* indicate the values of estimated *I*. Reproduced from [30] with permission of the American Institute of Physics

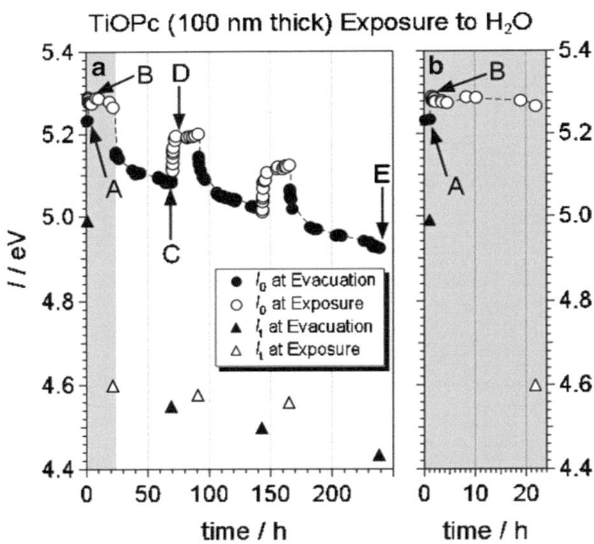


**Fig. 8.8** (**a**) The change of *I* of the TiOPc film in a time sequence with repeated cycles of the introduction and evacuation of $H_2O$ gas (770 Pa). The values of *I* determined by the cube-root plot of photoemission yield ($I_0$; *circles*) and the onset of the tailing part of the plot ($I_t$; *triangles*) are shown. *Filled circles* and *triangles* denote *I* at the evacuation stage (high vacuum of $1.0 \times 10^{-4}$ Pa was reached within 120 min), and open ones are *I* measured in $H_2O$ atmosphere. (**b**) The expansion of the data at the first exposure to $H_2O$ gas. Reproduced from [30] with permission of the American Institute of Physics

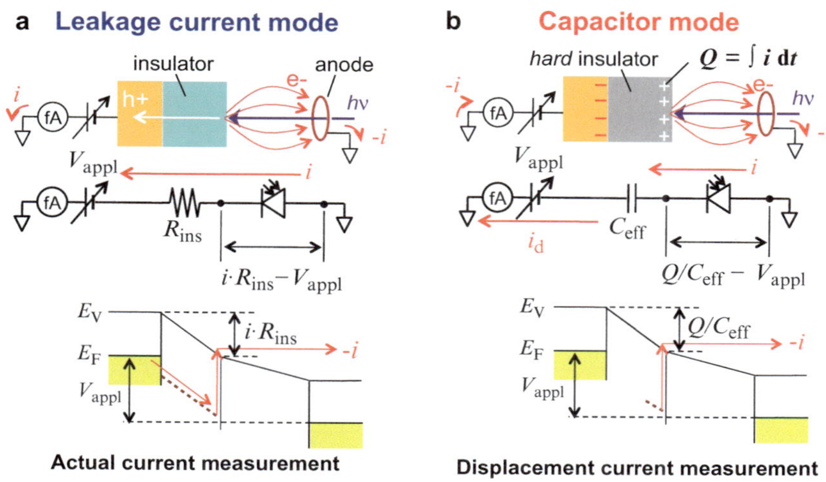


**Fig. 8.9** Two modes to avoid charge up during PYS measurements; (**a**) leakage current mode and (**b**) capacitor mode. Reproduced from [23] with permission of the American Institute of Physics

## *8.5.2 Application of PYS to Insulating Samples*

In contrast to conventional photoemission spectroscopy, *charge-up* problem is not serious in the case of PYS. This advantage is a *byproduct* of sample biasing to extract photoelectrons in atmospheric condition. In the case of insulating sample with negative biasing, the sample potential is gradually shifting downwards during photoemission event. Unless the vacuum level of the sample becomes below the surrounding potential (i.e. the vacuum level of the extracting electrode), all the photoelectrons can escape from the sample although the kinetic energy of photoelectrons is reduced. In the case of photoemission spectroscopy, such slow-down of photoelectron is critical, but it does not matter in PYS; only the total number of photoelectrons is essential. Thus, the relation between the sample bias and the potential drop due to charge up is important. There are two modes to induce the potential drop during charge up; (i) l*eakage current mode* and (ii) *capacitor mode* as shown in Fig. 8.9 [23].

In *leakage current mode*, the potential drop is related to the sample resistance to the ground, $R_{ins}$. Photoemission current $i$ induces the potential drop of $iR_{ins}$. If the sample bias ($V_{appl}$) is larger than $iR_{ins}$, all the photoelectrons can escape from the sample, i.e. no charge up. Thus we can avoid charge up and continue to measure PYS by reducing $i$ or $R_{ins}$; by reducing the photon intensity or reducing the thickness of sample film. This is the same *rule of thumb* to avoid charge up in conventional photoemission spectroscopy.

*Capacitor mode* to avoid charge up is unique to PYS. Let us consider a perfectly insulating film of thickness $d$. The area of light illumination is $S$. The dielectric constants of the vacuum and sample material are $\varepsilon_0$ and $\varepsilon_s$. If the amount of charge,

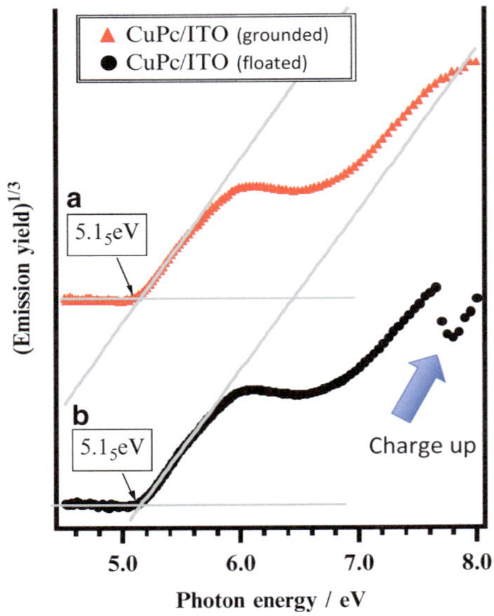


**Fig. 8.10** PYS spectra of CuPc thin film on ITO substrate. (**a**) The ITO electrode is grounded, (**b**) the ITO electrode is electrically floated

Q(t), is accumulated only at the film surface during the photoemission event, counter charge will be accumulated at the other side of the film. So, the potential drop across the sample film, $\Delta V$, is build up to form capacitor $C$. $\Delta V$ can be expressed by

$$\Delta V(t) = Q(t)/C = \frac{Q(t)d}{\varepsilon_0 \varepsilon_s S} = \frac{d}{\varepsilon_0 \varepsilon_s S} \int_0^t idt.$$

If $\Delta V$ can be kept to be less than $V_{\text{appl}}$, charge up is suppressed. In contrast to *leakage current mode*, charge up can be suppressed only for limited time. What we can do is to extend the persistence time. In other words, by using smaller thickness and larger illumination area to increase C, the time when the measurement is free of charge up can be extended.

Figure 8.10 shows the PYS measurement of copper phthalocyanine (CuPc) film on indium-tin-oxide (ITO) substrate in capacitor mode. If the ITO is grounded, the sample can be kept neutral as shown in the upper spectrum. If the ITO is not connected to the ground, i.e. electrically floated, PYS measurement can be performed in capacitor mode as in the lower spectrum. The photon energy was scanned from lower to higher energy side. Below 7.6 eV, the spectrum is the same as that of the grounded sample. At 7.7 eV, where the photon intensity shows the maximum (resonance line of $D_2$ lamp), the yield showed an abrupt decease due to sample charging. The similar results were obtained for gold film on mica substrates as shown in Fig. 8.11 [23]. An electrically floated gold thin film was prepared on a mica substrate, which is a good insulator. As shown in right panel, at the lower photon energy, the yield spectrum is similar to that of grounded film. At the photon

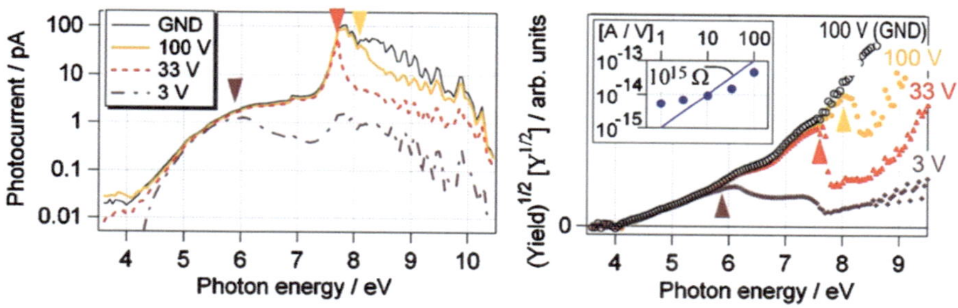


**Fig. 8.11** Photocurrent (*left*) and $Y^{1/2}$ (*right*) spectra of the Au films on a mica substrate. (*Inset*) Leakage current through the mica substrate plotted as a function of acceleration voltage. Reproduced from [23] with permission of the American Institute of Physics

energy marked by triangle, the yield showed an abrupt decrease due to charge up. With increasing sample bias voltage, the energy region without sample charge up is extended, demonstrating the validity of capacitance mode.

Next example of PYS is for an insulating polymer in relation to triboelectricity. When two different materials are contacted and then separated, one tends to be positively charged while the other negatively. Such triboelectricity is common phenomenon in dairy life and also applied in industry. The mechanism of the charging has been open question; there are many models on the mechanism (As a recent review, [31]). Despite of variation of the models, most models assume the existence of weak density-of-states (DOS) in HOMO-LUMO energy gap as a state to accept charges. The DOS and ionization energy of poly(ethylene terephthalate) (PET) film were investigated by PYS as shown in Fig. 8.12 [32]. The film thickness was 20 μm, which is too thick to perform conventional photoemission spectroscopy. This insulating polymer is literally *insulating* to keep charges in the film. Even for such insulating material, PYS can be easily applied. The spectrum shows two thresholds. By using $D_2$ light source, one threshold was found around 7.65 eV, which corresponds to the intrinsic ionization energy of PET. Even below the threshold, a long tail structure was observed, whose onset was found around 4.2 eV by Xe lamp. This lower threshold indicates that there are widely-distributed DOS within the HOMO-LUMO gap as shown in the energy diagram of Fig. 8.12b. These weak states should be investigated to discuss the mechanism of charging.

### *8.5.3 Application of PYS to Liquid Samples*

As described previously, one of the advantages of current-mode PYS is no-limitation of measurement environment; non-vacuum measurement is possible. That advantage can be applied to investigate the electronic structure of liquids with high vapor pressure. The photoemission spectroscopy has been applied to liquid system, but the micro-jet beam of samples should be introduced in vacuum

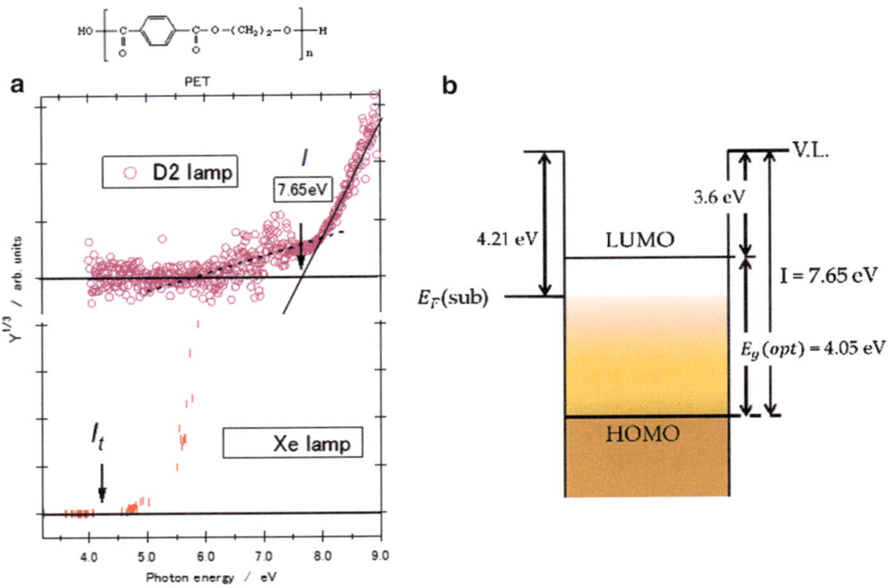


**Fig. 8.12** (**a**) PYS spectrum of PET film (20 μm thick) measured by using $D_2$ and Xe lamps. (**b**) The energy diagram of PET film obtained from PYS result

chamber, where the boiling surface of the beam is monitored (As a review paper, see [33]). PYS enable us to investigate the electronic structure of static liquid surface.

Figure 8.13 shows the PYS spectrum of 0.5 mol/L solution of $K_4[Fe(CN)_6]$ as a test sample. The $Y^{1/n}$ ($n = 2$, 2.5, 3) were plotted to check the $n$-dependence on ionization energy. The observed ionization energy was 5.2 eV in agreement with the previous study [34]. Next example is 9-(2-Ethylhexyl)carbazole (EHCz) in liquid form. Adachi and Xu have recently reported liquid organic light-emitting diodes using this liquid sample with doped rubrene [2]. The performance of the device is very poor, but it is attractive as future device due to an advantage of liquid form; for example, perfect flexibility, and easy replacement of active liquid material by injection to liquid cell. Figure 8.14 shows the PYS spectra of liquid EHCz, EHCz with doped rubrene, and rubrene film [35]. The spectrum of liquid EHCz showed threshold at 5.82 eV. This value of ionization energy is almost the same as poly(9-vinylcarbazole) (PVK) by PYS (5.8 eV [36]), suggesting good hole transporting property of EHCz comparable to PVK. If rubrene is doped into EHCz liquid, the new onset due to doped rubrene emerged at 5.42 eV in the spectra. Thus the energy location of HOMO of the doped rubrene molecule is 5.42 eV, which is 0.4 eV shallower than that of EHCz, suggesting the hole trapping nature of rubrene in EHCz. Thus, PYS measurement for liquid sample can give us useful information to discuss device performance.

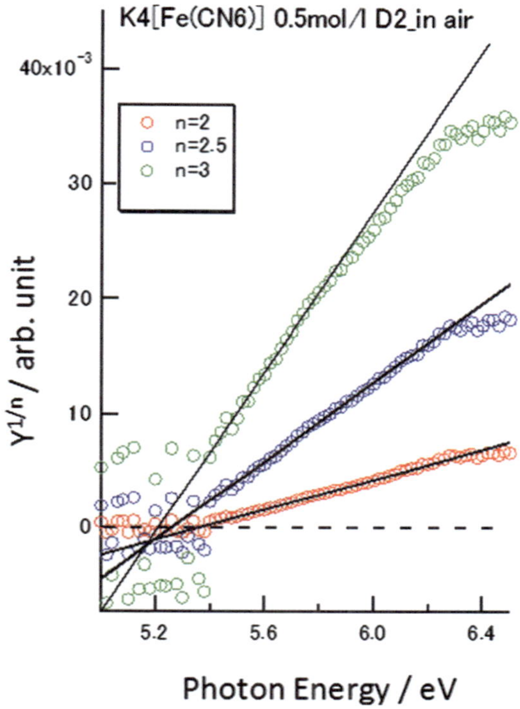


**Fig. 8.13** The PYS spectrum of 0.5 mol/L $K_4[Fe(CN)_6]$ solution. Function forms of $(h\nu - I)^n$ for $n = 2, 2.5, 3$ were used to fit the threshold region

### *8.5.4 Application of PYS to Organic Interfaces*

In the proceeding sections, the application of PYS to investigate the bulk electronic structures of various materials was described. As in the case of photoemission spectroscopy, PYS technique can be also applied to examine the interfacial electronic structures.

Figure 8.15 shows how to investigate the electronic structure of *metal on organic* interface by PYS. Here, small thickness region is investigated and band bending effect is neglected. At first, an organic layer is investigated by PYS, and from the ionization threshold, an energy diagram of organic layer can be drawn as Fig. 8.15a. Next metal layer is evaporated on the organic film. Here, the interfacial dipole layer is assumed to induce a vacuum level shift, Δ. The energy diagram of the system is shown in Fig. 8.15b. The Fig. 8.15c is overview of the expected spectra. The PYS of *metal on organic* system can be regarded as the superposition of the spectra from the metal layer and organic layer, leading to two threshold structures in the spectrum. For the spectral component from the metal layer, the threshold energy corresponds to the energy separation between the Fermi level of the metal overlayer and the vacuum level, i.e. the work function of the metal, $\Phi_m$. On the other hand, for the components from the organic layer, the threshold energy coincides to the energy separation between the vacuum level of the surface and the HOMO, i. e. $I + \Delta$.

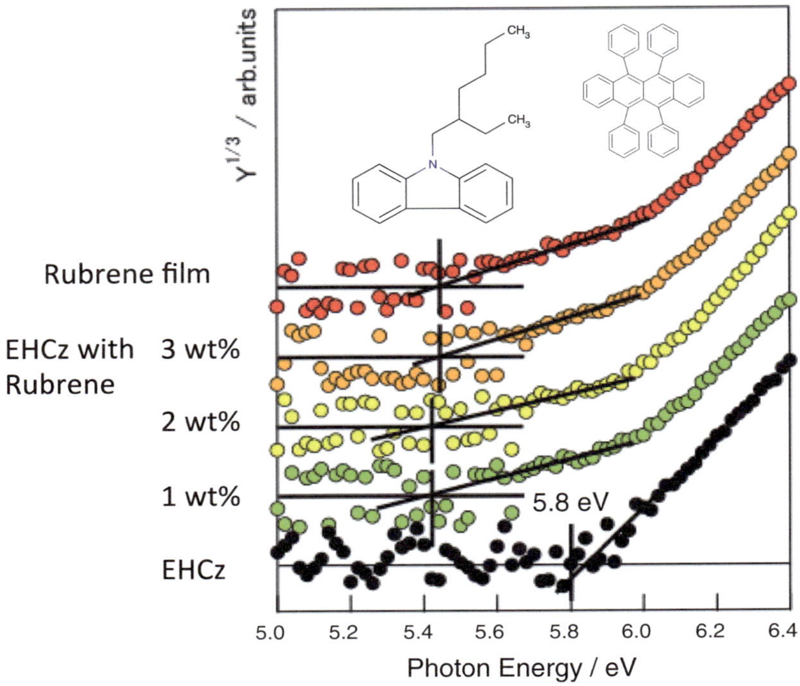


**Fig. 8.14** The PYS spectra of EHCz liquid (*bottom*), EHCz with doped rubrene, and rubrene film (*upper*). The chemical structures of EHCz and rubrene are in the *inset*

Thus, from the change of the higher threshold energy, the vacuum level shift, Δ, is determined, and from the energy separation between the lower and higher thresholds the hole injection barrier, $\varepsilon_V^F$, can be obtained by neglecting band bending effect. It should be noted that the energy reference in PYS is always the vacuum level of the sample film, while the Fermi level of sample holder (or electron analyzer) for photoemission spectroscopy.

Figure 8.16 shows an example for the case of *metal on organic* interface; Ag on rubrene single crystal. Figure 8.16a shows the variation of PYS spectra of rubrene single crystal with Ag overlayer. The bottom spectrum is for rubrene single crystal, indicating one threshold structure. With increasing Ag coverage, additional threshold structure appears on the lower energy side, which corresponds to the Fermi level of Ag layer. From the energy separation between the lower and higher thresholds, the hole injection barrier in vacuum can be estimated as 0.8 eV. This estimation has possible error, because there remains ambiguity to determine the higher threshold. The estimation by using data fitting is necessary, but here we demonstrate just the principle. After the measurement in vacuum, the PYS is performed in air. As shown in Fig. 8.16c, the spectrum is changed due to air atmosphere; the lower threshold shifted to higher energy side, while the higher threshold to lower. This result demonstrated that the hole injection barrier in air is 0.2 eV lower than that in

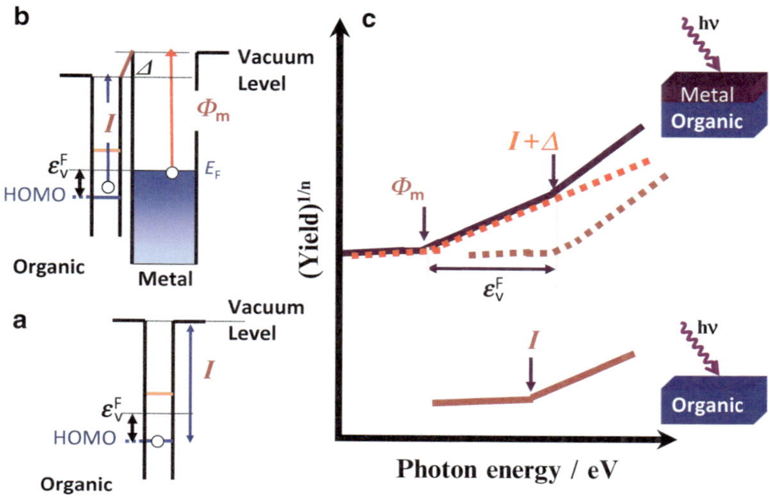


**Fig. 8.15** The principle of PYS measurement to investigate the electronic structure at *metal on organic* interface. (**a**) The energy diagram for an organic layer, (**b**) the energy diagram of *metal on organic* interface, (**c**) overview of the expected PYS spectra

vacuum; the energy diagram in air is shown in Fig. 8.16d. If the sample is evacuated and measured in vacuum again, the hole injection barrier increased a little bit. This indicates that direct measurement of interfacial electronic structure in air is necessary to examine the atmospheric effect to the electronic structure. The comparison between the spectra measured in vacuum before and after air-exposure, which is conventional method to examine atmospheric effect, cannot fully reflect the air effect.

This type of PYS measurement can be also applied to explore the electronic structure of *organic on metal* interface. Figure 8.17 shows how to investigate the electronic structure of *organic on metal* interface by PYS. First, a metal substrate is measured by PYS. The work function of the metal is determined from the threshold energy, and an energy diagram of metal substrate can be drawn as Fig. 8.17a. Next an organic layer is evaporated on the metal substrate. Here, the interfacial dipole layer is assumed to induce a vacuum level shift, Δ. The energy diagram of the system is shown in Fig. 8.17b. The Fig. 8.17c is overview of the expected spectra. The PYS of *organic on metal* system can be regarded as the superposition of the spectra from the organic layer and metal substrate, leading to two threshold structures in the spectrum. For the spectral component from the organic layer, the threshold energy corresponds to the energy separation between the HOMO of the organic layer and the vacuum level, i.e. the ionization energy, *I*. On the other hand, for the components from the metal substrate, the threshold energy coincides to energy separation between the vacuum level of the surface and the Fermi level of the substrate, i. e. $\Phi_m + \Delta$. Thus, from the shift of lower threshold energy, the vacuum level shift, Δ, is determined, and from the energy separation between the

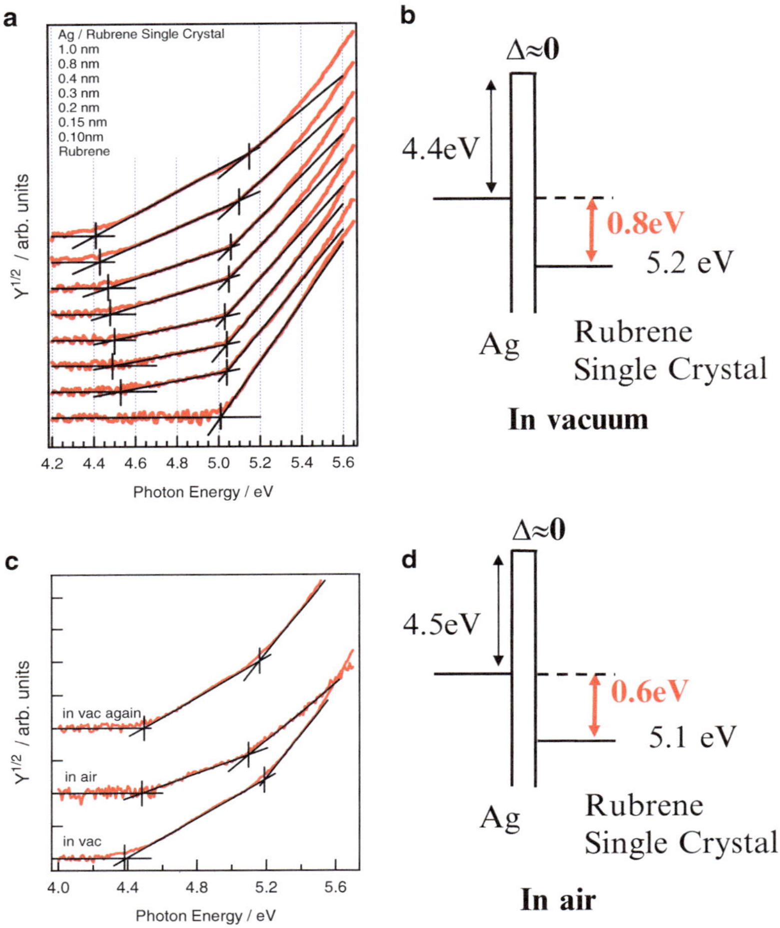


**Fig. 8.16** (**a**) The PYS spectra of Ag on rubrene single crystal for various coverage of Ag measured in vacuum. (**b**) The interfacial energy diagram obtained from the PYS in vacuum. (**c**) The spectral change of Ag on rubrene single crystal in vacuum and air. (**d**) The interfacial energy diagram estimated from the spectrum measured in air

lower and higher thresholds the hole injection barrier can be obtained by neglecting band bending effect.

Figure 8.18 shows an example for the case of *organic on metal* interface; evaporated rubrene film on Au substrate. Figure 8.18a shows the variation of UPS spectra of rubrene film on Au. The photon energy was 8.0 eV. With increasing thickness, the secondary cut-off showed 0.4 eV shift due to the formation of interfacial dipole. From the energy onset of HOMO peak (marked by an arrow) the hole injection barrier was determined as 1.2 eV. The obtained energy diagram was shown in Fig. 8.18b. PYS measurement was performed for the same sample,

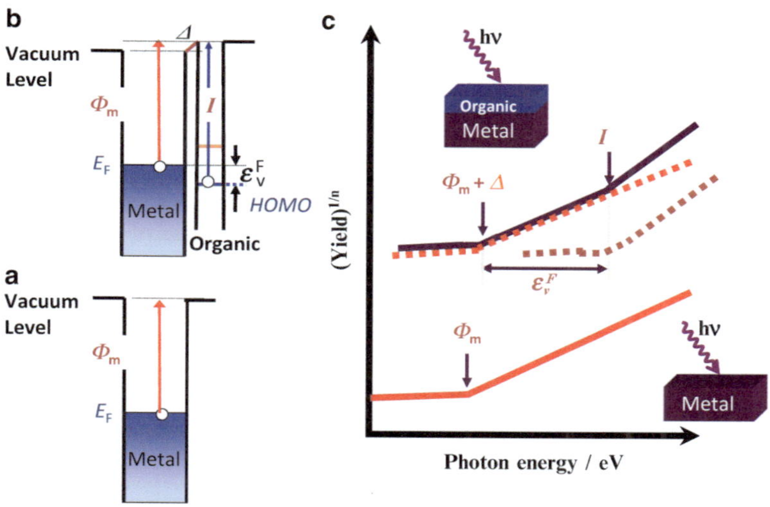


**Fig. 8.17** The principle of PYS measurement to investigate the electronic structure at *organic on metal* interface. (**a**) The energy diagram for a metal substrate, (**b**) the energy diagram of *organic on metal* interface, (**c**) overview of the expected PYS spectra

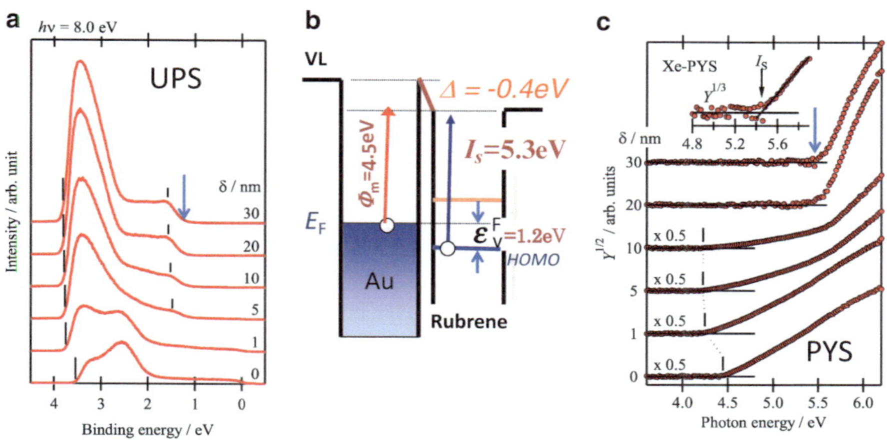


**Fig. 8.18** An example of analysis of PYS for *organic on metal* interface. (**a**) UPS spectra of rubrene on Au substrate, (**b**) Interfacial energy diagram obtained from UPS, (**c**) PYS spectra of rubrene on Au substrate. Reproduced from [24] with permission of the Japan Society of Applied Physics

and the spectra were shown in Fig. 8.18c. The bottom spectrum is for Au substrate. The onset corresponds to the initial work function of Au substrate. With increasing rubrene thickness, the onset suddenly shifted to lower energy side, indicating the downward shift of vacuum level due to the formation of interfacial dipole moment.

The amount of shift (ca. 0.3 eV) is similar to that in UPS. For above 5 nm thickness, the higher threshold became clear. The threshold is about 5.4 eV, which is close to the ionization energy obtained from UPS. Finally, the energy separation between the lower and higher thresholds, the hole injection barrier can be estimated as 1.2 eV, corresponding to that from UPS. This result demonstrated that PYS can be actually applied to investigate the interfacial electronic structures. Similar result was reported also for other organic system [37]. Although the precision to quantitatively determine the barrier height is limited, the advantage of PYS is useful to understand the actual organic electronics.

## 8.6 Analysis of an Organic Interface by Combining PYS and low Energy High Sensitivity UPS

So far, we described on the application of PYS to investigate the electronic structures of organic materials. By using electron energy analyzer, the combination of PYS and low energy high sensitivity UPS measurements can be easily realized. As described in the introduction, both methods are complimentary to each other, and the combination of UPS and PYS is useful. In addition, if the monochromatized light source has negligibly small contribution of stray photon, the high sensitivity photoemission measurement as described in Chap. 4 is also available. Here, we briefly describe an example by our PYS/UPS system to investigate extremely weak spectral feature of rubrene/$C_{60}$ interface.

Figure 8.19 shows the PYS and UPS spectra of $C_{60}$/rubrene interfaces [24]. In Fig. 8.19a, the bottom spectrum is for rubene film. $C_{60}$ was deposited upto 118 nm. The HOMO peak of rubrene shifted to the lower binding energy ($E_b$) side by 0.13 eV at 30 nm coverage of $C_{60}$. An additional peak derived from $C_{60}$ HOMO appeared at $E_b = 2$–4 eV at 2.5 nm coverage. Surprisingly, it hardly grew with increasing $C_{60}$ coverage up to 30 nm and also the intensity of the HOMO peak of rubrene did not correspondingly decrease. The HOMO peak of $C_{60}$ dramatically grew after further 30 nm deposition, indicating that $C_{60}$ molecules do not grow in the layer-by-layer mode on the rubrene film. Because the stray light from the excitation light source is negligible small, reliable spectra can be obtained even for very weak structure. As seen in the log scale graph [Fig. 8.19b], spectra have a feature around $E_F$ ($E_b = 0$). At the small coverage of $C_{60}$, the Fermi-edge appeared again and the intensity showed the maximum at the coverage of 1 nm (see inset). Further deposition of $C_{60}$ decreased the intensity of the Fermi-edge and it completely vanished at the coverage of 60 nm. This feature was also observed by PYS, as shown in Fig. 8.19c. The photoelectron threshold energy was around 4.4 eV and was smaller than the $I_s$ of rubrene at coverages of 1 and 2.5 nm. PYS and LE-UPS results strongly suggest that the photoemission signal from the Au substrate revived upon $C_{60}$ deposition onto the rubrene film, which implies enhancement of the surface roughness of the film. It should be emphasized that

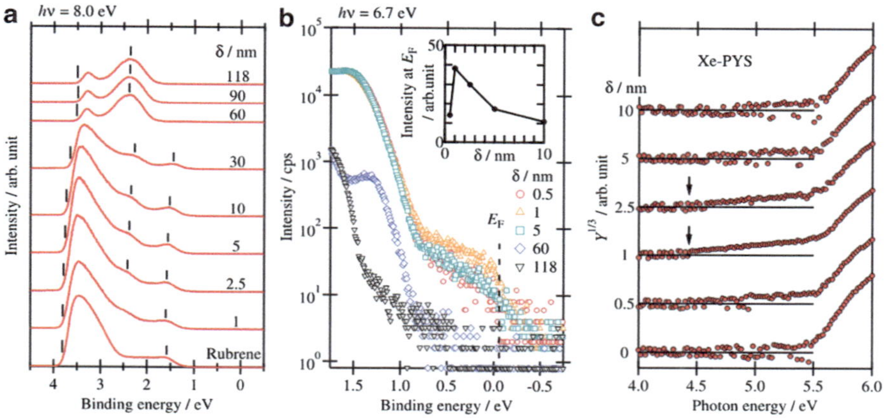


**Fig. 8.19** (**a**) LE-UPS spectra of the $C_{60}$/rubrene interface. The energy positions of the SECO and HOMO of rubrene and $C_{60}$ are indicated by *vertical bars*. (**b**) Log scale LE-UPS spectra around Fermi-edge. (*Inset*) Fermi-edge intensity determined from the least-squares fitting with Fermi–Dirac distribution function at each coverage. (**c**) PYS spectra of an identical sample. Reproduced from [24] with permission of the Japan Society of Applied Physics

this change is too small to see in the ordinary linear scale because the intensity of the Fermi-edge at the coverage of 1 nm $C_{60}$ is ca. 0.3 % of the initial one. Log-scalability by an extremely suppressed background level enables one to detect the small signal, which might otherwise not be noticed. The delayed evolution of the $C_{60}$-derived photoemission signal is ascribed to the finite diffusion of $C_{60}$ into the rubrene layer at the initial phase of interface formation. This type of high-sensitivity UPS and PYS measurements can be applied to examine other interfacial electronic structures, and the observed weak distribution of density-of-states can be related to injection property at organic semiconductor/electrode interface [38].

## 8.7 Summary

In this chapter, the basics and application of photoelectron yield spectroscopy were described. PYS is complimentary method to conventional photoemission spectroscopy. PYS can be available not only in vacuum but also in atmospheric condition. Because the performance of organic electronic devices is often affected by air, the direct observation of electronic structure of organic semiconductors in air is necessary to understand the practical device performance. Non-vacuum measurement can open the way to examine the electronic structures of liquid systems, where the investigation by photoemission spectroscopy has been much limited. PYS has another advantage; durability to sample charge-up. In principle, sample charging can be avoided in the case of PYS by applying a bias voltage to extract photoelectrons from sample. Even for electrically isolated film on mica, PYS measurement is possible. This advantage gives us a chance to examine more practical system such as organic film on gate insulator in organic transistor. We also demonstrated that PYS can be available to explore interface, too. Finally, the combination of PYS and UPS by using the same light source is fruitful to investigate various aspects of electronic structures of bulk and interface of organic semiconductors including weak density of states. We expect that PYS can be more widely-available and indispensable technique for organic semiconductor research.

**Acknowledgments** This work was supported by the Global-COE program at Chiba University (Advanced School for Organic Electronics, G-03, MEXT), KAKENHI (Grants No. 21245042, 22750167, 25288114), and the Funding Program for World-Leading Innovative R&D on Science and Technology (FIRST).

## References

1. Y.F. Liew, H. Aziz, N.X. Hu, H.S. Chan, G. Xu, Z. Popovic, Appl. Phys. Lett. **77**, 2650 (2000)
2. D. Xu, C. Adachi, Appl. Phys. Lett. **95**, 053304 (2009)
3. M. Faubel, in *Photoelectron Spectroscopy at Liquid Surfaces Chap. 12*, ed. by C-Y. Ng. Photoionization and Photodetachment (World Scientific, Singapore, 1999)
4. M.P. Seah, in *Practical Surface Analysis*, 2nd edn., vol. 1, ed. by D. Briggs, M.P. Seah (Wiley, New York, 1990), Appendix 2
5. J. Cazaux, J. Electron Spectros. Relat. Phenomena. **105**, 155 (1999)
6. M. Pope, C.E. Swenberg, *Electronic Processes in Organic Crystals and Polymers*, 2nd edn. (Oxford Science, New York, 1999). Chap. IV
7. C.N. Bergulund, W.E. Spicer, Phys. Rev. **136**, A 1030 (1964)
8. R.H. Fowler, Phys. Rev. **38**, 45 (1931)
9. C.R. Crowell, T.W. Kao, C.L. Anderson, V.L. Rideout, Surf. Sci. **32**, 591 (1972)
10. E.O. Kane, Phys. Rev. **127**, 131 (1962)
11. M. Kochi, Y. Harada, T. Hirooka, H. Inokuchi, Bull. Chem. Soc. Jpn. **43**, 2690 (1970)
12. J.M. Ziman, *Principles of the Theory of Solids* (Cambridge University Press, Cambridge, 1972)
13. C.A. Sebenne, II Nuvo Cimento **39**, 768 (1977)
14. R.P. Winch, Phys. Rev. **37**, 1263 (1931); Mendenhall, Science **73**, 107 (1931)
15. G.W. Gobeli, F.G. Allen, Phys. Rev. **127**, 141 (1962)
16. K. Winter, L. Ley, Phys. Rev. **B36**, 6072 (1987)
17. M. Sebastiani, L. Di Gapare, G. Capellini, C. Bittencourt, F. Evangelisti, Phys. Rev. Lett. **75**, 3352 (1995)
18. S. Miyazaki, T. Maruyama, A. Kono, M. Hirose, Microelectron Eng **48**, 63 (1999)
19. H. Kirihata, M. Uda, Rev. Sci. Instrum. **52**, 68 (1981)
20. T. Smith, J. Appl. Phys. **46**, 1553 (1975)
21. K. Inumaru, Y. Okubo, S. Yamanaka, Chem. Lett. **8**, 741 (1999)
22. S. Grigalevicius, G. Blazys, J. Ostrauskaite, J.V. Grazulevicius, V. Gaidelis, V. Jakauskas, E. Montrimas, Synth. Met. **128**, 127 (2002)
23. Y. Nakayama, S. Machida, D. Tsunami, Y. Kimura, M. Niwano, Y. Noguchi, H. Ishii, Appl. Phys. Lett. **92**, 153306-1 (2008)
24. S. Machida, Y. Ozawa, J. Takahashi, H. Tokairin, Y. Nakayama, H. Ishii, Appl. Phys. Express **6**, 025801 (2013)
25. Y. Nakayama, S. Machida, T. Minari, K. Tsukagoshi, Y. Noguchi, H. Ishii, Appl. Phys. Lett. **93**, 173305 (2008)

26. J. Takeya, M. Yamagishi, Y. Tominari, R. Hirahara, Y. Nakazawa, T. Nishikawa, T. Kawase, T. Shimoda, S. Ogawa, Appl. Phys. Lett. **90**, 102120 (2007)
27. Y. Harada, T. Takahashi, S. Fujisawa, T. Kajiwara, Chem. Phys. Lett. **62**, 283 (1979)
28. O. Mitrofanov, C. Kloc, T. Siegrist, D.V. Lang, W.-Y. So, A.P. Ramirez, Appl. Phys. Lett. **91**, 212106 (2007)
29. K. Komatsu, Master thesis, Nagoya University, 2003
30. M. Honda, K. Kanai, K. Komatsu, Y. Ouchi, H. Ishii, K. Seki, J. Appl. Phys. **102**, 103704 (2007)
31. S. Matsusaka, H. Maruyama, T. Matsuyama, M. Ghadiri, Chem. Eng. Sci. **65**, 5781 (2010)
32. T. Sato, Y. Nakayama, H. Ishii, to be published
33. B. Winter, M. Faubel, Chem. Rev. **106**, 1176 (2006)
34. I. Watanabe, Bunseki **3**, 102 (2006) (in Japanese)
35. T. Yamashita, T. Miyauchi, Y. Nakayama, H. Ishii, to be published
36. J. Kido, H. Shionoya, K. Nagai, Appl. Phys. Lett. **67**, 2281 (1995)
37. K. Kanai, M. Honda, H. Ishii, Y. Ouchi, K. Seki, Org. Electron. **13**, 309 (2012)
38. Y. Ozawa, H. Kinjo, T. Sato, Y. Nakayama, H. Ishii, to be published

# Part II
# Organic Devices and Their Properties

# Chapter 9
# Fabrication and Characterization of Organic Devices

**Kazuhiro Kudo and Masatoshi Sakai**

## 9.1 Basic Electrical Properties of Organic Films

The costs associated with inorganic single crystal semiconductors can be avoided by using organic semiconductors, which are suitable for low-cost electronic devices[1–3]. Organic semiconductors have enormous potential for optoelectronic applications, and would be significantly cheaper to produce than either single crystal and amorphous silicon (Si) equivalents. The development and encapsulation of organic electronic devices onto plastic substrates by large-scale and low-cost roll-to-roll production processes will allow their market implementation in numerous application areas, including organic light-emitting diode (OLED) displays, white lighting, solar cells, radio-frequency identification (RFID) tags, and bio-sensors, and pave the way to a new generation of unique applications. While the technology is novel, it can also be harnessed in many current applications, providing reduced cost and low energy consumption. Novel solution-processable organic semiconductor materials combined with innovative manufacturing techniques such as printing processes enable the production and development of many valuable and low-cost applications. Because of the lower cost and higher manufacturing throughput of organic-based electronic devices compared to today's Si-based devices, organic electronics also promise to expand the use of electronics technology in resource-limited areas of the world where supplies are limited or the necessary infrastructure is lacking. Our research

K. Kudo (✉) • M. Sakai
Graduate School of Engineering, Chiba University, 1-33 Yayoi-cho, Inage, Chiba 263-8522, Japan
e-mail: kudo@faculty.chiba-u.jp; sakai@faculty.chiba-u.jp



H. Ishii et al. (eds.), *Electronic Processes in Organic Electronics*,
Springer Series in Materials Science 209, DOI 10.1007/978-4-431-55206-2_9

focuses on exploring new device structures and fabrication processes such as vertical-type [4–6] and molecular wires [7, 8], as well as conventional planar-type organic field-effect transistors (OFET) [9, 10]. Device structures suitable for organic materials could lead to new designs for next-generation electronic devices, which could successfully replace traditional Si technologies. Our research employs different approaches to new functional devices, such as molecular wire devices and metal-insulator transition switching devices. We are able to demonstrate the new electronic functions of organic materials through these device physics. In addition, our in situ measurements verify their reliability and reproducibility on the basis of the agreement with measurements on actual devices from inventive molecular functions to practical large-area printed devices. Researchers initially used basic materials like pentacene and metallophthalocyanine, but then considered various alternatives that seemed likely to outperform them on account of their superior chemical and electrical properties. In this section, we will provide a brief overview of the fabrication process of organic devices, focusing mainly on organic transistor applications. First, we will report some of the latest advances in the fields of organic transistor structures and the fabrication and characterization of functional thin molecular films. Moreover, we will discuss the in situ optical and electrical monitoring and characterization of film fabrication. Finally, the potential for the development of organic electronic devices will be assessed to establish a framework for the achievement of future scientific and technological breakthroughs.

## 9.2 Fabrication Processes for Organic Films

In general, organic semiconductors are deposited from either the vapor or solution phase, depending on their vapor pressure and solubility. Device performance is governed by molecular structure and thin film morphology, which in turn are strongly influenced by various deposition conditions. To achieve high field-effect mobility, the orientation of the semiconducting molecules must be such that the $\pi$-$\pi$ stacking direction of the molecules is aligned with the current flow direction. Moreover, larger grain sizes and smoother grains tend to give better mobilities.

### *9.2.1 Physical Vapor Deposition*

The most widely used physical vapor deposition method is thermal evaporation using a resistive heating source under a vacuum of $10^{-6}$ to $10^{-4}$ Pa. A number of organic semiconductors have been deposited using the vacuum thermal evaporation method. The method is suitable for small organic molecules insoluble in common solvents, such as merocyanine [9], metallophthalocyanines [10, 11], acenes (pentacene and tetracene) [12, 13], and oligothiophene derivatives [14, 15]. Vacuum thermal evaporation has the advantage of forming highly uniform films with good reproducibility. Multilayer deposition and co-deposition of several organic semiconductors are possible. Other vacuum-based thin film deposition methods in use include organic vapor phase deposition, organic molecular beam deposition, and laser evaporation. Organic vapor phase deposition proceeds by evaporation of the molecular source material into a stream of hot inert carrier gas such as nitrogen or argon, which transports the vapor toward a cooled substrate.

### 9.2.2 Solution Deposition

Deposition methods based on solution-processable materials are compatible with large-area thin film fabrication and thus result in a lower production cost per device. Both spin-coating and solution casting are commonly used for solution deposition [16–18]. In the spin-coating method, the solvent dries relatively fast, allowing less time for molecular ordering compared to solution casting. However, film uniformity is usually better in the case of spin-coated films. It was found that for materials with a tendency to exhibit highly ordered molecular packing, even spin coating can achieve very high mobilities [19]. Solution deposition processes, such as screen printing and ink-jet printing [20–23], also allow printing of the active materials, whereby deposition and patterning are accomplished in a single step. In solution deposition, the quality of the resulting semiconductor films is strongly influenced by the semiconductor concentration, solvent evaporation rate, semiconductor solubility, and substrate surface properties. High mobility values on the order of 20–40 $cm^2/Vs$ are obtained with solution-deposited materials such as BTBT derivatives [23, 24]. One solution process, the electrospray deposition (ESD) method, is a simple process that uses a solution spray containing small drops formed by the electric field between the nozzle and conductive substrate [25, 26]. This method is suitable for fabricating uniform and large-area films, but cannot handle patterned deposition like ink-jet printing. Nevertheless, rough patterning is possible using a modified ESD method: By applying the electric field between the metal nozzle and selected patterned electrodes, selective deposition can be achieved around the gate lines on the substrate.

### 9.2.3 Thermal Press and Thermal Lamination Methods

Many kinds of printing methods have been developed for the fabrication of large-area flexible electronics. These printing methods, while considered promising for

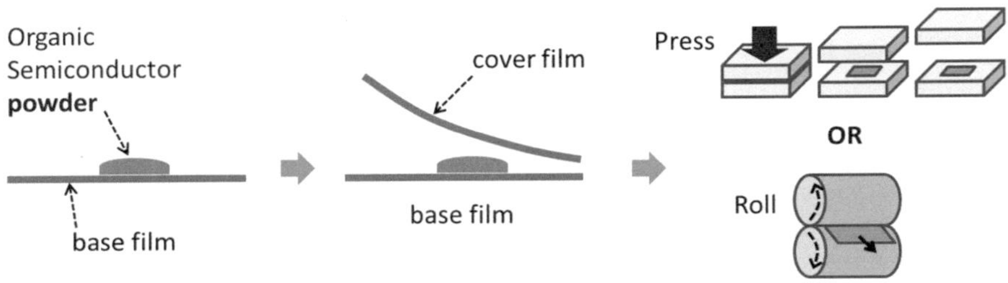


**Fig. 9.1** Basic concept of thermal press and thermal lamination methods

the industrial fabrication of flexible electronic devices, involve the use of toxic solvents. Thus, it has become necessary to implement toxicity reduction processes in order to dispose of these solvents and their vapors, which have a negative impact on the environment. The costs associated with such post-processing increases the total cost of industrial production. As such, a novel method for the fabrication of organic semiconductor devices that utilizes direct melting and subsequent recrystallization of the organic materials is proposed as thermal press [27] and thermal lamination methods [28], which can complement other industrial processes. The basic concept behind these methods is illustrated in Fig. 9.1. They requires no solvents to form the semiconductor layers and offer the advantage of imparting structural bending stability [29]. In addition, it is expected to allow large crystalline domain growth of organic materials in a two-dimensional confined gap (see paragraph after next). Moreover, materials suitable for this process are also suitable for high purification methods such as zone refining [30–34], which will be highly effective for obtaining device-grade materials. Furthermore, thermal lamination becomes applicable to high through-put roll-to-roll printing processes by exploiting toner technology to form a pattern of the organic powder on a plastic film.

In the thermal press [27], the two substrates (base and cover film in Fig. 9.1) with the organic semiconductor materials sandwiched between them are first clamped under a small load and heated to the melting point of the semiconductor. After the semiconductor melts, the sample is compressed under a pressure of 0.5–1 MPa to make a thin semiconductor layer. The first sample fabricated by the thermal press process was a $TTC_{18}$-TTF thin film. $TTC_{18}$-TTF FETs with glass substrates were fabricated to observe the growth of the crystal grains [27]. Figure 9.2 shows polarized optical micrographs of the grains grown. The open Nicols micrographs shown in Fig. 9.2a confirms that the $TTC_{18}$-TTF layer was formed homogeneously by melting and recrystallization. The crossed Nicols micrographs shown in Fig. 9.2b reveal that the organic layer is polycrystalline and has grains approximately $20 \times 5\,\mu m$ in size, which is larger than the crystal grains grown by the solution cast method. The thickness of the organic crystal was approximately 60–80 nm.

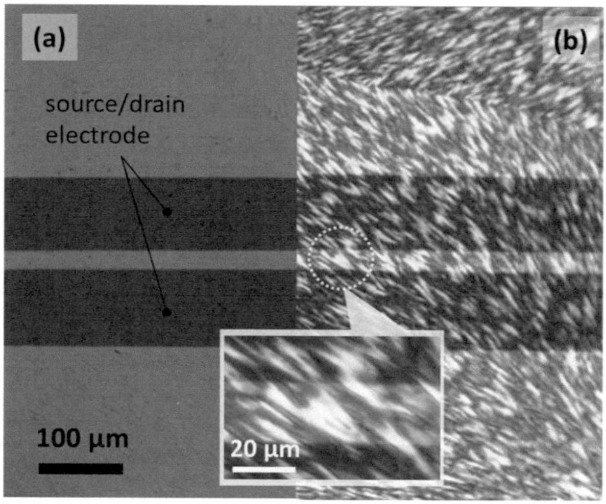


**Fig. 9.2** (**a**) Open and (**b**) crossed Nicols optical micrograph of $TTC_{18}$-TTF layer fabricated using the thermal press method. Copyright 2013, WILEY-VCH

**Fig. 9.3** Crossed Nicols optical micrograph of a perylene derivative crystalline layer fabricated by the thermal press method

As seen in the $TTC_{18}$-TTF case, large grain growth in a two-dimensional (2D) confined gap is expected to be possible. One significant example was obtained in a perylene derivative whose melting point was approximately 180°C. A large grain with a diameter of over 1 mm was obtained in a 2D confined gap (Fig. 9.3). Although the ranges of applied pressure and temperature are wide in the thermal press method, the throughput is not so high. When we consider practical organic device fabrication and the use of flexible substrates, having a high throughput and low energy input is rather important. Therefore, we proposed the thermal lamination process.

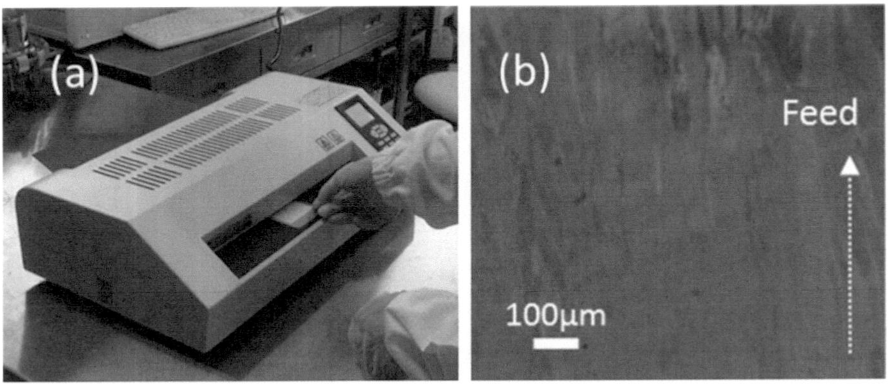


**Fig. 9.4** (**a**) The desktop laminator used in the thermal lamination method. (**b**) Crossed Nicols optical micrograph of grown $C_8$-BTBT crystalline films

In the thermal lamination process [28], we utilized an office-type desktop laminator (Fig. 9.4a), which had three pairs of rollers. The first two pairs were heat rollers, which served to melt the organic materials, while the third pair's function was to pull and cool the sample. The variable experimental parameters of the desktop laminator were the feed speed (0.4–1.5 m/min) and the roller temperature (80–160°C).

Figure 9.4b shows a crossed Nicols optical micrograph of the dioctylbenzothienobenzothiophene ($C_8$-BTBT) thin layer that was fabricated by the thermal lamination process. The sample was laminated under a heat roller temperature of 160°C and at a feed speed of 0.4 m/min. The feed direction is indicated by arrows in Fig. 9.4b. The needle-like $C_8$-BTBT grains are oriented parallel to the feed direction. This oriented growth may be due to one of two possible mechanisms: (i) crystal growth along the flow direction of the organic melt and (ii) crystal growth along the direction of movement of the melting zone, similar to zone melting methods. We concluded that the crystal orientation was mainly governed by the flow direction.

## 9.3 Characterization of Organic Films

### *9.3.1 Experimental Details of In Situ Field-Effect Measurements*

Generally, the electrical characteristics of organic semiconductors are strongly influenced by the measurement environment. For example, n-type characteristics disappear when the devices are measured in air after vacuum evaporation. Thus, the observed electrical properties are attributed to the experimental ambient, i.e., oxygen, humidity, etc [35]. For practical device applications, both p- and n-type semiconducting materials with high stability against air are desirable. As such, it is important to investigate the intrinsic electrical properties of organic semiconductor films before and after exposure to atmospheric gasses, especially oxygen. In situ field-effect measurements [9, 10] are promising for determining the conduction type (p or n), carrier mobility ($\mu$), electrical conductivity ($\sigma$), and carrier concentration ($N$) of evaporated films [8]. This method is advantageous in that it allows basic electrical parameters of even very thin films [7] to be evaluated, unlike time-of-flight (TOF) [36] measurements which require samples thicker than 1 $\mu$m, in the absence of ambient gases. The field-effect mobility of organic semiconductors determined by in situ field-effect measurements was for the first time reported for a merocyanine film [9]. Our research group investigated many kinds of organic semiconductor films using this method. Schematics of the in situ field-effect measurement system and the sample structure are shown in Figs. 9.5 and 9.6. The highly doped Si substrate, which acts as a gate electrode, was covered with a thermally grown $SiO_2$ film having a thickness of approximately 200 nm. The source and drain electrodes were fabricated on the substrate using standard vacuum evaporation and photolithographic techniques. The metallic components of the source and drain electrodes, i.e., Au for p-type organic films and In for n-type films, were chosen to make ohmic-like contacts to the organic materials. The typical channel length, $L$, and width, $W$, were 20 $\mu$m and 5 mm, respectively. The substrates were thoroughly cleaned with organic solvent in an ultrasonic bath and prebaked at 373 K for over 30 min in a vacuum chamber. Subsequently, organic semiconductor materials were evaporated as the active component of the FET. During evaporation, the substrate temperature, $T_G$, was adjusted according to the organic material from room temperature to 100°C (373 K). The typical thickness of the organic films was

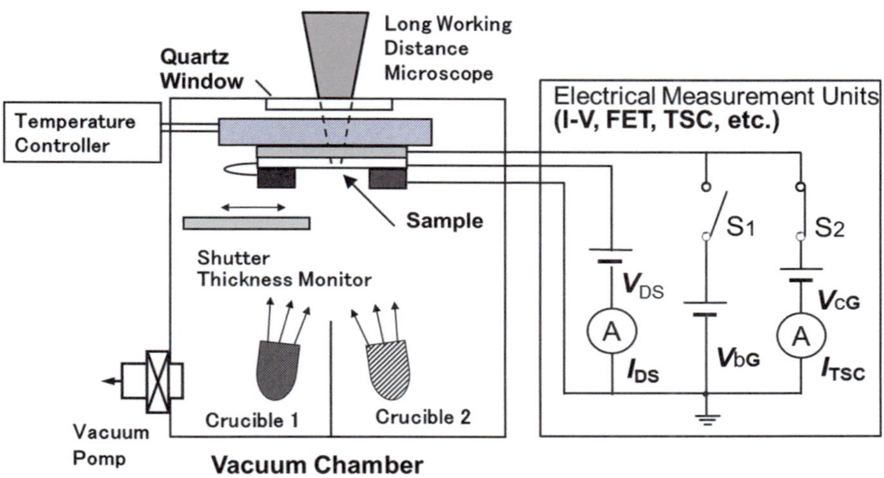


**Fig. 9.5** Schematic illustration of the in situ measurement system

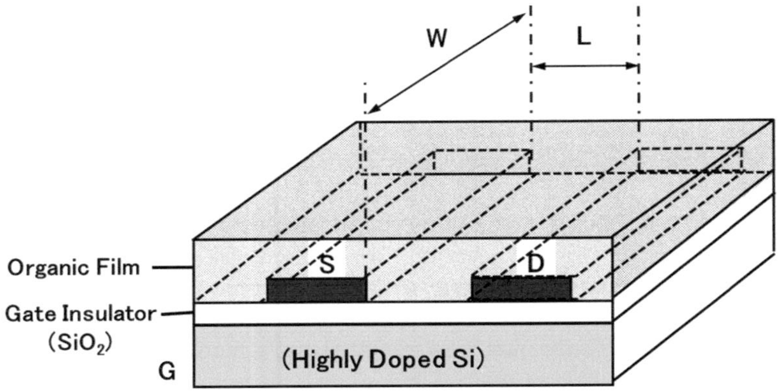


**Fig. 9.6** Device structure of OFET

approximately 200 nm. Field-effect measurements were performed immediately after the evaporation of the organic thin films, after a 5 h exposure to oxygen gas, and after a 1 h thermal anneal at 373 K in vacuum ($10^{-4}$ Pa). All electrical measurements were carried out in the dark. The characteristics of the source-drain current ($I_{DS}$) vs. source-drain voltage ($V_{DS}$) were measured under an applied gate voltage ($V_G$). $I_{DS}$ in the linear and saturation regions for a standard TFT is given by

$$I_D = \frac{\mu W C_i}{L} V_D \left( \frac{V_G - V_T - V_D}{2} \right), \tag{9.1}$$

$$I_D^{sat} = \frac{\mu W C_i}{2L} (V_G - V_T)^2, \tag{9.2}$$

respectively [37], where $C_i$ is the capacitance of the $SiO_2$ layer and $V_{th}$ is the threshold voltage. According to Eqs. (9.1) or (9.2), $\mu$ can be obtained from the transfer curve of the FET characteristics. The electrical conductivity, $\sigma$, was obtained from the slope of the $I_{DS}$-$V_{DS}$ plot at zero gate electric field ($V_G = 0$). Strictly speaking, $V_G$ should be set to the flat band potential because $\sigma$ is modified by the $V_{th}$ of FET. $\sigma$ is given by

$$\sigma = \frac{L}{Wd} \frac{I_{DS}}{V_{DS}}, \tag{9.3}$$

$\sigma$ is also expressible as

$$\sigma = qN\mu, \tag{9.4}$$

where q is the elementary electric charge and $N$ is the carrier concentration. $N$ was obtained from Eqs. (9.3) and (9.4). Phthalocyanine (Pc) films have been expected to find application as gas sensors [38, 39], and the coordinated metals in Pc are an important factor determining their electrical properties. It is necessary to investigate the relationship between their chemical structure and intrinsic electrical properties, especially in the absence of atmospheric gasses and impurities. As Pc derivatives, copper phthalocyanine (CuPc), lead phthalocyanine (PbPc), metal-free phthalocyanine ($H_2Pc$), and fluorophthalocyanine ($F_{16}CuPc$) were examined here, and the n-type behavior of $F_{16}CuPc$ was reported [40, 41].

Typical FET characteristics ($I_{DS}$ vs. $V_{DS}$ as a function of $V_G$) of a CuPc sample after deposition at a $T_G$ of 373 K are shown in Fig. 9.7a. $I_{DS}$ increases with negative $V_G$, and the CuPc FET operates in enhancement mode. This result indicates that negative gate voltages enlarge the conduction channel owing to the formation of a hole accumulation layer, so that CuPc films show p-type semiconducting properties. We have also investigated the effect of oxygen gas and annealing on the electrical properties of the films. Although $I_{DS}$ increases upon oxygen exposure, $I_{DS}$ decreases by the 1 h thermal treatment at 373 K in vacuum. It was also confirmed that $H_2Pc$ and PbPc showed p-type semiconducting properties in the absence of atmospheric gasses. However, the effect of oxygen gas on the FET characteristics depends on the molecular species. In terms of the FET characteristics of these p-type materials, the effect of oxygen on the PbPc film is significant, while that on the $H_2Pc$ films is small [10].

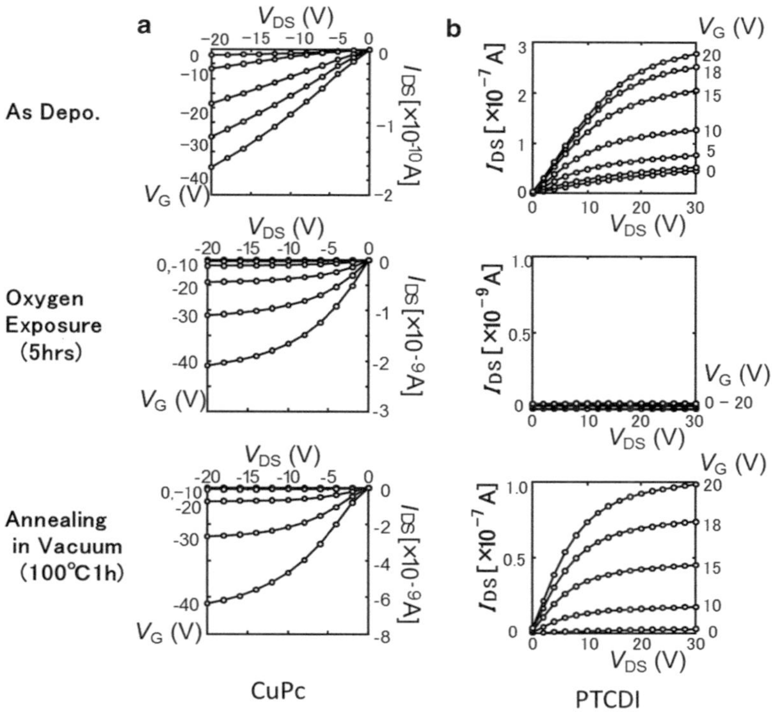


**Fig. 9.7** In situ field-effect measurement of CuPc and PTCDI evaporated films

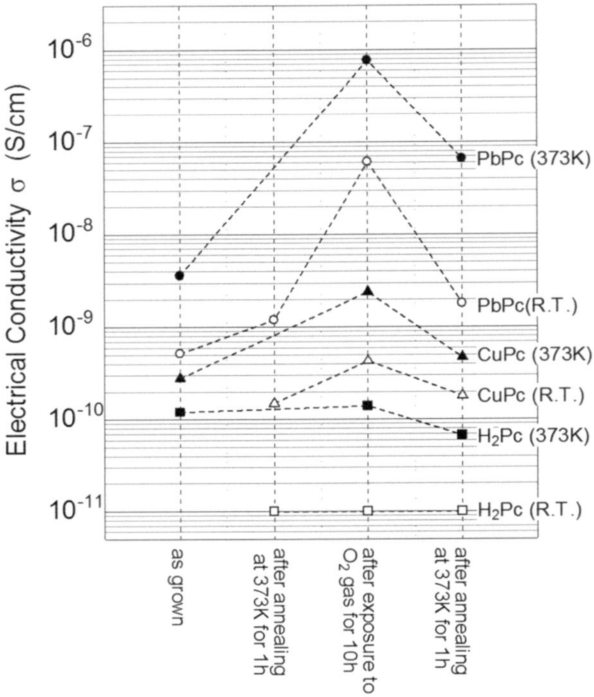


**Fig. 9.8** Variation of electrical conductivity in $H_2Pc$, CuPc and PbPc evaporated films. The substrate temperature during vacuum deposition is shown in the figure. Copyright 1997, The Japan Society of Applied Physics

Several kinds of perylene derivatives were reported as having n-type characteristics [42]. Figure 9.7b shows typical OFET characteristics of perylene derivative films. The channel conduction increases with positive $V_G$, and the OFET operates in enhancement mode. According to the results of field-effect measurements, the as-deposited films showed n-type semiconducting properties in the absence of atmospheric gasses. Oxygen gas decreases the conductivity in n-type films; the extent of the decrease depends on the molecular species. The FET characteristics obtained just after deposition were completely ruined by the exposure to oxygen gas and did not recover to the values of the as-grown sample. These findings can be explained as follows: Oxygen gas acts as an acceptor impurity and carrier compensation of the majority carriers (electrons) occurs in n-type films. The variation in electrical properties, however, depends on the molecular structure and growth conditions of the films. It is considered that electron transfer from Pc molecule to oxygen molecule and oxygen molecule is directly related to the composition of the central metal of the Pc molecule. This result is closely related to several reports that the interaction between $H_2Pc$ and oxygen is weak or that the adsorption site of $H_2Pc$ is different from that of metallophthalocyanine [35, 43]. The variation of electrical conductivity $\sigma$ in $H_2Pc$, CuPc, and PbPc films determined by in situ field-effect measurements is shown in Fig. 9.8. It is noteworthy that a significant increase in $\sigma$ occurs upon exposure of the PbPc film to oxygen gas and that $\sigma$ decreases upon annealing in vacuum. The effect of oxygen on $E_a(\sigma)$ depends on the chemical

**Table 9.1** Variation in thermal activation energy (eV) of PbPc, CuPc, and $H_2Pc$

| Material | As-grown | 1st annealing | After $O_2$ (5h) | 2nd annealing |
|---|---|---|---|---|
| PbPc | 1.12 | 0.53 | 0.23 | 0.54 |
| CuPc | 1.49 | 0.56 | 0.42 | 0.58 |
| $H_2Pc$ | 0.72 | 0.65 | 0.63 | 0.67 |

structure of the Pc molecules. While the $E_a$ of $H_2Pc$ remains almost constant, those of PbPc and CuPc decrease upon oxygen exposure. In particular, the $E_a(\sigma)$ of the PbPc film changed from 0.53 to 0.23 eV by introducing oxygen gas. After the thermal treatment in vacuum, $E_a(\sigma)$ became 0.54 eV, i.e., it nearly recovered to its initial value. Table 9.1 shows the variation in activation energy of PbPc, CuPc and $H_2Pc$ films. These results indicate that oxygen molecules interact with the PbPc molecules and an acceptor level is formed. Thus, oxygen gas acts as an acceptor impurity, increasing the net charge carriers. The ionization energies of PbPc, CuPc and $H_2Pc$ are reported to be 4.98, 5.17 and 5.20 eV, respectively [44]. This order agrees with the effect of oxygen gas on the electrical conductivity change obtained here.

## *9.3.2 Thermally Stimulated Current (Pyroelectric Current Measurement)*

Thermally stimulated current (TSC) is a conventional electrical measurement to investigate various polarization phenomena. There are many different types of TSC measurements. For example, pyroelectric current measurement is one variant that is used in ferroelectric materials to investigate the thermal dissociation of spontaneous polarization.

Electronic ferroelectricity due to charge-ordered phases has attracted considerable attention in a wide range of research areas [45–50]. In the quarter-filled charge-ordered phase, carriers are localized by on-site and inter-site Coulomb repulsion and form a stripe pattern of carrier distribution. Since this is a kind of electrically polarized electronic state, the charge distribution is modulated by an external electric field with increasing temperature.

The FET characteristics of $\beta'$- (BEDT-TTF)(TCNQ), known as an organic dimer Mott insulator [51, 52], were investigated in our previous study [53, 54]. The FET exhibited ambipolar FET characteristics at room temperature. We have also found that a ferroelectric-like transition exists in $\beta'$- (BEDT-TTF)(TCNQ). However, there has been no report of a ferroelectric-like phase transition in the $\beta'$- (BEDT-TTF)(TCNQ) crystal. Neither has there been any reports of the ferroelectric-like phase transition at 285 K for the two known polymorphs [55–57]. To elucidate the origin of the ferroelectric-like properties, a gate-induced pyroelectric current measurement was conducted.

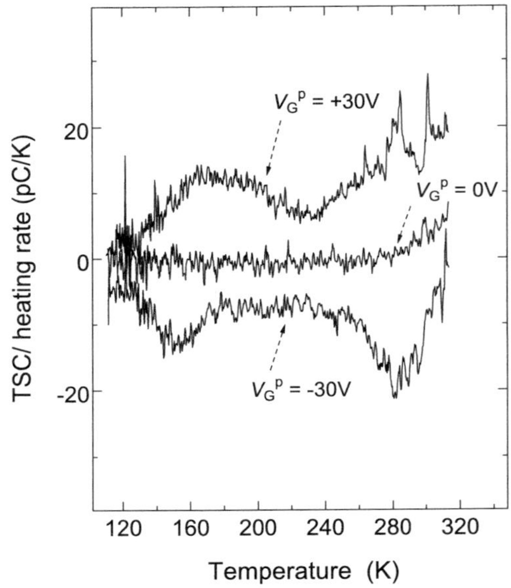


**Fig. 9.9** Gate-induced TSCs observed in the $\beta'$-(BEDT-TTF)(TCNQ) crystalline FET sample after application of a positive, negative and zero $V_G^p$

The source materials, substrate preparation, and crystal growth method were described in Refs [53, 54]. The high crystallinity, majority of the triclinic phase, and spontaneous oriented growth in the $a^*$-direction normal to the substrate surface were revealed by X-ray diffraction [53]. It is found that the $b - c$ conductive plane of the $\beta'$-BEDT-TTF layer is parallel to the $SiO_2$ surface and continuous from the source to drain electrode. The sample structure is not only a normal capacitor structure but also a bottom-contact FET structure.

A gate-induced thermally stimulated current was measured by the following procedure. First, the sample was placed in a vacuum chamber in which the base pressure was approximately $10^{-8}$ Pa, and the sample temperature was decreased to 280 K, which is lower than the observed ferroelectric-like phase transition temperature (285 K). A poling voltage ($V_G^p$) was applied at 280 K, and this poling bias was maintained for 30 min. After poling, the sample was cooled to 90 K under an applied $V_G^p$ to freeze the gate-induced polarization. After $V_G^p$ was turned to 0 V, an electrical current from the source electrode was measured under a constant sample heating rate of about 5 K/min. A thermally stimulated current was observed with the dissolution of the initially frozen polarization. A charge-ordered state is a polarized state whose polarization derives from the alternative localization of carriers. For a rigid charge-ordered phase, it is impossible to invert the polarization by an external electric field. By contrast, in a fluctuated charge-ordered state, weakly localized carriers become mobile under an external electric field.

Figure 9.9 shows observed gate-induced TSCs. Under a positive (negative) applied $V_G^p$, a positive (negative) TSC was observed. Under a zero applied $V_G^p$, no TSC peak was observed. The TSC observed under a nonzero $V_G^p$ exhibits a maximum at 285 K and broad peaks between 180 and 210 K. These minor and

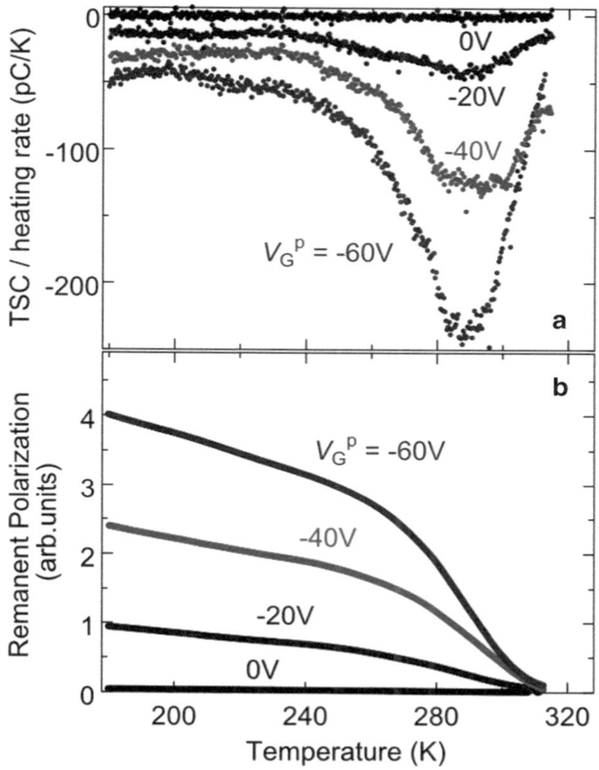


**Fig. 9.10** (**a**) $V_{\mathrm{G}}^{\mathrm{p}}$ dependence of the observed TSCs. (**b**) temperature and $V_{\mathrm{G}}^{\mathrm{p}}$ dependence of the remnant polarization charge. Copyright 2012, MDPI

broad peaks do not correspond to a pyroelectric current but rather to the detrapping of carriers because the peaks are not symmetric with respect to the polarization of an applied $V_{\mathrm{G}}^{\mathrm{p}}$. On the other hand, the major TSC observed above 210 K is due to a pyroelectric current because these peaks are symmetric with respect to the $V_{\mathrm{G}}^{\mathrm{p}}$, and the TSC corresponds to the dielectric response and $Q-V$ hysteresis observed in our previous work [54]. The $Q-V$ hysteresis and the divergent increase in the dielectric constant and loss are features of ferroelectricity.

We shall now discuss the TSC as a pyroelectric current. Figure 9.10a shows the $V_{\mathrm{G}}^{\mathrm{p}}$ dependence of the observed TSCs. These TSCs increase with increasing $V_{\mathrm{G}}^{\mathrm{p}}$ because the induced average polarization increases with increasing $V_{\mathrm{G}}^{\mathrm{p}}$. However, the TSC peak temperature shows little dependence on $V_{\mathrm{G}}^{\mathrm{p}}$. In this temperature region, the observed TSC is expected to consist mainly of the pyroelectric current because the temperature range is higher than that at which detrapping from the carrier traps is observed. On this basis, we integrated the observed TSCs. The relationship between the polarization and observed TSC can be expressed as

$$i(T) = -\frac{dQ_r}{dt} = -\frac{dQ_r}{dT}\frac{dT}{dt} = -s_{\mathrm{eff}}\frac{dP_r}{dT}\frac{dT}{dt}, \tag{9.5}$$

where, $s_{\mathrm{eff}}$ is the effective surface area of the sample and is defined for a parallel capacitor. $Q_{\mathrm{r}}$ and $P_{\mathrm{r}}$ are the remnant polarization charge and remnant polarization, respectively. In the case of the parallel capacitor, one can define the pyroelectric coefficient $p$(T) as

$$\frac{dP_r}{dT} = p(T). \tag{9.6}$$

Thus, one obtains:

$$i(T) = -s_{\mathrm{eff}} p(T) \frac{dT}{dt}. \tag{9.7}$$

If the heating rate ($\frac{dT}{dt}$) is constant, one can obtain the pyroelectric coefficient. Then, $P_{\mathrm{r}}$ is calculated as an integral of the pyroelectric current $i(T)$:

$$P_{\mathrm{r}}(T) = -\int p(T)dT = -\frac{1}{s_{\mathrm{eff}}} \int \frac{i(T)}{\frac{dT}{dt}} dT. \tag{9.8}$$

In our case, the remnant polarization charge is calculated via,

$$Q_{\mathrm{r}}(T) = -\int \frac{i(T)}{\frac{dT}{dt}} dT + Q_{ini}, \tag{9.9}$$

where $Q_{\mathrm{ini}}$ is a constant of integration that is fixed by the initial condition, i.e., the initial induced polarization. Figure 9.10b shows the integrated TSC, which corresponds to the temperature and $V_{\mathrm{G}}^{\mathrm{p}}$ dependence of the remnant polarization charge, ($Q_{\mathrm{r}}$). Although $Q_{\mathrm{ini}}$ is difficult to determine because of the non-uniform electric field in the crystal, we determined the $Q_{\mathrm{ini}}$ for each $V_{\mathrm{G}}^{\mathrm{p}}$ under the assumption that the initially induced $Q_{\mathrm{r}}$ is proportional to the $V_{\mathrm{G}}^{\mathrm{p}}$. Figure 9.10b indicates that $Q_{\mathrm{r}}$ gradually decreases with increasing temperature up to 260 K, and begins to decrease steeply at around 280–290 K. A small amount of $Q_{\mathrm{r}}$ remains at 320 K.

Figure 9.11 summarizes the physical parameters of the $\beta'$-(BEDT-TTF)(TCNQ) crystalline FET. There are clear relationships between the observed parameters; in particular, the ferroelectric-like transition, which is indicated in Figs. 9.11c and d, and the change in field-effect mobility (shown in Fig. 9.11b) occurred simultaneously. On the other hand, this ferroelectric-like transition hardly affected the bulk conduction shown in Fig. 9.11a. Thus, this correlation concerns the interface of the organic crystals and $SiO_2$ because both the field-effect mobility and the ferroelectric-like properties were observed around the interface, not in the bulk. Because the bulk $\beta'$-(BEDT-TTF)(TCNQ) crystal is known as a dimer Mott insulator, the ferroelectric-like phase below 285 K is exceptional.

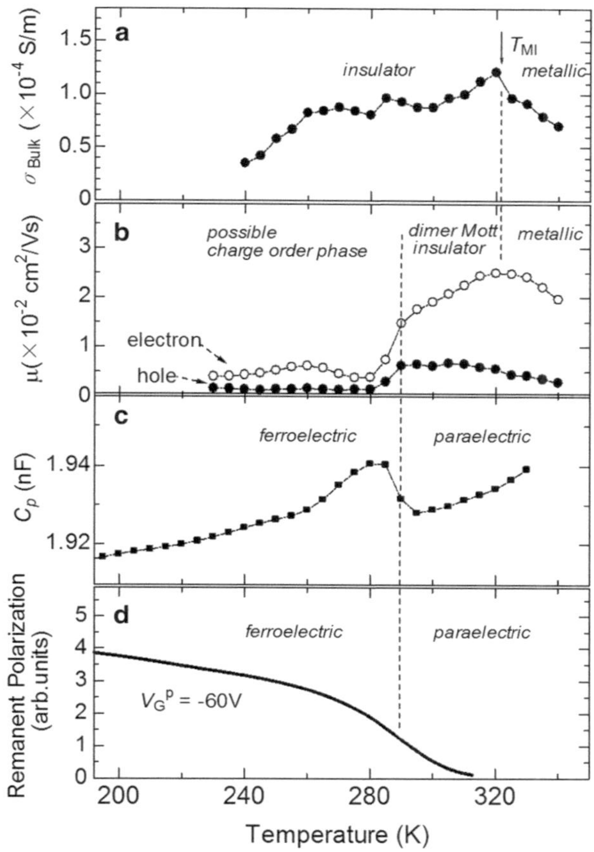


**Fig. 9.11** Summary of the physical and electrical parameters observed in the $\beta'$-(BEDT-TTF)(TCNQ) crystalline FET. Temperature dependence of (**a**) bulk conductance; (**b**) field-effect mobility; (**c**) gate capacitance, and (**d**) remnant polarization

The band filling of both the dimer Mott insulator and charge-ordered phase are quarter filling. The difference between the dimer Mott insulator and the charge-ordered phase is the degree of localization of correlated carriers. The carriers of the dimer Mott insulator and charge-ordered phase are localized in one dimer site and one BEDT-TTF site, respectively. If a carrier is localized in a BEDT-TTF site and thermal hopping to the intradimer neighboring BEDT-TTF site is allowed, spontaneous polarization derived from the biased carrier distribution in the dimer sites is induced by an external electric field. Under this assumption, the unusual temperature behavior shown in Fig. 9.11 can be explained by the abrupt change in the degree of localization of correlated carriers [58–60].

## 9.4 Organic Thin Film Transistors

### 9.4.1 Step-Edge Vertical-Channel OFET

OFETs have already shown promise for applications in electronic papers, sensors, and radio frequency identification cards (RFIDs). In order to be able to function, the OFETs in these devices must have high speed, high current, and low operational voltage. Conventional OFETs do not meet these demands owing to the low carrier mobility of organic semiconductors. One method for fabricating high-speed, high-power, and low-voltage OFETs is to shorten the channel length. A number of studies have attempted to shorten the channel length in order to achieve high-speed operation [61–69]. The operational speed (cut off frequency, $f_c$) is given by

$$f_0 \propto \frac{\mu(V_G - V_T)}{2\pi L^2} \propto \frac{\mu}{L^2}. \tag{9.10}$$

However, decreasing the channel length of conventional OFETs using a low-cost process is difficult because lateral-type OFETs usually have a planar structure where the source and drain electrodes are separated using shadow mask deposition or photolithography. An easy way to shorten the channel length is to form the channel in the vertical direction. Thus, there have been many reports of vertical-type organic transistors such as organic static induction transistors [4–6], charge injection controlled transistors [63, 64], metal-base vertical OFETs [65], and various edge-type vertical OFETs [66–69].

We fabricated step-edge vertical-channel OFETs (SVC-OFETs) based on pentacene or 6,13-bis(triisopropyl-silylethynyl) pentacene (TIPS-pentacene) and measured their static and frequency characteristics. SVC-OFET is a vertical-type OFET in which a submicron channel can easily be achieved in the film thickness level by using the step-edge structure. A short channel is formed in the vertical direction of the step-edge structure by a simple self-alignment process. A schematic of a bottom-contact (BC) SVC-OFET is shown in Fig. 9.12. First, an Al gate electrode with a line width of 100 $\mu$m was deposited by vacuum evaporation on a substrate. In this case, the thickness of the Al gate electrode, $D$, which was varied from 0.25 to 1 $\mu$m, corresponds to the channel length of the SVC-OFET, $L$. Subsequently, the substrate surface was covered with an insulating layer, and a gate dielectric layer was deposited so as to surround the Al gate electrode. The gate dielectric layer was

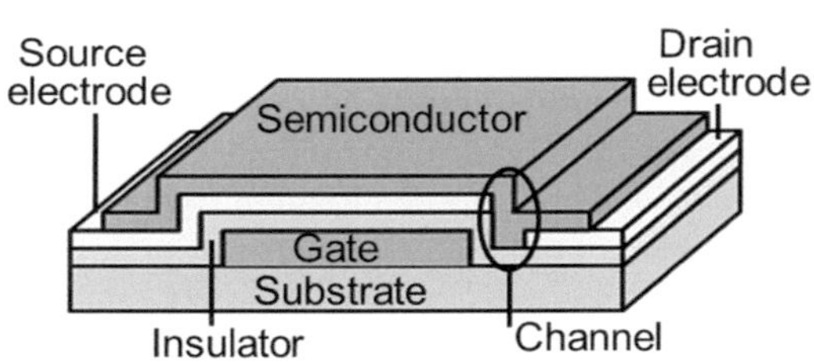


**Fig. 9.12** Device structure of SVC-OFET

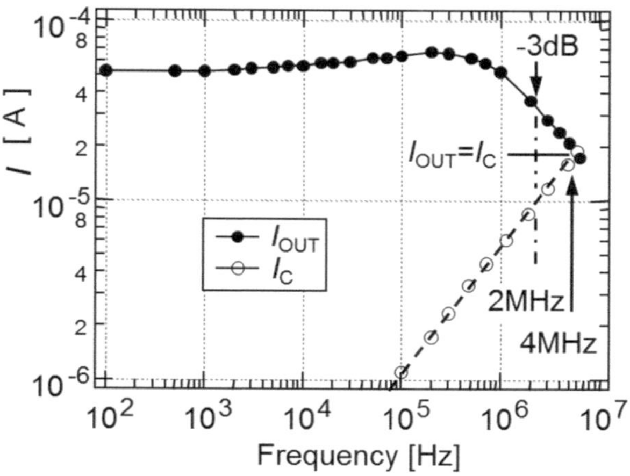


**Fig. 9.13** Frequency characteristics of a SVC-OFET. Copyright 2013, Trans. Mat. Res. Soc. Japan

composed of $SiO_2$ (200 nm) or parylene C, deposited by plasma-enhanced chemical vapor deposition (p-CVD). The surface of gate dielectric layer was treated with $UV/O_3$ for 10 min at room temperature and subjected to a silane coupling treatment with hexamethyldisilazane (HMDS) saturated vapor for 2 h at room temperature. Au was deposited by incline vacuum evaporation onto the gate-formed substrate to fabricate the upper and lower electrodes. The step edge of the gate electrode served as a shadow mask, and the Au electrode was separated by the wall of the step edge.

The active layer, composed of TIPS-pentacene, was formed by the drop casting method using a small volume from 1 wt% toluene solution and dried in toluene ambient. The substrate surface was coated with Pt to prevent charging during observation. A SVC-OFET with a 600-nm-thick $SiO_2$ film was used for the observation. The upper and lower electrodes were separated at the step edge and the channel length, $L$ (defined here as $L = D$), was approximately 1 $\mu$m. SVC-OFET can be fabricated by simple nanoimprint lithography and a self-aligned process [69].

The frequency characteristics of the SVC-OFET are shown in Fig. 9.13. The cutoff frequency ($f_c$) is defined here as the frequency where $I_{DS}$ (that is, $I_{out} - I_c$) at each frequency [$I_{DS}$(f)] is 3 dB lower than that at 100 Hz [$I_{DS}(100)$], i.e., 20 log [$I_{DS}$(fc)]/[$I_{DS}(100)$] $= -3$. SVC-OFETs showed excellent device performance and high $f_c$ values of approximately 2 MHz (defined as 3 dB down from the output current $I_{out}$ at low frequency) and 4 MHz (defined as output current equals gate charging current, $I_c$) were obtained [70]. This is presumably because of the shorter channel length of 250 nm.

### 9.4.2 *Self-Aligned Organic Nanochannel Transistors*

Electric field-induced oriented growth [71–75] and its in situ observation [76], selective growth [76, 77], and self-aligned wiring and nanogap formation are useful elemental technologies for fabricating organic nanodevices without the destruction

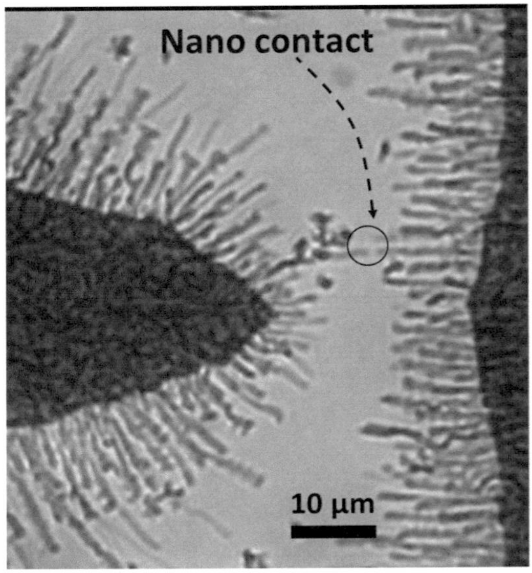


**Fig. 9.14** Optical micrograph of a nanocontact between two TTF-TCNQ conductive wires grown under a static electric field and quasi-thermal equilibrium conditions

of organic nanocrystals and modulation of the electronic state of the organic molecules. For example, organic nanochannel transistors were fabricated by the spontaneous formation of an active layer [78, 79] or by local Joule heating of connected TTF-TCNQ wires [80]. The spontaneous formation of an active layer occurs as a result of the absence of TTF at the growth tip [78]. Consequently, a small TCNQ region is formed at the junction, which was determined to be 500–700 nm by atomic force microscopic potentiometry (AFMP) [81]. These nanotransistors were demonstrated experimentally and exhibited a maximum electron mobility of 1–3 $cm^2/Vs$ [79], which is comparable to that of a single-crystal TCNQ device [82]. The high performance is mainly due to the high carrier injection efficiency because there is no carrier injection barrier at the TCNQ/TTF-TCNQ interface. The lowest unoccupied molecular orbital (LUMO) band of TCNQ and the TCNQ LUMO band of TTF-TCNQ are common, so that the LUMO band of TTF-TCNQ acts as a carrier reservoir for the TCNQ nanocrystal.

Figure 9.14 shows an optical micrograph of a spontaneously formed nanocontact between two organic conductive wires. This nanocontact is formed by self-aligned growth. If another organic semiconducting material, such as pentacene, is evaporated after making the nanogap, the nanogap is filled with the evaporated material.

Another nanofabrication method involves the self-termination mechanism [80] using self-concentrated Joule heating. A connected TTF-TCNQ wire has a highly resistant point at the junction just after making a connection. If an AC or DC current

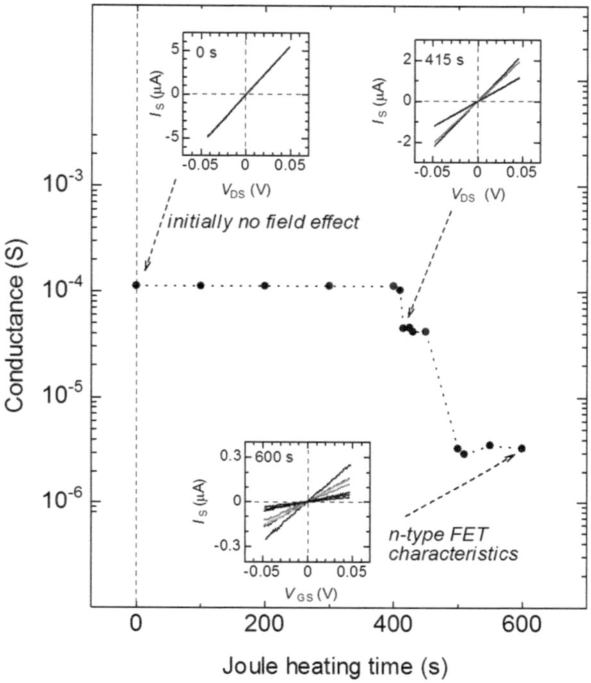


**Fig. 9.15** Joule heating time variation of the electrical conductivity for a connected TTF-TCNQ wire. A step-like decrease in electrical conductivity was observed upon increasing the Joule heating time. The FET characteristics, which were measured at each stage, indicate that an n-type nanochannel transistor was gradually formed by self-concentrated Joule heating

is applied, Joule heat is generated preferentially at this point of high resistance. The Joule heat causes TTF desorption from the high resistance point. During this process, the resistance of the initially high resistance point increases further and a non-doped semiconducting region is formed at the junction. This semiconductorization process by Joule heating ceases automatically because the higher resistance ($R$) limits the electric current. Therefore, the heat generated ($V^2/R$) decreases with increasing $R$ under constant-voltage operation. Figure 9.15 shows a typical example of the self-concentrated Joule heating process [80]. The initial total conductance of the connected TTF-TCNQ wire is approximately $10^{-4}$ S. During the first 400 s, the total conductance decreases slightly, and no gate-induced modulation of the source current ($I_S$) is observed at this stage. However, the conductance drops abruptly after 415 s, and a slight gate-induced modulation of $I_S$ appears. Another drop in conductance is observed after 500 s of Joule heating. After 600 s of Joule heating, the gate-induced modulation is significantly improved, and obvious n-channel FET characteristics are obtained.

## 9.5 Flexible Device Application

### 9.5.1 *Flexible Transistors*

The advantage of organic electronic devices is their flexibility, which is derived from their softness due to van der Waals interactions in their structure. In particular, placing the organic semiconductor layer at the neutral strain surface has been found to be very effective in improving the bending durability of organic devices [29]. The laminated sheet structure used in our thermal press and thermal lamination process (in Sect. 9.2.3) exploits this principle to improve the bending durability of the sheet device.

A flexible transistor was fabricated by the thermal lamination process [28] by the following procedure. The cover film was a thin (12 $\mu$m) polyimide film, with a 900-nm-thick parylene-C buffer layer and Au contact electrodes. The base film was also a 12-$\mu$m-thick polyimide film, with a Au gate electrode structure and a 900-nm-thick parylene-C gate insulating layer. The base and cover films had a symmetrical arrangement to cancel compressive and tensile strains in the organic layers during bending. A small amount of $C_8$-BTBT [83–85] powder was placed on the base film at the approximate device location, and was covered by the cover film. The pair of polyimide films, including the organic powder, was then inserted into the laminator.

Figure 9.16 shows the transfer characteristics of the laminated $C_8$-BTBT transistor and the results of sequential bending tests conducted under a bending radius of 1 mm. We carried out a sequential bending test starting from the flat device state, and then bent the channel region outward (so that the electrical current was parallel

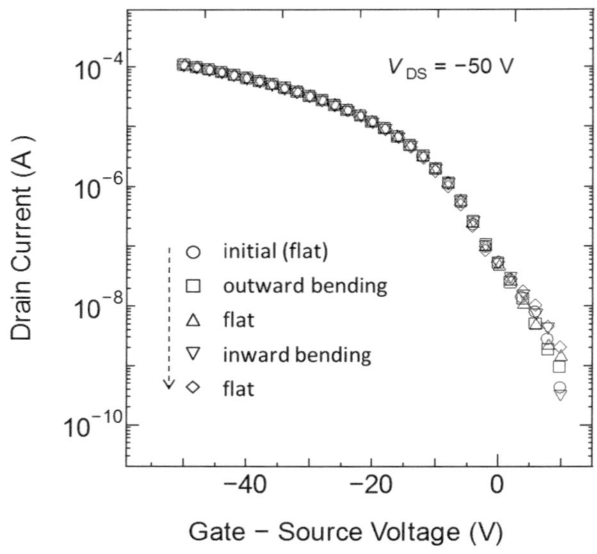


**Fig. 9.16** Transfer characteristics of laminated $C_8$-BTBT transistors during a sequential outward and inward bending test performed under a bending radius of 1 mm

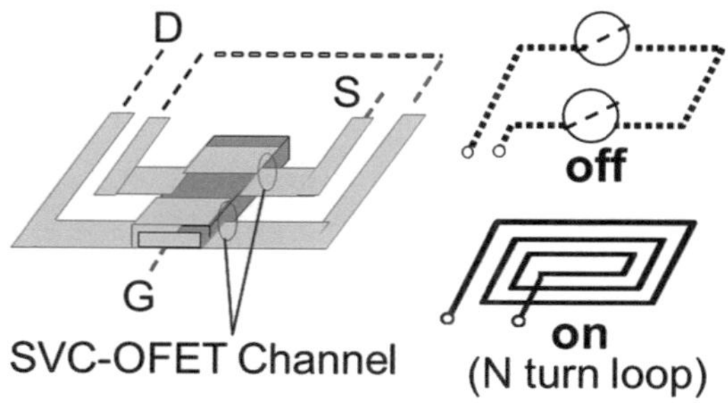


**Fig. 9.17** A typical example of fabricating an active antenna incorporating SVC-OFETs

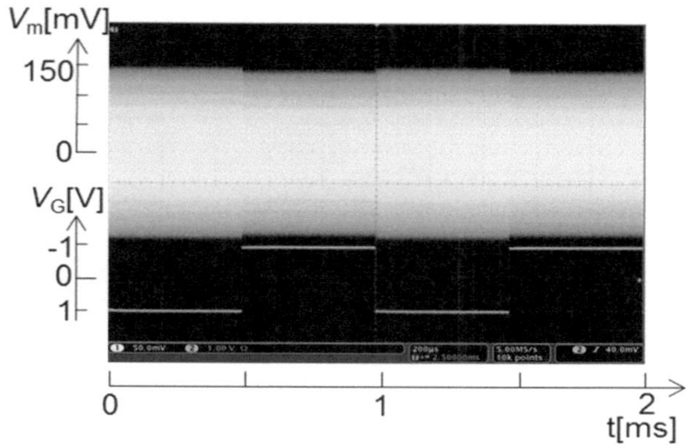


**Fig. 9.18** Wave form of $V_m$ observed in the modulation

to the strain, and the cover film was on the outside during outward bending); we then flattened the sample again, and bent the channel region inward, before we finally flattened the sample again. The transfer characteristics were measured at each stage. As seen in Fig. 9.16, the outward and inward bending process did not have any significant effect on the transfer curve. The effect of bending on the field-effect mobility was less than 2.6 %, and no irreversible degradation was observed. A bending radius of 1 mm was sufficient for reel-up type flexible sheet devices [3].

### 9.5.2 RFID Tag

Figure 9.17 shows a conceptual illustration of an active antenna using an OFET. In this case, a SVC-OFET is fabricated on one part of the loop antenna line-edge. The total impedance of an active antenna can be changed by operating SVC-OFETs. When the SVC-OFET is in the ‘on’ state, the active antenna forms an N-turn loop, whereas the loop is cut off when the SVC-OFET is in the ‘off’ state. Figure 9.18 shows a waveform of $V_m$ observed in the modulation monitor. The output voltage of the DC power source was − 25 V, the frequency of the carrier wave was 13 MHz, and the modulation frequency of the gate voltage was 1 kHz. Using the peak values

**Fig. 9.19** A typical example of an active-matrix OLET display. Copyright 2008, The Japan Society of Applied Physics

of $V_m$ at $V_G = 0$ V and $V_G = -20$ V, a modulation factor of 5.2 % was obtained from the results shown in Fig. 9.18. These results also demonstrate that the SVC-OFET is operating as a resistance circuit element around the RF region (13.56 MHz), although the transistor modulation of SVC-OFETs is restricted on the cutoff frequency (2 MHz). This means that the OFET operation is restricted by the gate charging and channel formation. Once the channel is formed by the low-frequency gate voltage, however, the resistance of the channel itself operates at a higher frequency like a doped organic material. Several reports demonstrate the resistance and diode operation (30–50 MHz) obtained using organic materials [86].

### 9.5.3 Active-Matrix OLED Display

Several vertical-type OFET drive OLEDs for flexible display applications have been reported. We have demonstrated the high performance of a novel vertical-type organic light-emitting transistor (OLET) and active-matrix OLET on a plastic film as well as on a glass substrate [70, 87–89]. Brightness values of approximately 400 and 50 cd/m$^2$ were obtained on glass and polyethersulfone (PES) substrates, respectively. The OLET display using organic materials is protected against mechanical stress, and can be bent while displaying an animation. Figure 9.19 shows an example of a flexible OLET display [89]. Figure 9.20 shows (a) the standard equivalent circuit of an OLED pixel and (b) a cross-sectional view of an SVC-OFET. The effective emission area (OLED part) in one pixel using a SVC-OFET (Fig. 9.20d) is larger than that for a conventional device layout (Fig. 9.20c). Because SVC-OFETs can be fabricated on active-matrix wires such as common, data, and scan lines, channels of switching and driving transistor of

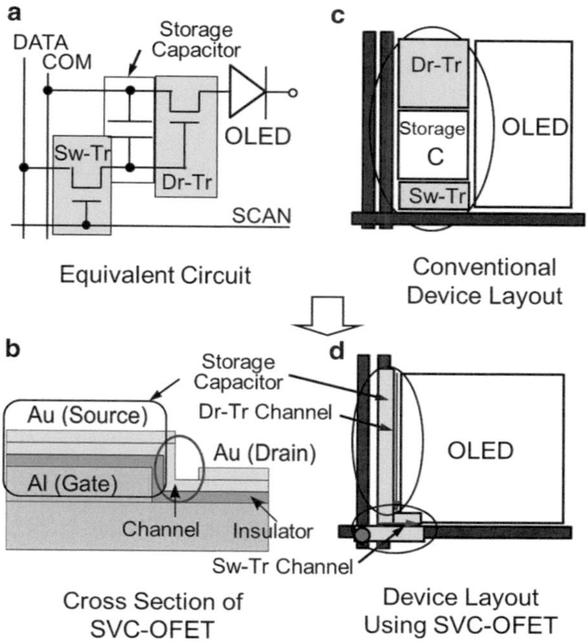


**Fig. 9.20** Device layout in one pixel of an active-matrix display.

active-matrix display are formed at each line edge. The storage capacitance can be controlled by adjusting the overlap area between the common line and the source electrode of the SVC-OFET.

As described above, SVC-OFETs are suitable not only for information tags, but also active-matrix display backplanes, owing to their high-performance operation and compact layout, because the capacitance between the gate and source (or drain) can be controlled by adjusting the gate line width and used as the matching capacitor of active antennae and storage capacitor of active-matrix display circuits.

**Acknowledgements** The authors would like to thank Dr. Shigekazu Kuniyoshi and Mr. Hiroshi Yamauchi of Chiba University for their assistance. This work was supported in part by a Grant-in-Aid for Science Research (Nos. 23360006, 24656008 and 24560006) and Global Center of Excellence Program 'Advanced School for Organic Electronics' from the Ministry of Education, Culture, Sports, Science, and Technology of Japan.

# References


1. C.D. Dimitrakopoulos, P.R.L. Malenfant, Adv. Mater. **14**, 99 (2002)
2. H. Klauk, D.J. Gundlach, J.A. Nichols, C.D. Sheraw, M. Bonse, T.N. Jackson, Solid State Tech. **43**, 63(2000)
3. S.R. Forrest, Nature, **428**, 911 (2004)
4. K. Kudo, D.X. Wang, M. Iizuka, S. Kuniyoshi, K. Tanaka, Thin Solid Films **331**, 51–54 (1998)

5. Y. Watanabe, H. Iechi, K. Kudo, Appl. Phys. Lett. **89**, 233509 (2006)
6. K. Fujimoto, T. Hiroi, K. Kudo, M. Nakamura, Adv. Mater. **19**, 525 (2007)
7. M. Sakai, M. Nakamura, K. Kudo, Appl. Phys. Lett. **90**(6), 62101 (2007)
8. M. Sakai, H. Miyata, K. Itami, M. Nakamura and K. Kudo, Appl. Phys. Express **1**, 081802 (2008)
9. K. Kudo, M. Yamashina, T. Moriizumi, Jpn. J. Appl. Phys. **23**, 130 (1984)
10. K. Kudo, T. Sumimoto, K. Hiraga, S. Kuniyoshi, and K. Tanaka, Jpn. J. Appl. Phys. **36**,6994 (1997)
11. Z. Bao, A.J. Lovinger, A. Dodabalapour, Appl. Phys. Lett. **69**, 3066 (1996)
12. H. Klauk, D.J. Gundlach, J.A. Nichols, C.D. Sheraw, M. Bonse, T.N. Jackson, Solid State Technol. **43**, 63 (2000)
13. D. Kinipp, R.A. Street, A.R. Volkel, Appl. Phys. Lett. **82**, 3907 (2003)
14. F. Garnier, A. Yassar, R. Hajlaoui, G. Horowitz, F. Dellofre, B. Servet, S. Ries, P. Alnot, J. Am. Chem. Soc. **115**, 8716(1993)
15. G. Horowitz, D. Fichou, X. Peng, Z. Xu, F. Garnier, Solid State Commun. **72**, 381 (1989)
16. H. Sirringhaus, P.J. Brown, R.H. Friend, M.M. Nielsen, K. Bechgaard, V.M.W. Langeveld, A.J.H. Spiering, R.A.J. Janssen, E.W. Meijer, P. Herwig, D.M. de Leeuw, Nature **401**, 685 (1999)
17. Z. Bao, J.A. Rogers, H.E. Katz, J. Mater. Chem. **9**, 1895 (1999)
18. V. Wagner, P. W?bkenberg, A. Hoppe, J. Seekamp, Appl. Phys. Lett. **89**, 243515 (2006)
19. Y.Y. Noh, N. Zhao, M. Caironi, H. Sirringhaous, Nature Nanotech. **2**, 784 (2007)
20. H. Sirringhaus, T. Kawase, R.H. Friend, T. Shimoda, M. Inbasekaran, W. Wu, E. P. Woo, Science **290**, 2123 (2000)
21. H. Klauk Chem. Soc. Rev. **39**, 2643 (2011)
22. H. Minemawari, T. Yamada, H. Matsui, J. Tsutsumi, S. Haas, R. Chiba, R. Kumai, T. Hasegawa, Nature **475**, 364 (2011)
23. T. Uemura, Y. Hirose, M. Uno, K. Takimiya, J. Takeya, Appl. Phys. Express **2**, 111501 (2009)
24. K. Nakayama, Y. Hirose, J. Soeda, M. Yoshizumi, T. Uemura, M. Uno, W. Li, M. J. Kang, M. Yamagishi, Y. Okada, E. Miyazaki, Y. Nakazawa, A. Nakao, K. Takimiya, J. Takeya, Adv. Mater. **23**, 1626 (2011)
25. M. Bongnitzki, H. Hou, M. Ishaque, T. Frese, M. Hellwig, C. Schwarte, A. Schaper, J.H. Wendorff, A. Greiner, Adv. Mater. **12**, 637 (2000)
26. T. Fukuda, T. Suzuki, R. Kobayashi, Z. Honda, N. Kamata, Thin Solid Films **518**, 575 (2009)
27. A. Inoue, T. Okamoto, M. Sakai, S. Kuniyoshi, H. Yamauchi, M. Nakamura, K. Kudo, Phys. Status Solidi A **210**, 1353 (2013)
28. M. Sakai, T. Okamoto, Y. Yamazaki, J. Hayashi, S. Yamaguchi, S. Kuniyoshi, H. Yamauchi, Y. Sadamitsu, M. Hamada, K. Kudo, Phys. Status Solidi RRL. doi:10.1002/pssr.201308118
29. T. Sekitani, S. Iba, Y. Kato, Y. Noguchi, T. Someya, and T. Sakurai, Appl. Phys. Lett. **87**, 173502 (2005)
30. A. Matsui, Y. Ishii, Jpn. J. Appl. Phys. **6**, 127 (1967)
31. M. Saleh, Jpn. J. Appl. Phys. **17**, 1031 (1978)
32. W. Warta, R. Stehle, N. Karl, Appl. Phys. A **36**, 163 (1985)
33. N. Karl, J. Marktanner, Mol. Cryst. Liq. Cryst. **355**, 149 (2001)
34. A.K. Tripathi, M. Heinrich, T. Siegrist, J. Pflaum, Adv. Mater. **19**, 2097 (2007)
35. J. Simon, J.J. Andre, *Molecular Semiconductors* (Springer, Berlin, 1985) p.116
36. R.G. Kepler, Phys. Rev. **119**, 1226 (1960)
37. S.M. Sze, *Physics of Semiconductor Devices* (Wiley, New York, 1969) p. 425
38. C. Hamann, A. Mrwa, M. Muller, W. Gopel, M. Rager, Sens. Actuat. B, **4**, 73 (1991)
39. A. Wilson, J.D. Wright, A.V. Chadwick, Sens. Actuat. B, **4**, 499(1991)
40. Z. Bao, A.J. Lovinger, A. Dodabalapur, Adv. Mater. **9**, 42 (1997)
41. Z. Bao, A.J. Lovinger, J. Brown, J. Am. Chem. Soc. **120**, 207 (1998)
42. P.R.L. Malenfant, C.D. Dimitrakopoulos, J.D. Gelorme, L.L. Kosbar, T.O. Graham, A. Curioni, W. Andreoni, Appl. Phys. Lett. **80**, 2517 (2002)
43. J.P. Contour, P. Lenfant, A.K. Vijh, J. Catal. **29**, 8 (1973)

44. R.O. Loutfy, Y.C. Cheng, J. Chem. Phys. **73**, 2902(1980)
45. K. Yamamoto, A.A. Kowalska, K. Yakushi, Appl. Phys. Lett. **96**, 122901 (2010)
46. K. Yamamoto, A.A. Kowalska, C. Nakano, K. Yakushi, Physica B, **405**, S363 (2010)
47. A.A. Kowalska, K. Yamamoto, C. Nakano, K. Yakushi, J. Phys. **132**, 012006 (2008)
48. K. Yamamoto, S. Iwai, S. Boyko, A. Kashiwazaki, F. Hiramatsu, C. Okabe, N. Nishi, K. Yakushi, J. Phys. Soc. Jpn. **77**, 074709 (2008)
49. M. Abdel-Jawad, I. Terasaki, T. Sasaki, N. Yoneyama, N. Kobayashi, Y. Uesu, C. Hotta, Phys. Rev. B **82**, 125119 (2010)
50. S. Niizeki, F. Yoshikane, K. Kohno, K. Takahashi, H. Mori, Y. Bando, T. Kawamoto, T. Mori, J. Phys. Soc. Jpn. **77**, 073710 (2008)
51. T. Mori, H. Inokuchi, Solid State Commun. **59**, 355 (1986)
52. Y. Iwasa, K. Mizuhashi, T. Koda, Y. Tokura, G. Saito, Phys. Rev. B **49**, 3580 (1994)
53. M. Sakai, H. Sakuma, Y. Ito, A. Saito, M. Nakamura, K. Kudo, Phys. Rev. B **76**, 045111 (2007)
54. M. Sakai, Y. Ito, T. Takahara, M. Ishiguro, M. Nakamura, K. Kudo, J. Appl. Phys. **107**, 043711 (2010)
55. T. Mori, H. Inokuchi, Bull. Chem. Soc. Jpn. **60**, 402 (1987)
56. H.M. Yamamoto, M. Hagiwara, R. Kato, Synth. Met. **133–134**, 449 (2003)
57. H.M. Yamamoto, N. Tajima, M. Hagiwara, R. Kato, J.-I. Yamaura, Synth. Met. **135–136**, 623 (2003)
58. H. Yoshioka, M. Tsuchiizu, H. Seo, J. Phys. Soc. Jpn. **76**, 103701 (2007)
59. H. Seo, Y. Motome, T. Kato, J. Phys. Soc. Jpn. **76**, 013707 (2007)
60. Y. Otsuka, H. Seo, Y. Motome, T. Kato, J. Phys. Soc. Jpn. **77**, 113705 (2008)
61. J. Collet, O. Tharaud, A. Chapoton, D. Vuillaume, Appl. Phys. Lett. **76**, 1941 (2000)
62. V. Wagner, P. Wobkenberg, A. Hoppe, J. Seekamp, Appl. Phys. Lett. **89**, 243515 (2006)
63. Y. Yang, A.J. Heeger, Nature, **372**, 344 (1994)
64. L. Ma, Y. Yang, Appl. Phys. Lett. **85**, 5084 (2004)
65. S. Fujimoto, K. Nakayama, M. Yokoyama, Appl. Phys. Lett. **87**, 133503 (2005)
66. T. Takano, H. Yamauchi, M. Iizuka, M. Nakamura, K. Kudo, Appl. Phys. Express **2**(7), 071501 (2009)
67. K. Kudo, T. Takano, H. Yamauchi, M. Iizuka, M. Nakamura, Jpn. J. Appl. Phys. **49**, 04DK03 (2010)
68. M. Uno, I. Doi, K. Takimiya, J. Takeya, Appl. Phys. Lett. **94**, 103307 (2009)
69. K. Kudo, H. Yamauchi, M. Sakai, Jpn. J. Appl. Phys. **51**, 11PD05 (2012)
70. K. Kudo, D. Tsutsumi, H. Yamauchi, S. Kuniyoshi, M. Sakai, Trans. Mat. Res. Soc. Japan, **38** (3), 369 (2013)
71. W.P. Hu, Y.Q. Liu, S.Q. Zhou, J. Tao, D.F. Xu, D.B. Zhu, Thin Solid Films **347**, 299 (1999)
72. E. Ina, N. Matsumoto, E. Shikada, F. Kannari, Appl. Surf. Sci. **127**, 574 (1998)
73. M. Sakai, M. Iizuka, M. Nakamura, and K. Kudo, Jpn. J. Appl. Phys. **43**, 2362 (2004)
74. N.A. Kato, M. Fujimura, S. Kuniyoshi, K. Kudo, M. Hara, K. Tanaka, Appl. Surf. Sci. **130–132**, 658 (1998)
75. M. Sakai, M. Iizuka, M. Nakamura, K. Kudo, Jpn. J. Appl. Phys. **42**, 2488 (2003)
76. M. Sakai, S. Kuniyoshi, H. Yamauchi, M. Iizuka, M. Nakamura, K. Kudo, J. Appl. Phys. **113**, 153513 (2013)
77. M. Sakai, M. Iizuka, M. Nakamura, K. Kudo, J. Appl. Phys. **109**, 054309 (2011)
78. M. Sakai, M. Iizuka, M. Nakamura, K. Kudo, J. Appl. Phys. **97**, 053509 (2005)
79. M. Sakai, M. Nakamura, K. Kudo, Appl. Phys. Lett. **90**, 062101 (2007)
80. M. Sakai, H. Miyata, K. Itami, M. Nakamura, K. Kudo, Appl. Phys. Express **1**, 081802 (2008)
81. M. Nakamura, M. Fukuyo, E. Wakata, M. Iizuka, K. Kudo, K. Tanaka, Synth. Met. **137**, 887 (2003)
82. E. Menard, V. Podzorov, S. -H. Hur, A. Gaur, M.E. Gershenson, J.A. Rogers, Adv. Mater. **16**, 2097 (2004)
83. H. Ebata, T. Izawa, E. Miyazaki, K. Takimiya, M. Ikeda, H. Kuwabara, T. Yui, J. Am. Chem. Soc. **129**, 15732 (2007)

84. T. Uemura, Y. Hirose, M. Uno, K. Takimiya, J. Takeya, Appl. Phys. Exp. **2**, 111501 (2009)
85. H. Minemawari, T. Yamada, H. Matsui, J. Tsutsumi, S. Haas, R. Chiba, R. Kumai, T. Hasegawa, Nature **475**, 364 (2011)
86. K. Kudo, S. Kuniyoshi, H. Yamauchi, M. Iizuka, M. Sakai, IEICE Trans. Electron. **E96-C(3)**, 340 (2013)
87. K. Kudo, Curr. Appl. Phys. **5**, 337 (2005)
88. K.Nakamura, T.Hata, A.Yoshizawa, K.Obata, H.Endo, K.Kudo, Appl. Phys. Lett. **89**, 103525 (2006)
89. K. Nakamura, T. Hata, A. Yoshizawa, K. Obata, H. Endo, K. Kudo, Jpn. J. Appl. Phys. **47**(3), 1889 (2008)

# Chapter 10
# Mobility Limiting Factors in Practical Polycrystalline Organic Thin Films

**Ryosuke Matsubara, Noboru Ohashi, Shi-Guang Li, and Masakazu Nakamura**

## 10.1 Overview of the Morphological Features of Pentacene Thin Films [5]

Pentacene is one of the candidates for active materials in organic thin-film transistors (OTFTs) having practical performance, because a relatively high mobility of >1 cm$^2$/Vs can be obtained even for vacuum evaporated thin films [1–4]. Active layers of OTFTs are generally used in polycrystalline state and transistor characteristics are strongly influenced by the film morphology. Therefore, knowledge of the correlation between film morphology and electronic properties is important to improve device performance [5–11]. At the beginning of this chapter, a variety of film morphology and crystallographic features of pentacene thin films grown on $SiO_2$ substrates by molecular beam deposition (MBD) is overviewed. Although

R. Matsubara • M. Nakamura (✉)
Graduate School of Materials Science, Nara Institute of Science and Technology, 8916-5 Takayama, Ikoma, Nara 630-0192, Japan

Graduate School of Engineering, Chiba University, 1-33 Yayoi-cho, Inage-ku, Chiba 263-8522, Japan
e-mail: mnakamura@ms.naist.jp

N. Ohashi
Graduate School of Engineering, Chiba University, 1-33 Yayoi-cho, Inage-ku, Chiba 263-8522, Japan

Department of Applied Chemistry School of Engineering, Tohoku University, 6-6-07 Aramaki Aza Aoba, Aoba-ku, Sendai, Miyagi 980-8579, Japan

S.-G. Li
Graduate School of Engineering, Chiba University, 1-33 Yayoi-cho, Inage-ku, Chiba 263-8522, Japan

Faculty of Automation and Information Engineering, Xi'an University of Technology, 5 South Jinhua Road, Xi'an, Shaanxi 710048, China

 
H. Ishii et al. (eds.), *Electronic Processes in Organic Electronics*, Springer Series in Materials Science 209, DOI 10.1007/978-4-431-55206-2_10

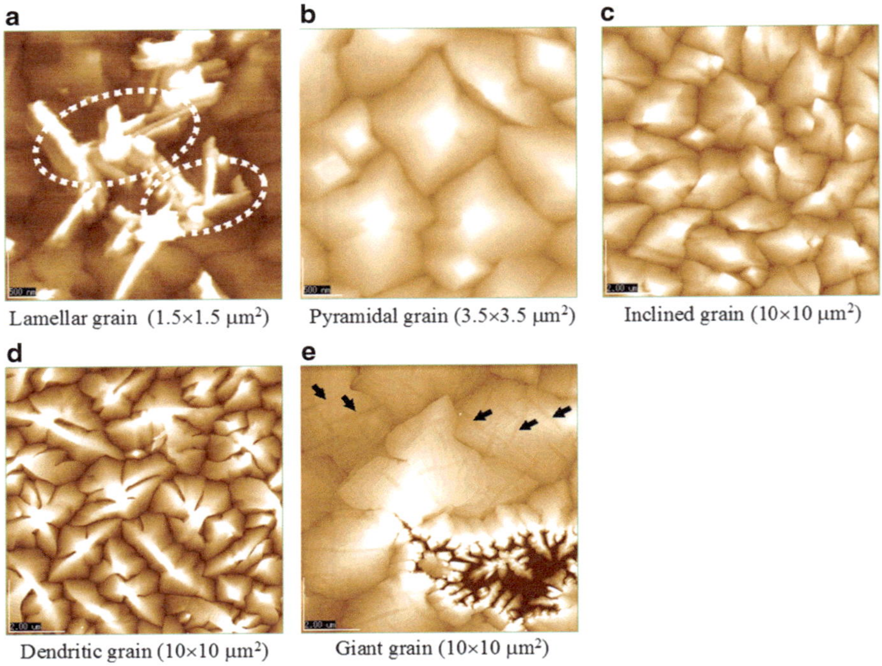


**Fig. 10.1** AFM images of pentacene films deposited under various growth conditions. Grain morphology was categorized into five groups: (**a**) lamellar, (**b**) pyramidal, (**c**) inclined, (**d**) dendritic and (**e**) giant grain. Average film thickness is within the range of 30–60 nm. The *arrows* in (**e**) point recessed regions which appear only in high temperature range

MBD is not a popular technique to fabricate practical devices, it has a great advantage in controlling growth rate and temperature broadly and precisely.

Morphology of pentacene crystalline grains is categorized into five groups, as shown in Fig. 10.1, according to their characteristic shapes. Figure 10.1a shows "lamellar grains" marked with dotted ovals, of which height tends to be higher than surrounding other grains. Figure 10.1b–d show typical morphologies of "pyramidal grains" which have a square pyramidal shape with relatively high symmetry, "inclined grains" which are similar to but more asymmetric than the pyramidal grains, and "dendritic grains" which have a branching structure as frequently seen in diffusion limited growth, respectively. Grains with irregular shape more than 5 μm in size are named as "giant grains" (Fig. 10.1e). Figure 10.2 summarizes a phase diagram of the grain morphology, or a morphodrom, against growth temperature and rate. Two overlapped markers indicate that the film is composed of two types of grains, a larger marker denotes a major component and a smaller a minor. Crystal growth was not observed on the growth conditions marked by crosses. Such a growth limit shifted to lower temperature by decreasing the growth rate. The lamellar structure appears frequently when the growth temperature is less

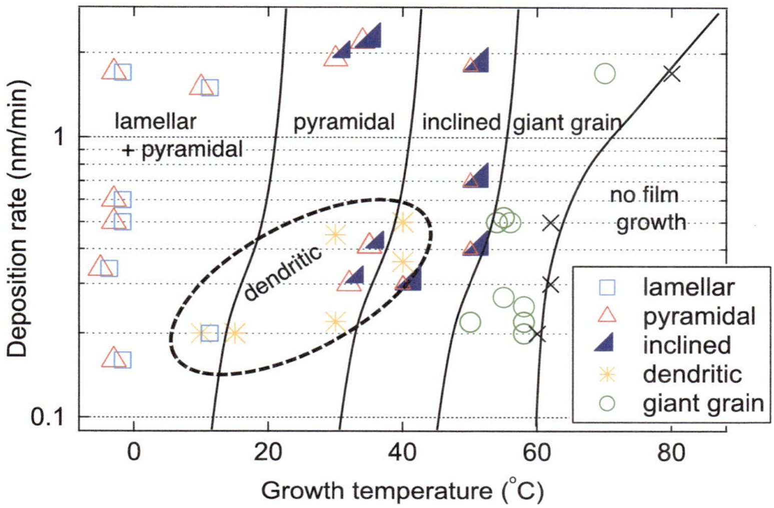


**Fig. 10.2** Morphological phase diagram of pentacene films on $SiO_2$. The vertical axis indicates not a flux of incident but deposition rate

than 10 °C. The dendritic region is isolated in the morphodrom, which would be related to the limited condition for supersaturation of migrating molecules [12].

Vacuum evaporated pentacene films are known to be composed of two crystalline phases, namely the "thin-film phase" and "bulk phase", and the ratio of the two phases varies depending on growth conditions [13]. The films consisting of thin-film phase generally exhibit higher mobility than those with a mixture of the two phases [14]. According to X-ray diffraction (XRD) measurements, all films shown in Fig. 10.1 are highly ordered showing sharp diffraction peaks. On any growth conditions, a diffraction peak indicating 1.54 nm interplanar spacing is the strongest, which indicates that the major component of these films is the thin-film phase. Since only the films in which lamellar-like grains exhibit a diffraction peak of which interplanar spacing is 0.46 nm, the lamellar grains are concluded to be composed of "flat-lying molecules" [15, 16]. A peak corresponding to the bulk phase with a 1.45 nm interplanar spacing is observed above 35 °C and gradually increased with increasing growth temperature. The bulk phase is, however, a minor component even near the growth limit. The intensity ratio between thin film and bulk phases are almost independent of the growth rate.

Above the growth temperature of 40 °C, a notable structure is observed on any film surface. Figure 10.3 shows the structure named "recessed regions" or "notches." An interesting point is that the recessed region is extended across the molecular steps without any perturbation on the step continuity. The boundary of the recess is obscure compared to the molecular steps. The wide range distribution of the recessed regions is also seen in Fig. 10.1e. The area density of the recessed

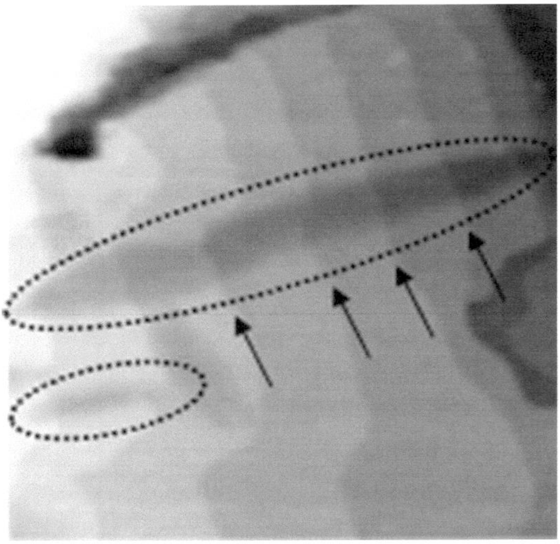

**Fig. 10.3** An AFM image ($2 \times 2\ \mu m^2$) showing recessed region (marked by *dotted ovals*) on a pentacene film surface. The recess is independent from molecular steps (some are marked by *arrows*)

region increases when approaching the growth limit and becomes almost negligible below 40 °C. Although the molecular scale structure has not been clarified, the mechanism forming these recessed structures is hypothesized as follows: When the growth temperature is above 40 °C, thermal expansion of the film is large enough to generate considerable tensile stress during the cooling process. To compensate the tensile stress, part of the film is transformed into bulk phase of which unit cell on (001) plane is larger than that of thin-film phase. Due to the structural transformation, interplanar spacing of (001) shrinks instead while keeping the relative position of the molecules almost unchanged. Such transformation would preserve the molecular steps formed during the film growth and appear with obscure boundary due to the gradual transition between thin-film and bulk phase. It should be noted that these recessed regions behave as large potential barriers against carrier transport, which is explained later in Sect. 10.2.

Figure 10.4 shows a contour plot of average grain size against growth temperature and rate. It is clear that the grain size tends to increase with increasing growth temperature and decreasing growth rate. Separation of the contour line becomes narrower when the deposition conditions approach the growth limit because the actual deposition rate steeply decreases in the near-limit region where the desorption rate of the molecules is comparable to adsorption rate. In the area where dendritic grains appear (a dotted oval), the monotonic variation of the grain size against the deposition conditions becomes irregular. This trend is considerable because the requirement for the dendritic growth meets with higher nucleus density.

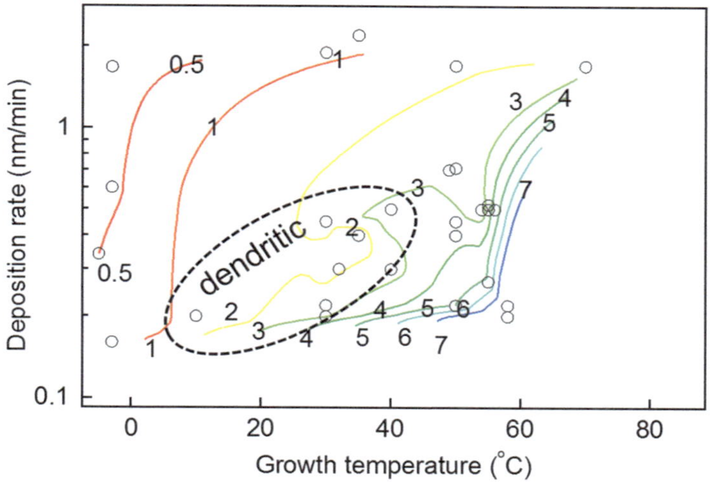


**Fig. 10.4** A contour plot of average grain size (μm) for various growth conditions. The area where dendritic grains appeared is marked by *dotted line*

## 10.2 Extrinsic Limiting Factors of Carrier Mobility in Pentacene Thin-Film Transistors [6]

### *10.2.1 Importance to Know the Influence of Defective Structures*

In many studies devoted to elucidating the mechanism of carrier transport or to improve device performance, the transfer characteristics of OTFTs themselves are employed to estimate carrier mobility. Then, the dependence of the mobility on various structural or chemical parameters is occasionally discussed without determining the true bottleneck of the carrier transport. However, the carrier mobility calculated from the transfer characteristics, namely apparent field-effect mobility, is not always an intrinsic physical parameter at the semiconductor/insulator interface but an extrinsic one showing merely the overall device performance. Apparent mobility often includes the influence of nonlinear resistances at metal/semiconductor contacts and ohmic series resistances, which may be modulated by gate voltage. Furthermore, even though rather intrinsic field-effect mobility is measured by a certain technique, the mobility may be limited mainly by local defective structures in the organic thin films.

Considering such a situation, we have been studying the limiting factors of carrier transport in OTFTs quantitatively using originally developed techniques, microscopic four-point probe FET measurement (MFPP-FET), and atomic-force-microscope potentiometry (AFMP) [7, 8, 17–20]. Carrier transport property in high-mobility OTFTs is generally determined at various stages (from the most intrinsic to the most extrinsic factors): (a) energy band structure of the perfect organic crystal, (b) modulation of the band structure and trap formation by the gate insulator surface, (c) bottleneck at the crystalline domain boundaries or line defects, and (d) carrier injection barriers at the organic/electrode interfaces and defective structures in the film. In this section, we comprehensively explain the cause and the influence of various extrinsic limiting factors of carrier transport, (c) and (d), on apparent mobility mainly focusing on pentacene OTFTs. From a practical point of view, understanding of the cause of extrinsic limiting factors helps us to obtain high-performance OTFTs reproducibly. Moreover, from a scientific point of view, it helps us to exclude unwanted bottlenecks in the carrier transport, and study rather intrinsic mechanisms.

### *10.2.2 Special Measurement Techniques*

In an MFPP-FET measurement (Fig. 10.5a), the sheet resistance of an organic semiconductor layer is measured by the four-point probe while modulating the carrier density by applying a gate voltage to a Si substrate [7, 21]. In this measurement, the well-known advantage of the four-point probe, i.e. the insensitivity to contact resistances, is utilized. Microscopic four-point probes (*CAPRES*) with 5–30 μm probe separations and Au coating were set in a vacuum-type atomic force microscope. The conductance $I_{14}/V_{23}$ in the linear regime, i.e. under sufficiently large gate voltage $V_G$, is described as

$$\frac{I_{14}}{V_{23}} = \frac{\mu\pi Ci}{\ln 2}\left(V_G - \frac{V_{24}+V_{34}}{2} - V_T\right), \tag{10.1}$$

where $\mu$ is the carrier mobility, $C_i$ is the $SiO_2$ capacitance per unit area and $V_T$ is the gate threshold voltage. According to Eq. (10.1), an intrinsic carrier mobility is calculated from the linear slope of the $I_{14}/V_{23}$ vs. $V_G$-$(V_{24}+V_{34})/2$ plot.

AFMP is composed of an AC-mode AFM equipped with a conductive cantilever and high-input-impedance voltage-measurement circuit integrated in a cantilever holder. The schematic diagram of AFMP is shown in Fig. 10.5b. This technique is capable of accurately measuring the potential distribution of working thin-film transistors [8, 18–20]. Kelvin probe force microscopy (KPFM) has also been used for a similar purpose [22–27]. In KPFM, tip potential is biased to make the electrostatic force between the sample and the tip zero, which indicates that vacuum levels are equalized. On the other hand, Fermi levels are equalized by the charge transfer through the contact in AFMP. Although KPFM is a powerful tool for the nano-scale mapping of work-function differences, it is not ideal for the precise potential mapping of semiconductor devices under an operational voltage application. A notable advantage of AFMP against KPFM is its accuracy in measuring the potential distribution near the electrode edges with a large applied bias voltage and grain boundaries with abrupt topographic features [28]. The spatial resolution of AFMP is less than 10 nm and the potential resolution ~100 μV.

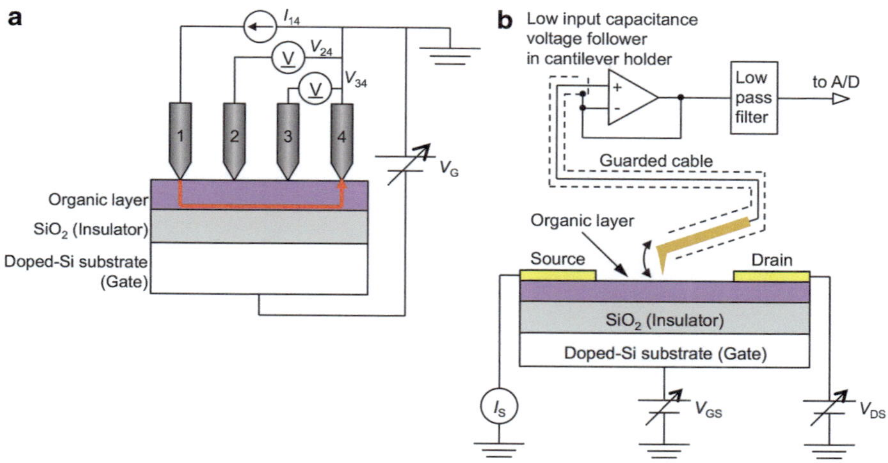


**Fig. 10.5** Schematic diagrams of the original measurement techniques used in this work: (**a**) MFPP-FET and (**b**) AFMP. (Reproduced from [6])

**Fig. 10.6** Ratio of carrier mobilities measured by TC-TFT (channel length: 20 μm) to those measured by MFPP-FET (probe separations: 25 μm). (Reproduced from [6])

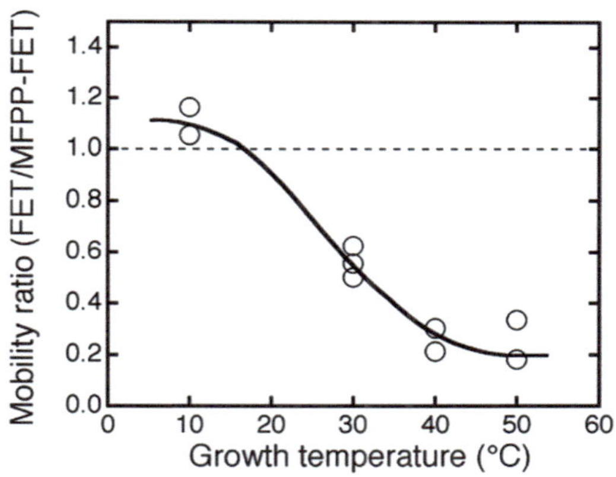


## 10.2.3 *Electrode Contacts*

In many cases, top-contact thin-film transistors (TC-TFTs) tend to exhibit higher apparent mobility than bottom-contact thin-film transistors (BC-TFTs). The contact resistance at source/drain electrodes in the top-contact configuration is, in general, smaller or almost negligible for relatively large channel-length OTFTs. However, even the top contact configuration is sometimes accompanied with extrinsic degradation of the apparent mobility depending on the fabrication processes. Figure 10.6 shows the ratio of field-effect mobilities measured by TC-TFT to those measured by MFPP-FET. Both measurements were carried out with the

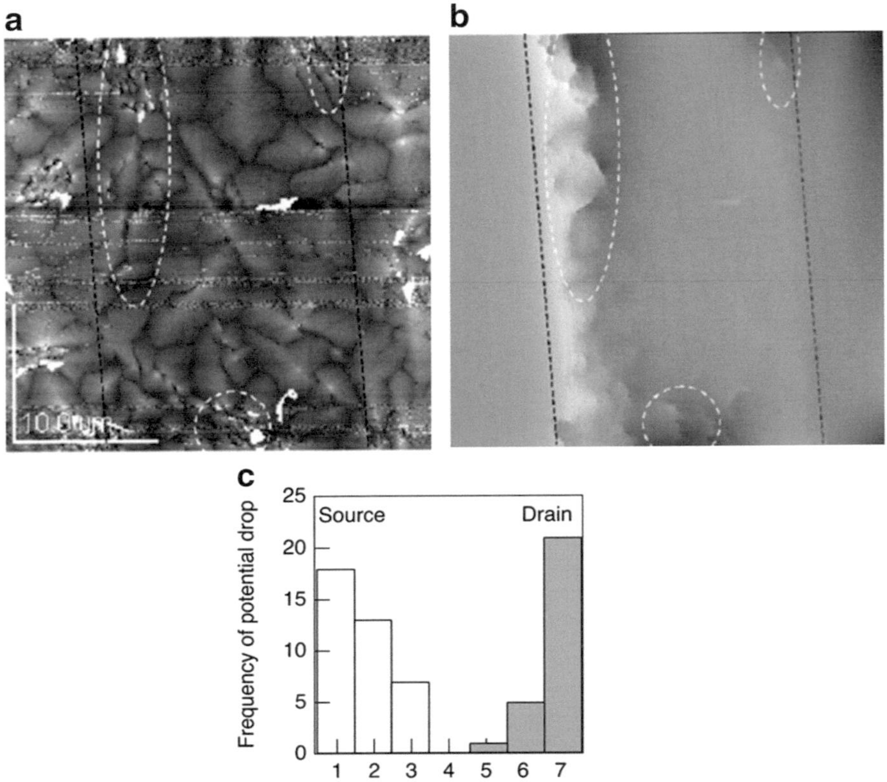


**Fig. 10.7** $30 \times 30\ \mu m^2$ AFMP images taken on a working pentacene TC-TFT (growth temperature: 50 °C, growth rate: 0.31 nm/min): (**a**) topographic and (**b**) high-pass filtered potential images. A source electrode is located on the *left side* and a drain on the *right side* of each image. The drain voltage $V_{DS}$ is −5 V and the gate voltage $V_G$ is −30 V. (**c**) Histogram of the position of major potential drops (>100 mV). The position is shown by the number of grains from source edge. (Reproduced from [6])

same electrode material, Au, and a similar channel length. The result implies that the organic films have been deteriorated by the top metal deposition. Accordingly, we have to be careful in analyzing the dependence of the apparent mobility on deposition conditions. Another notable finding is that the degree of degradation of the films grown at high temperatures is higher than that of the films grown at low temperatures. This is partly because the films grown at higher temperature tend to exhibit higher intrinsic mobilities and therefore the influence of the electrode-originated degradation becomes more apparent.

Figure 10.7a, b show AFMP images taken on a pentacene TC-TFT grown at relatively high temperature, 50 °C. The applied drain voltage $V_{DS}$ is −5 V and the gate voltage $V_G$ is −30 V, which correspond to the linear regime of the OTFT.

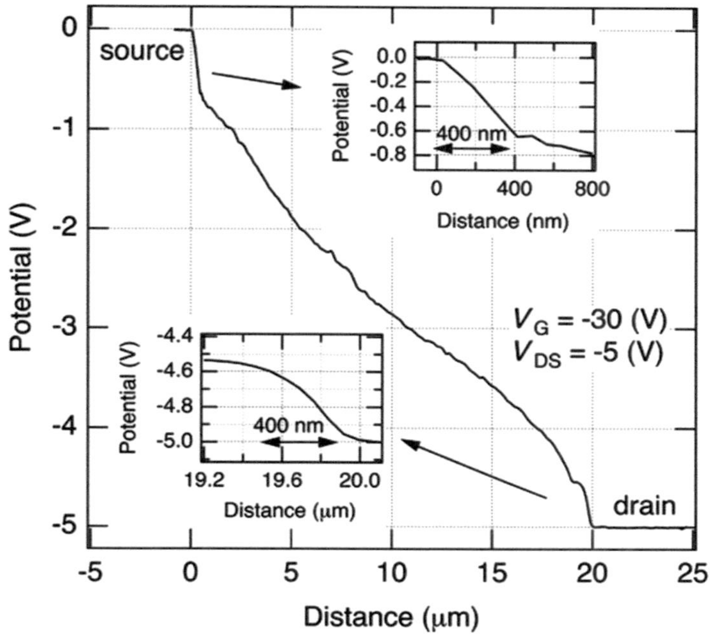


**Fig. 10.8** Potential profile across the channel of a working pentacene TC-TFT (growth temperature: 15 °C). *Insets* are magnified profiles at both ends of the channel. (Reproduced from [6])

The topographic and high-pass-filtered potential images indicate where the degradation has concentrated. In the potential image, potential drops are seen on certain parts of the grain boundaries encircled by dashed ovals. It seems difficult for carriers to flow straight without crossing these potential drops. A histogram of the position of large potential drops was obtained from many similar images, shown in Fig. 10.7c. The influence of the electrodes on the formation of potential drops is obvious from this distribution. Thus, the top contact causes serious electrical damage to the films grown at high temperatures, i.e. those with large grain sizes, because highly resistive grain boundaries are densely formed near the electrodes. As shown in Fig. 10.6, this damage possibly reduces the apparent field-effect mobility to only 10–30 % of the intrinsic one.

Such degradation is also seen when the growth temperature is relatively low. Figure 10.8 shows a potential profile through the channel region taken on a pentacene TC-TFT grown at a relatively low temperature, 15 °C. Steeper potential slopes near the source and drain edges are clearly observed. The width of the steeper slope is around 400 nm equally on each side, which means that it is too wide and symmetric to be caused by an injection barrier. The width corresponded well to the area with a trace amount of Au incident on it, of which the amount is too small to be detected by AFM topographic images. The Au is supposed to be scattered by the edge of the shadow mask or residual gas during the vacuum

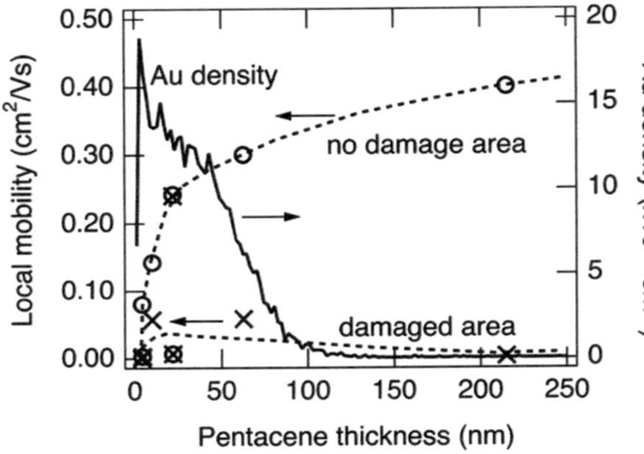


**Fig. 10.9** Pentacene-thickness dependence of local mobilities in damaged (*crosses*) and no-damage (*circles*) areas. The pentacene films were grown with the same conditions as Fig. 10.7. Au depth profile measured by SIMS (*solid line*) is also plotted on right axis. (Reproduced from [6])

evaporation. This steeper potential slope is therefore due to the electrical damage in pentacene induced by the trace amount of Au [18]. In this case, the apparent mobility was estimated to be about 75 % of the intrinsic one.

Depth of the Au-induced electrical damage to the pentacene films deposited at higher temperature was investigated by changing the pentacene thickness. Figure 10.9 shows pentacene-thickness dependence of the local field-effect mobilities in the damaged and no-damage areas calculated from potential profiles under FET operation. An Au depth profile measured by secondary ion mass spectrometry (SIMS) is also shown. According to the depth profile, Au penetrated approximately 100 nm into the pentacene layer. Such deep penetration would be mediated by the Au diffusion through grain boundaries, which has been confirmed by transmission electron microscopy for other organic polycrystalline films [29]. Since the output current flows at the pentacene/insulator interface, once the pentacene layer becomes thicker than the damaged layer, the local field-effect mobility is expected to increase. However, the low local mobility in the damaged area does not recover even when the pentacene thickness becomes much thicker than 100 nm as shown in Fig. 10.9. This result implies that the distribution of the electrical damage is deeper than that of the Au atoms themselves. One possible mechanism of the damage formation down to the bottom of the thick organic film, is the transformation of the crystal structure. Figure 10.10 shows the variation of XRD patterns of a pentacene thin film (grown at a moderate temperature) by the deposition of a thin Au film on it. By the Au deposition, the peak position of the thin-film phase (001) diffraction [12] shifts from 6.032° to 6.068°, which corresponds to the change of the lattice plane distance from 1.46 to 1.45 nm. Since the full width at half maximum (FWHM) of the peak becomes even narrower from 0.12° to 0.10° by the Au deposition, the peak shift is not due to the disordering of the pentacene crystal. These suggest that all of the pentacene molecules in the film have collectively inclined, i.e. became closer to the bulk phase. This transformation changes the molecular packing at the grain boundary from top to bottom of the film, and, as a result, makes the grain boundary more resistive due to the concentration of stress.

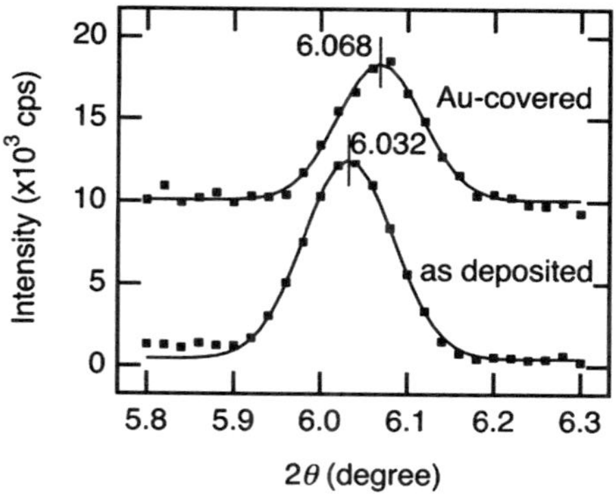


**Fig. 10.10** XRD patterns of an as-deposited pentacene film (growth temperature: 40 °C, thickness: 50 nm, average grain size: 3 μm) and the same film fully covered with a 0.5 nm-thick Au film. Cu-Kα radiation was used for the measurements. (Reproduced from [6])

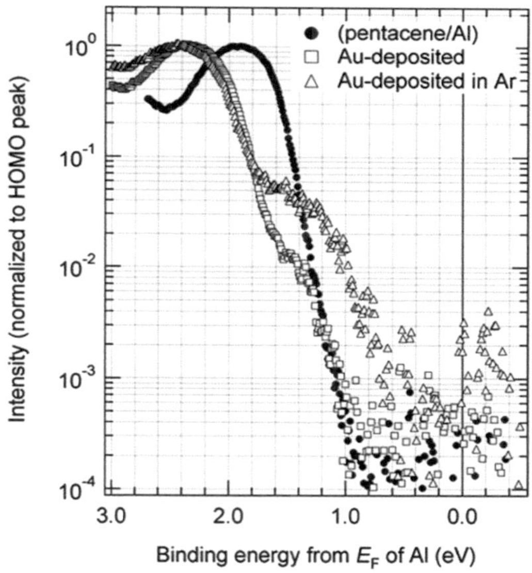


**Fig. 10.11** Photoelectron spectra of pentacene/Al samples: (*circle marks*) without Au deposition, (*square marks*) with Au deposited in vacuum, and (*triangle marks*) with Au deposited in 0.2 Pa Ar. (Reproduced from [31])

Another interesting fact seen in Figs. 10.7 and 10.8 is that no significant injection barrier is likely to exist at the source electrode/pentacene interface although a 0.85 eV barrier at the pentacene/Au interface is confirmed with electron spectroscopy [30]. The absence of the injection barrier may arise due to the carrier injection mechanism into the organic semiconductor with a large density of gap states mostly at grain boundaries, which is an advantage of the top-contact configuration. Figure 10.11 shows the influence of Au deposition on pentacene on the density-of-state (DOS) near the highest occupied molecular orbital (HOMO). These spectra were measured after the removal of the deposited Au layers from the pentacene films [31, 32]. By the Au deposition, the peak position of the pentacene HOMO shifts to

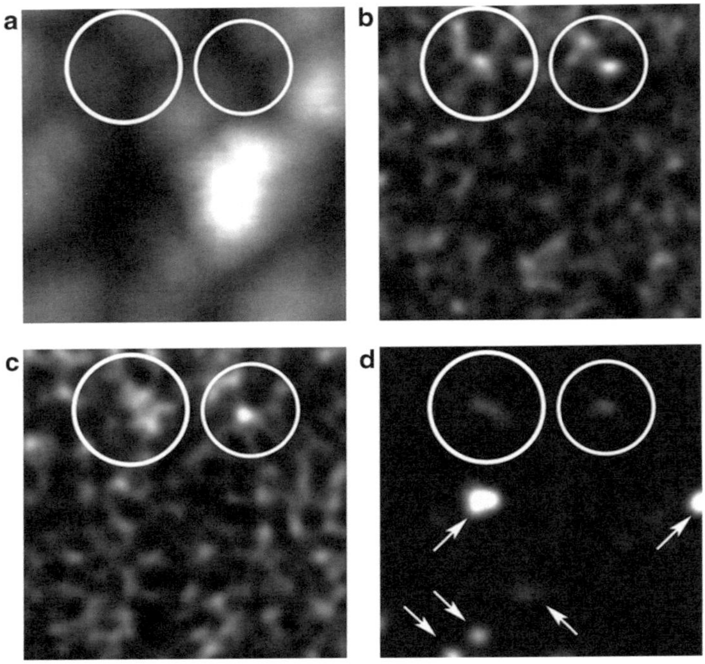


**Fig. 10.12** 500 × 500 $nm^2$ images taken in the same area of a pentacene/Al sample with 0.3 nm-thick Au deposited in 0.2 Pa Ar: (**a**) AFM topography, (**b**) absolute current at $V_{tip} = -2$ V, (**c**) at $V_{tip} = -3$ V, and (**d**) at $V_{tip} = 3$ V. Ranges displayed in the gray scale current images are 0.04 to −0.08 pA (more negative current is brighter) for (**b**) and (**c**), and −0.7 to 6.9 pA (more positive current is brighter) for (**d**). (Reproduced from [31])

the high binding energy side by ~0.5 eV. Moreover, another feature appears at approximately 1 eV above the HOMO level, demonstrating that the gap states appeared upon the Au deposition. The density of the gap states in the pentacene with low-energy Au deposited in Ar is much greater than that with Au deposited in vacuum. This fact indicates that the gap state is not due to the heat damage of pentacene but due to the incorporation of Au atoms in the pentacene film.

The spatial distribution of the large ohmic current flowing vertically through Au deposited pentacene films was studied with AFM current imaging [31]. Figure 10.12 shows (a) topographic and (b–d) current images taken in the same area of a Au (0.3 nm in Ar)/pentacene/Al sample using a Au-coated conductive probe. When the probe is biased negatively (Fig. 10.12b, c), i.e. reverse bias of the diode, electrical conduction appears only at a few specific spots, which are marked by white circles. The ohmic current spots are concluded to be located at the grain boundaries by comparing topographic and current images in Fig. 10.12. When the probe is biased positively (Fig. 10.12d), the forward thermionic current appears at several different points as marked by arrows. The thermionic spots are located even in the grains. Accordingly, the two current components do not uniformly flow but are

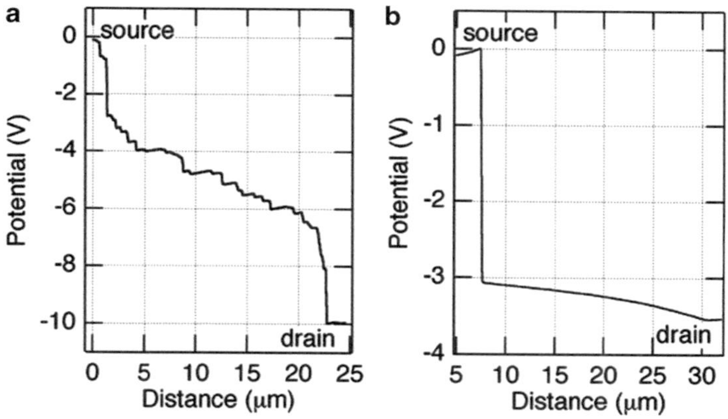


**Fig. 10.13** Potential profiles of working BC-TFTs: (**a**) a pentacene OTFT with large contact resistance and (**b**) a CuPc OTFT with large injection barrier. (Reproduced from [6])

concentrated at different spots in the polycrystalline pentacene film. These images are instructive for study on metal/organic interfaces because they indicate that such singular spots possibly dominate the current conduction. On the other hand, many spectroscopic techniques measure averaged information over the entire film. This difference frequently causes the quantitative disagreement between the electrical and spectroscopic results [32].

In contrast to the small contact resistance of TC-TFTs, there frequently appears large contact resistance, or a carrier injection barrier, when the bottom-contact configuration is used. Figure 10.13a shows an example of the potential profile obtained on an inferior performance BC-TFT with a pentacene active layer. One can see abrupt potential slopes at both sides of the channel region of which width is approximately 1 μm. These are due to small crystal grains of which nucleation and growth is strongly influenced by the edge structure of the source/drain electrodes. In this case, approximately 2/3 of the applied $V_{DS}$ is consumed at the narrow low mobility area near the source/drain electrodes. We have to note that this resistive area does not behave as a fixed resistance but as a very low mobility semiconductor. Since 10 % of the channel length is filled with such a very low mobility semiconducting film, the apparent mobility becomes approximately 1/3 of the intrinsic one. This possibly appears as the $V_G$ dependent "contact resistance" when transfer line method is used for BC-TFTs [33].

When there exists a large injection barrier at the electrode/organic interface, a large abrupt potential drop appears only on the source side as shown in Fig. 10.13b. In this case, only 1/7 of the applied $V_{DS}$ is used for the drift motion of the carriers in the channel region and the rest is consumed to inject the carriers from the source electrode. Furthermore, the effective gate electric field becomes much smaller than expected because the average channel potential is much closer to the negatively biased drain electrode, which results in poor accumulation of carriers and extremely low apparent field-effect mobility, 1/10 to 1/100 of the intrinsic value.

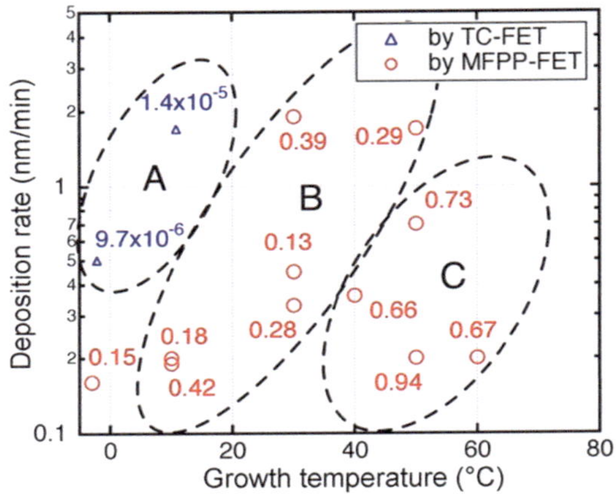


**Fig. 10.14** Carrier mobilities in units of cm$^2$/Vs summarized in the plane of growth temperature and deposition rate. Numbers with *triangle marks* are the apparent mobilities measured by TC-TFT, indicating that these samples are difficult to measure by MFPP-FET. (Reproduced from [6])

### *10.2.4 Crystal Grains and Grain Boundaries*

The problems originated in the source/drain electrodes discussed above are practically the major issue when the reproducibility of the device characteristics is poor and the insufficient characteristics are suspected to depend on the detail of the fabrication methods. In turn, one of the next major issues is the grain boundary of the organic layers. Dependence of the intrinsic field-effect mobility, which means that it does not contain the influence of the contact resistance, on the growth condition is summarized in Fig. 10.14. There is a general tendency that the carrier mobility becomes higher when the film is deposited at higher temperature and lower deposition rate. Here, we classify the films into A, B and C groups as indicated in Fig. 10.14. From the standpoint of film morphology, group A is characterized by the frequent existence of the lamellar grains (brighter parts in Fig. 10.1a), where molecules lie flat against the substrate, and group B mainly contains the pyramidal or the dendritic grains (Fig. 10.1b–d). Group C is composed of the giant grains with many defective structures, such as cracks between grains and depressions in the grains (Fig. 10.1e). Not only the cracks in the film but also the lamellar grains are confirmed to behave as insulators by AFMP analyses, of which a typical potential profile is shown in Fig. 10.15. Molecules lie flat in lamellar grains, while they are nearly perpendicular to the substrate in other types of grains. From this difference, the overlap of π-orbitals between the lamellar and other grains is much smaller than that between the grains with similar molecular orientations. The apparent field-effect mobility is therefore strongly affected by the insertion of these insulating structures into the organic layers especially in group A.

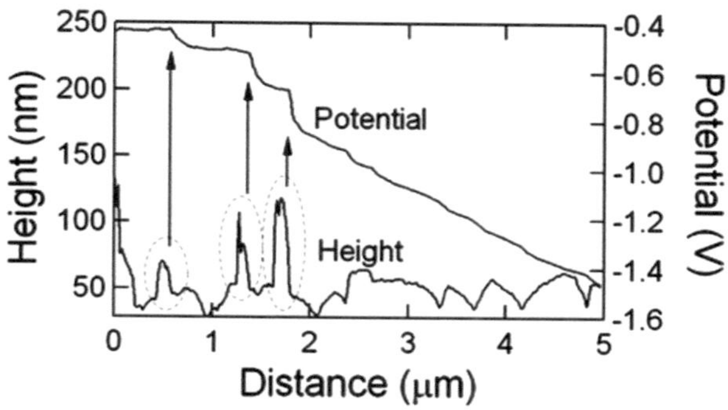


**Fig. 10.15** Height and potential profiles under FET operation taken on a sample including lamellar grains grown at 0 °C. (Reproduced from [6])

## 10.3 Intrinsic Limiting Factors of Carrier Mobility in Pentacene Thin-Film Transistors [6–8]

### *10.3.1 Grain-Boundary–Limited Carrier Transport in Polycrystalline Films*

Figure 10.16 shows the intrinsic carrier mobility values plotted against the average diameter of pentacene grains. The mobility is similarly proportional to the grain size as several have reported [1, 15, 34], except in a sample with extremely large grains, 8 μm in average, grown near the growth limit conditions and containing a large amount of cracks as shown in Fig. 10.1e. This implies that the mobility is limited by the domain boundaries.

Similar results have been obtained in the field-effect measurement of oligothiophene by Horowitz and Hajlaoui [34, 35]. They tried to analyze the property by a conventional "polycrystalline model" where carriers pass over back-to-back Schottky barriers, which are generated by traps at grain boundaries, via thermionic emission. Although the proportionality to the domain size is well explained by the polycrystalline model, the analysis is not so satisfactory because the calculated mean thermal velocity of carriers is too small as a thermal velocity. In the case of pentacene, the mean free path is almost equal to the molecular distance, and typical Debye length of lightly doped pentacene films, e.g. those exposed to the air, is several tens of nanometers. Accordingly, application of the diffusion theory is more appropriate for the analysis of pentacene polycrystalline films. The band diagram used in the polycrystalline model with the diffusion theory is shown in Fig. 10.17. In this model, the externally applied drain-source voltage, $V_{DS}$, is divided by the number of grain boundaries, and thus $V_b = V_{DS}/(L/l)$, where $L$ is the channel length and $l$ the grain size, is applied to a single grain boundary. At a very low $V_b$ condition, $V_b << 4\phi_b$ and $V_b << 2k_BT/q$, the current density flowing through the boundary is described, by using linear approximation of the diode equation, as

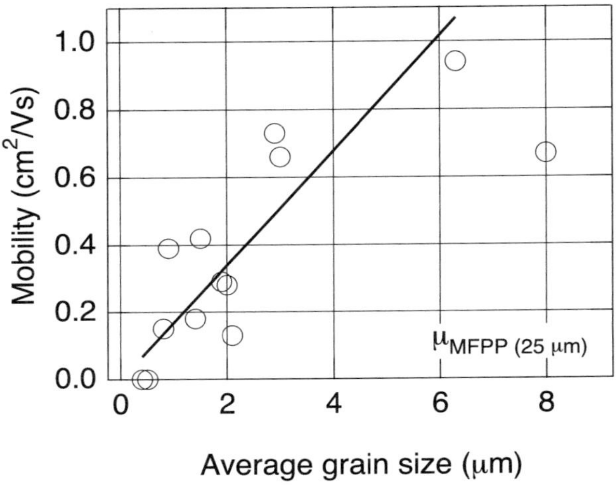


**Fig. 10.16** Plot of carrier mobility obtained with MFPP-FET vs. average diameter of the grains. The probe spacing used was 25 μm. (Reproduced from [6])

$$J = qn_{\mathrm{h}}\mu_{\mathrm{h}}\sqrt{\frac{qN_{\mathrm{A}}\phi_{\mathrm{b}}}{2\varepsilon_{\mathrm{S}}}}\exp\left(-\frac{q\phi_{\mathrm{b}}}{k_{\mathrm{B}}T}\right)\frac{qV_{\mathrm{b}}}{2k_{\mathrm{B}}T}, \tag{10.2}$$

where $q$ is the elementary charge, $n_{\mathrm{h}}$ the hole density in the grain, $\mu_{\mathrm{h}}$ the mobility in the grain, $N_{\mathrm{A}}$ the acceptor density, $\varepsilon_{\mathrm{S}}$ the dielectric constant, $\phi_{\mathrm{b}}$ the barrier height at the grain boundary, $k_{\mathrm{B}}$ the Boltzmann constant, and $T$ the absolute temperature. Then, considering that the current density is macroscopically expressed as $J = qn\mu(V_{\mathrm{DS}}/L)$, the apparent mobility, $\mu$, is given by

$$\mu = \frac{ql}{2k_{\mathrm{B}}T}\mu_{\mathrm{h}}\sqrt{\frac{qN_{\mathrm{A}}\phi_{\mathrm{B}}}{2\varepsilon_{\mathrm{S}}}}\exp\left(-\frac{q\phi_{\mathrm{B}}}{k_{\mathrm{B}}T}\right). \tag{10.3}$$

This model is valid when the true drift mobility $\mu_{\mathrm{h}}$ is relatively high and the grain boundary is resistive enough, where most of the applied drain–source voltage is distributed to the depletion region near the grain boundaries, as depicted in Fig. 10.17. From Eq. (10.3), one may notice that we should increase $l$ and $N_{\mathrm{A}}$, and decrease $\phi_{\mathrm{B}}$ to increase the apparent mobility. The influence of $N_{\mathrm{A}}$ is due to the reduction of the depletion region width near the grain boundary, and it could be a considerable cause of the increase of apparent mobility by oxygen or other dopant species in relatively high mobility materials [36, 37].

### *10.3.2 Estimation of Potential Barriers and In-Grain Mobility*

In some organic thin films, the morphological crystal grain, which is frequently determined by electron microscopy or AFM as a granular cluster, does not necessarily correspond to a single crystalline domain. A pentacene grain in OTFTs

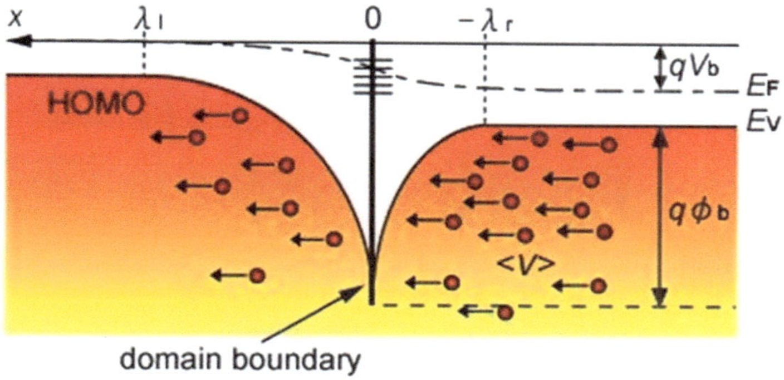


**Fig. 10.17** Band diagram near a grain boundary used for the polycrystalline model with diffusion theory. (Reproduced from [7])

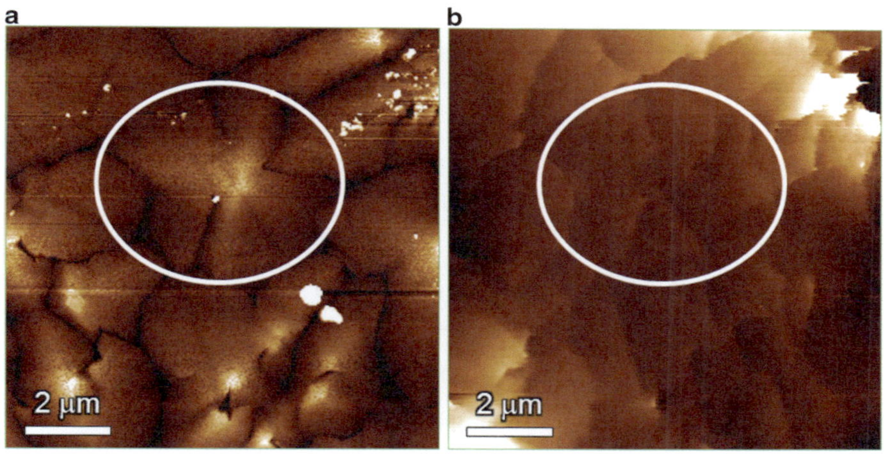


**Fig. 10.18** AFMP images taken at the center of the channel area of a working pentacene OTFT: (**a**) topographic and (**b**) potential (average gradient is subtracted) images. Operation condition of the OTFT is −8 V for $V_{DS}$ and −40 V for $V_G$. (Reproduced from [7])

also possibly contains several differently oriented crystalline domains [38, 39]. Figure 10.18 shows AFMP images taken at the center of the channel area of a working pentacene OTFT. Many potential drops are seen in the potential image. They are the barriers against carrier transport. By comparing the topographic and potential images, many barriers located within the morphological grains can be found. Some of these are locate in the deep notches as seen at the center of the topographic image, but many are at shallow depressions, which are better viewed in Fig. 10.1e (marked by arrows). The latter has a relatively smooth surface but is considered a boundary between differently oriented crystal domains [38]. By the analysis of the $V_G$ dependence of these potential drops, they have been concluded to

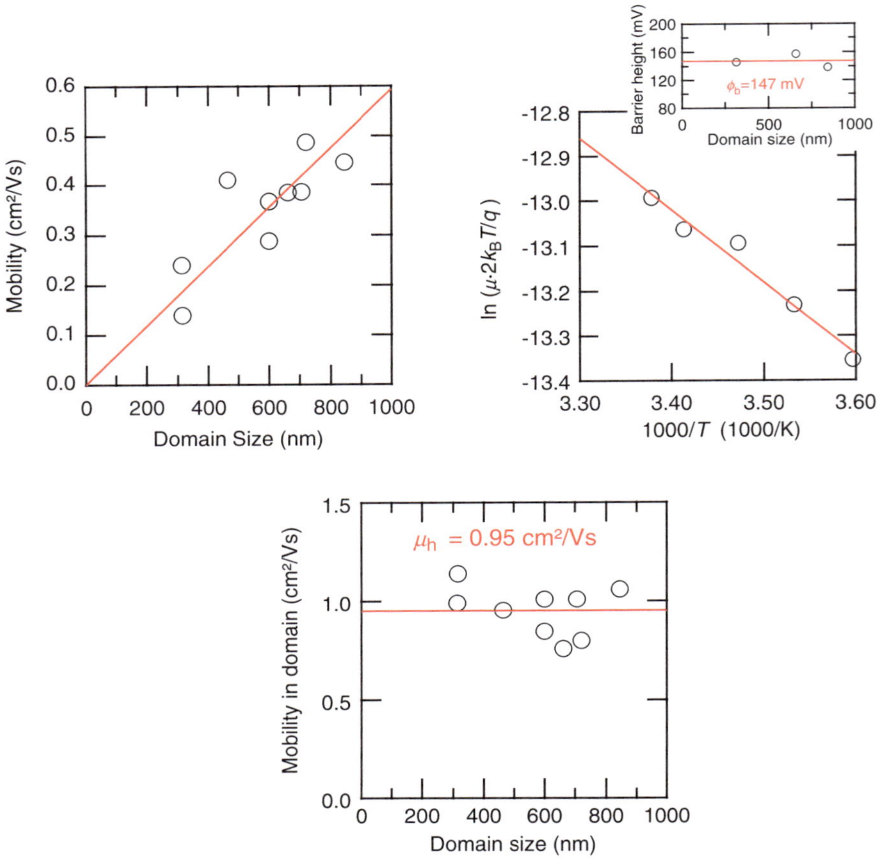


**Fig. 10.19** (**a**) Domain size dependence of the mobility at 296 K measured by MFPP-FET. (**b**) Temperature dependence of mobility in a sample with a domain size of 850 nm. *Inset* shows domain size dependence of the barrier height. (**c**) Domain size dependence of the "mobility in domain" calculated according to the diffusion theory. (Reproduced from [7])

be the same type of potential barriers as the morphological grain boundaries. Therefore, in the analysis of organic polycrystalline films using Eq. (10.3), we have to use the crystalline domain size for $l$ instead of the morphological grain size. In the case of pentacene, domain size is approximately half of the morphological grain size according to the statistical occurrence of the notches [7].

Figure 10.19a shows the re-plotted domain size dependence of the mobility in the films deposited at 296 K. In this graph, only the results showing moderate mobility are used to exclude the influence of the extrinsic limiting factors as explained in Sect. 10.2. Figure 10.19b shows an example of $\ln(\mu \cdot 2k_BT/q)$ vs. $1/T$ plot of the temperature dependence. The plot is linear in this temperature range as expected from Eq. (10.3). Although regular Arrhenius-type and $\ln(\mu \cdot 2k_BT/q)$ vs. $1/T^2$ plots were also examined, the $\ln(\mu \cdot 2k_BT/q)$ vs. $1/T$ plot exhibited the best linearity. In addition, the estimated barrier height in this temperature range is almost independent of domain size as seen in the inset of Fig. 10.19b, and is 150 mV in average.

Using the slope of the $\mu$–$l$ plot in Fig. 10.19a, acceptor density of each sample calculated from the mobility, a film conductivity at $V_G = 0$, and dielectric constant calculated from capacitance–voltage ($C$–$V$) measurement, $4.6\varepsilon_0$, the mobility in crystalline domain, $\mu_h$, can be estimated by Eq. (10.3). Figure 10.19c shows a plot of the estimated $\mu_h$ against the domain size. $\mu_h$ is again independent of domain size, and the average value is $0.95 \pm 0.04$ cm$^2$/Vs.

Let us compare this value with those in other reports. Field-effect mobility of 15–40 cm$^2$/Vs is reported for carefully fabricated pentacene single crystal FETs with organic gate insulators [39]. Over 10 cm$^2$/Vs mobility is also expected from the hole effective masses estimated from band calculation for single crystals [40] and angle-resolved ultraviolet photoemission spectroscopy against single-crystal-like monolayers on atomically flat substrates [41, 42]. These values are over ten times as large as that obtained in this work. This disagreement indicates that, even in a polycrystalline pentacene film on $SiO_2$, there exists another limiting factor of carrier transport in addition to the barriers at domain boundaries. This will be explained in the next subsection.

### *10.3.3 Potential Fluctuation Within a Crystalline Domain [8, 9]*

Figure 10.20a–c show potential images obtained at three different areas in a working pentacene OTFT. One may find that the crystal domains of pentacene commonly contain small potential fluctuations in addition to large potential drops at the domain boundaries, marked by dotted lines in the figure. Similar wavy shapes are observed in almost all the domains. However, not all fluctuations are due to the channel potential, because some of the images show surface-structure-induced potential fluctuations that are superimposed on the channel potential, as shown in Fig. 10.20c. We therefore attempted to eliminate such influence by measuring a profile including a large flat molecular terrace, which is defined as an area with a roughness within a monolayer height (about 1.5 nm). Figure 10.20d shows a topographic image that contains a large molecular terrace elongated in the scanning direction. Topographic and potential profiles were then extracted along the black line in Fig. 10.20d. In the potential profile, small fluctuations are frequently seen, even in the molecular-terrace region. The amplitude of the fluctuations is approximately 12 mV$_{P-P}$, which is much higher than the noise level of 2.5 mV$_{P-P}$ estimated in the electrode area. Furthermore, the fluctuations are not correlated with the sub-monolayer noise in the topographic profile.

Potential fluctuations themselves do not have a physical meaning. We therefore converted the potential profile into a conductivity profile. The relationship between conductivity and potential is

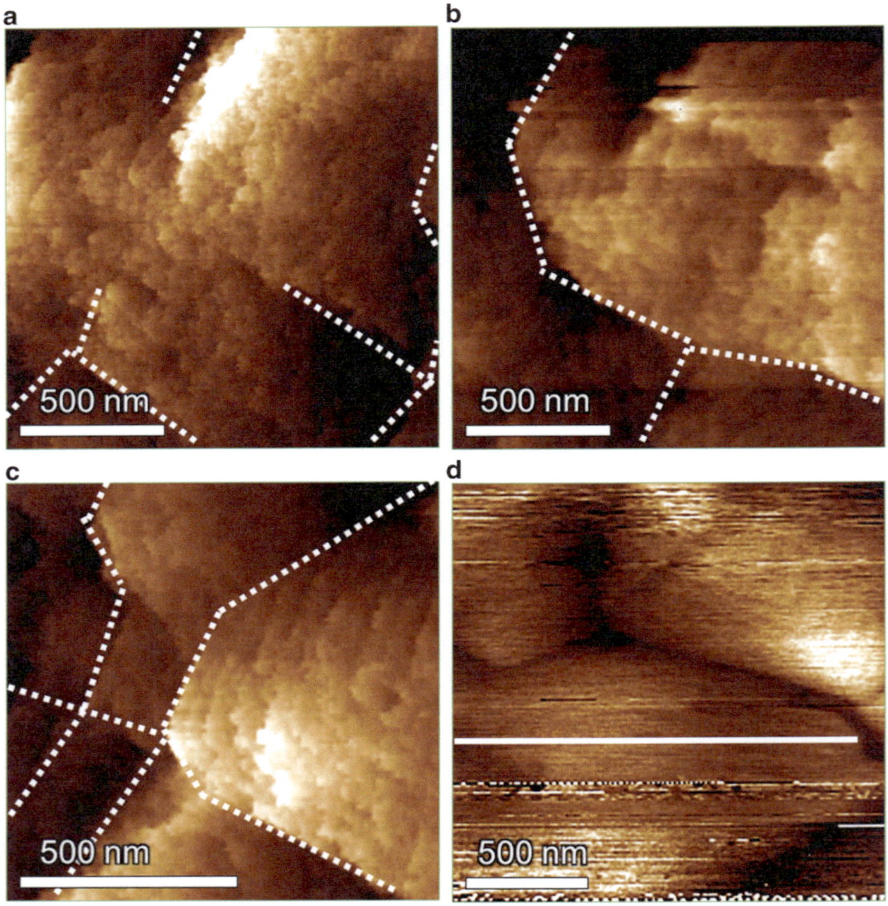


**Fig. 10.20** (**a**)–(**c**) Potential images obtained at different areas in a working OTFT. To show potential fluctuations within domains clearly, the average potential gradient $V_{DS}/L$ was subtracted as a background. Images (**a**) and (**b**) are examples showing potential fluctuations in the channel region, while (**c**) shows molecular-step-induced potential fluctuation. (**d**) Topographic image with a large molecular terrace elongated in the scanning direction (marked by a *solid line*). (Reproduced from [8])

$$\sigma(x) = \frac{J}{E(x)} = \frac{J}{dV/dx}, \tag{10.4}$$

where $J$ is the two-dimensional (2D) current density in the channel, $E$ the electric field in the lateral direction, and $x$ the distance. The current density perpendicular to the source/drain electrodes can be regarded as homogeneous throughout the channel, because the potential gradient in the channel of this OTFT was almost homogeneous, except for small potential drops at domain boundaries and for the

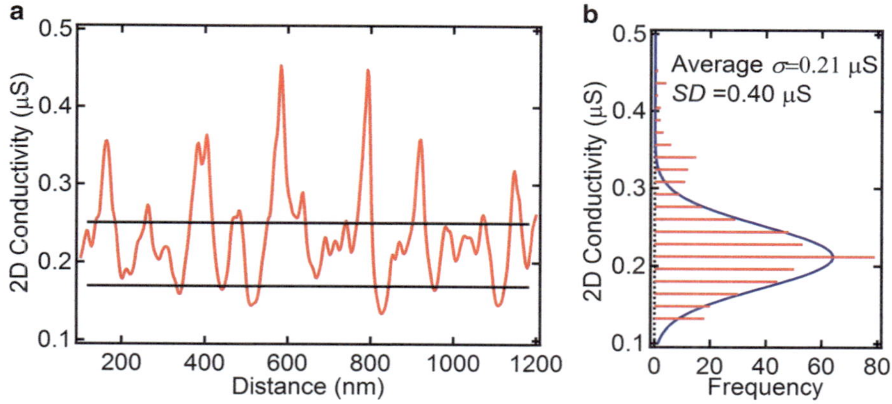


**Fig. 10.21** 2D conductivity in a crystalline domain: (**a**) profile and (**b**) histogram. The two *horizontal lines* in (**a**) represent $\pm SD$ obtained from the histogram in (**b**). (Reproduced from [8])

much smaller fluctuations discussed in this work. Therefore, the current density can be calculated as the source current divided by the channel width. Figure 10.21 shows the converted 2D conductivity profile. Most parts of the 2D conductivity, except for some irregularly higher peaks, are within $\pm SD$ of the mean conductivity (*SD* is the standard deviation). Because of these higher values, the histogram has a broad shoulder on the upper side of the peak. This type of profile cannot be caused naturally by scattered defects in the domain. Therefore, we assume that it originates from the energy fluctuation of the HOMO band edge rather than energetic barriers. The inset of Fig. 10.22a shows a schematic diagram of the band-edge fluctuation in a domain. Band-edge fluctuation may change both carrier density and mobility. First, we assume that the mobility is constant, and later discuss the validity of this assumption.

The 2D drift current in the channel region with band-edge fluctuations is described as

$$J = \sigma(x)E(x) = q\mu N_{\mathrm{v}}\exp\left(-\frac{\varepsilon_{\mathrm{f}} - \varepsilon_{\mathrm{v}}(x)}{kT}\right)E(x), \tag{10.5}$$

where $\mu$ is the mobility, $N_{\mathrm{v}}$ the 2D effective density of states in the HOMO band, $\varepsilon_{\mathrm{f}}$ the Fermi level, and $\varepsilon_{\mathrm{v}}$ the top level of the HOMO band. By rearranging Eq. (10.5), $\varepsilon_{\mathrm{v}}$ can be extracted from the potential profile as follows:

$$\ln[1/E(x)]kT = \left(\ln\frac{q\mu N_{\mathrm{v}}}{J}kT - \varepsilon_{\mathrm{f}}\right) + \varepsilon_{\mathrm{v}}(x). \tag{10.6}$$

Figure 10.22a shows the extracted band fluctuation in the domain. The fluctuation profile has become symmetric about the energy axis as in Fig. 10.22b and become closer to that of a normal distribution. This result strongly indicates that the

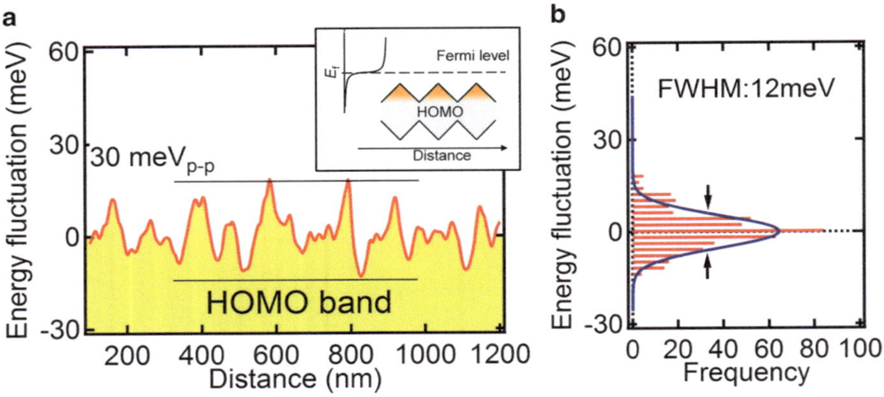


**Fig. 10.22** Fluctuation of HOMO-band edge in a crystalline domain calculated using Eq. (10.6): (**a**) profile and (**b**) histogram. The *inset* of (**a**) shows a schematic diagram of the band fluctuation. (Reproduced from [8])

major origin of the potential fluctuation is the fluctuation of the HOMO-band edge as we assumed, because the histogram of a random event usually follows a normal distribution. Here, the maximum amplitude within the domain must be much smaller than the barrier height of 110 meV at the domain boundaries [7] since the mobility of pentacene OTFTs shows thermally active behavior with a 110 meV barrier. The maximum amplitude of 30 meV and the FWHM of 12 meV obtained in this study are consistent with this requirement.

Here, let us discuss whether or not mobility can be regarded as a constant. Mobility is proportional to the bandwidth according to band theory. Therefore, band fluctuation may change the local mobility if the band fluctuation originates from the HOMO bandwidth. On the basis of the model shown in the inset of Fig. 10.22a, we can estimate a possible maximum ratio for the local-mobility variation in the domain from the bandwidth of the single crystal, 210 meV [41], and the maximum amplitude of the band fluctuation, 30 meV. The maximum ratio obtained is 1.3, which is much smaller than that of the fluctuating conductivity, 3.2. The influence of mobility variation on the conductivity fluctuation is therefore concluded to be minor.

According to these results and the discussion above, the potential fluctuation observed in the domains must originate from the HOMO-band-edge fluctuation. In such a situation, the band fluctuation notably degrades the in-domain mobility, $\mu_h$, in Eq. (10.3). We propose two possible degradation mechanisms: (a) If the band fluctuation behaves as shallow traps, carrier transport within a domain follows a multiple-trapping and release mechanism, [43] and (b) if current flow meanders through the high-conductance region for small $V_{DS}$, carrier transport within the domain follows a percolation mechanism [44]. The conductance in a domain is proportional to the power of the space-filling factor in this case. According to the THz-wave absorption

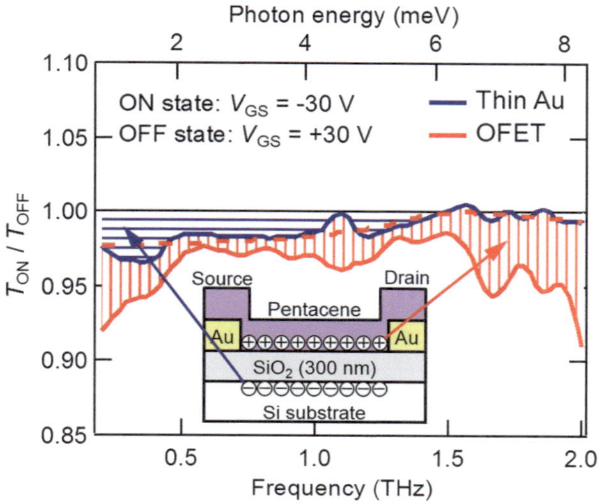


**Fig. 10.23** THz-wave modulation absorption spectra of a pentacene OTFT measured by time-domain spectroscopy. The *red thick line* indicates the decrease of THz-wave transmission by the accumulation of carriers in an OTFT, and the *blue thin line* that of a control sample where only the absorption by the carriers in the Si substrate appears. *Dashed line* is a theoretical curve obtained using the Drude model. (Reproduced from [9])

spectra, which are explained later, it is highly probable that the multiple-trapping and release mechanism dominantly determines the typical carrier transport phenomena in OTFTs. Either of these mechanisms results in the reduction of the mean carrier velocity in a domain. It thus appears as if $\mu_h$ is much smaller than, approximately 1/10 at room temperature, that of perfect single crystals.

The potential fluctuation, of which root-mean-square amplitude is around 10 meV, is relatively small and may not be clearly noticed by most of spectroscopic analyses at room temperature. However, it strongly influences the carrier motion even in a short distance range. Figure 10.23 shows THz absorption spectra by gate-field-induced electrons in a single-crystal silicon (horizontally hatched area) and gate-field-induced holes in an OTFT-grade pentacene thin film (vertically hatched area) [8]. Although the spectral shape of the electron in silicon is well explained by the Drude model, which classically describes the dynamics of free electrons in solids, that of the holes in pentacene cannot. The absorption increases by increasing frequency from 0.5 to 2.0 THz, while a Drude-like tail appears at the low frequency end. The Drude component appears only after sufficiently high on-state gate voltage is applied, as seen in Fig. 10.24.

The potential environment for the holes in pentacene as depicted in Fig. 10.22a could explain this shape. When the frequency becomes closer to 2.0 THz, which corresponds to 8.27 meV photon energy, holes at the bottom of the fluctuated band become able to overcome the surrounding potential barriers. Since the local barrier

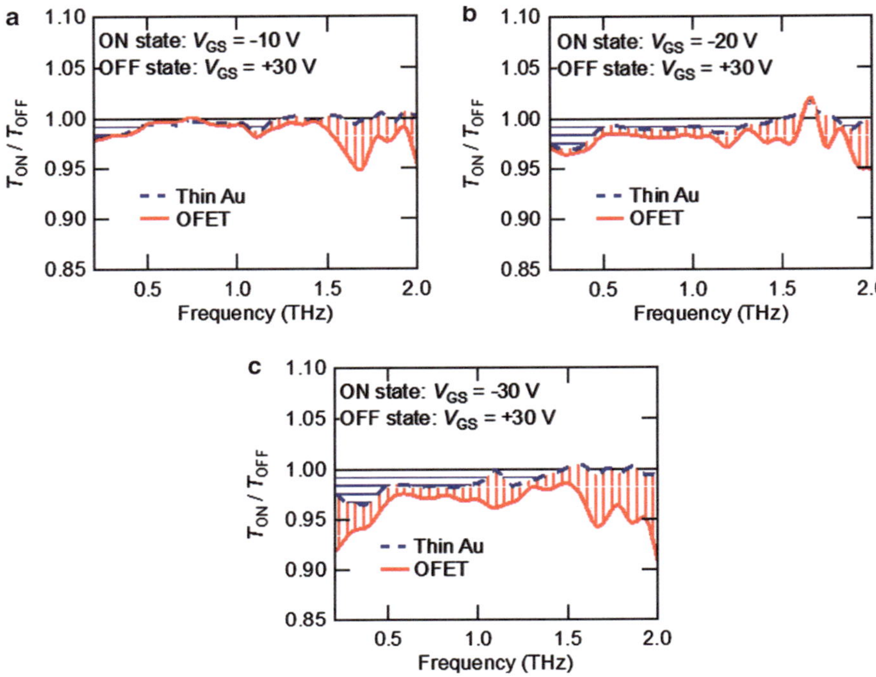


**Fig. 10.24** Variation of modulation absorption spectra of a pentacene OTFT (*red thick line*) and a control sample (*blue thin line*) for on-state gate bias [(**a**): −10, (**b**): −20, and (**c**): −30 V]. *Vertically hatched area* denotes absorption by holes accumulated in pentacene. (Reproduced from [9])

height is distributed from a few to 20 meV and the experiment is done at room temperature, increasing the photon energy in this measurement range gradually increases the probability to accept the photon energy by the weakly-restricted holes, and they gain the kinetic energy to go across the barriers. During the usual OTFT operation, this activation process must be made by the thermal energy, i.e. the mobility in the domain becomes a thermal-activation type and must be smaller than the maximum mobility that is expected by the effective mass and the relaxation time by scattering. The increase of Drude components with the application of higher gate voltage can be explained by the formation of a large carrier pool at the fluctuated band edge. The accumulated holes in the large pool may move freely without significant restriction, as in the single-crystal silicon. However, the free motion of such holes would not contribute to the field-effect carrier mobility, which is the index of the long-distance carrier transport, unless the carrier pool grows as large as the channel length.

## 10.4 A Model to Express the Apparent Carrier Mobility in Practical Thin-Film Transistors

In this section, we summarize a model to account for the apparent and 'intrinsic' carrier mobility in practical OTFTs where polycrystalline films are used as active layers. Here, the word 'intrinsic' does not mean that the semiconductor layer is a perfect single crystal but that the overall carrier mobility is not influenced by many extrinsic limiting factors that are explained in Sect. 10.2. Although this model is derived from the experimental results on pentacene thin films grown on $SiO_2$ substrate, it could be universally used for most of high-mobility organic polycrystalline thin films.

When the resistance of the grain (domain) boundary is high enough to consume almost all of the applied drain–source voltage, the boundary-limited mobility is well explained by the modified polycrystalline model with diffusion theory and is expressed as

$$\mu_{\rm BL} = \mu_{\rm DI}\sqrt{\frac{qN_{\rm A}\phi_{\rm b}}{2\varepsilon_{\rm S}}}\exp\left(-\frac{q\phi_{\rm b}}{k_{\rm B}T}\right)\frac{ql}{2k_{\rm B}T}, \tag{10.3'}$$

where $\mu_{\rm DI}$ is the mobility for the domain interior which is formally expressed as the mobility in the grain (domain), $q$ the elementary charge, $l$ the grain size, $N_{\rm A}$ the acceptor density, $\varepsilon_{\rm S}$ the dielectric constant, $\phi_{\rm b}$ the barrier height at the grain boundary, $k_{\rm B}$ the Boltzmann constant, and $T$ the absolute temperature.

Then, $\mu_{\rm DI}$ is not the one obtained in the single crystal and notably reduced by the potential fluctuation in the grain (domain). As shown in Fig. 10.22, potential distribution of the fluctuated band edge can also be regarded as an exponential distribution of the tail state of which decay energy is around 10 meV. The influence of such a tail state on carrier transport is well studied for inorganic amorphous semiconductors, e.g. a-Si or a-Ge, where the gap-state distribution is frequently referred as an "Urbach tail". Various models have been proposed to precisely explain the carrier transport in amorphous semiconductors. There exists however a quantitative difference between the tail states in inorganic semiconductors and pentacene. The decay energy of the inorganic tail states is two to three times larger than that of pentacene and, furthermore, their spatial period of band-edge fluctuation is much shorter than carrier mean free path in contrast to the much longer period in pentacene. The "detrapping and drift" phenomenon in pentacene crystal grain (domain) is therefore expected to be much closer to a simple thermally activated process. Nevertheless, as long as the temperature dependence is noted, Arrhenius-like temperature dependence is observed for both inorganic [45] and organic [46] semiconductors. In this chapter, we therefore assume that the mobility in domain interior can be simply expressed by

$$\mu_{\rm DI} = \mu_0\exp\left(-\frac{q\phi_{\rm f}}{k_BT}\right), \tag{10.7}$$

where $\mu_0$ can be now regarded as the mobility in single crystal and $\phi_f$ the mean barrier height in the fluctuating band. The temperature dependences of Eqs. (10.3') and (10.7) are slightly different due to the pre-exponential factors in Eq. (10.3'). It would however be difficult to separate these parameters, $\phi_b$ and $\phi_f$, only by the temperature dependence of apparent mobility, if they are close to each other.

When the voltage consumption by the grain (domain) interior is not negligible, we have to consider the series connection of the two equivalent registers: total resistance of the part where the carrier mobility is limited by the grain (domain) boundary, and that limited by the band-edge fluctuation. The overall mobility of such a polycrystalline film is now expressed by

$$\frac{1}{\mu} = \frac{\alpha}{\mu_{BL}} + \frac{1-\alpha}{\mu_{DI}}, \tag{10.8}$$

where $\alpha$ is the ratio between the total length of the grain (domain) boundary region and the channel length.

## 10.5 Crystal Order in Pentacene Crystals on $SiO_2$ and its Influence on Electronic Band Structure [10]

### *10.5.1 GIXD Analysis of Pentacene Films Grown at Various Temperatures*

As explained in the previous part of this chapter, the band-edge fluctuation within a crystalline domain is the major factor reducing the apparent mobility. In this section, the structural origin of the potential fluctuation will be discussed by using glazing-incidence X-ray diffraction (GIXD) analyses.

There are two possible origins of the potential fluctuations:

(1) The crystalline domain itself is a single crystal but contains random strains, or
(2) The crystalline domain is a mosaic of smaller crystallites and potential fluctuations are generated due to the crystallite boundaries.

Therefore, the microstructures in pentacene thin films need to be precisely evaluated to determine the major cause of the band-edge fluctuation. Some groups have reported the correlation between out-of-plane crystallite sizes of pentacene films on $SiO_2$ and their mobility [8, 47]. However, there have been a limited number of reports that mention the in-plane microstructures in pentacene thin films [48, 49], even though they are more important for carrier transport in a TFT channel rather than out-of-plane ones. This is mainly because crystallite sizes of organic polycrystalline materials tend to be larger than those of inorganic ones, which obliges us to use narrower-width slits or an analyzer crystal to measure very narrow FWHM of the diffraction peaks, and also because organic materials are composed mainly of

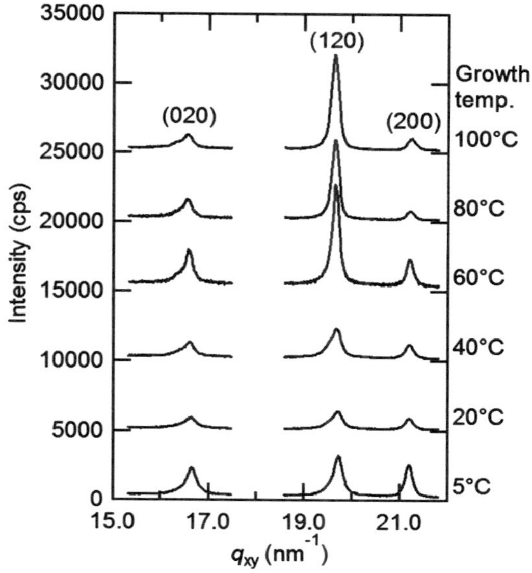


**Fig. 10.25** In-plane GIXD patterns of pentacene thin films grown on $SiO_2$ substrates at 5, 20, 40, 60, 80, and 100 °C. The growth rate was 0.3 nm/min. The indexes of a thin-film phase pentacene crystal are labeled. (Reproduced from [10])

light elements. Since both of them significantly weaken the intensity of diffracted X-ray, it is almost impossible to use a laboratory-scale GIXD instrument for this purpose. Therefore, the use of very bright X-ray from a synchrotron radiation source is necessary for the GIXD analyses of in-plane microstructures in organic thin films.

Pentacene films grown on $SiO_2$ substrates at 5, 20, 40, 60, 80, and 100 °C were used in this analysis. By increasing growth temperature, average morphological grain size is widely increased from 0.4 to 2.5 μm. Figure 10.25 shows GIXD patterns measured with the pentacene films. Among many detectable in-plane diffraction peaks, these three peaks shown in the figure, corresponding to the thin-film phase [50], were selected for further analysis because they exhibited clear peak shapes with sufficient signal-to-noise ratios. It should be noted that the (020) and (120) peaks have asymmetric shoulders. By fitting two Voigt functions to these peaks, the lattice constant obtained from the major components was equal to that of the thin-film-phase pentacene, and the constant from the minor peaks was in between those of the thin-film and the bulk phases, of which composition ratio increases while increasing the thickness [51]. This suggests that the minor component is an intermediate state wherein the crystal structure of the pentacene changes from a thin-film phase to a bulk phase. In the structural analyses, only the major thin-film-phase components were used because the minor component is judged not to affect carrier transport in the film.

As mentioned previously, the peak widths of XRD patterns are broadened not only by the size of crystallite but also by the random strain in the crystal structure. The Williamson-Hall (W-H) method can separate these two factors from the widths

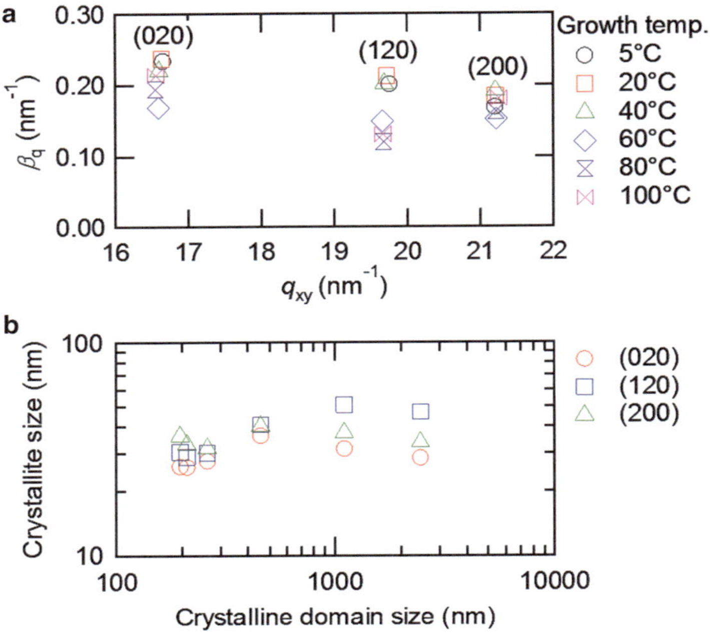


**Fig. 10.26** (**a**) Williamson-Hall plot of the major GIXD peaks in Fig. 10.25. (**b**) Relationship between crystallite size estimated using the Scherrer equation and the polycrystalline domain size of the pentacene film. A Scherrer constant value of 0.95 was used to calculate the crystallite sizes. (Reproduced from [10])

of the XRD peaks [52]. In this method, the relationship between these structural parameters and peak width is described as follows:

$$\beta_q = \frac{2\pi K_S}{\langle D \rangle} + \varepsilon_{rms} q_{xy}, \tag{10.9}$$

where $\beta_q$ is the FWHM of a peak, $K_S$ is the Scherrer constant, $<D>$ is the average crystallite size, $\varepsilon_{rms}$ is root mean square of the random strain, and $q_{xy}$ is the scattering vector. The Scherrer constant is determined depending upon the composition and type of crystal. For organic crystals, values of 0.87–1.05 are frequently used [53–56]. In this work, we adopt $K_S = 0.95$ which is in the middle of the range used for organic crystals. By fitting a straight line to the $\beta_q$ versus $q_{xy}$ plot, the random strain and crystallite size can be estimated from the positive slope and the positive intercept of the fitted line, respectively.

Figure 10.26a shows W-H plots of all major peaks in Fig. 10.25. It is to be noted that no sample has a clear positive slope in its W-H plot. This result indicates that the FWHMs of the X-ray diffraction peaks from the pentacene films are determined by their finite crystallite sizes and the contribution of random strain is negligible.

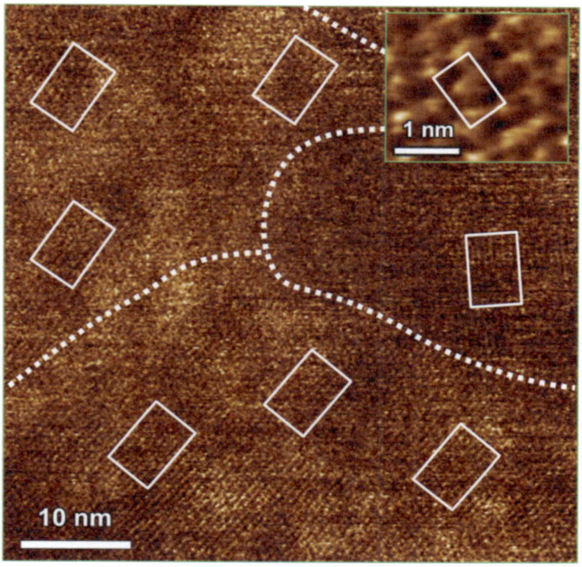


**Fig. 10.27** A molecular lattice image obtained with an FFM on a flat molecular terrace of a pentacene film. The *inset* shows a magnified image to show a unit cell. *Rectangles* in the wide image show directions of unit cells in each area and the *broken lines* show boundaries between differently oriented domains. (Reproduced from [10])

Therefore, hereafter, we only discuss the crystallite size estimated for each peak by the Scherrer equation,

$$\langle D \rangle = \frac{2\pi K_{\mathrm{S}}}{\beta_{\mathrm{q}}}. \quad (10.10)$$

In Fig. 10.26b, the crystallite size calculated from each peak is plotted against the polycrystalline domain size of the film. Although polycrystalline domain sizes are varied more than ten times by changing the growth temperature, surprisingly, the crystallite sizes are unchanged. The estimated crystallite size is within the range of 25–50 nm and one- to two-orders-of-magnitude smaller than the domain size.

To verify the small crystallite size estimated from GIXD, scanning probe microscopy is a good complementary technique. We observed molecular lattice images on flat terraces of a 5-nm-thick (3–4 molecular layers) pentacene film using high-resolution frictional force microscopy (FFM) [57, 58]. Figure 10.27 shows a $50 \times 50\ \mathrm{nm}^2$ FFM image along with a magnified one indicating the unit cell structure of thin-film-phase pentacene. White rectangles denote directions of unit cells in each area. Shorter and longer sides correspond to *a* and *b* of the unit cell, respectively, and its direction was determined by the period of the molecular rows. Judging from the direction of molecular rows, there exist at least three differently oriented crystallites within this area. Similar images are observed for many different terraces in different grains, which strongly support the small (25–50 nm) crystallite size estimated by GIXD.

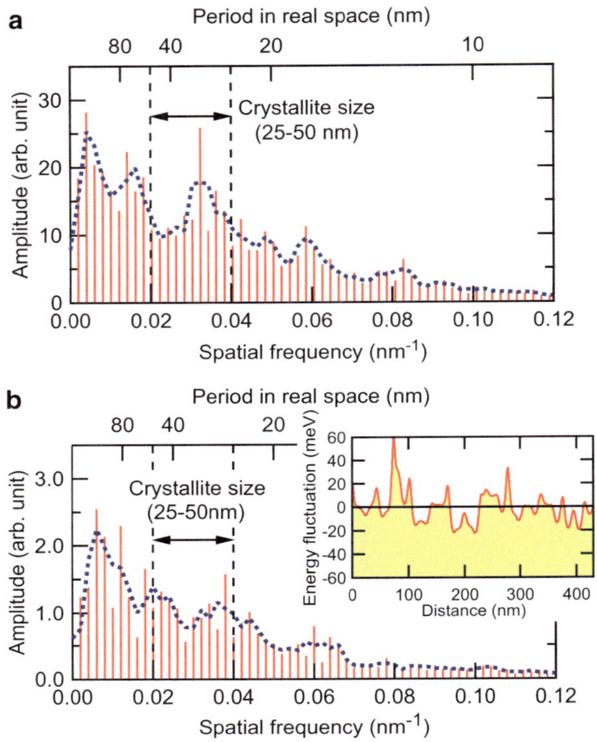


**Fig. 10.28** FFT spectrums of (**a**) the surface topography of the $SiO_2$ substrate (average of eight spectra) and (**b**) the HOMO-band edge profile of the sample grown at 60 °C. Two pairs of *dashed vertical lines* show the range of crystallite size estimated in this work. *Thick dashed lines* are smoothing results of the FFT spectra using the 5-point Savitzky–Golay method. The *inset* in (**b**) is a typical HOMO-band-edge profile of the same sample taken on a flat terrace of pentacene. (Reproduced from [10])

### *10.5.2 Origin of the Temperature Independent Small Crystallite Size and Band-Edge Fluctuation*

As shown in Fig. 10.4, the size of a polycrystalline domain generally increases by elevating the growth temperature. This is explained by the decrease in the nucleation density due to the increased desorption rate of the molecules from the growing nuclei. Since film growth of pentacene is strongly two-dimensional in this thickness range, the nucleation density at the initial stage of the growth, during the growth of the first or second molecular layer, determines the polycrystalline domain size. On the other hand, the size of the crystallite is determined by the generation probability of stacking faults during horizontal crystal growth, which interrupts the coherency of the molecular lattice.

Figure 10.28a shows a fast-Fourier-transform (FFT) spectrum of the surface topography of the $SiO_2$ substrate. A clear peak is seen within the range of the crystallite size (25–50 nm), while totally three peaks are prominent in the spectrum. Two peaks seen in the lower frequency side are components with double and quadruple periods of the highest frequency one. Minor features in the higher frequency side also indicate the surface topography unless the period does not become smaller than the tip radius, less than 10 nm. Their amplitudes are, however, much smaller than those of the major peaks. Consequently, the highest spatial frequency of the characteristic surface corrugation of the substrate is at around 0.03 $nm^{-1}$ (30 nm in period), which agrees well with the crystallite size. Based on the above fact, it is highly probable that the crystallite size is regulated by the corrugation of the substrate surface, which induces stacking faults during the lateral growth of the first or second layer (In the pentacene thin-film growth, the second layer possibly determines the entire crystal structure of the thicker film [59]). Once these stacking faults are introduced, successive growth of the upper layers will follow the underlying crystallographic structure. Thus, crystallite size cannot be increased even by elevating the growth temperature.

Single crystal pentacene exhibits HOMO band dispersion with a bandwidth of a few hundred meV, and its carrier mobility reaches 40 $cm^2/Vs$ [39–42]. These characteristics are produced by high coherency of the molecular lattice and a relatively large transfer integral of the π orbitals. Under this condition, both random strain in the molecular lattice and disruption of lattice coherency lead to modulation of the band structure. According to the GIXD results, the lattice coherency of the pentacene thin films on $SiO_2$ is mainly interrupted by the small crystallite size rather than the random strain. Therefore, the mosaic structure of the polycrystalline domain with small crystallites would be the most probable cause of the band-edge fluctuation. To confirm the validity of this assumption, characteristic period of the band-edge fluctuation was also compared to the crystallite size. Inset of Fig. 10.28b shows a typical line profile of the band-edge fluctuation obtained with AFMP on a crystalline domain [7] and Fig. 10.28b shows its corresponding FFT spectrum. Although AFMP measures a completely different physical property from the topography of the substrate, the FFT spectrum has similar prominent peaks. Based on the same discussion as for Fig. 10.28a above, the highest spatial frequency of the characteristic band-edge fluctuation is concluded to be around 0.03 $nm^{-1}$ (30 nm in period), which again agrees well with the crystallite size. The characteristic periods of the band-edge fluctuation are nearly constant among many pentacene films grown at different temperatures. Accordingly, it is highly probable that the mosaic structure with small crystallites causes the band-edge fluctuation.

### *10.5.3 Overview of HOMO-Band-Edge and Crystallographic Structures in Pentacene Thin Films*

Figure 10.29 schematically summarizes the energy diagram and film structure of a pentacene polycrystalline film based on the knowledge introduced in Sects. 10.1–10.5. A pentacene polycrystalline film consists of hierarchical structures, including (i) the micrometer-scale morphological grain, (ii) the polycrystalline domain surrounded by large energy barriers for carrier transport, (iii) the ten-nanometer-scale crystallite, (iv) the unit cell of the crystal, and (v) the molecule. A morphological grain of pentacene on $SiO_2$ has its nucleus at around the center and four branches grow from the nucleus toward four directions. Each branch finally becomes the crystalline

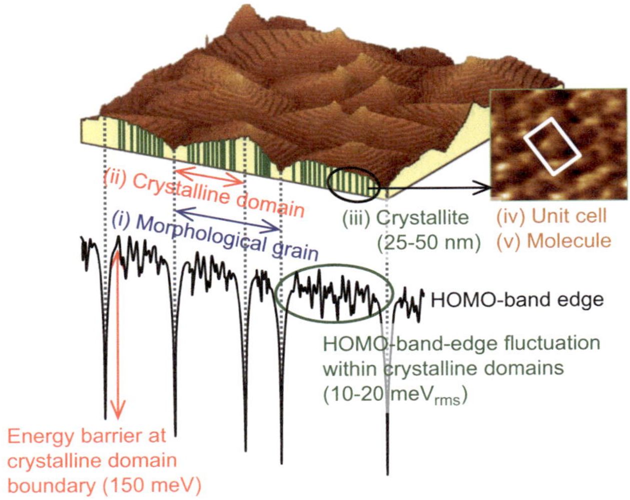


**Fig. 10.29** Schematic illustration showing (*top*) crystallographic and (*bottom*) HOMO-band-edge structure in a pentacene polycrystalline film. A pentacene polycrystalline film consists of hierarchical structures, including *i*) the μm-scale morphological grain, (*ii*) the polycrystalline domain surrounded by large energy barriers for carrier transport, (*iii*) the 10-nm-scale crystallite, (*iv*) the unit cell of the crystal, and (*v*) the molecule. (Reproduced from [10])

domain (ii). During film growth, the branches coalesce so as to form, in most cases, a smooth pyramid-shaped morphological grain (i) and the boundary of (ii) is invisible in an AFM image. The boundaries of (i) and (ii) are however almost equivalent and crucial limiting factors for carrier transport because their barrier height is more than 100 meV. Moreover, (ii) is still not a single crystal and the boundary of (iii) introduces 10-meV-scale fluctuations of the band edge. The barrier height at the boundary of (ii) is much higher than that of (iii) because crystal orientations are more different between two neighboring domains than between connected crystallites due to the different orientation of triclinic crystals. Overall apparent carrier mobility of the film is greatly reduced by the band-edge fluctuation by (iii) because it is proportional to the mobility in (ii) (see Eq. (10.3')).

## 10.6 Influence of Chemical Composition of Insulator Surface on Carrier Mobility [11]

### *10.6.1 Surface Chemical Treatments to Improve Carrier Mobility*

Among various techniques to enhance mobility, changing surface chemical composition of a gate insulator surface using self-assembled monolayers (SAM) such as hexametyl-disilazane (HMDS) [60, 61] and octadecyltrichlorosilane (OTS) [1, 2, 15, 62],

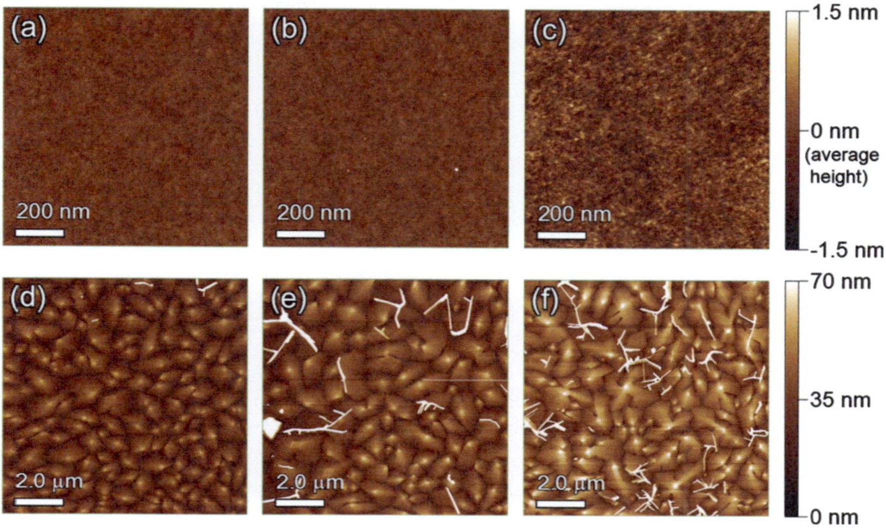


**Fig. 10.30** AFM height images of the substrate surface of (**a**) bare $SiO_2$, (**b**) HMDS-treated $SiO_2$, (**c**) polyimide (PI), and 30-nm-thick pentacene films grown on (**d**) bare $SiO_2$, (**e**) HMDS-treated $SiO_2$, (**f**) PI. Growth temperatures was 60 °C, and deposition rate was fixed at 0.3 nm/min

or polymer dielectrics such as polyimide (PI) [2, 63, 64] and polymethyl methacrylate (PMMA) [65–67], are the most frequently used techniques because of their simplicity and high reproducibility. However, despite their usefulness, it has been a mystery how the surface treatments increase the carrier mobility. Some groups present the correlation between surface free energy or wettability of a gate insulator surface and the mobility [1, 65, 68]. Other groups argue that an increase in the adhesion of molecules to dielectrics or adjacent molecules enhances the mobility [15, 69]. In this section, influence of the surface treatments of the gate-insulator on the carrier mobility is elucidated based on the knowledge of mobility limiting factors discussed in the previous part of this chapter. Three types of pentacene TFTs with different gate-insulator surfaces, bare, HMDS-treated, and PI-coated $SiO_2$ are compared to confirm which limiting factor is the major cause of the mobility increase.

### *10.6.2 Influence on Film Morphology and Crystallographic Structure*

Since growth of pentacene polycrystalline films is significantly influenced not only by the surface chemical treatment but also by adsorbates or contaminations on the substrate surface, we have to eliminate these unintentional factors to evaluate the influence of surface treatment correctly. Figure 10.30 shows AFM images

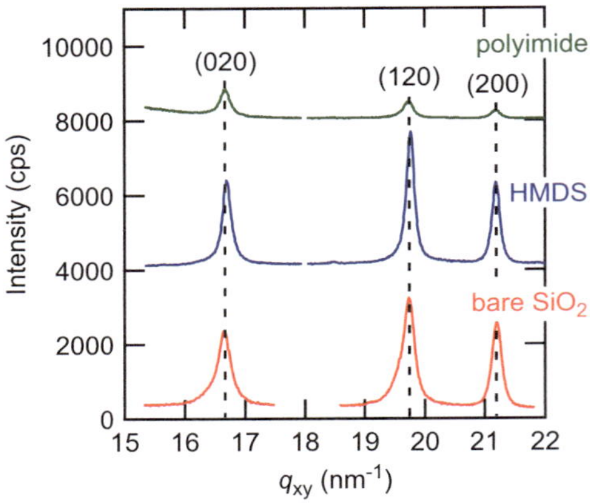


**Fig. 10.31** GIXD patterns of pentacene film grown on each substrates grown at 60 °C

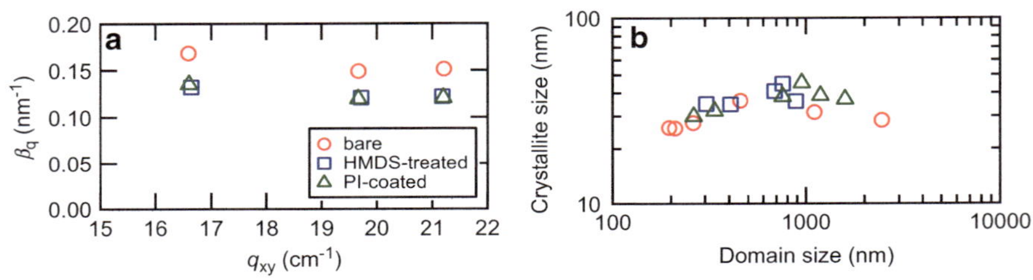


**Fig. 10.32** (**a**) Williamson–Hall plots of the GIXD patterns obtained by the pentacene films grown at 60 °C and (**b**) relationship between average crystallite size and crystalline domain size

of the substrate surface used in this work and pentacene films grown on these substrates. As shown in Fig. 10.30a–c, there exists very little difference in surface roughness and no observable adsorbate or particle can be seen on any substrates. Universality of the similar images as in Fig. 10.30a–c is confirmed in other areas over several tens of μm$^2$.

Morphological shape of the pentacene crystals grown on these substrates does not show any notable difference except their grain sizes as shown in Fig. 10.30d–f. The difference of grain size must be due to the difference in the chemical composition of the surfaces. The average domain sizes on the bare, the HMDS-treated, and the PI-coated $SiO_2$ are estimated to be 455, 753, and 945 nm, respectively. On each substrate, pentacene films with various domain sizes were prepared by changing the growth temperature.

Figure 10.31 shows GIXD patterns from the same films in Fig. 10.30d–f. The positions of the peaks obtained are the same for all three samples and correspond to the diffraction of typical thin-film phase pentacene [12, 13]. Figure 10.32a shoes W-H plots of the GIXD patterns observed from the pentacene films. The W-H plots are nearly horizontal for all samples. These results indicate that only the crystallite size determines the width of GIXD peaks and the contribution of

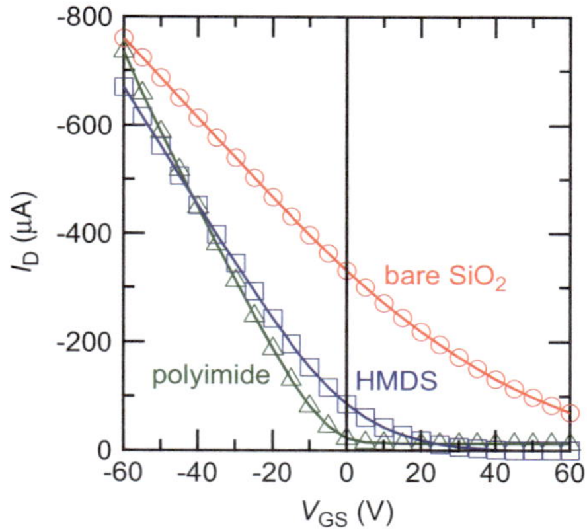


**Fig. 10.33** Transfer characteristics of pentacene TFTs fabricated on three different substrates. The growth temperature of pentacene films was 60 °C and the drain-source voltage during the measurement was set to −10 V

random strain is negligible also on HMDS-treated and PI-coated $SiO_2$ as well as on bare $SiO_2$. Figure 10.32b shows the estimated crystallite size of pentacene lattice along (020) direction plotted against crystalline domain size. Even on HMDS-treated and PI-coated $SiO_2$ samples, the crystallite size is at least one-order-of-magnitude smaller than the crystalline domain size and almost constant against the growth temperature. In Sect. 10.5, we proposed that the crystallite size is determined by nanometer-scale surface corrugation of the substrate [10]. It is therefore understood that the crystallite sizes on these substrates do not show notable difference because the spatial periods of surface corrugations on these substrates are nearly the same.

### *10.6.3 Influence on Transistor Characteristics*

Figure 10.33 shows linear-regime transfer characteristics of OTFTs fabricated on the three gate insulators. Gate threshold voltages are quite dependent on the insulators, which indicates that deep-trap states at the interfaces between pentacene and insulator are quite different [70]. The density of interface traps decreases when HMDS-treated or PI-coated $SiO_2$ is used as an insulator. This would be one of the major benefits to select adequate chemical composition of the gate insulator. However, since we focus on field-effect mobility, which is estimated in a range of gate voltage where the deep trap states are filled, we do not discuss the influence on deep trap states in this chapter. Mobilities calculated from the $I_D$-$V_{DS}$ relationship in a linear regime [71] are 0.35, 0.51, and 0.61 $cm^2/Vs$ for the OTFTs on bare, HMDS-treated, and PI-coated $SiO_2$, respectively.

Figure 10.34 shows HOMO-band edge profiles within a domain for three insulators. Standard deviations of the fluctuation amplitude are also indicated in the figures. Comparing the band-edge profiles, we can conclude that the surface

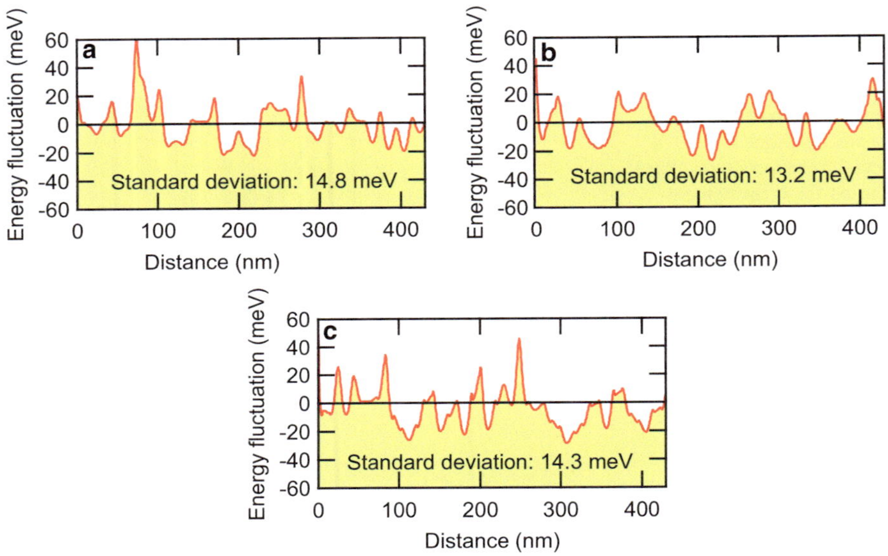


**Fig. 10.34** HOMO-band-edge profiles within a pentacene crystal and their FFT spectra measured for OTFTs on (**a**), (**b**) bare $SiO_2$, (**c**), (**d**) HMDS-treated $SiO_2$, and (**e**), (**f**) PI-coated $SiO_2$. The growth temperature of pentacene films is 60 °C

treatments do not affect the HOMO-band-edge fluctuations as predicted by the GIXD results. This implies that their mobility in domain ($\mu_{DI}$) is also unchanged by the surface treatments.

### *10.6.4 Mobility Limiting Factors Influenced by the Chemical Composition of Insulator Surface*

As explained in the previous sections, there exist hierarchical structures in pentacene thin films and corresponding limiting factors of carrier mobility. From molecular to the macroscopic scales, carrier mobility is determined by: (A) transfer integral between neighboring molecules in the molecular lattice, (B) perfectness of the crystallite, (C) crystallite boundary, i.e. crystallite size, (D) boundary of crystalline domain and morphological grain, i.e. domain size, and (E) extrinsic factors. Factors (A) and (B) determine $\mu_0$ in Eq. (10.7), (C) introduces $\phi_f$ in Eq. (10.7) and decreases $\mu_{GI}$ in Eq. (10.3'), (D) introduces $\phi_b$ and determines $l$ in Eq. (10.3'). Among them, the factors (A), (B) and (C) are the same for all of the OTFTs discussed in this section because the GIXD analysis tells us that the lattice constants, negligible random strain, crystallite sizes, and their insensitivity to the growth temperature are the same. Factor (E) always makes the reproducibility of the transistor characteristics inferior and increases the ambiguity in understanding

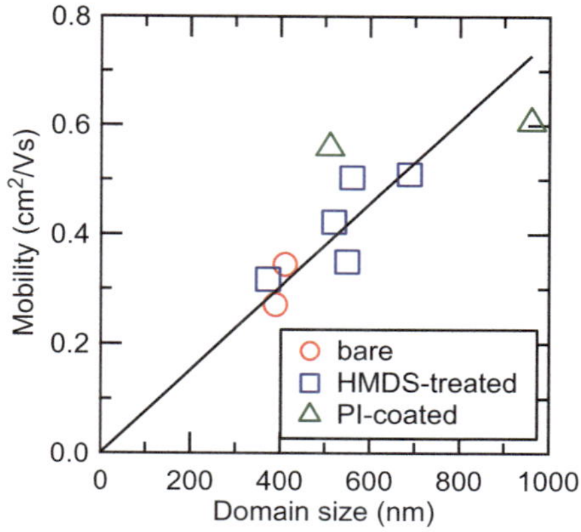


**Fig. 10.35** Relationship between the field-effect mobility and average size of crystalline domain for all samples analyzed in this work

the physics of OTFTs. In the experiments in this chapter, we have understood the origins of factor (E) and made all possible efforts to eliminate them. Consequently, we can conclude that the increase of the field-effect mobility is due to the improvement in factor (D), i.e. increase of the domain size and/or decrease of the height of the potential barrier at domain boundaries.

Here, we discuss the influence of the surface treatments on these two factors. As shown in Fig. 10.30d–f, domain size increases by the surface treatments. According to Eq. (10.3'), mobility should be proportional to domain size. Figure 10.35 shows mobility values of all samples analyzed in this section plotted against their average domain sizes. It clearly shows that the mobility values can be fitted by one straight line with zero intercept and the slope is independent of the surface treatments. On the other hand, the barrier height at domain boundary $\phi_b$, can be estimated from the linear slope of the $\ln(\mu 2k_B T/q) - 1/T$ plot as in Fig. 10.19, and $\mu_{DI}$ based on Eq. (10.3') because both $\mu/l$ and $\mu_{DI}$ are independent of the surface treatments. Estimated $\phi_b$ values are within 139–143 meV. The deviation of the barrier heights by the surface treatment is less than 3 %, which causes only a 0.5 % deviation in mobility. Therefore, the difference in the barrier height is almost negligible. Based on the above discussion, we can finally conclude that the influence of the surface treatments must be due only to the increase of crystalline domain size.

The increase in domain size by HMDS treatment or PI coating can be explained with the decrease in the nucleation density. According to the theory of crystal growth, adsorbed molecules diffuse on a substrate surface and form nuclei, and then a thin film starts its continuous growth. In this process, surface free energy is an important parameter because it governs the diffusion length of molecules on a substrate surface and the Gibbs free energy of the nuclei. The longer diffusion length and smaller free energy of the substrate surface result in the decrease of nucleation probability. Then, a thin film with larger grains become able to grow. While surface roughness also plays an important role on the determination of nucleation density, it is not the cause of the difference in this work because surface roughness is the same for all substrates.

As mentioned in Sect. 10.6.1, some groups have reported a correlation between surface free energy or wettability of a gate insulator surface and the mobility [65, 68]. Wettability of the surface is generally governed by the surface chemical structures with larger bond dipole or hydrogen-bonding functional groups, and a surface with lower free energy becomes, in general, more hydrophobic. Accordingly, experimental results with mobility increase relating to surface wettability can be attributed to the increase of grain size as in this work. However, some other groups have reported that mobility increases, even though grain size does not increase or rather decreases [1, 15, 68, 69, 72]. These results are probably due to the quality of the surface treatment. In the process of surface treatment, especially those using SAMs with trichlorosilyl head groups, the molecules easily polymerize due to their high chemical reactivity in the vicinity of water. The polymerized clusters of SAM molecules adsorbed on a substrate would become nucleation sites, and induce the thin-film growth with smaller grains. We therefore suspect that the quality of the surface may not be strictly controlled in such experiments.

## 10.7 Summary

In this chapter, we strived to elucidate all of the extrinsic and intrinsic limiting factors of carrier mobility in organic polycrystalline films according to our comprehensive study on pentacene thin films. We can find thousands of papers that discuss parts of this issue. However, as long as we compare imperfect OTFTs fabricated by different techniques and conditions, the argument will never reach a consensus. This is the reason why we have paid continuous efforts to investigate all extrinsic and intrinsic factors using the OTFTs similarly fabricated with the same material, techniques, and conditions. As a result, we believe that the mobility limiting factors in pentacene OTFTs are almost completely classified and quantitatively understood. Physical or chemical origins of some of the limiting factors, e.g. physical origin of the bad-edge fluctuation, chemical origin of the traps at the domain boundary inducing carrier depletion, etc., are still open questions. Nevertheless, the knowledge and methodology in this series of works should be generally applicable to any polycrystalline OTFTs.

In Sect. 10.1, surface morphology and crystallographic structures of precisely grown pentacene films on $SiO_2$ were classified. Five types of grain morphology are observed in pentacene films, which is probably one of the most complicated morphological variations among the many organic small molecules. In Sect. 10.2, extrinsic limiting factors of carrier mobility, defective structures or carrier injection barriers near or at the organic/electrode interfaces, irregular structures in the polycrystalline film, metal-induced damage, were demonstrated. Without this sort of survey work, a study discussing semiconductor physics would be interfered by these irregular factors.

In Sect. 10.3, relatively intrinsic factors in polycrystalline semiconductors were demonstrated. Various "polycrystalline models" have been proposed and used for inorganic TFTs. However, we have to be careful when borrowing a theory or a model from device physics for inorganic semiconductors. In the case of organic semiconductor, ballistic motion of carriers seldom plays a role. By using modified polycrystalline models, characteristics of domain-boundary-limited mobility are well explained. The mobility in the crystalline domain obtained according to the model is only around 1 $cm^2/Vs$. The band-edge fluctuation within a crystalline domain, which has been discovered by AFMP measurements, is the cause of the small mobility in domain. The root-mean-square amplitude of the fluctuation is only around 10 meV, but large enough to reduce the mobility in domain to nearly 1/10 of that of the single crystals. According to the findings explained in Sect. 10.3, equations reproducing the overall mobility in high-quality polycrystalline semiconductor films are summarized in Sect. 10.4.

In Sect. 10.5, the crystallographic order of pentacene thin films grown on $SiO_2$ was introduced according to the analyses using in-plane GIXD. The coherency of the molecular lattice is disturbed not by the random strain but by the size of the crystallites. The crystallite size, 25–50 nm, is surprisingly smaller than the morphological grain size. Since Fourier-transformed spectra of the band-edge fluctuation and substrate surface corrugation exhibited the same period as crystallite size, the surface roughness of the substrate is considered to cause the interruption of the crystallite growth and, as a result, introduce the potential fluctuation in the carrier-transporting band. The limiting factors and related phenomena in this section would be the particular nature of pentacene and differ material by material. For example, crystal order of some other molecules is known to be determined only by random strain. Even so, it is highly probable that the crystalline domains of most organic materials have potential fluctuation due to the structural disorder, charges on the substrate surface, deviation of local dielectric relaxation, etc.

In Sect. 10.6, the influence of the chemical composition of the gate-insulator surface on the mobility limiting factors was demonstrated. We concluded that the increase of carrier mobility by surface chemical modification of the gate insulator is only due to the increase in grain size, which is determined by the nucleation density at the initial stage of film growth. Both band-edge fluctuation and barrier height at the domain boundary remain unchanged. Despite this minor influence on the carrier mobility, surface chemical modification plays a more important role in the decrease of deep-trap states at the interfaces between organic films and insulator surfaces. This reduces the gate threshold voltage and its short- to long-term instabilities.

**Acknowledgments** This work is supported in part by the Chiba University Global COE Program ''Advanced School for Organic Electronics'' and Grant-in-Aid for Scientific Research (No. 21350099). Synchrotron radiation experiments were performed in the SPring-8 with the approval of the Japan Synchrotron Radiation Research Institute (JASRI) (Proposal No. 2008A1813, 2009A1827, 2010A1874, 2012B1443, and 2014A1593).

## References


1. D. Knipp, R.A. Street, A. Volkel, J. Ho, J. Appl. Phys. **93**, 347 (2003)
2. H. Klauk, M. Halik, U. Zschieschang, G. Schmid, W. Radlik, W. Weber, J. Appl. Phys. **92**, 5259 (2002)
3. R. Schroeder, L.A. Majewski, M. Grell, Appl. Phys. Lett. **83**, 3201 (2003)
4. D. Knipp, R.A. Street, B. Krusor, R. Apte, J. Ho, J. Non Cryst. Solids **299**, 1042 (2002)
5. H. Yanagisawa, T. Tamaki, M. Nakamura, K. Kudo, Thin Solid Films **464–465**, 398 (2004)
6. M. Nakamura, H. Ohguri, N. Goto, H. Tomii, M.-S. Xu, T. Miyamoto, R. Matsubara, N. Ohashi, M. Sakai, K. Kudo, Appl. Phys. A **95**, 73 (2009)
7. R. Matsubara, N. Ohashi, M. Sakai, K. Kudo, M. Nakamura, Appl. Phys. Lett. **92**, 242108 (2008)
8. N. Ohashi, H. Tomii, R. Matsubara, M. Sakai, K. Kudo, M. Nakamura, Appl. Phys. Lett. **91**, 162105 (2007)
9. S.-G. Li, R. Matsubara, T. Matsusue, M. Sakai, K. Kudo, M. Nakamura, Org. Electron. **14**, 1157 (2013)
10. R. Matsubara, M. Sakai, K. Kudo, N. Yoshimoto, I. Hirosawa, M. Nakamura, Org. Electron. **12**, 195 (2011)
11. R. Matsubara, T. Nomura, Y. Sakai, M. Sakai, K. Kudo, Y. Majima, M. Nakamura, Influence of substrate-surface treatments on carrier transport and crystal structures in pentacene films, in *Abstracts of the International Conference on Science and Technology of Synthetic Metals 2010*, Kyoto, Japan, 4–9 July 2010, p. 291.
12. U. Nakaya, *Snow Crystals, Natural and Artificial* (Harvard University Press, Cambridge, 1954)
13. I.P.M. Bouchoms, W.A. Schoonveld, J. Vrijmoeth, T.M. Klapwijk, Synth. Met. **104**, 175 (1999)
14. C.D. Dimitrakopoulos, A.R. Brown, A. Pomp, J. Appl. Phys. **80**, 2501 (1996)
15. M. Shtein, J. Mapel, J.B. Benziger, S.R. Forrest, Appl. Phys. Lett. **81**, 268 (2002)
16. J.H. Kang, X.-Y. Zhu, Appl. Phys. Lett. **82**, 3248 (2003)
17. M. Nakamura, M. Fukuyo, E. Wakata, M. Iizuka, K. Kudo, K. Tanaka, Synth. Met. **137**, 887 (2003)
18. M. Nakamura, N. Goto, N. Ohashi, M. Sakai, K. Kudo, Appl. Phys. Lett. **86**, 122112 (2005)
19. M. Nakamura, H. Ohguri, H. Yanagisawa, N. Goto, N. Ohashi, K. Kudo, in *Proceedings of the International Symposium on Super-Functionality Organic Devices*, IPAP Conference Series, vol. 6 (2005), p. 130
20. M.-S. Xu, A. Ohno, S. Aramaki, K. Kudo, M. Nakamura, Org. Electron. **9**, 439 (2008)
21. H. Ohguri, T. Tamaki, H. Yanagisawa, M. Nakamura, M. Iizuka, K. Kudo, in Technical Report of IEICE, OME2003-22 (in Japanese)
22. L. Bürgi, H. Sirringhaus, R.H. Friend, Appl. Phys. Lett. **80**, 2913 (2002)
23. J.A. Nichols, D.J. Gundlach, T.N. Jackson, Appl. Phys. Lett. **83**, 2366 (2003)
24. K.P. Puntambekar, P.V. Pesavento, C.D. Frisbie, Appl. Phys. Lett. **83**, 5539 (2003)
25. T. Miyazaki, K. Kobayashi, K. Ishida, S. Hotta, T. Horiuchi, H. Yamada, K. Matsushige, Jpn. J. Appl. Phys. **42**, 4852 (2003)
26. D.M. Taylor, Thin Solid Films **331**, 1 (1998)
27. H.O. Jacobs, A. Stemmer, Surf. Interface Anal. **27**, 361 (1999)
28. M. Nakamura, T. Yamada, Roadmap of Scanning Probe Microscopy: Ch. 6, in *Electrostatic Force Microscopy*, ed. by S. Morita (Springer, Berlin, 2007)
29. K. Suemori, M. Yokoyama, M. Hiramoto, J. Appl. Phys. **99**, 036109 (2006)
30. N. Koch, A. Kahn, J. Ghijsen, J.-J. Pireaux, J. Schwartz, R.L. Johnson, A. Elschner, Appl. Phys. Lett. **82**, 70 (2003)
31. T. Sawabe, K. Okamura, T. Miyamoto, T. Sueyoshi, N. Ueno, K. Kudo, M. Nakamura, Appl. Phys. A **95**, 225 (2009)
32. T. Sugiyama, T. Sasaki, S. Kera, N. Ueno, T. Munakata, Appl. Phys. Lett. **89**, 202116 (2006)

33. C. Vanoni, S. Tsujino, T.A. Jung, Appl. Phys. Lett. **90**, 193119 (2007)
34. G. Horowitz, M.E. Hajlaoui, Adv. Mater. **12**, 1046 (2000)
35. G. Horowitz, M.E. Hajlaoui, R. Hajlaoui, J. Appl. Phys. **87**, 4456 (2000)
36. X.-J. Yan, H. Wang, D.-H. Yan, Thin Solid Films **515**, 2655 (2006)
37. S. Verlaak, V. Arkhipov, P. Heremans, Appl. Phys. Lett. **82**, 745 (2003)
38. J.T. Sadowski, G. Sazaki, S. Nishikata, A. Al-Mahboob, Y. Fujikawa, K. Nakajima, R.M. Tromp, T. Sakurai, Phys. Rev. Lett. **98**, 046104 (2007)
39. O.D. Jurchescu, M. Popinciuc, B.J. van Weeks, T.T.M. Palstra, Adv. Mater. **19**, 688 (2007)
40. K. Hummer, C. Ambrosch-Draxl, Phys. Rev. B **72**, 205205 (2005)
41. H. Kakuta, T. Hirahata, I. Matsuda, T. Nagao, S. Hasegawa, N. Ueno, K. Sakamoto, Phys. Rev. Lett. **98**, 247601 (2007)
42. H. Yamane, E. Kawabe, D. Yoshimura, R. Sumii, K. Kanai, Y. Ouchi, N. Ueno, K. Seki, Phys. Status Solidi B **245**, 793 (2008)
43. G. Le Comber, W.E. Spear, Phys. Rev. Lett. **25**, 509 (1970)
44. D. Stauffer, A. Aharony, *Introduction to percolation theory,* Rev. 2nd edn. (Taylor & Francis, London, 1994)
45. Q. Wang, H. Antoniadis, E.A. Schiff, Phys. Rev. B **47**, 9435 (1993)
46. M.C.J.M. Vissenberg, M. Matters, Phys. Rev. B **57**, 12964 (1998)
47. W.Y. Chou, Y.S. Mai, H.L. Cheng, C.Y. Yeh, C.W. Kuo, F.C. Tang, D.Y. Shu, T.R. Yew, T.C. Wen, Org. Electron. **7**, 445 (2006)
48. Y. Natsume, T. Minakata, T. Aoyagi, Synth. Met. **159**, 338 (2009)
49. K. Hong, S.Y. Yang, C. Yang, S.H. Kim, D. Choi, C.E. Park, Org. Electron. **9**, 864 (2008)
50. S.E. Fritz, S.M. Martin, C.D. Frisbie, M.D. Ward, M.F. Toney, J. Am. Chem. Soc. **126**, 4084 (2004)
51. T. Kakudate, N. Yoshimoto, Y. Saito, Appl. Phys. Lett. **90**, 081903 (2007)
52. G.K. Williamson, W.H. Hall, Acta Metall. **1**, 22 (1953)
53. R. Hussain, D. Mohammad, Turk. J. Chem. **28**, 725 (2004)
54. Y. Iwanami, H. Kamiishi, K. Matsuura, A. Sano, S. Yokoyama, US Patent 5,002,714, 1991
55. H.S. Soliman, M.M. El Nahass, A.M. Farid, A.A.M. Farag, A.A. El Shazly, Eur. Phys. J. AP **21**, 187 (2003)
56. F.Y. Tsai, T.N. Blanton, D.R. Harding, S.H. Chen, J. Appl. Phys. **93**, 3760 (2003)
57. M. Nakamura, H. Tokumoto, Surf. Sci. **377–379**, 85 (1997)
58. M. Nakamura, H. Tokumoto, Surf. Sci. **398**, 143 (1998)
59. V. Kalihari, E.B. Tadmor, G. Haugstad, C.D. Frisbie, Adv. Mater. **20**, 4033 (2008)
60. I. Yagi, K. Tsukagoshi, Y. Aoyagi, Appl. Phys. Lett. **86**, 103502 (2005)
61. A. Zen, D. Neher, K. Silmy, A. Holländer, U. Asawapirom, U. Scherf, Jpn. J. Appl. Phys. **44**, 3721 (2005)
62. Y.-Y. Lin, D.J. Gundlach, S.F. Nelson, T.N. Jackson, IEEE Electron Device Lett. **18**, 606 (1997)
63. J.-H. Shim, L.-Y. Jung, S.-W. Pyo, Y.K. Kim, Thin Solid Films **441**, 284 (2003)
64. Y. Kato, S. Iba, R. Teramoto, T. Sekitani, T. Someya, H. Kawaguchi, Appl. Phys. Lett. **84**, 3789 (2004)
65. S. Uemura, M. Yoshida, S. Hoshino, T. Kodzasa, T. Kamata, Thin Solid Films **438–439**, 378 (2003)
66. J. Puigdollers, C. Vos, A. Orpella, R. Quidant, I. Martin, M. Vetter, R. Alcubilla, Org. Electron. **5**, 67 (2004)
67. F. De Angelis, S. Cipolloni, L. Mariucci, G. Fortunato, Appl. Phys. Lett. **86**, 203505 (2005)
68. S.C. Lim, S.H. Kim, J.H. Lee, M.K. Kim, D.J. Kim, T. Zyung, Synth. Met. **148**, 75 (2005)
69. P.S. Abthagir, Y.-G. Ha, E.-A. You, S.-H. Jeong, H.-S. Seo, J.-H. Choi, J. Phys. Chem. B **109**, 23918 (2005)
70. S. Yogev, R. Matsubara, M. Nakamura, U. Zschieschang, H. Klauk, Y. Rosenwaks, Phys. Rev. Lett. **110**, 036803 (2013)
71. S.M. Sze, *Physics of Semiconductor Devices*, 3rd edn. (Wiley-Interscience, New York, 2006)
72. W.-Y. Chou, C.-W. Kuo, H.-L. Cheng, Y.-R. Chen, F.-Y. Yang, D.-Y. Shu, C.-C. Liao, F.-C. Tang, Jpn. J. Appl. Phys. **45**, 7922 (2006)

# Chapter 11
# Materials for Organic Light Emitting Diode (OLED)

**Takashi Karatsu**

## 11.1 OLED and Phosphorescent Cyclometalated Iridium (III) Complexes

Iridium complexes have a wide range of applications such as photocatalysts to reduce carbon dioxide [1], imaging reagents for living cells [2], and oxygen sensors [3]. In particular, organic light emitting diodes (OLED) is one of important industrial application for iridium complexes, due to their high phosphorescent efficiency at ambient temperature [4, 5]. OLEDs have many advantages, including self-emission (no backlight required), an almost 180° wide view angle, light weight, thin (<2 mm), quick response (1,000 times faster than LCD), high contrast, and can be fabricated on flexible plastic substrates.

In 1987, Tan et al. reported the potential of OLEDs using tris (8-hydroxyquinolinato) aluminium (III) ($Alq_3$) [6]. Before them, the OLED device has simple configuration that has single organic crystal sandwiched by two electrodes. They introduced concept of OLED device configuration composed of multiple thin layers (Fig. 11.1). Here, each layer has an exact function, such as charge transporting and emitting abilities. The external efficiency of this OLED device was 1 %, which meant an internal efficiency of 5 %, because the output efficiency from the device was approximately 20 % [7]. After charge (hole and electron) injection into the emitting layer, 25 % singlet and 75 % triplet excitons are generated by charge recombination. Therefore, the only usable amount of fluorescence is 25 %. If phosphorescent materials can be used, then the 75 % triplet excitons are usable. In addition, the triplet excited state has lower energy than the singlet excited state; therefore, there is a chance that intersystem crossing of singlet excited state to the triplet excited state could occur.

T. Karatsu (✉)
Department of Applied Chemistry and Biotechnology, Graduate School of Engineering, Chiba University, 1-33, Inage-ku, Chiba 263-8522, Japan
e-mail: karatsu@faculty.chiba-u.jp



H. Ishii et al. (eds.), *Electronic Processes in Organic Electronics*,
Springer Series in Materials Science 209, DOI 10.1007/978-4-431-55206-2_11

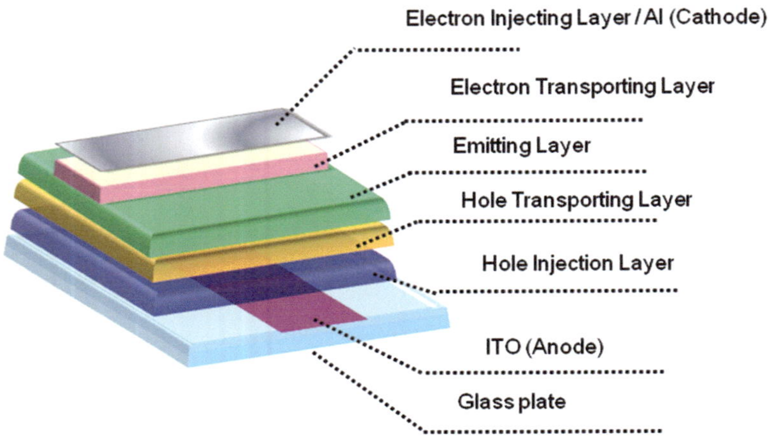


**Fig. 11.1** Configuration of OLED device composed of multiple thin layers

In 1998, Thompson and co-workers reported a device that employed 2,3,7,8,12,13,17,18-octaethyl-12*H*,23*H*-porphine platinum(II) (PtOEP), which had an external quantum efficiency of 4 % [8, 9]. In 1999, the efficiency of PHOLED was improved to 7.5 % using phosphorescent *fac*-tris (2-phenylpyridinato,$NC^{2'}$)iridium(III) (*fac*-Ir(ppy)$_3$) [10, 11]. These reports boosted the research activity in this field. The efficiency was jumped up to 29 %, using also *fac*-Ir(ppy)$_3$, as reported by Kido and co-workers in 2007 [5]. This meant internal emission quantum efficiency reached 100 %. Here, not only triplet excitons (75 %), but also singlet excitons (25 % singlet excitons also generate triplet excitons after intersystem crossing) were used after charge recombination. Phosphorescent triscyclometalated iridium complexes have advantage to obtain wide variety of the emission colors by changing structure of cyclometalated ligands.

Recent development of OLED materials has been focused on the maximizing substance ability to each function, such as charge injection and transport ability, emitting ability, exciton confinement ability, and so on. Designing of the materials has wide varieties based on variation of organic molecules. In this chapter, it is focused on the materials forming the emitting layer, especially, the nature of phosphorescent triscyclometalated iridium (III) complexes. Selection of cyclometalated ligands from the wide variation of organic molecules provides fine tuning of phosphorescence color. The typical triscyclometalated iridium (III) complexes consist of three bidentate ligands to make octahedral structure surrounding iridium atom, and symmetry generated by these coordination provides *fac* and *mer* geometrical isomers. Each isomer has $\Delta$ and $\Lambda$ optical isomers. Their selective preparation and characteristic in the excited state including isomerization between their isomers are the main part in the next section [12–14]. In addition, comparing with blue, green, and red phosphorescent complexes, blue phosphorescent complexes have difficulties for their emission color purities and materials stabilities. Therefore, one section is spent to explain recent development of blue phosphorescent iridium complexes [15, 16]. Iridium complexes have been used as dopant in an emitting layer, therefore, host materials are also important. Host materials require high abilities of charge transport and confinement of triplet energy of dopant exciton. In the last part, our attempt to develop wet processable host materials for green to blue phosphorescent complexes has been mentioned. Wet process is one of the key processes to reduce process cost and improve quality of large area devices by climbing over difficulties of vacuum sublimation method [17–19].

## 11.2 *Meridional* and *Facial* Isomers of Iridium Complexes and Their Photochemical Isomerization

$Ir(ppy)_3$, (tris(4,6-difluorophenyl)pyridinato,$NC^{2'}$) iridium (III) ($Ir(F_2ppy)_3$), and other triscyclometalated iridium (III) complexes have $d^6$ electron configuration, and have octahedral structure (Fig. 11.2). $Ir(ppy)_3$ has three 2-phenylpyridine bidentate ligands and coordinated at 2-phenyl anionic carbon and pyridyl nitrogen atoms. This is quite different from nitorogen atom coordinated ruthenium (II) trisbipyridine ($Ru(bpy)_3$). Symmetry generated *C* and *N* atom coordinations makes *fac* and *mer* isomers. Here in the *fac*-isomer, three pyridyl nitrogen atoms locate vertexes of a triangle make up of octahedron. On the other hand, three pyridyl nitrogen atoms locate on the meridian of *mer*-isomer. Up-to-date, photophysics and photochemistry of the *fac*-isomer have been caught attention and studied extensively, however, those of the *mer*-isomer have been caught limited attention. The reasons are synthetic easiness, chemical stability, and high emission ability of the *fac*-isomer. In addition, each *fac*- and *mer*-isomer has $\Delta$- and $\Lambda$ – optical isomers, and those characteristics have also been studied minimal. In this section, characters of those isomers and photochemical isomerization between those isomers are described. Radiative process is important as a radiative material, however, understanding of nonradiative process that is a complemental process to

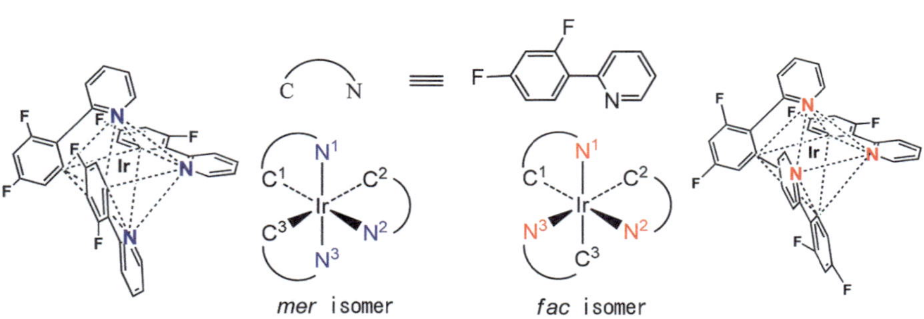


**Fig. 11.2** Structures of *mer* and *fac* isomers of $Ir(F_2ppy)_3$, and their abbreviated structures

the radiative process is also important for the understanding of radiative materials and chemical stability of the materials.

For syntheses of the iridium complexes, Nonoyama method through Ir(III) chlorine $\mu$ dimer complex is widely used (Eq. (11.1)) [20]. For this reaction and following reaction introducing the third ligand are typically quantitative and are performable under a moderate reaction condition. R. J. Watts and co-workers reported 10 % formation of $Ir(ppy)_3$ beside formation of Cl-$\mu$ dimer complex [21].

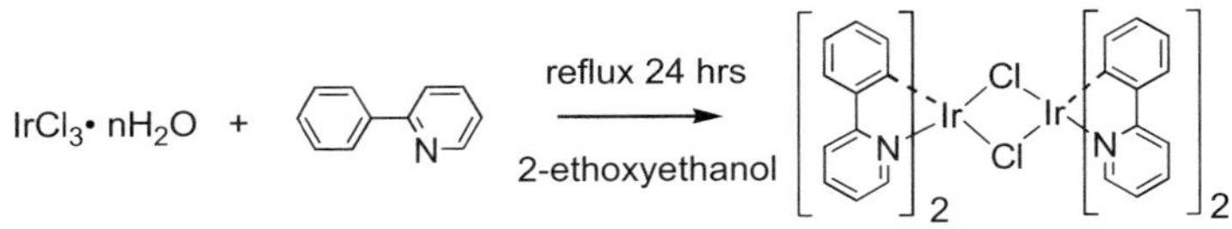


(11.1)

The reaction yield from $\mu$-dimer complex was improved by using silver triflate (trifluoromethanesulfonate) as using leaving group and also dechlorination reagent [22]. The synthesis of the triscyclometalated complex has been reported to synthesize by one-step reaction from tris(acetylacetonato) Ir(III) complex [23]. For the preparation and purification methods for *mer*-isomer have not been well studied and have not established. Recently, selective preparation method of the *fac*- and *mer*-isomer has been reported by controlling reaction temperature. The *mer*-isomer has been prepared 65–80 % reaction yield under mild condition in the presence of base at ambient temperature to 150 °C [24, 25] (Eq. (11.2)). In addition, a method through bis(phenylpyridinato)acetylacetonato iridium (III) complex synthesized from $\mu$-dimer complex, and then acetylacetonato ligand was converted to the third ligand improves reaction yield of the *mer*-isomer [12]. In general, reaction under the mild condition gives *fac*-*mer* mixture, and [*mer*]/[*fac*] ratio is increased and total reaction yield is decreased by decreasing reaction temperature. Ligands exchange reaction proceeds through thermodynamic controlled mechanism giving thermodynamic stable *fac*-isomer as a product at high temperature, and kinetic favored *mer*-isomer is produced at ambient temperature. Usually, *fac*-isomer is almost 10 kcal mol$^{-1}$ more stable than the corresponding *mer*-isomer [13].

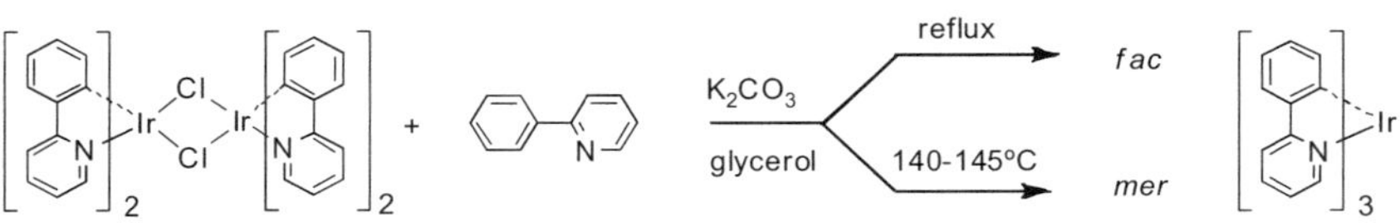


(11.2)

For the synthesis expressed in Eq. (11.2), glycerin used as typical solvent even hard to reach its boiling point. In such case, use of microwave reactor is useful method, and this shortened reaction time. On the other hand, synthesis for heat sensitive complexes, decomposition of the complex becomes problem. In such cases, use of *mer*→*fac* photochemical isomerization after the synthesis of the *mer*-isomer is very useful method [22, 23].

The isomers were separated by a column chromatography and purified by the crystallization, and analyzed by HPLC, NMR, elemental analysis, X-ray crystallography. Especially, by means of H NMR, signals from three ligands are appeared in equivalent for the *fac*-isomers, but inequivalent for *mer*-isomers. For $^{13}C$ NMR, signals of the *mer*-isomer's are triple than those of the *fac*-isomer's.

There are some reports for structures of the complexes by a single crystal X-ray crystallography. All of the crystals were racemic crystal consist of 1:1 $\Delta$- and $\Lambda$-enantiomers. There have been reported many data for the *fac*-isomers [13, 24], and not so many for the *mer*-isomers [12, 24]. Therefore, there are limited cases we can compare those structures. The *fac*-isomers have higher symmetry than the *mer*-isomers. All of *trans* positions of Ir–N bond are Ir–C bonds for *fac*-isomer. On the other hand, combination of the atoms of *trans* positions of the *mer*-isomer are N–Ir–N, N–Ir–C, and C–Ir–C. These bond lengths reported for tris(1-phenylpyrazolato, $NC^{2'}$)iridium(III) (Ir(*ppz*)$_3$) are compared between *fac*- and *mer*-isomers (Table 11.1). In case of the *fac*-isomer, bond lengths of three pairs of Ir–C and Ir–N bonds are almost same. Mean values of Ir–C and Ir–N bond lengths are 2.02 and 2.12 Å, respectively, and always Ir–N bonds are longer than Ir–C bonds. On the other hand, for the *mer*-isomer, bond length has a large variation, and difference in bond length of Ir–N and Ir–C is small. Relation Ir–N > Ir–C accomplished in the *fac*-isomer, does not hold in the *mer*-isomer. Bond lengths of Ir–$C^2$ and Ir–$C^3$, those are in *trans* position each other, are long and similar *trans* position of Ir–$N^1$ and Ir–$N^3$ bond lengths are short. This kind of relation in bond length is generally found for other iridium (III) complexes (Table 11.1).

Measurments of the absorption spectra and emission spectra gave excitation energies, emission lifetimes ($\tau_p$) and emission quantum yields ($\Phi_p$) and then radiative rate constant ($k_r = \Phi_p/\tau_p$) and nonradiative rate constant ($k_{nr} = (1-\Phi_p)/\tau_p$) as shown in Table 11.2. For example, absorption spectrum of *fac*-Ir($F_2$ppy)$_3$ is very similar to that of *fac*-Ir(ppy)$_3$ [21, 26–29], however, spectrum of *mer*-Ir($F_2$ppy)$_3$ was quite different from the following two points. In the short wavelength region, $\pi$–$\pi$* absorption band splitted to two bands for *fac*-isomer was single in the case of *mer*-isomer, and decrease of molar extinction coefficient was observed for $^1$MLCT band around 350 nm. Phosphorescence band of the *mer*-isomer generally appeared at longer wavelengths than that of the corresponding *fac*-isomer, and degree of shift is between a couple of nanometers and up to 50 nm dependent on the ligand. $\Phi_p$ values of *mer*-isomer were extremely smaller than that of *fac*-isomer. Emission lifetime of *mer*-isomer measured by a single photon counting method was shorter than that of the *fac*-isomer. As a result, no significant difference between $k_r$ of *mer*- and *fac*-isomers and difference in $\Phi_p$ and $\tau_p$ was caused by difference of $k_{nr}$.

The significantly low ability of phosphorescence of *mer*-isomer was partly due to the photochemical isomerization. Irradiation of UV light induced *mer* → *fac* isomerization under deaerated solution, and emission spectrum and lifetime were identical with those of the *fac*-isomer, and product was finally identified as the *fac*-isomer by chemical analytic methods such as $^1$H NMR and mass spectrometry. On the other hand, irradiation of the *fac*-isomer brought no chemical change at the same reaction condition. This means this isomerization is *mer* → *fac* one-way (Fig. 11.3).

**Table 11.1** Key bond lengths and bond angles of *mer* and *fac* isomers obtained by single-crystal X-ray diffraction method

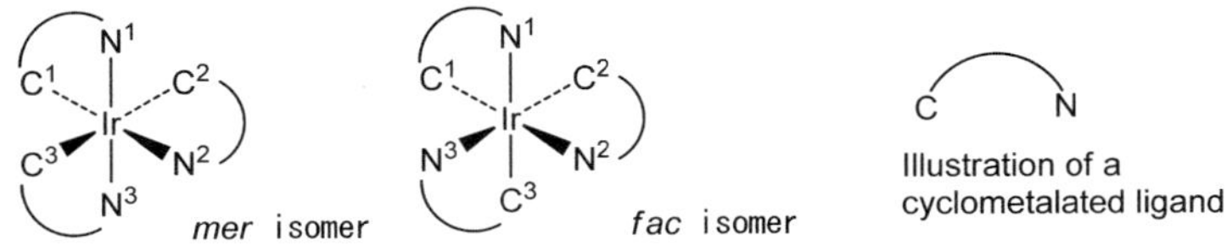


| Bond length (Å) or angle (°) | $Ir(ppz)_3$ | | $Ir(CF_3ppz)_3$ | | $Ir(F_2ppy)_3$ | |
|---|---|---|---|---|---|---|
| | *mer* | *fac* | *mer*[a] | *fac*[a] | *mer* | *fac* |
| $Ir–N^1$ (Å) | 2.013 | 2.135 | 2.016 | 2.113 | 2.018 | 2.118 |
| $Ir–N^2$ | 2.049 | 2.121 | 2.104 | 2.111 | 2.142 | 2.128 |
| $Ir–N^3$ | 2.025 | 2.118 | 2.014 | 2.108 | 2.041 | 2.116 |
| $Ir–C^1$ (Å) | 1.994 | 2.027 | 1.999 | 2.017 | 2.030 | 2.011 |
| $Ir–C^2$ | 2.061 | 2.021 | 2.078 | 2.014 | 2.077 | 2.002 |
| $Ir–C^3$ | 2.053 | 2.016 | 2.078 | 2.019 | 2.067 | 1.997 |
| $N^1–Ir–C^1$ (°) | 79.63 | 79.50 | 80.32 | 79.02 | 81.12 | 79.32 |
| $N^2–Ir–C^2$ | 79.13 | 79.62 | 78.27 | 79.26 | 78.69 | 79.47 |
| $N^3–Ir–C^3$ | 78.49 | 78.73 | 79.13 | 79.19 | 80.32 | 79.33 |
| $N^1–Ir–N^3$ (°) | 171.52 | | 172.25 | | 173.24 | |
| $C^1–Ir–N^2$ | 172.17 | | 170.18 | | 175.30 | |
| $C^2–Ir–C^3$ | 172.68 | | 169.62 | | 172.60 | |
| $N^1–Ir–C^3$ (°) | | 170.56 | | 170.93 | | 172.50 |
| $N^2–Ir–C^1$ | | 170.05 | | 169.57 | | 174.16 |
| $N^3–Ir–C^2$ | | 171.32 | | 168.60 | | 170.41 |

[a]Average of Δ and Λ isomers

**Table 11.2** Phosphorescence quantum yields ($\Phi_p$), lifetime ($\tau$), radiative ($k_r$) and nonradiative rate constants ($k_{nr}$) of iridium (III) triscyclometalate complexes [12, 13, 24]

| Complex | $\Phi_p$ | $\tau$/μs | $k_r/s^{-1}$ | $k_{nr}/s^{-1}$ |
|---|---|---|---|---|
| *fac*-$Ir(ppy)_3$ [24] | 0.4 | 1.9 | $2.1 \times 10^5$ | $3.2 \times 10^5$ |
| *mer*-$Ir(ppy)_3$ [24] | 0.036 | 0.15 | $2.4 \times 10^5$ | $6.4 \times 10^6$ |
| *fac*-$Ir(tpy)_3$ [24] | 0.5 | 2 | $2.5 \times 10^5$ | $2.5 \times 10^5$ |
| *mer*-$Ir(tpy)_3$ [24] | 0.051 | 0.26 | $2.0 \times 10^5$ | $3.6 \times 10^6$ |
| *fac*-$Ir(F_2ppy)_3$ [12] | 0.43 | 1.6 | $2.7 \times 10^5$ | $3.6 \times 10^5$ |
| *mer*-$Ir(F_2ppy)_3$ [12] | 0.053 | 0.21 | $2.5 \times 10^5$ | $4.5 \times 10^6$ |
| *fac*-$Ir(ppz)_3$[a][13] | 1 | 28 | $3.3 \times 10^4$ | $3.3 \times 10^6$ |
| *mer*-$Ir(ppz)_3$[a][13] | 0.81 | 14 | $6.4 \times 10^4$ | $6.4 \times 10^6$ |
| *fac*-$Ir(tpy)_2(ppz)$ [13] | 0.35 | 1.5 | $2.3 \times 10^5$ | $4.1 \times 10^5$ |
| *mer*-$Ir(tpy)_2(ppz)$ [13] | 0.012 | 0.064 | $1.9 \times 10^5$ | $1.5 \times 10^7$ |
| *fac*-$Ir(tpy)(ppz)_2$ [13] | 0.37 | 1.8 | $2.1 \times 10^5$ | $3.6 \times 10^5$ |
| *mer*-$Ir(tpy)(ppz)_2$ [13] | 0.068 | 0.4 | $1.7 \times 10^5$ | $2.5 \times 10^6$ |

[a]Rate constants, $k_{nr}$ is estimated on the assumption that $k_r$ at 77 K and that at ambient temperature are equal, because those complexes did not give phosphorescence at ambient temperature in solution

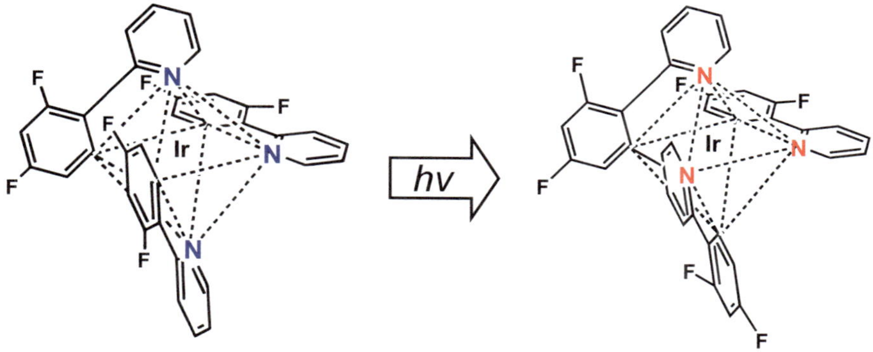


**Fig. 11.3** Photochemical *mer* → *fac* one-way isomerization of $Ir(F_2ppy)_3$ complex

**Table 11.3** Quantum yield of photochemical *mer* → *fac* isomerization ($\Phi_{isom}$) in deaerated solution ($CH_3CN$)

| Complex | $\Phi_{isom}$ |
|---|---|
| *mer*-$Ir(F_2ppy)_3$ | 0.067 |
| *mer*-$Ir(tpy)_3$ | 0.00018 |
| *mer*-$Ir(tpy)_2(ppz)$ | 0.00028 |
| *mer*-$Ir(tpy)(ppz)_2$ | 0.12 |
| *mer*-$Ir(ppz)_3$ | 0.072 |

Photochemical *mer* → *fac* one-way isomerization quantum yield were between 0.12 and $10^{-4}$ in deaerated acetonitrile at ambient temperature as shown in Table 11.3 [12, 13]. An activation energy of this isomerization was determined to be 15.2 kJ $mol^{-1}$ by the Arrhenius plot of the isomerization rate constants for Ir $(F_2ppy)_3$. Both emission and isomerization were quenched by the triplet quencher azulene that has lowest triplet energy $E_T = 163$ kJ $mol^{-1}$. This indicated that the emission was phosphorescence and the isomerization also occurred via a triplet excited state. $\Phi_{isom}$ values showed quite small <0.1 quantum yield in the temperature range between 283 and 313 K. On the other hand, temperature dependence of $\Phi_P$ showed large dependence on temperature, and $\Phi_P$ increased intensity by decreasing measurement temperature. At temperature $T \leqq 243$ K, *mer*-isomer showed equivalent to the *fac*-isomer's $\Phi_P$ at ambient temperature. Therefore, the isomerization and phosphorescence do not act as complementally that indicates the isomerization do not occurred through $^3$MLCT state (phosphorescent state).

This isomerization require activation energy. The reaction quantum yields ($\Phi_{react}$) decreases by decreasing reaction temperature. Comparing temperature dependence of isomerization of *mer*-$Ir(F_2ppy)_3$ [12] and carbonyl ligand and solvent acetonitrile exchange reaction of renium carbonyl complex showed similarity [30]. Emission quantum yields ($\Phi_{em}$) and $\Phi_{react}$ behave complementary, therefore, $\Phi_{em}$ are increasing by decreasing temperature, however, $\Phi_{react}$ are decreasing by decreasing temperature. For the exchange reaction, $\Phi_{react} + \Phi_{em}$ was almost constant in the measured temperature range, however, $\Phi_{react} + \Phi_{em}$

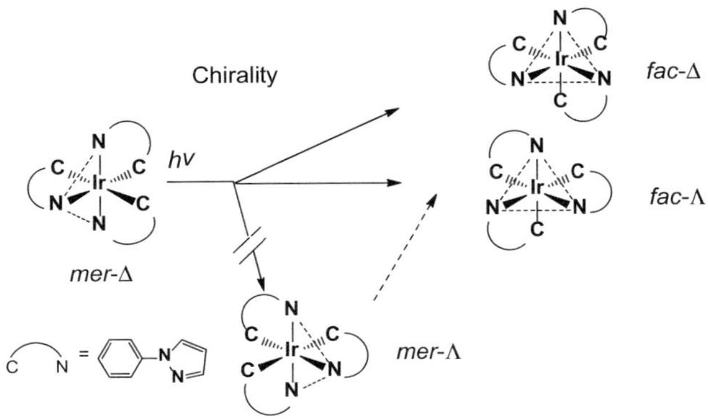


**Fig. 11.4** Chirality in the photochemical isomerization of *mer*-$\Delta$-Ir(ppz)$_3$ in solution. This reaction produced no *mer*-$\Lambda$ isomer, but *fac*-$\Delta$ and -$\Lambda$ isomers. The *fac*-$\Delta$ isomer was produced in excess amount

was not constant for the isomerization implying existence of another deactivation path. From the above results, a plausible reaction mechanism is proposed as follows. This isomerization occurs through thermally accessible $^3$d* state, which is in equilibrium between emissive $^3$MLCT state.

Ir(ppz)$_3$ complex is interesting because this complex is not only *mer*-isomer but also *fac*-isomer also does not give phosphorescence at ambient temperature in solution. Reason of these phenomena must be *mer* → *fac* one-way isomerization or enantiomeric isomerization between $\Delta$- and $\Lambda$-isomers through a twist mechanism or ligand dissociation-association mechanism. We optical resolved $\Delta$- and $\Lambda$-isomers of Ir(ppz)$_3$, and examined their photochemical isomerization (Fig. 11.4) [14].

Through the twist mechanism of both *mer* and *fac* isomers cause no *mer-fac* geometrical isomerization but only cause $\Delta - \Lambda$ optical isomerization. *Mer* and *fac* isomers cause optical isomerization and geometrical isomerization by dissociation-recoordination mechanism. For example, irradiation of *mer*-$\Delta$ isomer of Ir(ppz)$_3$ produced *fac*-$\Delta$ and *fac*-$\Lambda$ isomers, however, no production of the *mer*-$\Lambda$ isomer (Fig. 11.4). Here, *fac*-$\Delta$ and $\Lambda$ isomers were produced 59 and 41 %, respectively. This means *fac*-Δ isomer is produced in 18%ee (enantio excess). Plausible reaction mechanism is trigonal bipyramidal (TBP) mechanism as shown in Fig. 11.5.

There are two important points to consider this isomerization mechanism. The first point is identification of photochemically active ligand, here, there is Adamson's empirical rule [31–36]. There is argument for the application limit of this rule, however, photochemistry of iridium complexes fits very well. According to this rule, individual Ir–N bond in axial position dissociates of *mer*-$\Delta$ isomer, Ir–N$^1$ bond dissociation (Route A) gives *fac*-$\Delta$ isomer (chirality retains) and Ir–N$^3$ bond dissociation (Route B) gives *fac*-$\Lambda$ (chirality inversion) isomer. The second point is that C–Ir–N axis (not C–Ir–C axis) is kept in two diastereomeric TBP intermediates. If these two hypothesis hold in the mechanism, we can rationally explain experimental results. Reason why enantio-excess arise is explained by the different efficiency over come diastereomeric transition states (Fig. 11.6).

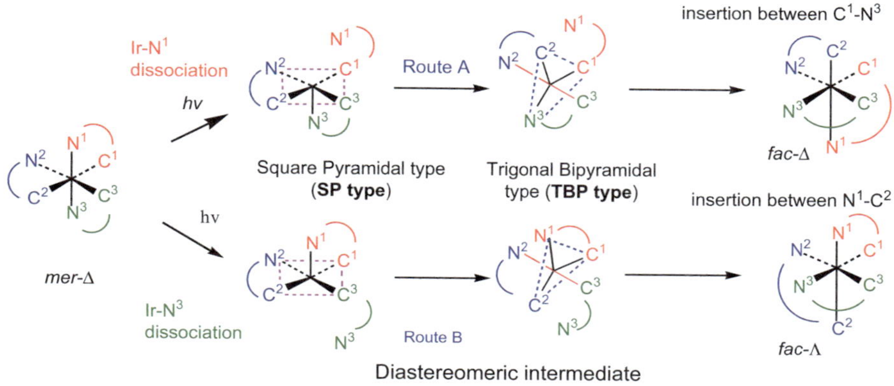


**Fig. 11.5** Isomerization mechanism through dissociation and recoordination (TBP mechanism)

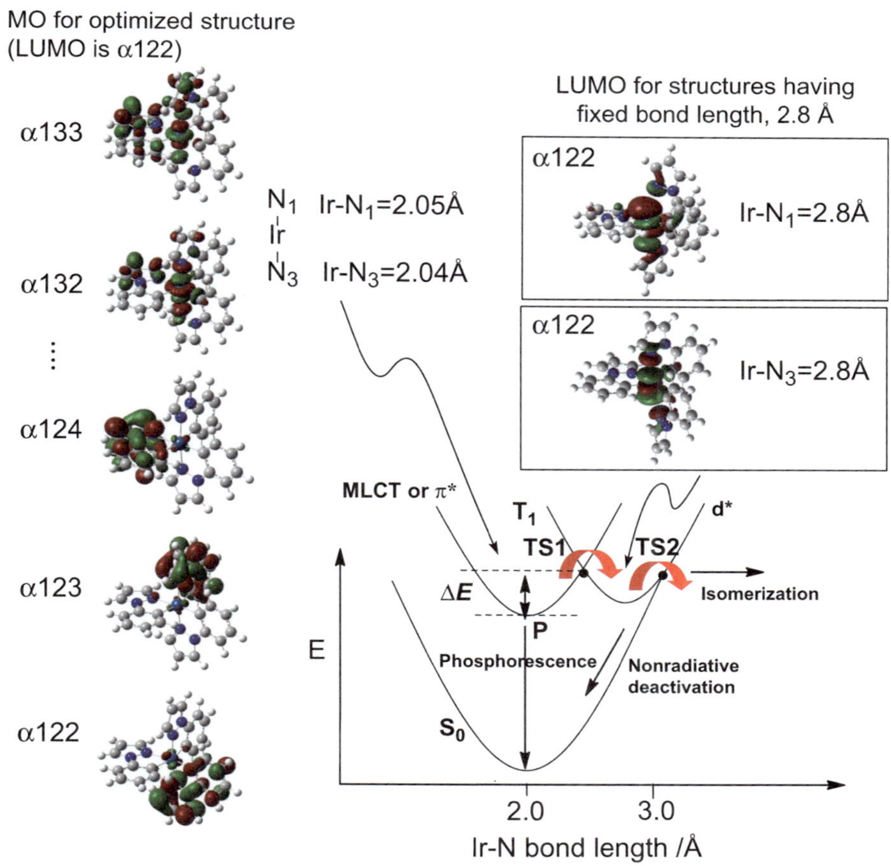


**Fig. 11.6** Potential energy surfaces of ground state and excited state and their MO

Lowest triplet excited state generated by the excitation gives $^3$MLCT state. Structures for this state, elongation of Ir–N$^1$ or Ir–N$^3$ bond gives d* state through the transition state caused by energy surface crossing (TS1). In this structure, axial bond shows anti-bonding nature. Difference in activation energies ($\Delta E_a$) formed by bond elongation of Ir–N$^1$ or Ir–N$^3$ determines enantioexcess value.

Whether this isomerization occurs or not in solid phase, *fac*- and *mer*-Ir(ppy)$_3$ gave equivalent PL efficiency in EPA at 77 K, and PL lifetimes were 2.83 and 2.65 μs for *mer*- and *fac*-isomer, respectively. In addition, relative PL intensities of powder *mer*- and *fac*-isomer were equivalent. Therefore, isomerization is not important, however, there is a chance that this isomerization may take place during vacuum sublimation process.

OLED device composed of *mer*-Ir(ppy)$_3$ gave equivalent efficiency as a device of *fac*-isomer [37]. Actually, radiative rate constants of the *mer*-isomers are almost identical with those of the corresponding *fac*-isomer, this is quite reasonable considering that molecular motion and structural transformation are prohibited in the solid state device. On the other hand, efficiency of the emission decreased for the device composed of *fac*-isomer doped with small amount of the *mer*-isomer. This is explained by the trapping of the excitation energy at *mer*-isomers. *Mer*-isomer which has smaller excitation energy than that of the *fac*-isomer acts as a trap site. In addition, *mer*-isomer may have smaller emission efficiency than corresponding *fac*-isomer [38]. PL of *fac*-Ir(ppy)$_3$ doped in CBP, in the case of no *mer*-isomer doped, i.e. [*mer*-isomer] = 0 %, $\Phi_{PL}$ was 92 %, and this value decreased to 48 and 37 % when doped *mer*-isomer concentrations were 30 and 46 %, respectively.

This *mer* → *fac* one-way isomerization is often observed for blue phosphorescent Ir(III) complexes, since energy level of phosphorescent state is high and it locates nearby d* state that is responsible for the isomerization [12–14]. However, recent our report showed that no *mer* → *fac* isomerization occurred for blue phosphorescent Ir(III) complexes having carbene type ligand such as tris[2-(4-X-phenyl)-3-butyl-[1,3]-imidazolinato-$C^2$, $N^1$]iridium (III) X=fluoro, *mer*-Ir(Fpim)$_3$; X=trifuluoromethyl, *mer*-Ir(CF$_3$pim)$_3$; and X=trifluoromethoxy, *mer*-Ir (OCF$_3$pim)$_3$ [15].

## 11.3 Blue Phosphorescent Cyclometalated Iridium (III) Complexes and Their Nonradiative Deactivation

For PHOLED, red and green phosphorescent iridium complexes have been used in commercial devices [39]. However, compared to the significant success of green and red phosphorescent materials, there are still difficulties with blue phosphorescent materials, which is the barrier to achieve total phosphorescent full color display and white color lighting.

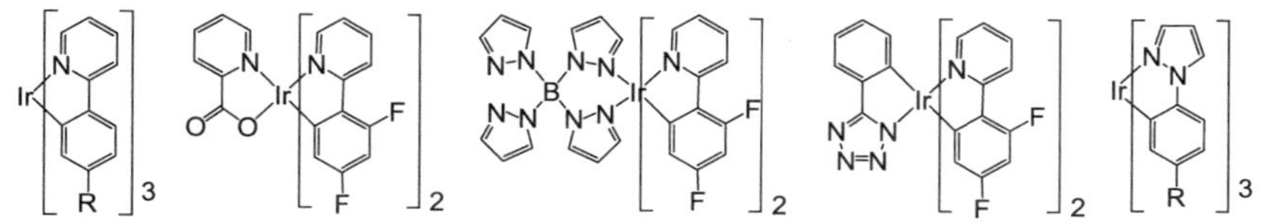


**Fig. 11.7** Typical blue phosphorescent Ir (III) complexes

There are three difficulties with blue Phosphorescent OLEDs. The first is that they do not have sufficient color purity. The National Television Standards Committee (NTSC) determined that the Commission Internationale de l'Eclairage (CIE) coordinates for blue are (x, y = 0.14, 0.08). However, in typical blue phosphorescent complexes, such as iridium (III) bis(4,6-difluorophenylpyridinato)picolinate (FIrpic) [40], iridium (III) bis(4,6-difluorophenylpyridinato)tetrakis(1-pyrazolyl) borate (FIr6) [41], and iridium (III) bis(4,6-di-fluorophenylpyridinato)-5-(pyridine-2-yl)-1*H*-tetrazolate (FIrN4) [42], the sum of the x, y coordinates is more than 0.3, and the color is called sky blue. The second difficulty is insufficient emission efficiency (Fig. 11.7). A blue phosphorescent complex has a higher emission state ($^3$MLCT, metal-to-ligand charge-transfer) than other color complexes, due to the large transition energy. This enables thermal activation to the metal-centered excited state (d*), which promotes nonradiative deactivation. For example, *fac*-Ir(ppz)$_3$ has strong blue phosphorescence ($\Phi_{PL} = 1.0$) at 77 K, but has almost no emission at 298 K ($\Phi_{PL} = 0.001$) [24]. This is explained by thermal excitation to the upper $^3$d* state, which is responsible for nonradiative deactivation from the phosphorescent $^3$MLCT* state at 298 K [43]. In addition, the development of host and carrier transport materials for blue phosphorescent materials is also very important, because the high triplet excitation energy of blue phosphorescent materials is becoming difficult to confine [40]. This accompanies tuning of the HOMO–LUMO level of the host and carrier transport materials, and causes a decrease of the carrier transport efficiency, which has an influence on device lifetime. For example, a device composed of FIrpic produced a large amount of defluorinated product (detected by mass spectrometry (MS)) after application of voltage–current at 10 V, 100 mA/cm$^2$ for 24 h [44]. Substitution with the strong electron-withdrawing difluoro group on the phenyl ring is an effective method to decrease the x + y value, and has thus been one of the main methods. However, there may be a shift toward the development of fluorine-free materials [45].

Computational investigation of *fac*-Ir(ppy)$_3$ revealed that the HOMO is mainly localized at the iridium d-orbital and phenyl moiety, and the LUMO is localized at the pyridyl moiety. Therefore, control of the HOMO–LUMO energy gap has been attempted by the modification of green phosphorescent *fac*-Ir(ppy)$_3$ using the following three strategies.

(1) Stabilization of HOMO by the introduction of electron withdrawing groups (EWGs) on a phenyl ring: Thompson and co-workers reported an OLED device

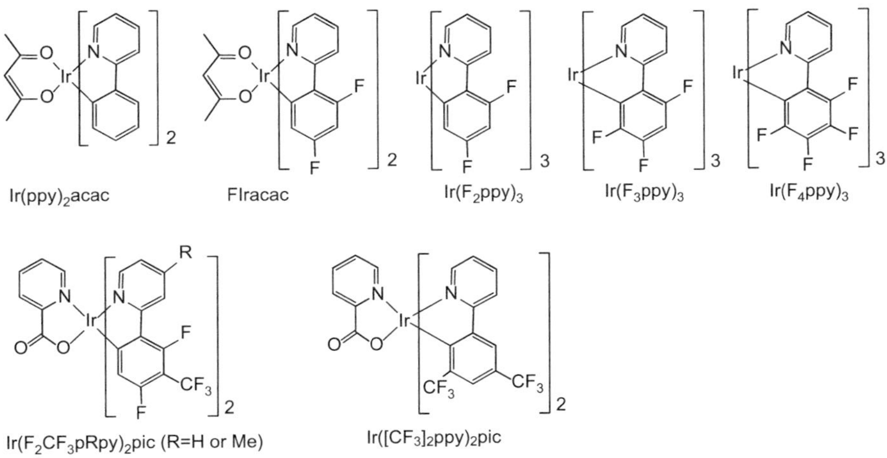


**Fig. 11.8** Complexes with electron withdrawing groups

doped with *fac*-Ir(ppy)$_3$ in 1999, and thereafter reported several homoleptic complexes. A typical preparation is conducted through a $\mu$-Cl dimer complex; however, there are difficulties involved. Thompson and co-workers also reported the synthesis and use of a diketonate complex through a dimer complex in 2001 [46]. These diketonate complexes have equivalent or slightly lower emission efficiency than the homoleptic complexes; however, they are easily synthesized. Many green and red phosphorescent diketonate complexes have also been reported [47]. Iridium (III) bis(4,6-difluorophenylpyridinato) acetylacetonate (FIr(acac)) and FIrpic were reported in 2001 (Fig. 11.8) [40]. Both complexes exhibited very high efficiency emission; $\lambda_{max}$ for FIr(acac) is blue-shifted 40 nm compared with that for iridium (III) bis(2-phenylpyridinato) acetylacetonate (Ir(ppy)$_2$(acac)), and $\lambda_{max}$ for FIrpic is also blue-shifted 20 nm compared with that for FIr(acac). These results indicate that the introduction of fluoro groups on the phenyl stabilizes the HOMO, because the HOMO is partly localized at the phenyl moiety. In addition, changing the ancillary ligand to the EWG picolinate also stabilized the HOMO partly localized at the d-orbital of the metal center.

Thompson and co-workers also reported *fac*-Ir(F$_2$ppy)$_3$ in 2003 [24]; however, $\lambda_{max}$ for this complex was the same as FIrpic (468 nm). In 2006, De Cola and co-workers reported a complex with trifluoro groups introduced on the phenyl rings of Ir(ppy)$_3$, *fac*-iridium (III) tris(3,4,6-trifluorophenylpyridinate) (*fac*-Ir(F$_3$ppy)$_3$), in addition to a complex with tetrafluoro groups, *fac*-iridium (III) tris(3,4,5,6-tetrafluorophenylpyridinate) (*fac*-Ir(F$_4$ppy)$_3$). These complexes had $\lambda_{max}$ at 459 and 468 nm, respectively [48]. Interestingly, the increased number of F substitution, from three to four, resulted in red-shift of $\lambda_{max}$ (decrease of the HOMO–LUMO energy gap). In the same year, Yamashita and co-workers reported trifluoromethyl substituted FIrpic, iridium

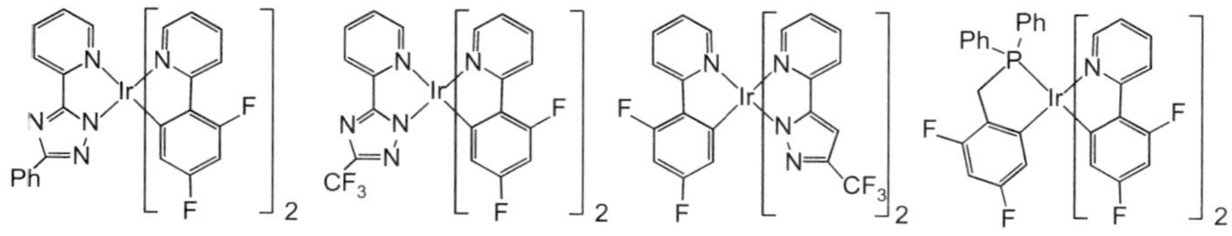


**Fig. 11.9** Complexes with electron withdrawing ancillary ligands

(III) bis(2,4-difluoro-3-(trifluoromethyl)phenylpyridinato)picolinate (Ir($F_2CF_3$pRpy)$_2$pic), and iridium (III) bis(2,4-difluoro-3-(trifluoromethyl)phenyl-4- methylpyridinato)picolinate, which had $\lambda_{max}$ at 457 and 454 nm, respectively [49]. Che and co-workers reported $\lambda_{max}$ for iridium (III) bis(3,5-bis (trifluoromethyl)phenylpyridinato)- picolinate (Ir([$CF_3$]$_2$ppy)$_2$(pic)) as 471 nm in 2008 [50]. These attempts demonstrate the effectiveness and limitations of design for blue phosphorescent Ir complexes by tuning of the HOMO energy level.

(2) Using strong EWG ancillary ligands: Many types of ancillary ligands have been added to the complexes after FIrpic exhibited a shorter $\lambda_{max}$ than FIracac. In 2003, Thompson and co-workers reported FIr6 with borate as a ancillary ligand, and this complex had $\lambda_{max}$ at 457 nm [41]. In 2004, De Cola and co-workers reported iridium (III) bis(4,6-difluorophenylpyridinato)pyridyltriazole with $\lambda_{max}$ at 461 nm ($\Phi_{PL} = 0.27$) [51]. At this stage, pyridyl-azole type ligands became popular. Chi and co-workers reported in 2005 that iridium (III) bis(4,6-difluorophenylpyridinato)-3-(trifluoromethyl)-5-(pyridine-2-yl)-1,2,4-triazolate and FIrN4 had $\lambda_{max}$ at 460 and 459 nm, respectively (Fig. 11.9) [46]. Then in 2007, iridium (III) bis(2-pyridyl-3-trifluoromethylpyrazolato)-4,6-difluorophenylpyridinate, which has pyridyltrifluoromethylpyrazole as main ligands, had $\lambda_{max}$ at 450 nm ($\Phi_{PL} = 0.50$) [52]. Chi and co-workers reported a phosphine type ligand in 2009 [53]; iridium (III) bis(4,6-difluorophenylpyridinato)-(2,4-difluorobenzyl)diphenylphosphinate had $\lambda_{max}$ at 457 nm ($\Phi_{PL} = 0.19$) [54].

(3) Breakaway from phenyl pyridinato complexes, using phenylheterocycles: Complexes with many types of ancillary ligands have been synthesized; however, the shift of $\lambda_{max}$ was only 10 nm, and color clarity of the blue was less than satisfactory. Further blue-shift is very difficult using difluorophenylpyridine as a ligand; therefore, an approach to tune the resonance stabilization energy by changing the pyridyl group to other heterocycles was attempted. In 2003, complexes with phenyl pyrazole type ligand were reported [24]. *fac*-Ir(ppz)$_3$ and the 4,6-difluorophenyl derivatives *fac*-Ir($F_2$ppz)$_3$ exhibited phosphorescence at 77 K with $\lambda_{max}$ at 414 and 390 nm, respectively. This was a significant blue-shift compared to phenyl pyridine based complexes such as *fac*-Ir($F_2$ppy)$_3$ ($\lambda_{max} = 450$ nm). However, these complexes in solution were poorly phosphorescent at room temperature ($\Phi_{PL} < 0.001$). Thompson and co-workers

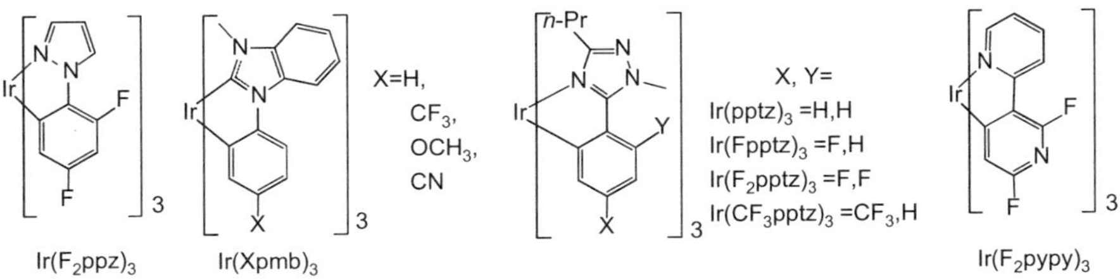


**Fig. 11.10** Complexes having characteristic heterocyclic rings

proposed that this was due to the bond weakness between iridium and the ligand nitrogen atoms, and they therefore synthesized a complex with a carbene type ligand [43]. In 2010, we examined the substituent effect of the tris(1-phenyl-3-methylbenzimidazolin-2-ylidene) iridium (III) (Ir(pmb)$_3$). *fac*-Ir(CF$_3$pmb)$_3$ and *fac*-Ir(CH$_3$Opmb)$_3$ exhibited deep blue phosphorescence with $\lambda_{max}$ at 396 and 403 nm ($\Phi_{PL} = 0.84$ and 0.76), respectively (Fig. 11.10) [15]. *fac*-Ir(pmb)$_3$ had poor solubility, which was improved to some extent using *N*-butyl substitution. However, there are no appropriate host and charge carrier materials for such large band gap complexes; therefore, the external quantum efficiency of the OLED device was only 2.6 % [55].

Samuel and co-workers reported a series of phenyl triazole type complexes in 2006 [56]. The *fac*-iridium (III) tris(1-methyl-5-phenyl-3-propyl-[1,2 4] triazolate) (Ir(pptz)$_3$ in Fig. 11.10), *fac*-iridium (III) tris(1-methyl-3-propyl-5-(4-fluorophenyl)-1*H*-[1,2,4]triazolate) (Ir(Fpptz)$_3$), and *fac*-iridium (III) tris (1-methyl-3-propyl-5-(4,6-difluorophenyl)-1*H*-[1,2,4]triazolate) (Ir(F$_2$pptz)$_3$) complexes had $\lambda_{max}$ at 449 ($\Phi_{PL} = 0.66$), 428 ($\Phi_{PL} = 0.27$), and 425 nm ($\Phi_{PL} = 0.03$), respectively. $\Phi_{PL}$ decreased by the increase of the number of substituted fluorine atoms; however, it is interesting that the chromaticity coordinate of Ir(pptz)$_3$ (0.16, 0.20) was deeper than the device composed of FIr6 (0.16, 0.26).

In 2009, Kang and co-workers reported that the phenyl ring of phenyl-pyridine was also converted to a heterocycle. *fac*-tris(2′,6′-difluoro-2,3′-bipyridinato-*N*,$C^{4'}$) iridium(III) (*fac*-Ir(F$_2$pypy)$_3$) exhibited $\lambda_{max}$ at 438 nm ($\Phi_{PL} = 0.71$) [57].

Research on blue phosphorescent iridium complexes is currently performed based on approaches (1)–(3). Complex that have generally high efficiency have long emission lifetimes that are inadequate for device fabrication [45], and complexes with $\lambda_{max}$ shorter than 450 nm have very small quantum efficiency (for example complex Ir(F$_2$pptz)$_3$).

There have been reports of complexes having phenylpyrazole and phenyltriazole type ligands; however, only a few reports of phenylimidazole derivatives. In 2009, Grätzel and co-workers reported a diketonate complex, iridium (III) bis(1-methyl-2-phenylimidazolato)acetylacetonate (N966) [58], which gives a broad emission between 440 and 800 nm that is applicable to a single molecular white lighting OLED.

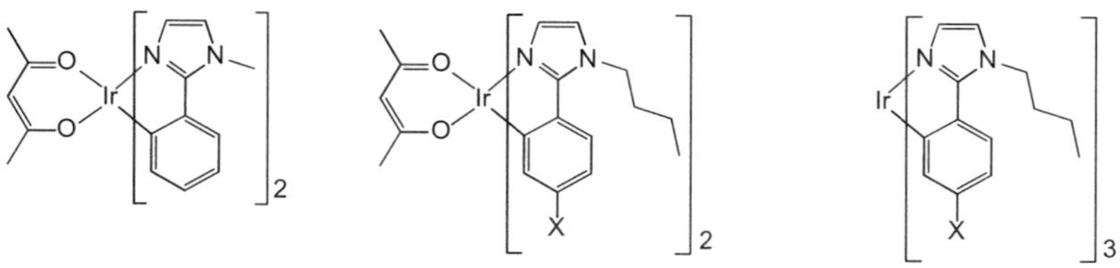

Ir(*N*-Mepim)$_2$acac

Ir(Xpim)$_2$acac

*fac*-Ir(Xpim)$_3$

Ir(pim)$_2$acac: X=H
Ir(OCH$_3$pim)$_2$acac: X=OCH$_3$
Ir(Fpim)$_2$acac: X=F
Ir(CF$_3$pim)$_2$acac: X=CF$_3$
Ir(OCF$_3$pim)$_2$acac: X=OCF$_3$

*fac*-Ir(Fpim)$_3$ :X=F
*fac*-Ir(CF$_3$pim)$_3$ :X=CF$_3$
*fac*-Ir(OCF$_3$pim)$_3$ :X=OCF$_3$

**Fig. 11.11** Homoleptic and heteroleptic phenylimidazolynato complexes

Grätzel and co-workers also reported continuous substituents and ancillary ligand effects for N966 in 2011 [59]. They examined the effect of methyl and phenyl groups as *N*-substituents, and also examined chloro substitution on a phenyl ring. For example, iridium (III) bis(4,5-dimethyl- 1,2-diphenylimidazolato)-4-(dimethylamino)picolinate (Ir(Ph$_2$pim)$_2$dmapic) has $\lambda_{max}$ at 539 nm ($\Phi_{PL} = 0.22$) and iridium (III) bis(4-methyl-1,2,5-triphenylimidazolato)- acetylacetonate (Ir(pmppim)$_2$acac) has $\lambda_{max}$ at 551 nm with an efficiency of $\Phi_{PL} = 0.95$. The aim of this research was also to achieve white phosphorescence. In the same year, Perumal and co-workers also reported substituent and solvent effects on the emission of diketonate complexes [60]. For example, iridium(III) bis(4,5-dimethyl-1-(3,5-dimethylphenyl)-2-fluorophenylimidazolato)acetylacetonate (Ir(pmdmpim)$_2$acac) had a green emission of $\lambda_{max}$ at 514 nm ($\Phi_{PL} = 0.56$). Both reports of phenylimidazolinato complexes do not deal with the blue phosphor, and there is no report of homoleptic complexes. However, there are some patents of phenylimidazolinato complexes [61–66], such as that applied by Konica–Minolta in 2006 [61].

Phenylimidazolinato complexes are one of the important candidates for blue phosphorescent materials. In particular, we have synthesized phenylimidazolinato complexes with various substituents introduced on the phenyl ring (Fig. 11.11), and examined their effects on the photophysical properties and frontier orbitals (HOMO and LUMO).

MO analysis of *fac*-Ir(ppy)$_3$ indicated that the HOMO is mainly localized in the d-orbital of iridium atom (46.4–56.0 %) and phenyl moiety (30.7–39.8 %) [56] as shown in Table 11.4. Similar behavior has been observed in *fac*-Ir(ppz)$_3$[67]. On the other hand, detailed analysis of the MOs of *fac*-Ir(CF$_3$pim)$_3$, *fac*-Ir(OCF$_3$pim)$_3$, and *fac*-Ir(Fpim)$_3$ revealed the HOMO contribution of the phenyl moiety was 30.7–32.8 % [16], which is smaller than that of *fac*-Ir(ppy)$_3$ (38.9 %). Accordingly, the contribution of the imidazole ring part (13.3–15.0 %) is higher than that of the pyridine part (8.2 %) in *fac*-Ir(ppy)$_3$.

The LUMO localized on the ligand (nearly 100 %) is the same result as that for *fac*-Ir(ppy)$_3$ [56] and *fac*-Ir(ppz)$_3$ [67]. These results strongly indicate a

**Table 11.4** Calculated contribution of iridium metal (Ir), phenyl (Ph), and heterocyclic (Hetero) moieties to HOMO and LUMO

| | HOMO | | | LUMO | | |
|---|---|---|---|---|---|---|
| | Ir | Ph | Hetero[a] | Ir | Ph | Hetero[a] |
| *fac*-Ir(ppy)$_3$ [56] | 52.8 | 38.9 | 8.2 | 0.2 | 25.9 | 73.5 |
| *fac*-Ir(pmb)$_3$ [43] | 30.2 | 47.7 | 22.1 | 2.5 | 2.5 | 95 |
| *fac*-Ir(CF$_3$pmb)$_3$ [15] | 40.3 | 38.5 | 21.2 | 3.2 | 19.7 | 77.1 |
| *fac*-Ir(pptz)$_3$ [56] | 56.6 | 31.8 | 10.7 | 0 | 49.2 | 46.7 |
| *fac*-Ir(CF$_3$pptz)$_3$ [56] | 58.6 | 30.8 | 9.7 | 0.1 | 56.1 | 36.8 |
| *fac*-Ir(Fpim)$_3$ [16] | 52.4 | 32.8 | 14.8 | 0.6 | 53.6 | 45.8 |
| *fac*-Ir(CF$_3$pim)$_3$ [16] | 56 | 30.7 | 13.3 | 0.4 | 64.1 | 35.5 |
| *fac*-Ir(OCF3pim)$_3$[16] | 53.2 | 31.8 | 15 | 0.5 | 57 | 42.5 |

[a]Hetero means heterocyclic moiety, thus, pyridyl, benzoimidazolyl, triazolyl, or imidazolyl group including alkyl substituents on it

HOMO–LUMO transition with MLCT character. Details of the LUMO in the phenyl (Ph) moieties (53.6–64.1 %) of homoleptic *fac*-Ir(CF$_3$pim)$_3$, *fac*-Ir (OCF$_3$pim)$_3$, and *fac*-Ir(Fpim)$_3$ have higher contribution than those of the imidazolyl (Im) parts (35.5–45.8 %) [16]; therefore, these populations are quite different from the LUMO of *fac*-Ir(ppy)$_3$ [56] with 73.5 % pyridyl (Py) and 25.9 % phenyl moieties.

The calculated results indicate that the HOMO of iridium phenylimidazolinate complexes have similar or slightly smaller contributions at the phenyl moiety, as with *fac*-Ir(ppy)$_3$; however, much larger localization of the LUMO at the phenyl moiety. Therefore, substitution of the phenyl group by EWGs affects not only the HOMO, but also the LUMO [16]. This type of HOMO–LUMO relation has also been reported for iridium phenyltriazolinate complexes [56].

The contributions of phenyl ring and heterocyclic ring parts of diketonate complexes were similar to those of *fac*-Ir(ppy)$_3$; however, the contribution of the iridium d-orbital had smaller values 46–48 % and these decreases appeared as increase of acetylacetone parts (4.6–5.8 %). Therefore, in *fac*-Ir(CF$_3$pim)$_3$, *fac*-Ir (OCF$_3$pim)$_3$, and *fac*-Ir(Fpim)$_3$, the substitution of EWGs on the phenyl ring is less effective to stabilize the HOMO than in the case of *fac*-Ir(ppy)$_3$ (Table 11.5) [16].

The LUMO of diketonate complexes also has a higher coefficient at the phenyl moiety (51.2–62.3 %) than at the imidazole moiety (35.7–46.3 %), similar to the homoleptic complexes. However, there are two exceptions, Ir(OCH$_3$pim)$_2$acac and Ir(pim)$_2$acac, where the LUMO is localized at acetylacetone (93.6 and 83.8 %, respectively). In addition to these two extreme cases, other diketonate complexes have a LUMO + 1 (MHacac and Ir(Fpim)$_2$acac) or LUMO + 2 (Ir(CF$_3$pim)$_2$acac and Ir(OCF$_3$pim)$_2$acac) localized at acetylacetone. Complexes with trifluoromethyl substituents, *fac*-Ir(CF$_3$pim)$_3$ and Ir(CF$_3$pim)$_2$acac, have LUMO highly localized at the phenyl moiety with 64.1 and 62.3 %, respectively.

Substitution of the acetylacetonate complex has a significant effect on the emission quantum efficiency. This can be explained from the energy difference ($\Delta E$) between the LUMO and MO localized on the acetylacetone moiety. Table 11.5

**Table 11.5** Calculated energy levels of the HOMO, LUMO, LUMO + 1, and LUMO + 2 and contribution of iridium metal (Ir), phenyl (Ph), imidazolyl (Im), and acetylacetonate (acac) moieties [16]

| | | Ir | Ph | Im | acac | $E$ (eV) |
|---|---|---|---|---|---|---|
| $Ir(Fpim)_2acac$ | LUMO + 2 | 3.6 | 42.3 | 39.7 | 14.4 | 0.74 |
| | LUMO + 1 | 1.4 | 7.5 | 7.4 | 83.7 | 0.87 |
| | LUMO | 2.1 | 51.9 | 45.6 | 0.4 | 0.88 |
| | HOMO | 46.4 | 38.4 | 10.2 | 5.0 | 4.86 |
| $Ir(CF_3pim)_2acac$ | LUMO + 2 | 2.2 | 1.2 | 1.0 | 95.6 | 1.05 |
| | LUMO + 1 | 2.8 | 60.3 | 34.3 | 2.6 | 1.41 |
| | LUMO | 1.6 | 62.3 | 35.7 | 0.4 | 1.54 |
| | HOMO | 48.8 | 36.8 | 8.6 | 5.8 | 5.16 |
| $Ir(OCF_3pim)_2acac$ | LUMO + 2 | 2.8 | 10.5 | 8.8 | 77.9 | 1.03 |
| | LUMO + 1 | 2.2 | 43.0 | 34.5 | 20.3 | 1.15 |
| | LUMO | 1.9 | 55.4 | 42.3 | 0.4 | 1.25 |
| | HOMO | 47.6 | 37.0 | 9.9 | 5.5 | 5.16 |
| $Ir(OCH_3pim)_2acac$ | LUMO + 2 | 3.0 | 46.4 | 46.2 | 4.4 | 0.31 |
| | LUMO + 1 | 2.0 | 50.9 | 46.8 | 0.3 | 0.43 |
| | LUMO | 1.6 | 2.2 | 2.6 | 93.6 | 0.54 |
| | HOMO | 46.4 | 38.9 | 10.0 | 4.6 | 4.43 |
| $Ir(pim)_2acac$ | LUMO + 2 | 3.6 | 42.2 | 39.8 | 14.4 | 0.50 |
| | LUMO + 1 | 2.1 | 51.7 | 45.8 | 0.4 | 0.64 |
| | LUMO | 1.4 | 7.4 | 7.4 | 83.8 | 0.64 |
| | HOMO | 47.0 | 39.8 | 8.4 | 4.8 | 4.48 |

shows the percentage MO (LUMO to LUMO + 2) contribution of iridium metal (Ir), phenyl (Ph), imidazolyl (Im), and acetylacetonate (acac) moieties, and the MO energy levels ($E$) calculated by the DFT method. For example, in the case of $CF_3$acac, $\Delta E$ is 0.49 eV ($E_{\text{LUMO}}$–$E_{\text{LUMO+2}}$ = 1.54–1.05), and for $OCF_3$acac, $\Delta E$ is 0.22 eV ($E_{\text{LUMO}}$–$E_{\text{LUMO+2}}$ = 1.25–1.03). This indicates that a larger $\Delta E$ gives a larger $\Phi_{\text{PL}}$; therefore, simple MO energy levels calculated for optimized structure are useful to understand nonradiative deactivation processes through the ancillary ligand. The weak and broad emission of $Ir(CF_3pim)_2acac$ and $Ir(OCF_3pim)_2acac$ at 298 K is caused by the quenching of excitation energy by the acetylacetone part. The inefficiencies of other diketonated complexes at 298 K are due to thermal activation to the upper excited state responsible for nonradiative deactivation. Therefore, these thermal activations are prohibited at 77 K. The smallest $\Phi_{\text{PL}}$ of $Ir(OCH_3pim)_2acac$ and $Ir(pim)_3acac$ among the diketonate complexes at 77 K is explained by the LUMO being localized on acetylacetone [16].

In contrast, the homoleptic complexes showed efficient emission, not only at 77 K, but also at 298 K ($\Phi_{\text{PL}}$ = 0.40–0.60). No significant difference of $k_r$ between the homoleptic and diketonate complexes was evident, whereas $k_{nr}$ become smaller for the homoleptic complexes than for the diketonate complexes. For example, in the case of *fac*-$Ir(Fpim)_3$ and $Ir(Fpim)_2acac$, $k_r$ is almost the same (1.1 and $1.7 \times 10^5\ s^{-1}$); however, $k_{nr}$ decreased by almost 1/300.

OLED devices were fabricated for *fac*-Ir($OCF_3$pim)$_3$, which showed efficient luminescence in 2MeTHF, and FIrpic was used as a reference. Device was composed of ITO/PEDOT: PSS/mB-4Cz/BCP/CsF/Al. The wet-processable m-terphenyl derivative, 3,3″,5,5″-tetra(9H-carbazol-9-yl)-1,1′:3′,1″-terphenyl (mB-4Cz), was synthesized and used as a host material [17, 68]. The performance of the fabricated devices were determined from plots of current density ($J$: mA/cm$^2$), luminance ($L$: cd/m$^2$), and current efficiency ($\eta$: cd/A) versus applied voltage ($V$: V).

EL spectra of both devices were red-shifted by 2–3 nm and had smaller 0–0 bands than the 0–1 band, compared with the spectra measured in 2MeTHF. Therefore, the CIE coordinates (x and y) were slightly increased; the coordinate of FIrpic is more blue than *fac*-Ir($OCF_3$pim)$_3$; however, there was no significant difference observed by visual check.

The maximum luminance for *fac*-Ir($OCF_3$pim)$_3$ and FIrpic was 889 and 3,490 cd/m$^2$, respectively. The lower luminance of *fac*-Ir($OCF_3$pim)$_3$ is explained by inefficient carrier injection into the emitting layer. This is supported by both the inefficient current density and larger driving voltage. The HOMO–LUMO energy levels of *fac*-Ir($OCF_3$pim)$_3$ were shifted by approximately 0.7 eV to higher energy than those for FIrpic.

From the energy diagram, the HOMO–LUMO level of *fac*-Ir($OCF_3$pim)$_3$ is moved by 0.7 V almost parallel to those of FIrpic in the anodic direction. Therefore, charge injection from the charge conducting layer to the emitting layer became difficult in the case of *fac*-Ir($OCF_3$pim)$_3$. The HOMO level of *fac*-Ir($OCF_3$pim)$_3$ is higher than that of PEDOT:PSS. The HOMO–LUMO energy levels of imidazole have been reported to take high values among some nitrogen-containing cyclic compounds, according to ab initio calculations [69].

The FIrpic device has the same configuration with the mB-4Cz host materials. A hole only device (device fabricated without a PEDOT: PSS layer) and an electron only device (device fabricated with no BCP layer) were fabricated and the *J-V* characteristics were measured. The hole and electron only devices showed current densities of 237 and 64 mA/cm$^2$, respectively, at an applied voltage of 10 V [68]. Therefore, the high hole and low electron transferability of mB-4Cz is partly responsible for the moderate performance of these devices.

## 11.4 Wet Processable Host Materials for Phosphorescent OLED

Importance of the materials for PHOLED is directly affected to the performance of the devices, as I already mentioned in earlier section. In addition, host materials and device fabrication process has been caught attentions. Because of their high quantum efficiency and low-cost processing, considerable research has been made on solution processable materials for PHOLED during the last decade [70–78].

However, most of these solution processable materials for PHOLED are based on the polymer materials, which have an inherent number of problems, including lower carrier mobility of host materials, shorter lifetime of EL devices, especially their difficult syntheses and purifications [78, 79]. Recently, soluble small molecule-based PHOLEDs have been reported [78–85], which can be overcoming above problems of polymer hosts. On the other hand, because of their high triplet energy gap and bipolar nature of carrier transport, carbazole units frequently have been used in phosphorescence host materials. For example, 4,4′-bis(9-carbazolyl)-2,2′-biphenyl (CBP), 1,4-di(9H-carbazolyl)benzene (CCP) [86], 4,4′-N-N′-dicarbazoleterphenyl (CTP) [87], 1,3-bis(9-carbazolyl)benzene (mCP) [88], and 2,2′- di(9H-carbazole-9-yl)biphenyl (o-CBP) [89]. These small-molecule hosts could be great host materials in vacuum deposited device, however not be obtained approving EL performance in solution processed device. This is because these small-molecule hosts possess the lower glass transition temperature ($T_g$), lead to the occurrence of crystallization upon drying of emitting layer, thus cannot form a uniform amorphous thin film [80]. Therefore, to obtain better performance of EL device great needs to increase the $T_g$ for small-molecular hosts in solution processed device.

It is conceivable that above carbazole-based hosts can be easily used to solution process through the relatively simple molecular design. Along this respect, in this paper, we report two small-molecule host materials for solution processing PHOLED, 1,4-bis(3,6-di([1,1′:3′,1″:3″,1‴:3‴,1⁗-quinquephenyl]-5″-yl)-9*H*-carbazol-9-yl)benzene (P-mPCCP) and 1,4-bis(3,6-bis(4,4″-di-tert-butyl-[1,1′:3′,1-″-terphenyl]-5′-yl)-9*H*-carbazol-9-yl) benzene (T-mPCCP) (Fig. 11.12). We have

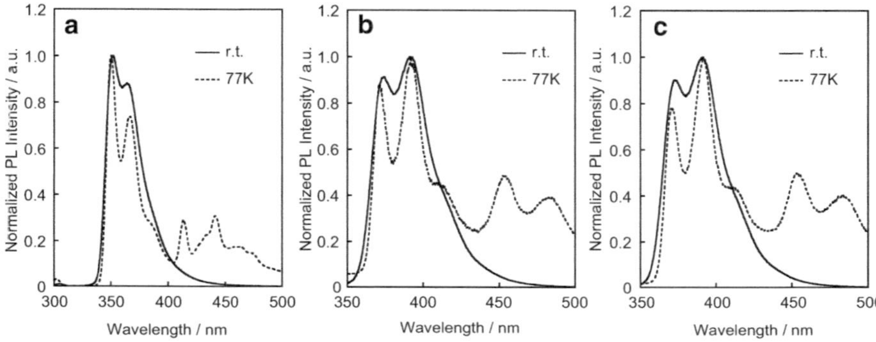


**Fig. 11.12** Molecular structures (*above*) and (**a**) PL spectra of CCP, (**b**) P-mPCCP, and (**c**) T-mPCCP at room temperature (*solid line*) and 77 K in Me-THF (*dotted line*)

chosen the CCP as the molecular core because of its good solubility and high triplet energy relative to order carbazol-based hosts, such as well-known CBP. In order to advance thermal stability of host materials, the *m*-terphenyl group attached to the CCP molecule. Since possessing of high $E_T$ and high carrier mobility, *m*-terphenyl group has been successfully applied to phosphorescent host designing [78, 90, 91]. However, combining with *m*-terphenyl group generally molecules become very rigid and difficult to dissolve in common organic solvents. Actually, host material, 1,4-bis(3,6-bis([1,1′:3′,1″-terphenyl]-5′-yl)-9*H*-carbazol-9-yl)benzene (mPCCP) synthesized in this study was also insoluble in any solvent used. To improve the solubility, the both approach of increasing of molecular free volume and introduction of *tert*-butyl group was adopted to synthesized P-mPCCP and T-mPCCP, respectively [17].

P-mPCCP and T-mPCCP were prepared as described below. First, compound 5″-bromo-1,1′:3′,1″:3″,1‴:3‴,1⁗-quinquephenyl was prepared from 3-biphenylboronic acid and 1,3,5-tribromobenzene by $Pd(PPh_3)_4$ in THF, which then converted to the arylboronic ester of 2-([1,1′:3′,1″:3″,1‴:3‴,1″-″-quinquephenyl]-5″-yl)-4,4,5,5-tetramethyl-1,3,2-dioxaborolane (M1). The same procedure for compound M2 was applied to give compound 2-(4,4″-di-*tert*-butyl-[1,1′:3′,1″-terphenyl]-5′-yl)-4,4,5,5-tetramethyl-1,3,2-dioxaborolane (M2). Next, the Suzuki cross-coupling reaction of compound M1 and compound M2 with 1,4-bis(3,6-diiodo-9*H*- carbazolyl)benzene [92] led to P-mPCCP and T-mPCCP, respectively. The model compound mPCCP was synthesized from CCP and m-terphenyl group. The chemical structures were characterized by $^{1}$H NMR, $^{13}$C NMR and elemental analysis.

mPCCP showed poor solubility and not soluble in any solvents. Contrast to mPCCP, P-mPCCP and T-mPCCP can readily dissolve in common organic solvents, such as chloroform, 1,2-dichloroethane, toluene, chlorobenzene. For P-mPCCP molecule, although without alkyl group also showed good solubility due to increasing of molecular free volume through additional aromatic ring.

The thermal properties of both compared are evaluated by differential scanning calorimetry (DSC, SII, extar 600). P-mPCCP and T-mPCCP possess high $T_g$ of 161 and 185 °C, respectively, which is significantly higher than the commonly used host materials of CBP (66 °C) and mCP (60 °C). The morphologies of P-mPCCP and T-mPCCP were characterized by AFM (SII, SPA-400). For P-mPCCP and T-mPCCP films doped with 6 wt% $Ir(ppy)_3$, the root-mean square (RMS) values are 0.57 and 0.48 nm, respectively. This indicates that allowing both form morphologically stable and uniform amorphous films in solution process.

P-mPCCP and T-mPCCP have a similar absorption and emission spectra in chloroform and film state, the maximum absorption and emission peak at around 260 and 395 nm, respectively. The absorption band gap was 3.40 eV from the onset of optical absorption in both compounds, which is 0.1 eV lower than that of CCP. The emission spectrum at 77 K is also measured for estimation of the triplet energy (Fig. 11.12). The spectra show the first phosphorescence peak at 452 nm in P-mPCCP and T-mPCCP, corresponding to a triplet energy of 2.70 eV, which is higher than the triplet energy of *fac*-$Ir(ppy)_3$, $E_T = 2.41$ eV [93]. This indicates that

**Table 11.6** Physical properties of hosts molecules

| Compounds | $T_g$ (°C) | UV–vis/nm | $PL_{\lambda max}$/nm | HOMO[a]/eV | LUMO[b]/eV | $\Delta E^c$/eV | $E_T{}^d$/eV |
|---|---|---|---|---|---|---|---|
| CCP | – | 293 | 378 | −5.46 | −1.94 | 3.52 | 3 |
| P-mPCCP | 161 | 254 | 397 | −5.56 | −2.16 | 3.4 | 2.7 |
| T-mPCCP | 185 | 260 | 395 | −5.51 | −2.11 | 3.4 | 2.7 |

[a]HOMO is derived from electrochemical oxidation potentials
[b]LUMO = HOMO + $\Delta E$
[c]$\Delta E$ is obtained from the absorption band gap
[d]$E_T$ values estimated from phosphorescence 0,0 band in Fig. 11.12

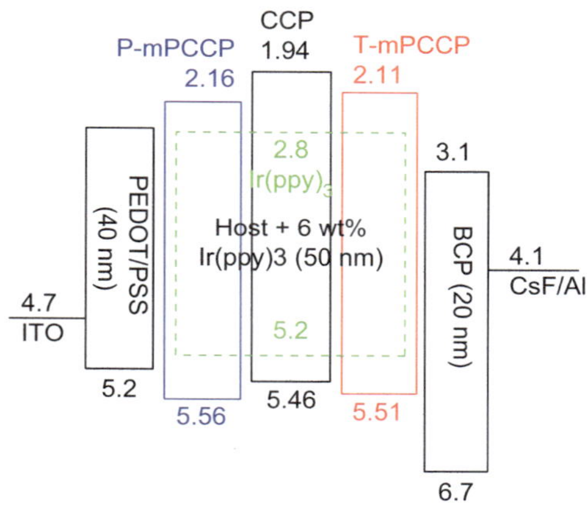


**Fig. 11.13** Energy level diagram for OLED devices

allowing P-mPCCP and T-mPCCP to serve as appropriate host for $Ir(ppy)_3$. The triplet energy of CCP is calculated to be 3.0 eV, which is 0.3 eV higher than that of P-mPCCP and T-mPCCP. This is because the addition of aromatic ring at 3 or 6 position of carbazole will significantly reduced the triplet energy by expanded the π-electrons of delocalization [94]. The performance of photophysical properties are summarized in Table 11.6.

The electrochemical oxidation potentials of P-mPCCP and T-mPCCP were measured at 0.76 and 0.71 V vs ferrocene/ferricinium ion in $CH_2Cl_2$; thus, using ferrocene ionization potential of −4.8 eV led to HOMO energy levels of P-mPCCP and T-mPCCP to be −5.56 and −5.51 eV, respectively. We can be deduced that the favorable hole injection from the PEDOT: PSS layer to the emitting layer in EL device. The LUMO energy levels were calculated from HOMO levels and optical band gaps obtained from their absorption spectra to be −2.16 and −2.11 eV, respectively.

To investigate the OLED properties of P-mPCCP and T-mPCCP composing devices, EL device with a structure of ITO/PEDOT:PSS (40 nm)/hosts + 6 wt% Ir $(ppy)_3$ (50 nm)/BCP (20 nm)/CsF (2 nm)/Al (100 nm) was fabricated (Fig. 11.13). As shown in Fig. 11.14a, the EL spectra of both devices are almost similar and showed a maximum emission peak at 510 nm, with the emission of $Ir(ppy)_3$,

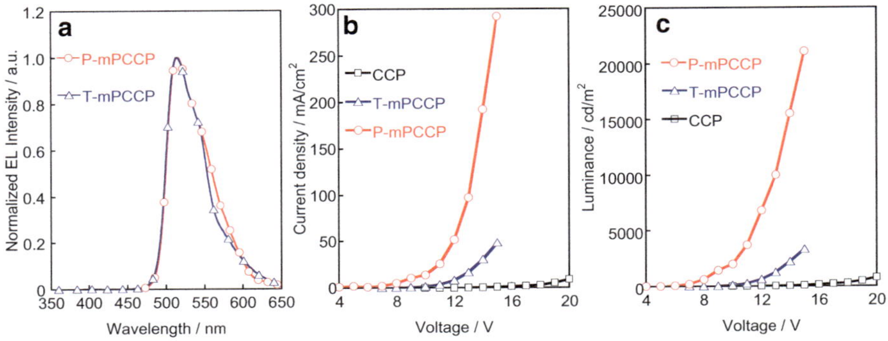


**Fig. 11.14** (**a**) EL spectra, (**b**) *I-V* characteristic and (**c**) *L-V* characteristic of OLED devices

without trace of host emission at around 400 nm. The PL spectra of doped Ir(ppy)$_3$ in P-mPCCP and T-mPCCP were similar to their EL spectra and intensities at their maxima had no significant difference. This indicated that an efficient energy transfer occurred from both host to guest Ir(ppy)$_3$. The properties of these devices are compared to that of CCP composing device. The current–voltage (*I-V*) and luminance-voltage (*L-V*) characteristics are showed Fig. 11.14b, c, respectively. The performance of P-mPCCP and T-mPCCP composing device were much higher than that of CCP composing device. The turn-on voltage of P-mPCCP and T-mPCCP composing device was 4.0 and 7.0 eV (corresponding to 1 cd/m$^2$) and showed a $L_{max}$ of 21,100 and 3,290 cd/m$^2$ and a maximum luminance efficiency, $\eta_{c, max}$ of 15.0 and 7.6 cd/A, respectively. In contrast, the CCP composing device was turned on at 10.0 V and showed a $L_{max}$ of only 771 cd/m$^2$ and a $\eta_{c, max}$ of 5.1 cd/A. This is indicated that introducing *m*-terphenyl derivatives of molecular design is very successful for improve the device performance in solution processed device. The device composed of CCP showed very low performance probably because of insufficient smoothness by the spin coating and poor thermal stability of the emitting layer. The device coated with CCP, crystallization phenomenon was found in the next day.

Although both hosts have a same triplet energy, HOMO and LUMO levels, but the P-mPCCP composing device possessed much higher device performance than that of T-mPCCP composing device, which was maximum luminance of sixfold and maximum luminance efficiency of twofold higher than that of T-mPCCP composing device. We attributed to reducing of carrier mobility by introducing a *t*-butyl alkyl in T-mPCCP based on emitting layer. We fabricated the hole transport only device (ITO/PEDOT: PSS/emitting layer/Al) and electron transport only device (ITO/emitting layer/TPBI/LiF/Al) to investigate the conductivity of emitting layer. As a result, P-mPCCP composing device showed higher the current density of hole and electron, especially electron current density was tenfold higher than that of T-mPCCP.

We also fabricated OLED devices by similar method using FIrpic as phosphorescent dopant, however, those devices performance was poor. Considering triplet energy of FIrpic ($E_T = 2.62$ eV [77]), those devices satisfied triplet energy requirement, however, there is other factors to be solved. One possible factor is that HOMO–LUMO energy levels of FIrpic are 5.72 and 3.07 eV, respectively. Therefore, HOMO energy level of FIrpic is much higher than those of host materials examined. Selection of materials such as dopant and host molecules is somehow requires tailor-made manner.

**Acknowledgments** This work was supported by Grants-in-Aids for Scientific Research (No. 20550056 and 26288111) and the G-COE Program (Advanced School for Organic Electronics) from the Ministry of Education, Culture, Sports, Science and Technology (MEXT), and COE Start-up Program from Chiba University.

## References

1. K.A. Belmore, R.A. Vanderpool, J.C. Tsai, M.A. Khan, K.M. Nicholas, J. Am. Chem. Soc. **110**, 2004 (1988)
2. K.K.-W. Lo, C.-K. Chung, T.K.-M. Lee, L.-K. Lui, K.H.-K. Tsang, N. Zhu, Inorg. Chem. **42**, 6886 (2003)
3. M.C. DeRosa, D.J. Hodgson, G.D. Enright, B. Dawson, C.E.B. Evans, R.J. Crutchley, J. Am. Chem. Soc. **126**, 7619–7626 (2004)
4. H. Yersin (ed.), *Highly Efficient OLEDs with Phosphorescent Materials* (Wiley VCH, Berlin, 2007)
5. D. Tanaka, H. Sasabe, Y.-J. Li, S.-J. Su, T. Takeda, J. Kido, Jpn. J. Appl. Phys. **46**, L10–L12 (2007)
6. C.W. Tang, S.A. Vanslyke, Appl. Phys. Lett. **51**, 913 (1987)
7. R. Schlaf, B.A. Parkinson, P.A. Lee, K.W. Nebesny, G. Jabbour, B. Kippelen, N. Peyghambarian, N.R. Armstrong, J. Appl. Phys. **84**, 6729–6736 (1998)
8. M.A. Baldo, D.F. O'Brien, Y. You, A. Shoustikov, S. Sibley, M.E. Thompson, S.R. Forrest, Nature **395**, 151 (1998)
9. D.F. O'Brien, M.A. Baldo, M.E. Thompson, S.R. Forrest, Appl. Phys. Lett. **74**, 442 (1999)
10. M.A. Baldo, S. Lamansky, P.E. Burrows, M.E. Thompson, S.R. Forrest, Appl. Phys. Lett. **75**, 4 (1999)
11. M.A. Baldo, M.E. Thompson, S.R. Forrest, Nature **403**, 750 (2000)
12. T. Karatsu, T. Nakamura, S. Yagai, A. Kitamura, K. Yamaguchi, Y. Matsushima, T. Iwata, Y. Hori, T. Hagiwara, Chem. Lett. **32**, 886 (2003)
13. T. Karatsu, E. Ito, S. Yagai, A. Kitamura, Chem. Phys. Lett. **424**, 353 (2006)
14. K. Tsuchiya, E. Ito, S. Yagai, A. Kitamura, T. Karatsu, Eur. J. Inorg. Chem. **2104** (2009)
15. K. Tsuchiya, S. Yagai, A. Kitamura, T. Karatsu, K. Endo, J. Mizukami, S. Akiyama, M. Yabe, Eur. J. Inorg. Chem. 926 (2010)
16. T. Karatsu, M. Takahashi, S. Yagai, A. Kitamura, Inorg. Chem. **52**, 12338 (2013)
17. X. Lin, S. Yagai, A. Kitamura, D.-R. Hwang, S.-Y. Park, Y.-S. Park, J.-J. Kim, T. Karatsu, Synth. Met. **162**, 303 (2012)
18. X. Lin, T. Hattori, S. Yagai, A. Kitamura, T. Karatsu, J. Photopolym. Sci. Tech. **24**, 299 (2011)
19. G. Zhou, W.-Y. Wong, S. Suo, J. Photochem. Photobiol. C Photochem. Rev. **11**(133) (2010)
20. M. Nonoyama, Bull. Chem. Soc. Jpn. **47**, 767 (1974)
21. K.A. King, P.J. Spellane, R.J. Watts, J. Am. Chem. Soc. **107**, 1431 (1985)
22. K. Dedeian, P.I. Djurovich, F.O. Garces, G. Garlson, R.J. Watts, Inorg. Chem. **30**, 1685 (1991)
23. M.G. Colombo, T.C. Brunold, T. Riedener, H.U. Güdel, M. Förtsch, H.-B. Bürgi, Inorg. Chem. **33**, 545 (1994)

24. A.B. Tamayo, B.D. Alleyne, P.I. Djurovich, S. Lamansky, I. Tsyba, N.N. Ho, R. Bau, M.E. Thompson, J. Am. Chem. Soc. **123**, 7377 (2003)
25. V.V. Grushin, N. Herron, D.D. LeCloux, W.J. Marshall, V.A. Petrov, Y. Wang, Chem. Commun. 1494 (2001)
26. J.N. Demas, G.A. Crosby, J. Am. Chem. Soc. **92**, 7262 (1970)
27. J.N. Demas, G.A. Crosby, J. Am. Chem. Soc. **93**, 2841 (1971)
28. R.J. Watts, G.A. Crosby, J. Am. Chem. Soc. **94**, 2606 (1972)
29. R.J. Watts, G.A. Crosby, J.L. Sansregret, Inorg. Chem. **11**, 14 (1972)
30. K. Koike, N. Okoshi, K. Takeuchi, O. Ishitani, H. Tsubaki, I.P. Clark, M.W. George, F.P.A. Johnson, J.J. Turner, J. Am. Chem. Soc. **124**, 11448 (2002)
31. A.W. Adamson, J. Phys. Chem. **71**, 798 (1967)
32. L.G. Vanquickienborne, A. Ceulemans, J. Am. Chem. Soc. **99**, 2208 (1977)
33. L.G. Vanquickienborne, A. Ceulemans, Inorg. Chem. **17**, 2730 (1978)
34. L.G. Vanquickienborne, A. Ceulemans, Coord. Chem. Rev. **48**, 157 (1983)
35. E. Larsen, G.N. La Mar, J. Chem. Educ. **51**, 633 (1974)
36. G.L. Miessler, D.A. Tarr, *Inorganic Chemistry*, 3rd edn. 2004 (Prentice Hall Inc, New Jersey)
37. E.J. Nam, J.H. Kim, B.-O. Kim, S.M. Kim, N.G. Park, Y.S. Kim, Y.K. Kim, Y. Ha, Bull. Chem. Soc. Jpn. **77**, 751 (2004)
38. N. Ide, N. Matsusue, T. Kobayashi, H. Naito, Thin Soild Films **509**, 164 (2006)
39. L. Xiao, Z. Chen, B. Qu, J. Luo, S. Kong, Q. Gong, J. Kido, Adv. Mater. **23**, 926 (2011)
40. C. Adachi, R.C. Kwong, P. Djurovich, V. Adamovich, M.A. Baldo, M.E. Thompson, S.R. Forrest, Appl. Phys. Lett. **79**, 2082 (2001)
41. R.J. Holmes, B.W. D'Andrade, S.R. Forrest, X. Ren, J. Li, M.E. Thompson, Appl. Phys. Lett. **83**, 3818 (2003)
42. S.-J. Yeh, M.-F. Wu, C.-T. Chen, Y.-H. Song, Y. Chi, M.-H. Ho, S.-F. Hsu, C.H. Chen, Adv. Mater. **17**, 285 (2005)
43. T. Sajoto, P.I. Djurovich, A. Tamayo, M. Yousufuddin, R. Bau, M.E. Thompson, R.J. Holmes, S.R. Forrest, Inorg. Chem. **44**, 7992 (2005)
44. V. Sivasubramaniam, F. Brodkorb, S. Hanning, H.P. Loebl, V. van Elsbergen, H. Boerner, U. Scherf, M. Kreyenschmidt, J. Fluor. Chem. **130**, 640 (2009)
45. C.-H. Lin, Y.-Y. Chang, J.-Y. Hung, C.-Y. Lin, Y. Chi, M.-W. Chung, C.-L. Lin, P.-T. Chou, G.-H. Lee, C.-H. Chang, W.-C. Lin, Angew. Chem. Int. Ed. **50**, 3182 (2011)
46. S. Lamansky, P. Djurovich, D. Murphy, F. Abdel-Razzaq, R. Kwong, I. Tsyba, M. Bortz, B. Mui, R. Bau, M.E. Thompson, Inorg. Chem. **40**, 1704 (2001)
47. S. Lamansky, P. Djurovich, D. Murphy, F. Abdel-Razzaq, H.-E. Lee, C. Adachi, P.E. Burrows, S.R. Forrest, M.E. Thompson, J. Am. Chem. Soc. **123**, 4304 (2001)
48. R. Ragni, E.A. Plummer, K. Brunner, J.W. Hofstraat, F. Babudri, G.M. Farinola, F. Naso, L. De Cola, J. Mater. Chem. **16**, 1161 (2006)
49. S. Takizawa, H. Echizen, J. Nishida, T. Tsuzuki, S. Tokito, Y. Yamashita, Chem. Lett. **35**, 748 (2006)
50. M.L. Xu, R. Zhou, G.Y. Wang, Q. Xiao, W.S. Dua, G.B. Che, Inorg. Chim. Acta **361**, 2407 (2008)
51. P. Coppo, E.A. Plummerab, L. De Cola, Chem. Commun. 1774 (2004)
52. C.-H. Yang, Y.-M. Cheng, Y. Chi, C.-J. Hsu, F.-C. Fang, K.-T. Wong, P.-T. Chou, C.-H. Chang, M.-H. Tsai, C.-C. Wu, Angew. Chem. Int. Ed. **46**, 2418 (2007)
53. Y.-C. Chiu, C.-H. Lin, J.-Y. Hung, Y. Chi, Y.-M. Cheng, K.-W. Wang, M.-W. Chung, G.-H. Lee, P.-T. Chou, Inorg. Chem. **48**, 8164 (2009)
54. J.-Y. Hung, Y. Chi, I-H. Pai, Y.-C. Yu, G.-H. Lee, P.-T. Chou, K.-T. Wong, C.-C. Chenc, C.-C. Wu, Dalton Trans., 6472 (2009)
55. R.J. Holmes, S.R. Forrest, T. Sajoto, A. Tamayo, P.I. Djurovich, M.E. Thompson, J. Brooks, Y.-J. Tung, B.W. D'Andrade, M.S. Weaver, R.C. Kwong, J.J. Brown, Appl. Phys. Lett. **87**, 243507 (2005)

56. S.-C. Lo, C.P. Shipley, R.N. Bera, R.E. Harding, A.R. Cowley, P.L. Burn, I.D.W. Samuel, Chem. Mater. **18**, 5119 (2006)
57. S.J. Lee, K.-M. Park, K. Yang, Y. Kang, Inorg. Chem. **48**, 1030 (2009)
58. H.J. Bolink, F. De Angelis, E. Baranoff, C. Klein, S. Fantacci, E. Coronado, M. Sessolo, K. Kalyanasundaram, M. Graetzel, Md. K. Nazeeruddin, Chem. Commun. 4672 (2009)
59. E. Baranoff, S. Fantacci, F. De Angelis, X. Zhang, R. Scopelliti, M. Graetzel, M.K. Nazeeruddin, Inorg. Chem. **50**, 451–462 (2011)
60. J. Jayabharathi, V. Thanikachalam, N. Srinivasan, M.V. Perumal, J. Fluoresc. **21**, 1585–1597 (2011)
61. T. Oshiyama, M. Nishizeki, N. Sekine, H. Kita, Japan Patent, P2006-60198A (2006)
62. J. Adler, A. Kanitz, G. Schmid, O. Freydenzon, A. Maltenberger, U.S. Patent, US2010/0213824 A1 (2010)
63. O. Molt, C. Lennartz, K. Dormann, E. Fuchs, T. Geβner, N. Langer, S. Watanabe, C. Schildknecht, G. Wagenblast, U.S. Patent, US2010/0253988 A1 (2010)
64. C.-H. Cheng, W.-H. Liao, H.-H. Shih, M.-J. Huang, T.-W. Tang, U.S. Patent, US 8,062,767 B2 (2011)
65. F. Okuda, T. Iwakuma, K. Yamamichi, C. Hosokawa, U.S. Patent, US 2011/0017984 A1 (2011)
66. S. Otsu, M. Nishizeki, E. Katoh, T. Oshiyama, U.S. Patent, US 2011/0073851 (2011)
67. T. Fei, X. Gu, M. Zhang, C. Wang, M. Hanif, H. Zhang, Y. Ma, Synth. Met. **159**, 113 (2009)
68. T. Hattori, Master Thesis, Chiba University, 2012
69. K. Tamao, M. Uchida, T. Izumizawa, K. Furukawa, S. Yamaguchi, J. Am. Chem. Soc. **118**, 11974 (1996)
70. X.W. Chen, J.L. Liao, S.A. Chen, J. Am. Chem. Soc. **125**, 636 (2002)
71. S. Tokito, M. Suzuki, F. Sato, Org. Electron. **4**, 105 (2003)
72. A.V. Dijken, J.J.A.M. Bastiaansen, K. Brunner, J. Am. Chem. Soc. **126**, 7718 (2004)
73. X.H. Yang, D.C. Muller, K. Meerholz, Adv. Mater. **18**, 948 (2006)
74. Y.C. Chen, G.S. Huang, S.A. Chen, J. Am. Chem. Soc. **128**, 8549 (2006)
75. Y.M. You, S.H. Kim, S.Y. Park, Macromolecules **39**, 349 (2006)
76. S.P. Huang, T.H. Jen, S.A. Chen, J. Am. Chem. Soc. **130**, 4699 (2008)
77. Z.L. Wu, Y. Xiong, Y. Cao, Adv. Mater. **20**, 2359 (2008)
78. W. Jiang, L. Duan, Y. Qiu, J. Mater. Chem. **20**, 6131 (2010)
79. K.S. Yook, S.E. Jang, J.Y. Lee, Adv. Mater. **22**, 4479 (2010)
80. J.H. Jou, W.B. Wang, J. Mater. Chem. **20**, 8411 (2010)
81. Z.Y. Ge, T. Hayakawa, M. Kakimoto, Chem. Mater. **20**, 2532 (2008)
82. Y.T. Tao, Q. Wang, D.G. Ma, J. Mater. Chem. **18**, 4091 (2008)
83. J.Q. Ding, B.H. Zhang, L.X. Wang, Adv. Mater. **21**, 4983 (2009)
84. J.J. Park, T.J. Park, J.H. Kwon, Org. Electron. **10**, 189 (2009)
85. W. Jiang, L. Duan, Y. Qiu, J. Mater. Chem. **21**, 4918 (2011)
86. D.K. Rayabarapu, J.P. Duan, C.H. Cheng, Adv. Mater. **17**, 349 (2005)
87. J.Y. Jeon, T.J. Park, J.Y. Lee, Chem. Lett. **36**, 1156 (2007)
88. R.J. Holmes, S.R. Forrest, M.E. Thompson, Appl. Phys. Lett. **82**, 2422 (2003)
89. J. He, H.M. Liu, D.G. Ma, J. Phys. Chem. C **113**, 6761 (2009)
90. S.C. Lo, G.J. Richards, P.L. Burn, Adv. Funct. Mater. **15**, 1451 (2005)
91. H. Sasabe, Y.J. Pu, J. Kido, Chem. Commun. **19**, 6655 (2009)
92. Q. Zhang, J.S. Chen, L.X. Wang, J. Mater. Chem. **14**, 895 (2004)
93. J.X. Jing, C.Y. Jiang, Y. Cao, Macromolecules **38**, 4072 (2005)
94. K. Brunner, A.V. Dijken, B.M.W. Langeveld, J. Am. Chem. Soc. **126**, 6035 (2004)

# Chapter 12
# DNA Electronics and Photonics

**Norihisa Kobayashi and Kazuki Nakamura**

## 12.1 Structure and Properties of DNA Functional Molecule Complexes

DNA is one of the oldest naturally occurring polymers and its history goes back to the origins of the most primitive life forms. The structure of the DNA molecule was famously described as a double helical structure by Watson, Crick, and Wilkinson in Nature in 1953 [1]. The 'monomer' unit of DNA consists of a base, which is covalently bonded to a sugar molecule, which is then covalently bonded to a phosphate group that makes up the backbone of the DNA 'polymer.' A schematic representation of this structure is shown in Fig. 12.1a. Four different base molecules make up the DNA: Adenine (A), Thymine (T), Guanine (G), and Cytosine (C). Each base has a conjugated ring structure, as shown in Fig. 12.1b.

The Watson–Crick model (Fig. 12.2a) was the first to describe double-stranded DNA, which consists of two single-stranded DNA molecules arranged in a double helix and connected by hydrogen bonding between bases, known as base pairs [1, 2]. Because of the complex geometry of double-stranded DNA molecules, bases can only hydrogen bond to form base pairs of Adenine-Thymine and Guanine-Cytosine. This can be seen in Fig. 12.1b, which shows that Adenine and Guanine are both double-ringed structures, and Thymine and Cytosine are single-ringed. Having two double-ringed bases would not allow the DNA to form the low energy configuration of a double helix. The hydrogen bonding is shown in Fig. 12.2b.

Recently, the application of DNA as a biomacromolecule has been attracting additional attention from researchers in diverse areas of science [3–7]. As a result, DNA is regarded as an ideal macromolecule for creating new functional materials, such as optical amplifiers [8], conductive and semiconductive nanowires [9],

N. Kobayashi (✉) • K. Nakamura
Graduate School of Advanced Integration Science, Chiba University,
Inage-ku, Chiba 263-8522, Japan
e-mail: koban@faculty.chiba-u.jp



H. Ishii et al. (eds.), *Electronic Processes in Organic Electronics*,
Springer Series in Materials Science 209, DOI 10.1007/978-4-431-55206-2_12

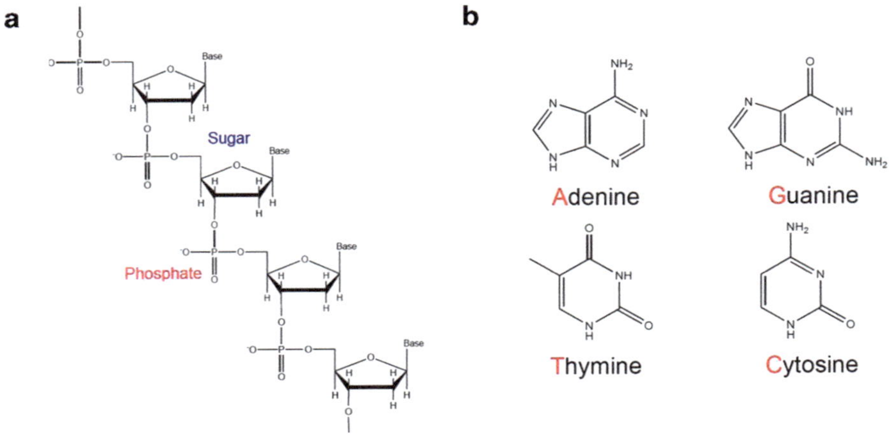


**Fig. 12.1** Chemical structure of (**a**) DNA backbone, and (**b**) DNA bases

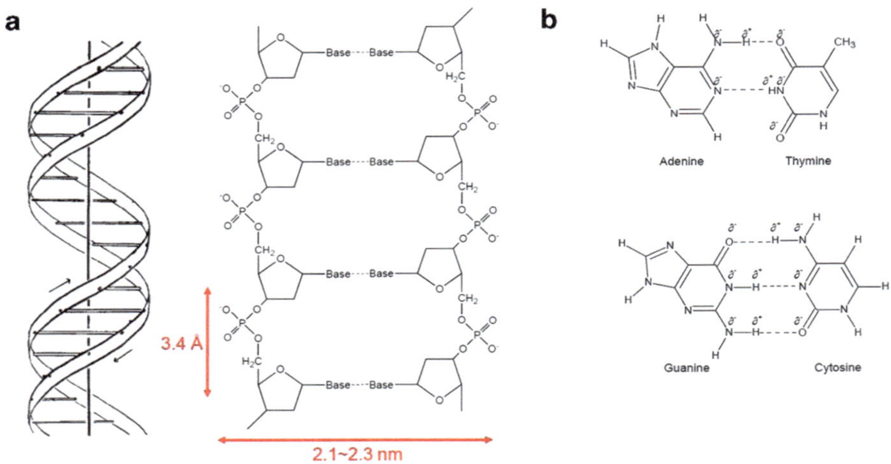


**Fig. 12.2** (**a**) Schematic illustrations of DNA double helix, and (**b**) base pair structures

transistors [10, 11], and biosensors [12]. Many experimental reports have been published in the last two decades, expanding the research field and making it attractive enough to obtain more attention from scientists. The DNA molecule possesses three unique features for creating functional complexes by interacting with other materials: being a polyelectrolyte for electrostatic connection, having selective affinity for small molecules by intercalation, and binding specific molecules into its grooves (Fig. 12.3).

The electrostatic property of DNA as a highly charged polyelectrolyte is important for functional complex creation, and it has been widely used for exploring various DNA-based applications. A DNA molecule consists of two

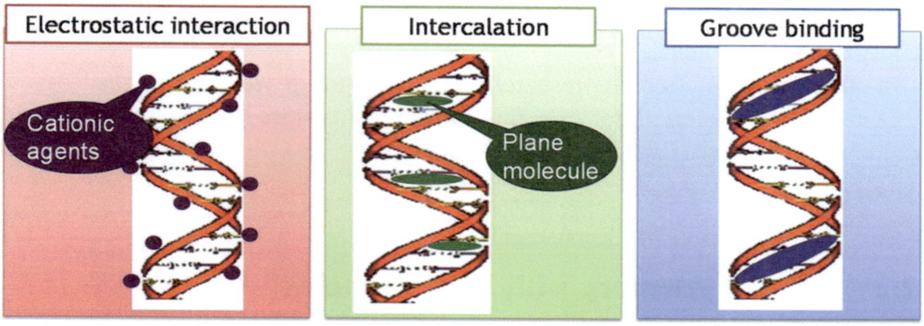


**Fig. 12.3** Schematic illustrations of the interaction between DNA and functional materials

polynucleotide strands coiled around each other in a helical structure, with a diameter of approximately 2 nm. Moreover, the double helix chains of DNA are negatively charged by the phosphate groups that are regularly arranged in the two backbones. Therefore, DNA can be an ideal template to fabricate highly ordered nanostructures by binding cationic agents such as metal ions, cationic surfactants, and polycationic agents through electrostatic interactions.

The second feature of DNA is its selective incorporation properties for small molecules. The most common DNA structure is the B-DNA type, in which the stacked bases are regularly spaced at 0.34-nm intervals. Some small planar molecules can intercalate into the spaces between the stacked bases, but the interaction patterns are highly selective regarding the structure of the small molecules.

The third feature of DNA is its groove-binding property. The helical structure of DNA possesses a wide, major groove and a narrow, minor groove of approximately the same depth, enabling binding of specific small or large molecules into the grooves between the two backbones. By these special affinities, DNA can be used as an environmental material for selective detection of certain molecules and removal of toxic pollutants, or as a framework to arrange functional molecules.

Currently, various designed DNA sequences can be synthesized by automated synthesis systems and amplified by polymerase chain reaction (PCR). However, synthetic DNA is still expensive and not easy to produce in large quantities. Therefore, laboratories usually use it at microgram or milligram levels. This makes it difficult to apply the synthetic DNA for uses requiring much larger quantities. On the other hand, another type of DNA that is extracted from natural products, referred to as natural DNA, has substantial potential for mass-production. A typical case is salmon milt in fisheries, which is often used for livestock feedstuff. The salmon milt contains over 10 % dry weight DNA, and it is very easy to manufacture. Because salmon fishing is an expanding industry with a worldwide supply exceeding 2.4 million tons per year, it is estimated that about 3,000 t of salmon DNA are available per year. Other resources include herring milt, scallop testis, and even artificial microorganisms produced by biological methods. Not only is this material bio-organic, but it is also biodegradable. Since it comes from waste material, it is inherently cost effective. Undoubtedly, such simple and renewable resources will effectively promote further DNA research by a variety of scientific fields.

In this chapter, recent research by our group regarding the use of DNA-functional materials for organic thin film transistors (BiOTFT) and organic light emitting diodes (BiOLED) are described.

## 12.2 BiOTFT Memory with DNA Complex as Gate Dielectric

### *12.2.1 OTFT Using Bio-Related Materials and DNA-Surfactant Complexes*

Printing processes are attractive for the fabrication of electronic devices such as radio-frequency identification (RFID) tags [13], sensors [14], and active matrix displays [15] on a plastic substrate. Since organic compounds can be dissolved in a solvent, electric materials are promising candidates for use as inks in the printing process [16–18]. Further, functional molecules can be oriented along the highly ordered structure of biopolymers, such as DNA, proteins, and polysaccharides. On this basis, biopolymers have also attracted attention for their possible use in the fabrication of electronic devices; such device fabrication using the ordered structure of DNA has been reported, as mentioned above.

Printable non-volatile memories also have high applicability to electronic devices, but there has been little research on promising materials using a low-temperature process. Since an organic thin film transistor (OTFT) was fabricated using ferroelectric material as a gate dielectric and exhibited storage properties, the development of new ferroelectric material has great significance for the potential fabrication of all printing devices.

Polyvinylidene fluoride and its derivatives are typical ferroelectric materials used as dielectric layers in OTFT memories because of their high ferroelectricity [19, 20]. It has also been reported that an OTFT memory was fabricated using biopolymers, and we reported the fabrication of a printable OTFT memory using a-helical poly(g-methyl-L-glutamate) (PMLG) as a ferroelectric layer in a previous study [21]. PMLG exhibited ferroelectric properties without any special treatment, such as poling, annealing, and stretching, possibly because of its rod-like, a-helical structure. DNA also has a rod-like secondary structure. Hence, OTFTs fabricated using DNA are also expected to exhibit novel storage properties. However, an OTFT memory typically has a low on/off ratio and short memory retention time.

We reported that the BiOTFT memory device prepared by natural DNA as the gate dielectric shows poor device performance, such as high OFF current because the presence of sodium ions probably contributed to ion conduction [11]. Okahata et al. reported preparation of the DNA-cetyltrimethylammonium (CTMA) complex, which is soluble in organic solvents and was effective for reducing the

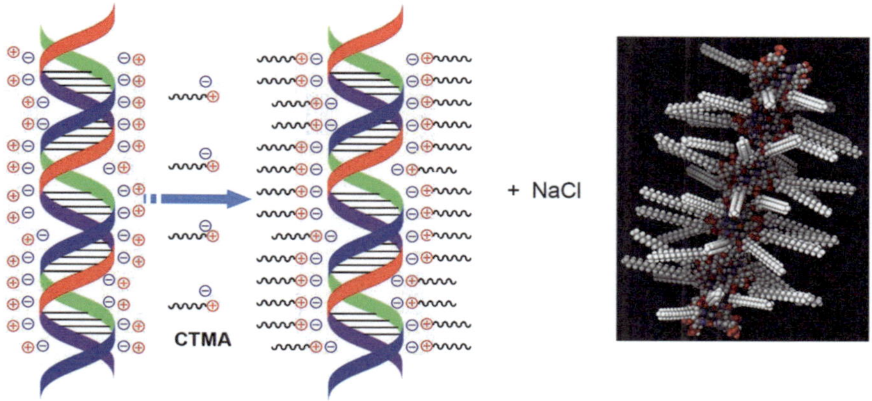


**Fig. 12.4** Schematic illustration of DNA-CTMA complex

mobility of counter ions [22]. CTMA is a quaternary ammonium compound that results in a positive charge on the nitrogen (cation) due to the delocalization of the lone pair of electrons, with chlorine as the counter anion. The negatively charged phosphate groups in the DNA backbone and corresponding sodium cations in the aqueous solution form electrostatic charge pairs. The positive charge on the nitrogen of the CTMA replaces the sodium cation and ionically binds to the DNA molecule. The displaced sodium cation then ionically binds with the chlorine anion of CTMA, forming NaCl as the byproduct, as schematically shown in Fig. 12.4.

The cationic surfactant complexation of DNA with CTMA in the laboratory setting is a simple processing step. Since both materials are water soluble, they are dissolved in deionized water, and the CTMA solution is titrated into the DNA solution. The DNA-CTMA material is precipitated out, while the NaCl byproduct remains in solution. The mixture is filtered to obtain the DNA-CTMA, which is thoroughly washed with water to rinse out any unreacted CTMA and remaining NaCl. DNA-CTMA is water-insoluble, but is soluble in organic solvents such as methanol, ethanol, isopropyl alcohol, and butanol. This lends the material to use in solvent fabrication techniques for thin films such as casting, dip coating, and spin coating. After this novel finding, in order to exclude the influence of DNA sodium ions and to prepare high-quality thin films, researchers utilized DNA-surfactants for application of electronic devices, including OTFT.

In this section, in order to investigate the influence of surfactant structure on BiOTFT device performance, the physicochemical properties of various DNA-surfactant complexes in solution and film state have been studied. Further, BiOTFT device performance, as inferred by aspects such as transfer properties, has also been examined, and the possible explanation of device performance is discussed.

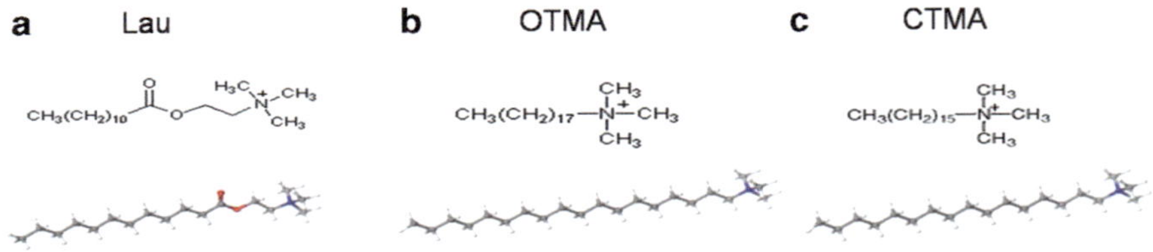


**Fig. 12.5** Chemical structures of (**a**) Lau, (**b**) OTMA, (**c**) CTMA

**Fig. 12.6** UV–vis spectra of DNA complex solution with high molecular weight

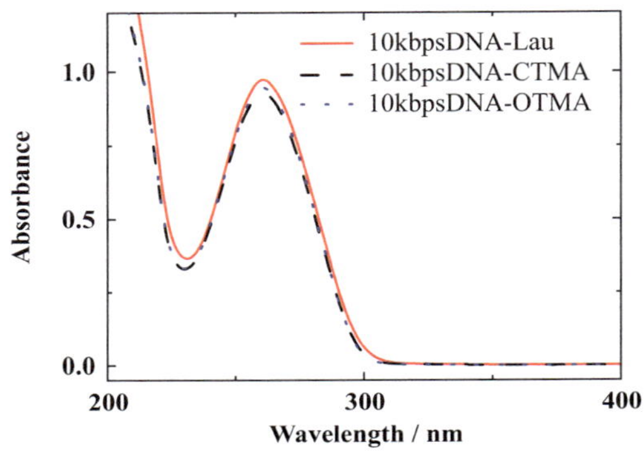


## 12.2.2 *Photo-Electrical Properties of DNA-Surfactant Complexes*

DNA-surfactant complexes were prepared by adding 10 mmol/L of DNA (the concentration of the phosphate group) aqueous solution into 10 mmol/L of the surfactant (CTMA, OTMA, Lau, Fig. 12.5) aqueous solution, and then the precipitate was filtered, washed with ultrapure water, and dried in vacuo. The DNA-surfactant complexes were obtained as a white powder, and the yield was over 95 %. Then, in order to obtain DNA-surfactant films, the 100 mmol/L DNA-surfactant solution was spin coated on ITO glass substrates. The thicknesses of the DNA and DNA-surfactant complex films were approximately 6 and 2 μm, respectively.

The photophysical property of UV absorption by DNA is an important property for determining whether π–π stacking of nucleobases occurs. The DNA double helix in aqueous solution had a specific absorption band from 220 to 300 nm, with a $\lambda_{max}$ at 260 nm [23]. In order to compare the effect of different DNA molecular weights with different cationic surfactants on structural regularity, the UV-visible spectra were measured. Figures 12.6 and 12.7 show the UV-visible spectra of DNA-Lau, DNA-OTMA, and DNA-CTMA in butanol solutions. It was clear that both high (10 kbps) and low (100 bps) molecular weights of different DNA complexes have the same peak absorption wavelength at 260 nm, which was assigned as the characteristic absorption of DNA nucleobases. The molecular

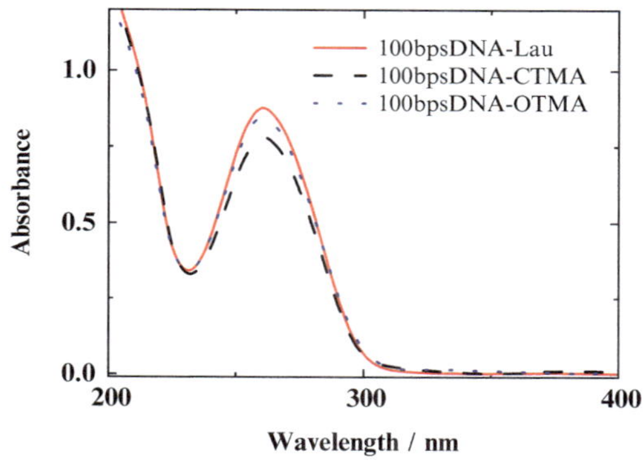


**Fig. 12.7** UV–vis spectra of DNA complex solution with low molecular weight

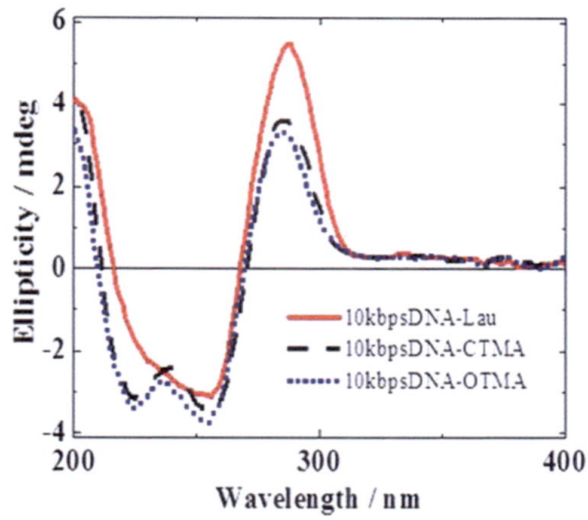


**Fig. 12.8** CD spectra of DNA complex solution with high molecular weight

weight had no relationship with the UV spectra. The π–π stacking of DNA nucleobases did not change when the molecular weight was altered, indicating that both the high and low DNA molecule weights retained the double helix structure after an ion exchange reaction in the butanol solution.

Circular dichroic (CD) analysis is one of the most useful techniques for probing the conformation of DNA complexes in many kinds of solutions, including gels, films, and fibers [24]. In order to compare different DNA molecular weights with different surfactants and to analyze its effect on the double helix structure, both the solution and film have been investigated. Figures 12.8, 12.9, and 12.10 show CD spectra of the solution and film. All of the complexes in either butanol solution or film state showed positive and negative CD signals. In the case of the solution, a positive Cotton effect at about 280 nm and a negative Cotton effect at about 225 and 245 nm have been observed, which is similar in shape to natural DNA. Meanwhile, the A form of the DNA complexes in the butanol solution appeared to transform into the B form in the film state. Additionally, with the alkyl chain increase, the CD signal decreased. This indicates the decrease of structure regularity, which would

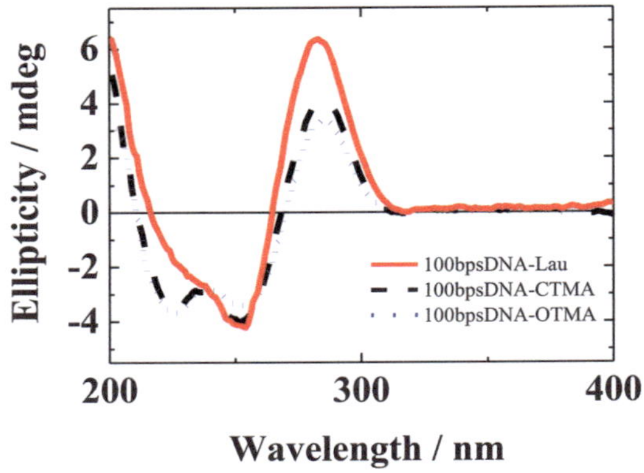


**Fig. 12.9** CD spectra of DNA complex solution with low molecular weight

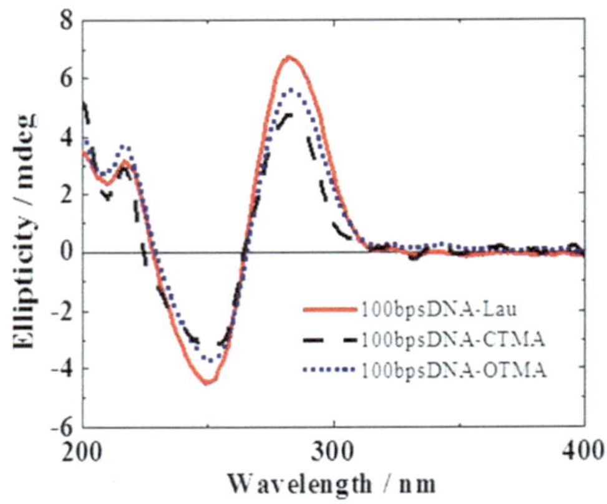


**Fig. 12.10** CD spectra of DNA complexes in film formation

be caused by the hydrophobic properties of the alkyl chain and steric hindrance between the long alkyl chains.

We also evaluated DNA complex resistivity because it plays an important role in the BiOTFT memory when used as the insulator layer. For comparison, the current (I)-voltage (V) characteristics of DNA alone with different chain lengths are also shown in Fig. 12.11. Figures 12.12 and 12.13 show that DNA complex resistivity increased significantly compared to DNA alone (Fig. 12.11) with either long or short chain lengths. This is caused by the DNA, which is one of the poly-anions that could interact with cationic surfactants through the ion-exchange reaction, leading to the decrease of movable ions, such as sodium ions, and improved resistivity. Additionally, Figures 12.12 and 12.13 also show that the resistivity increased as the alkyl chain length increased, because the longer alkyl chains provided a higher proportion of the insulative alkyl chains in the DNA complex film, which would improve the insulating properties.

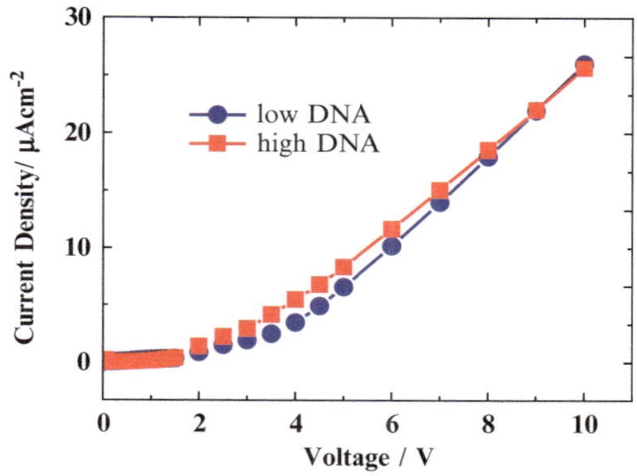


**Fig. 12.11** I-V curves of DNA films with different chain lengths

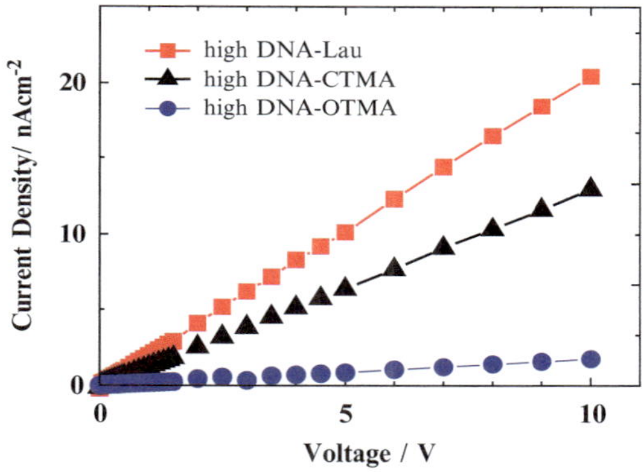


**Fig. 12.12** I-V characteristics of ITO/high DNA/Au cells

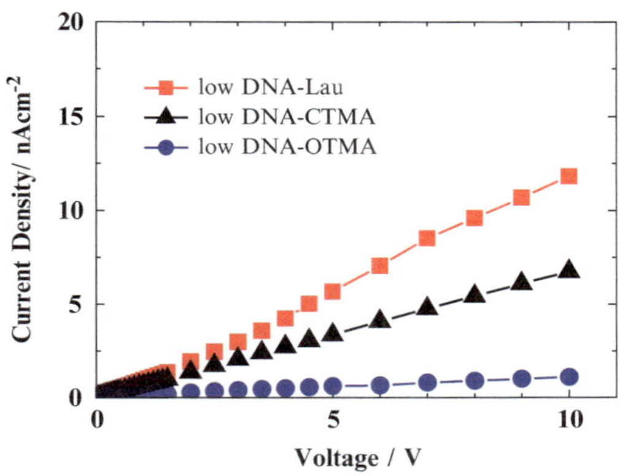


**Fig. 12.13** I-V characteristics of ITO/low DNA/Au cells

Furthermore, compared to the high-molecular weight DNA, low-molecular weight DNA showed better resistivity when the same surfactants were applied. This was due to the decrease of the carrier conductive pathway along the DNA chains when the DNA chain length was short. In conclusion, the kinds of surfactant

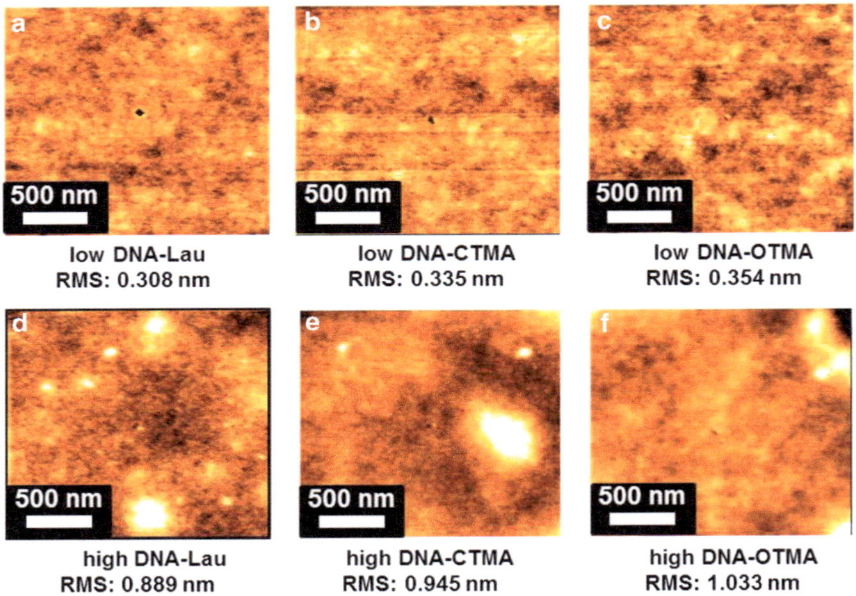


**Fig. 12.14** AFM topographical images of (**a**)(**d**)DNA-Lau, (**b**)(**e**)DNA-CTMA, and (**c**)(**f**)DNA-OTMA films

and DNA chain length have substantial effects on film resistivity. The resistivity of the low molecular weights of DNA-Lau, DNA-CTMA, and DNA-OTMA were $1.41 \times 10^{13}\ \Omega\cdot\text{cm}$, $8.2 \times 10^{13}\ \Omega\cdot\text{cm}$, and $1.51 \times 10^{14}\ \Omega\cdot\text{cm}$, respectively. Thus, it is reasonable to use them as dielectric material in the OTFT memory.

In general, thin film surface roughness and structure play crucial roles in the use of any insulating dielectric film for OTFT memory. This stems from the fact that charge transport takes place at the interface between dielectric and semiconductor films [25]. Figure 12.14 shows the $2 \times 2\ \mu\text{m}$ surface morphology of different DNA complexes with high and low molecular weights. The RMS of the high molecular weight DNA complex was estimated to be 0.8–1.0 nm. In contrast, the RMS of the low-molecular weight DNA complex was lower, at about 0.3 nm. Additionally, in comparison with the morphology of the low molecular weight DNA complex, some bump structures were found in the higher molecular weight DNA complex. It is possible that this occurred because the longer chain length would lead to entanglement of the DNA complex. Therefore, smoothness and film formability of the shorter DNA complex were substantially higher than that of the longer DNA complex. This indicates that low-molecular weight DNA is favorable for use in the dielectric layer of the OTFT memory.

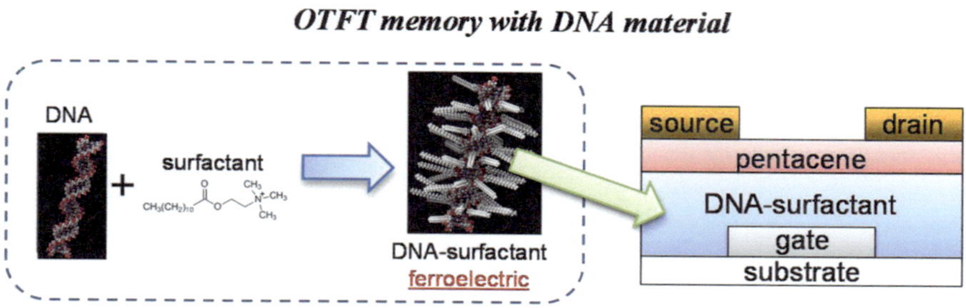


**Fig. 12.15** Schematic illustrations of the BiOTFT memory device configuration

**Fig. 12.16** Transfer characteristics of BiOTFTs with DNA complexes as gate dielectrics

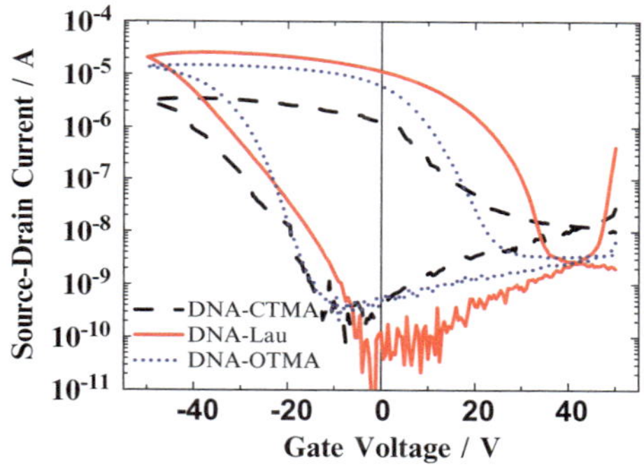


## 12.2.3 *Fabrication of BiOTFT Devices Using DNA-Surfactant Complexes*

Next, we fabricated the OTFT memories using the following method. On the prepared ITO/DNA or ITO/DNA complex films, a pentacene layer (film thickness = 50 nm) was an active layer at a pressure of $2 \times 10^{-3}$ Pa and an evaporation rate of 0.2–0.4 $Ås^{-1}$. The Au layer was the source and drain electrodes (W/L = 5 mm/20 μm) were deposited by vacuum evaporation on the pentacene film. The BiOTFT structure using a top contact and gate bottom geometry is schematically depicted in Fig. 12.15.

Figure 12.16 shows the transfer characteristics, which represent the transistor current $I_{ds}$, plotted as a function of the gate voltage at constant $V_d$ of the BiOTFTs fabricated by using DNA-OTMA, DNA-CTMA, and DNA–Lau films as a gate dielectric layer. Figure 12.17 also shows the transfer characteristics when the DNA film was used as the dielectric layer.

In all of these BiOTFTs, hysteresis behaviors were observed, indicating that both the DNA and DNA complex films work as memory layers. Because of the poor resistivity of the DNA film, the OFF current, which equals $3.8 \times 10^{-7}$ A, was too high to use as the insulator in the OTFT memory. In contrast, when the DNA complex was applied as the dielectric layer, the OFF current decreased remarkably

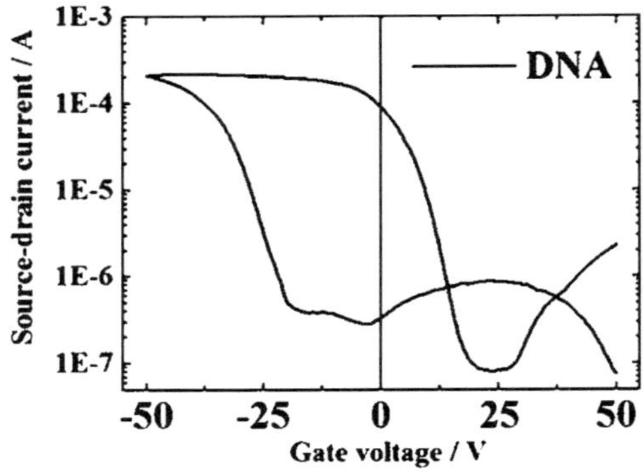


**Fig. 12.17** Transfer characteristics of BiOTFTs with DNA film as the gate dielectric

to about $10^{-9}$ A because of the resistivity improvement. At the same time, hysteresis of the BiOTFT memory with the DNA complex significantly increased, and the ON/OFF current ratio at $V_G = 0$ V was as high as $10^4$.

Moreover, compared to DNA-OTMA and DNA-CTMA, when the DNA-Lau was used as the dielectric layer, the higher ON current and restrain of the OFF current were observed. This improvement of device performance might be effected by varying the composition and structure of the DNA complex, because the DNA-Lau film roughness was the lowest and had the best structure regularity when compared to the DNA-CTMA and DNA-OTMA complex films.

The appearance of hysteresis in the DNA-based BiOTFT was thought to be due to the accumulation of mobile ions at the interface. However, more hysteresis was observed in the DNA complex, indicating that the accumulation of mobile ions is not the dominant factor, but instead, ferroelectricity of DNA complex and/or the interface are thought to be possible explanations. Research is now underway to provide a better understanding of this memory performance.

### *12.2.4 Summary*

To summarize this section, DNA and various DNA complex films were used as gate dielectric materials of the BiOTFT memories. In order to exclude the influence of sodium ions and other impurities, it is necessary to make the DNA complex with cationic surfactant through the ion-exchange reaction. At the same time, the molecular weight of DNA also plays an important role in the DNA complex film resistivity and morphology. In this study, the most favorable surfactant for preparing the BiOTFT is DNA-Lau, as compared to DNA-OTMA and DNA-CTMA. The ON current and restrain of the OFF current have been improved extensively by using DNA-Lau as the gate dielectric.

## 12.3 BiOLED with DNA Complexes

### *12.3.1 Application of DNA Complexes as Charge Conductive Material*

In the previous section, we described the application of DNA-functional molecule complexes to organic electronics, such as BiOTFT memory. The DNA-surfactant complexes acted as ferroelectric materials in the OTFT devices. On the other hand, DNA has sparked interest as a functional material since the stacked aromatic bases of DNA may act as a "π-way" for efficient charge transfer [26]. This is possibly because of application to novel electronic devices, such as highly integrated circuits with nano-structures.

The electron transfer and conductivity of DNA have been extensively debated. While the actual magnitude of these conductive properties is unclear because of different measurement conditions and apparati [27–30], attempts to improve DNA conductivity have been carried out by modifying DNA with other functional materials [31–33]. In this section, we demonstrate opt-electronic application of DNA complexes, which contain conductive polymers and luminescent metal complexes in the structure.

Conducting polymers such as poly(acetylene) have been gaining interest regarding applicability for electronic devices. These polymers are generally prepared by oxidative polymerization. If conducting polymers are prepared by the photoinduced electron-transfer process, which enables vectorial electron transfer (redox reaction) between photocatalysts and other molecules by illumination, patterning with conducting polymers is possible at any place and on any geometry. This process would allow the possibility of fabricating molecular electronic and/or optical devices, and for microprocessing. On the basis of this information, photopolymerization of pyrrole using tris(2,2-bipyridyl)ruthenium ($Ru(bpy)_3{}^{2+}$) as a photocatalyst has already been performed, resulting in conductive polypyrrole.

We have already reported photopolymerization of dimeric aniline by the photoinduced electron transfer between $Ru(bpy)_3{}^{2+}$ and methylviologen ($MV^{2+}$) [34–36], because polyaniline (PAn) is one of the most promising conductive polymers due to its high environmental stability in air. We also demonstrated its application in imaging and micropatterning [37, 38]. However, for the development of electronic devices, the obtained PAn does not seem to be sufficient in its conjugation length, because the PAn photopolymerized in homogeneous systems involves a branched and/or "compact coil" structure. The use of a polyelectrolyte as a polymerization template has been reported for the polymerization of aniline by an enzyme in the presence of hydrogen peroxide in order to minimize branching [39]. We also carried out polymerization in the presence of clay minerals [40] and micelles [41], in order to utilize its specific structure as a template and to improve physicochemical properties of the photopolymerized PAn. Since template polymerization is expected to result in a characteristic structure reflecting template structure, template variations are advantageous for different applications.

In this section, structurally ordered DNA was employed as a template to prepare the DNA/PAn complex. The relevance of polymerization and its structure for electronic material have been discussed because DNA works as a rigid, straightforward template, due to its rod-like and double helical structure. Further, the DNA/PAn complex should include $Ru(bpy)_3^{2+}$ in its matrix even after purification, because $Ru(bpy)_3^{2+}$ could be electrostatically bound on the free anionic sites of the DNA and remained after complexation with PAn.

$Ru(bpy)_3^{2+}$ works as a light-emitting molecule as well as photosensitizer. In the complex, PAn should be p-type conductor. If materials such as $Ru(bpy)_3^{2+}$ showed electronic conduction, the DNA/PAn complex containing $Ru(bpy)_3^{2+}$ was expected to be a photovoltaic and light-emitting molecular system, where electrons and holes were injected from electrodes through DNA and PAn, respectively, and combined at $Ru(bpy)_3^{2+}$ to emit red-light. We therefore analyzed the light-emitting properties of the device, which was fabricated from a novel, processable, and water-soluble DNA/photopolymerized PAn complex containing $Ru(bpy)_3^{2+}$. Further, a green-light-emitting $Alq_3$ complex layer was deposited on the DNA/PAn complex. The device showed voltage-controlled emission color tenability. The mechanism of multi-emission was also investigated.

## *12.3.2 Structure of Template Photopolymerized PAn/DNA Complex*

We first analyzed the structure of DNA associated with phenylenediamine (PPD, dimeric aniline) and $Ru(bpy)_3^{2+}$ (photocatalyst) in order to obtain a highly ordered complex. The kinds of structures that can be obtained by the association of DNA with these chemicals are important for template photopolymerization. The DNA-based functional complex binding the PAn chain in its groove, DNA/PAn, was prepared as follows. The aerated hydrochloric acid (HCl) aqueous solution (pH 3.0–6.5) containing $6 \times 10^{-5}$ M $Ru(bpy)_3^{2+}$, 1.0 mM PPD, and a given concentration of DNA (the concentration of phosphate group) was illuminated with a 500 W xenon lamp through a 420–600 nm filter. The light intensity was adjusted to 15 mW $cm^{-2}$ at 450 nm.

Then, we investigated the interaction between the $Ru(bpy)_3^{2+}$ photocatalyst and DNA in aqueous solution by analyzing the emission behavior of $Ru(bpy)_3^{2+}$. Since the polymerization of aniline derivatives is commonly carried out in acidic aqueous solutions, we used aqueous solutions with pH 3.1. The emission behavior of Ru $(bpy)_3^{2+}$ is different in a polyanionic environment compared to solutions without polyanions.

The emission intensity at about 600 nm from excited $Ru(bpy)_3^{2+}$ ($6.0 \times 10^{-5}$ M) increased with DNA concentration, and then reached constant emission intensity at a DNA concentration of $1 \times 10^{-3}$ M. Further, the increase in DNA melting temperature was also observed in the solution at pH 3.1 in the presence of $Ru(bpy)_3^{2+}$.

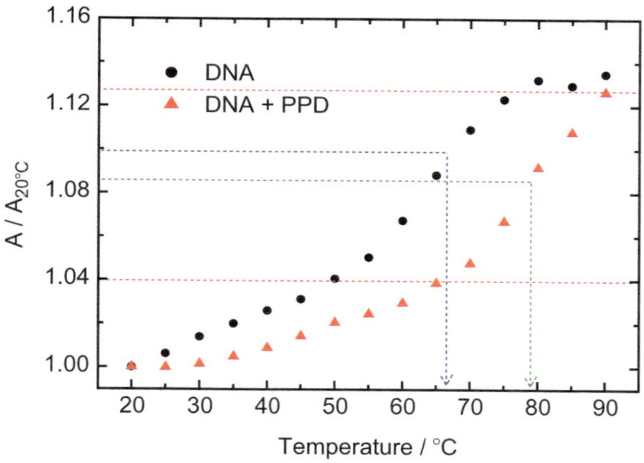


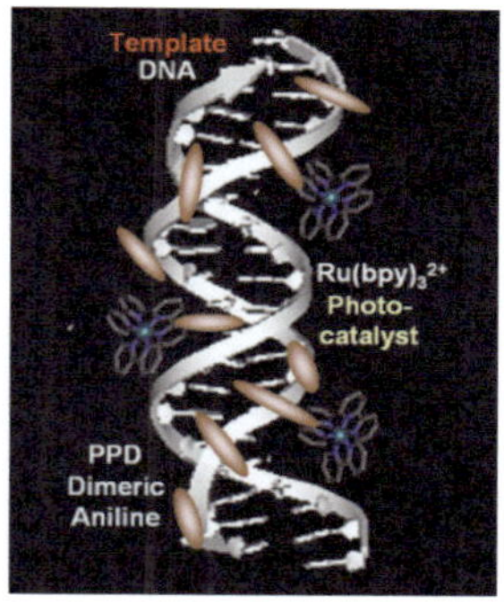


**Fig. 12.18** (**a**) The helix melting curves of $1.0 \times 10^{-5}$ M DNA in the presence and absence of $1.0 \times 10^{-5}$ M PPD (pH 3.1). (**b**) Schematic representation of the structure of DNA associated with PPD and $Ru(bpy)_3^{2+}$

pH-titration measurements indicated that the pKa of the DNA was ca. 4–4.5. We suggest that electrostatic interaction between $Ru(bpy)_3^{2+}$ and DNA minimally takes place at pH 3.1. Rather, $Ru(bpy)_3^{2+}$ is associated with duplex DNA by specific interaction, such as by intercalation at pH 3.1.

The absorption of PPD ca. 280 nm in HCl aqueous solution (pH 3.1) containing DNA underwent a slight red shift and exhibited 25 % hypochromicity by increasing the concentration of DNA from $7.5 \times 10^{-5}$ M to $1.25 \times 10^{-4}$ M. Further, increase in the melting temperature of DNA was also observed in the solution at pH 3.1 in the presence of PPD, as shown in Fig. 12.18a. However, the absorption of PPD at pH 6.5 did not show hypochromicity. These results suggest that PPD, protonated by the primary amino group at pH 3.0 ($pK_{a1} = -0.1$, $pK_{a2} = 5.72$), is associated with duplex DNA by intercalation and/or electrostatic interaction. Considering the above-mentioned results, DNA is associated with PPD and the $Ru(bpy)_3^{2+}$ photocatalyst, with a structure as schematically represented in Fig. 12.18b. This kind of architecture, a so-called supramolecular polymer, can be easily obtained by dissolving only DNA, PPD, and $Ru(bpy)_3^{2+}$ in the aqueous solution.

When an aqueous solution (pH 6.5) containing $6 \times 10^{-5}$ M $Ru(bpy)_3^{2+}$, $1.0 \times 10^{-3}$ M PPD, and $1.0 \times 10^{-3}$ M DNA was illuminated with a xenon lamp through a 420–600 nm filter, absorption peaks ca. 400 and 800 nm emerged, which are assignable to the polaron band of PAn [42]. The mechanism of photocatalytic polymerization has already been discussed [35]: protonated PPD is oxidized by excited $Ru(bpy)_3^{2+}$, and the formed $Ru(bpy)_3^{+}$ is oxidized back to $Ru(bpy)_3^{2+}$ by an acceptor (e.g., oxygen in an aerated solution). Polymerization proceeds with the reaction of the protonated form of doubly-oxidized PPD with unoxidized PPD, followed by the successive reaction at the end of the chain via photoinduced electron transfer. When polymerization was carried out at pH 6.5 in the absence of DNA, the absorption peaks were observed ca. 300 and 600 nm, which are

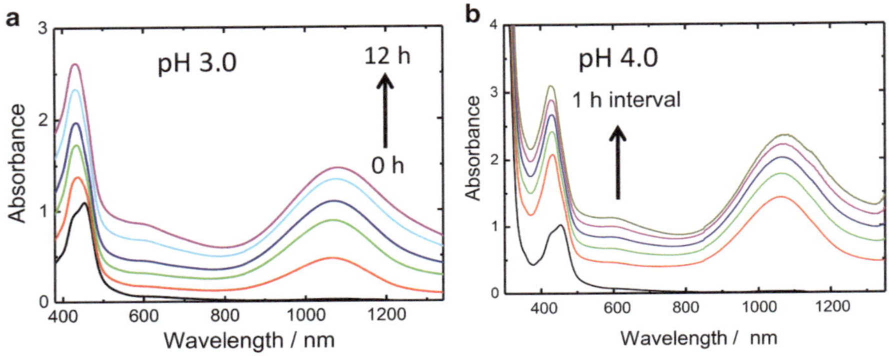


**Fig. 12.19** Change in absorption spectra of the aqueous solution containing 1 mM PPD, $6.0 \times 10^{-5}$ M $Ru(bpy)_3^{2+}$, and 1 mM DNA at (**a**) pH 3.0 and at (**b**) pH 4.0 upon visible light illumination. Figure 12.16 Transfer characteristics of BiOTFTs with DNA complexes

assignable to π–π* transition of the benzenoid ring and exciton absorption of the PAn quinoid ring, respectively [13, 14]. These results strongly suggest that the photopolymerized PAn is acid-doped (i.e., protonated) by DNA, and that DNA phosphate groups provide a local, lower pH environment.

Figure 12.19 shows change in the absorption spectra of the HCl aqueous solution containing $6 \times 10^{-5}$ M $Ru(bpy)_3^{2+}$, $1.0 \times 10^{-3}$ M PPD, and $1.0 \times 10^{-3}$ M DNA at (a) pH 3.0 and (b) pH 4.0 upon visible light illumination. The polaron absorption also emerged at 600–800 nm at pH 3.0 in the absence of DNA. Its absorption maxima ($\lambda_{max}$), however, significantly shifted to the longer wavelength side (i.e., ca. 200 nm), in the presence of DNA. The longer-wavelength shift indicates the formation of PAn with a longer π-electron conjugation. MacDiarmid et al. reported that the shift is attributable to the delocalization of electrons in the polaron band of PAn due to conformational change of PAn from "compact coil" to "expanded coil." The shift, therefore, is also explained by the "expanded coil" formation of PAn photopolymerized at pH 3.0 in the presence of DNA.

When polymerization was carried out at pH 3.0 in the presence of DNA, a larger amount of PPD electrostatically interacted with DNA than at pH 6.5 because of PPD protonation. In other words, while DNA at pH 6.5 provides the local, lower-pH environment required for photopolymerization, the template effect of DNA on polymerization was insufficient. Differential CD spectra of the solution containing DNA and PPD and that containing only DNA at pH 3.0 showed a negative CD band at 280 nm assignable to π–π* PPD transition. However, the negative CD band was not found at pH 6.5.

It is well known that the helical structure of DNA is deeply affected by the pH of aqueous solutions, and these results indicate that PPD in the DNA solution at pH 3.0 is aligned along the main chain of DNA through intercalation and/or electrostatic interaction, leading to the formation of an "expanded-coil" PAn structure in the PAn/DNA complex. DNA studied in these experiments was about 60 % unwounded at pH 3.0. We carried out the same experiments at pH 4.0, where more than 90 % of

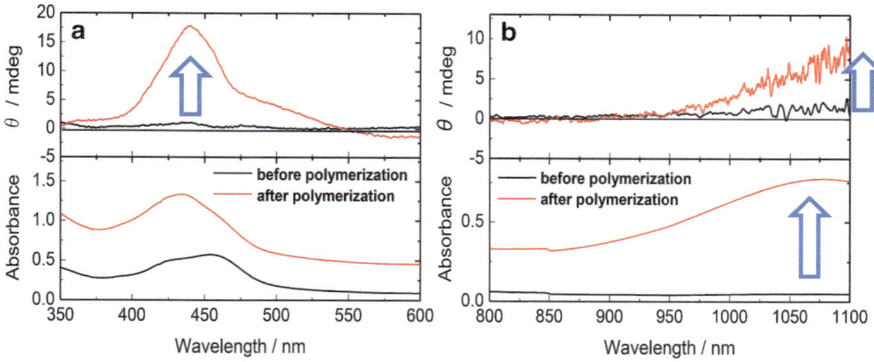


**Fig. 12.20** UV–vis and CD spectra of the aqueous solution (pH 4.0) containing DNA and PPD and $Ru(bpy)_3^{2+}$ before and after photopolymerization

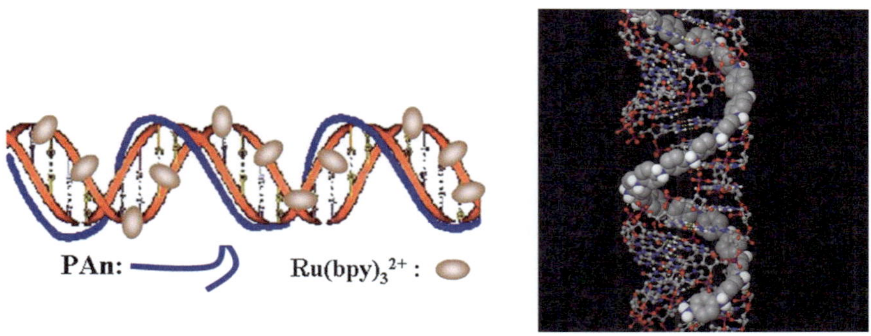


**Fig. 12.21** Schematic representation (*left*) and calculated structure using the molecular dynamics method (*right*) of the PAn/DNA complex

the helical structure remained. As can be seen in Fig. 12.19b, a similar absorption spectral change was obtained at pH 4.0. This indicates that DNA works well at pH 4.0 as a template for the PAn photopolymerization.

Figure 12.20 shows CD spectra of the PAn/DNA complex in an aqueous solution at pH 4.0. CD signal emerged after photopolymerization at around 440 and 1,100 nm, which were assignable to the absorption bands of PAn in the PAn/DNA complex. However, PAn in the absence of DNA did not show a CD signal between 350 and 1,100 nm. Similar spectra were obtained in the solution at pH 3.0, but the intensity of the CD signal at pH 4.0 was larger than that at pH 3.0 ($\theta$ = ca. 3 mdeg at 440 nm and ca. 2 mdeg at 1,100 nm). This is due to a higher double helical structure content of DNA at pH 4.0. This clearly indicates that PAn was associated with DNA, and PAn in the complex had a helical structure along the DNA main chain [6]. Taking the above results into account, schematic representation of the structure of the PAn/DNA complex is depicted in Fig. 12.21. PAn was associated with DNA and wound around the duplex DNA, just like a triple helix. The complex contains photocatalysts even after purification. Therefore, the resulting so-called triple complex is a highly ordered structure composed of DNA, PAn, and the photocatalyst, $Ru(bpy)_3^{2+}$. DNA as a template plays an important role to form the PAn/DNA complex.

### 12.3.3 EL Properties of $Ru(bpy)_3^{2+}$-Based DNA BiOLED

As mentioned in the Introduction, the DNA/PAn complex included $Ru(bpy)_3^{2+}$ in its matrix even after purification, because $Ru(bpy)_3^{2+}$ could be electrostatically bound on the free anionic sites of DNA, remaining after complexation with PAn. $Ru(bpy)_3^{2+}$ works as a light-emitting molecule as well as photocatalyst (photosensitizer). In this complex, PAn should be a p-type conductor. If materials such as $Ru(bpy)_3^{2+}$ regularly binding to the PAn/DNA complex enabled electronic conduction, the PAn/DNA complex containing $Ru(bpy)_3^{2+}$ would be expected to be a photovoltaic and light-emitting molecular system, where electrons and holes were injected from electrodes through DNA and PAn, respectively, and combined at $Ru(bpy)_3^{2+}$ to emit red-light, as schematically shown in Fig. 12.22.

The DNA/PAn complex was prepared by template photopolymerization of dimeric aniline, as mentioned above, and was purified by reprecipitation with acetone. Since $Ru(bpy)_3^{2+}$ should work as an emitter and the concentration of Ru $(bpy)_3^{2+}$ on the DNA/PAn complex was not enough to fabricate an organic light emitting diode (OLED), a solution containing $4.2 \times 10^{-3}$ g/mL of DNA/PAn complex and $3.2 \times 10^{-4}$ g/mL of $Ru(bpy)_3^{2+}$ was prepared by adding an appropriate amount of $Ru(bpy)_3^{2+}$ to the solution. An indium-tin oxide (ITO)-coated glass electrode was dipped in the solution to prepare the DNA/PAn complex containing $Ru(bpy)_3^{2+}$ on the electrode. The film thickness was estimated to be 30–50 nm.

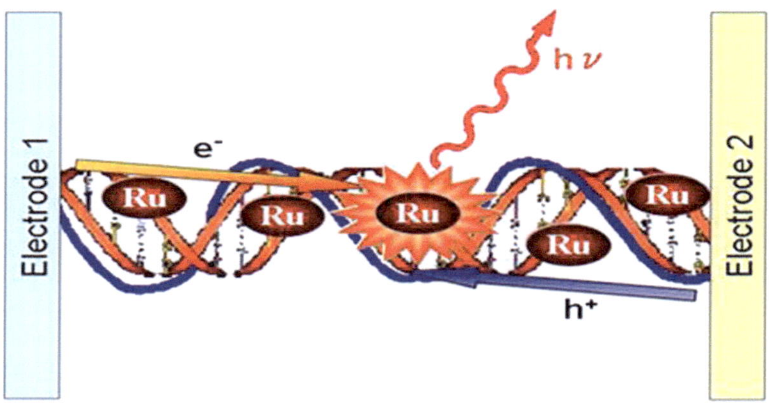


**Fig. 12.22** Schematic representation of the PAn/DNA light-emitting molecular system structure

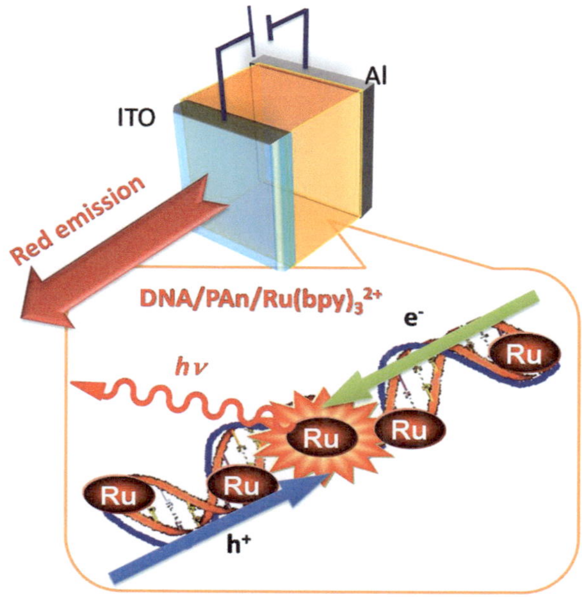


**Fig. 12.23** Schematic representation of the cell structure of PAn/DNA-based OLED

The precipitate and crystal corresponding to $Ru(bpy)_3{}^{2+}$ itself were not found in the film micrograph, indicating that the incorporated $Ru(bpy)_3{}^{2+}$ was homogeneously dispersed in the film, possibly due to electrostatic interaction between the DNA phosphoric group and $Ru(bpy)_3{}^{2+}$. The Al top electrode was finally deposited at $4.0 \times 10^{-6}$ Torr with a thickness of 100 nm. The emitting area was $0.2 \times 0.2$ cm$^2$. We then analyzed the light-emitting properties of the cell, which was fabricated from a novel, processable, and water-soluble DNA/photopolymerized PAn complex containing $Ru(bpy)_3{}^{2+}$. The Al electrode DNA/PAn complex/ITO cell with a DNA/PAn thickness of 30–50 nm was fabricated and the voltage was applied as schematically shown in Fig. 12.23 [43].

The cell I-V curve is shown in Fig. 12.24a. The current abruptly increased above a bias voltage of 7 V, and a large current of 1,600 mA/cm was obtained at 11 V. The current over 7 V seems to obey the space-charge-limited current (SCLC) mechanism. The current at 11 V is relatively higher than that of devices commonly fabricated by vacuum evaporation of low-molar mass organic molecules. This is due to the employment of conductive PAn in the complex. With regard to the I-V characteristics, light was emitted from the cell above the bias voltage of 7 V, and the luminance of the device increased with bias voltage as shown in Fig. 12.24b.

A maximum luminance of 1,500 cd/m$^2$ was observed at 11 V. As can be seen by comparison between Fig. 12.24a, b, the luminance seems to increase linearly with injected charges. Turn-on response of the cell luminance was very fast for our device. Although we could not measure its quantitative response time, we do not doubt that the turn-on response was much faster than 1 s. We confirmed flicker due to emission-on and -off from the cell when an AC bias voltage of 9 V was applied to

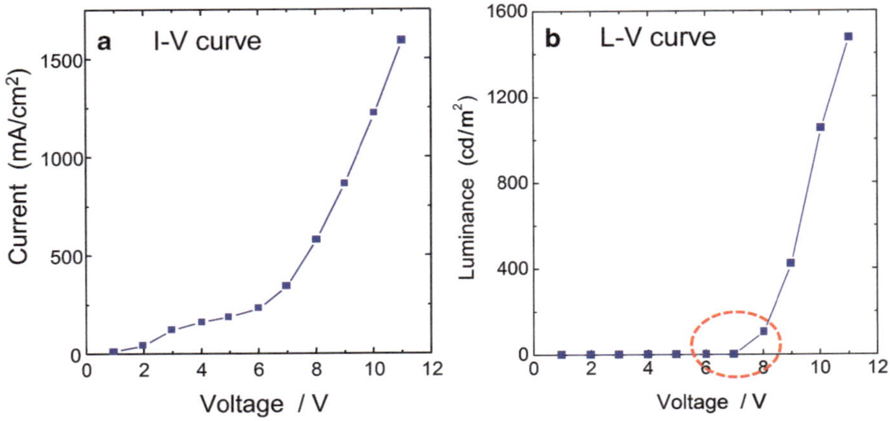


**Fig. 12.24** (**a**) Current–voltage characteristics and (**b**) luminance-voltage characteristics of a device fabricated with DNA/PAn complex containing $Ru(bpy)_3^{2+}$. Those figures were reused from Kobayashi et al. [43]. Copyright 2001, The Royal Society of Chemistry 2001

the cell at a frequency of 30 Hz. This indicates that the turn-on and -off response should be faster than 30 ms. We fabricated a cell composed of DNA/$Ru(bpy)_3^{2+}$ without PAn, and the emission-time response of the cell was measured with a photomultimeter by applying 0–10 V of rectangular wave to the cell at 10 Hz [44]. The luminance increased rapidly just after turning on the bias and reached a constant value within ca. 70 mse. This behavior seems to be different from that of previously reported Ru complex-based devices. This indicates that the emitting mechanism is not electrochemical, but is commonly observed in an electroluminescence (EL) device fabricated by vacuum-evaporation of organic molecules.

The EL spectrum of the cell and photoluminescence (PL) spectrum of $Ru(bpy)_3^{2+}$ are shown in Fig. 12.25 (left), and the photo images of the emission from the cell are shown in Fig. 12.25 (right). The PL spectrum was measured for $Ru(bpy)_3^{2+}$ in an aqueous solution. The EL spectrum was found to be almost identical to PL spectrum, indicating that the red-light was emitted from $Ru(bpy)_3^{2+}$ in the DNA/PAn matrix. However, the emission peaks were found to red-shift from 610 nm for PL in the solution to 620 nm for the OLED cell, and the EL spectrum was slightly broader than the PL spectrum. This is due to the electronic interaction between $Ru(bpy)_3^{2+}$ molecules, which increases as the separation between $Ru(bpy)_3^{2+}$ molecules decreases.

As described above, the role of DNA in this cell is interesting to examine. Whether or not DNA is an electronic conductor is unclear, because we do not have any quantitative evidence that electrons or holes pass through DNA chains. Taking into account the presence of electroactive $Ru(bpy)_3^{2+}$ in the complex and the very low conductivity ($<10^{-10}$ S/cm) of DNA film, electron transfer through Ru $(bpy)_3^{2+}$ from the cathode is plausible in the EL cell.

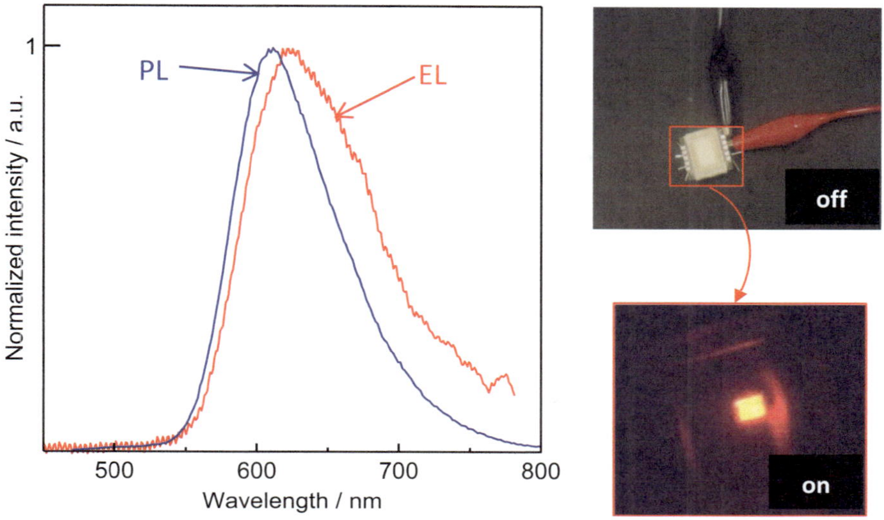


**Fig. 12.25** PL spectrum of $Ru(bpy)_3^{2+}$ in an aqueous solution and EL spectrum of the cell (*left*), and digital camera images of cell emission (*right*). This figure was reused from Kobayashi et al. [43]. Copyright 2001, The Royal Society of Chemistry 2001

When poly(aniline sulfonic acid) (SPAn) was employed instead of the DNA/PAn complex to fabricate the EL device, no emission was observed from the device, even at the bias voltage of 11 V. This is due to the increase of ineffective current for emission, resulting in the decrease of recombination probability between holes and electrons at $Ru(bpy)_3^{2+}$ in the complex by employing more conductive SPAn as a matrix.

When poly(vinyl sulfate) (PVS) was employed instead of DNA, the PVS/PAn complex was also obtained by the template photopolymerization of PPD. A similar solution to the PAn/DNA complex containing $Ru(bpy)_3^{2+}$ was prepared for the PVS/PAn complex for EL device fabrication. The ITO glass electrode, however, shed the solution differently from the DNA/PAn complex solution, and it was difficult for us to prepare a thin, smooth, and homogeneous complex film on the electrode. The film surface was qualitatively confirmed by the AFM technique, indicating that the DNA/PAn complex film on the ITO glass electrode was smooth and homogeneous.

Further, when DNA/$Ru(bpy)_3^{2+}$ without PAn was employed to fabricate the Al/DNA/$Ru(bpy)_3^{2+}$/ITO-based cell, a maximum luminance of 50 cd/$m^2$ was observed at the bias voltage of 15 V[20]. The cell also showed fast turn-on response, as mentioned above. In this experiment, a brand new cell was prepared in each measurement to avoid potential formation from previous electrochemical reactions. These results clearly suggest that DNA plays an important role for at least OLED cell fabrication.

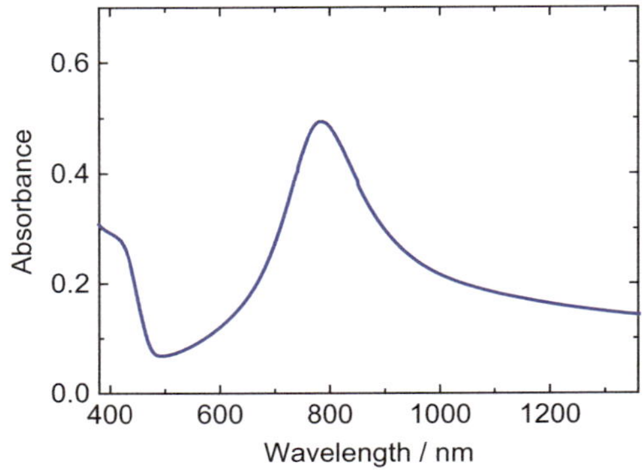


**Fig. 12.26** Vis-NIR absorption spectrum of the DNA/PAn complex in an aqueous solution

### 12.3.4 Color Tunable OLED Based on the DNA/Pan/Ru $(bpy)_3{}^{2+}$ Complex

OLEDs have a wide spectral range, require low driving voltage, and are lightweight. Therefore, they are suitable for use in portable display devices such as full-color flat panel displays, as described above. Color tunable OLEDs are of particular interest because they would enable control of picture elements in large screen displays and improve resolution and color quality. These devices can generate two or more colors depending on device parameters such as driving voltage, current, and local temperature [45, 46]. Voltage-controlled multicolor OLEDs are widely used in the fabrication of color-variable devices. These OLEDs are classified into several types based on their components: a dye-dispersed polymer emissive layer, a single quantum-well layer inserted in the emissive layer, two emissive layers separated by a carrier blocking layer, or two emissive parts stacked on a transparent electrode.

In this section, the color tunability of a DNA-based OLED was successfully demonstrated for the first time. The OLED consists of a DNA/PAn/Ru$(bpy)_3{}^{2+}$ complex layer (hole transport layer, red-light-emitting material) and a Tris (8-hydroxyquinolinato) aluminum layer ($Alq_3$, electron transport layer, green-light-emitting material). Owing to this structure, the OLED exhibited multicolor emission—ranging from green to yellow to red—upon application of different voltages.

We first prepared the DNA/PAn complex with a simple complexation procedure as follows. Five mmol/L of the DNA/PAn complex solution ([DNA]/[PAn] = 10/1) was prepared by mixing suitable amounts of a DNA aqueous solution, PAn/NMP solution, and dilute HCl (pH 3.8). This solution was purified by reprecipitation with acetone to remove NMP, and the precipitated DNA/PAn complex was obtained by filtering the solution. The vis-NIR absorption spectra of the DNA/PAn complex measured in aqueous solution are shown in Fig. 12.26.

A large absorption band was observed around 750 nm, which was assignable to the localized polaron structure of PAn [47]. While PAn cannot dissolve in water by itself, the DNA/PAn complex can. This result indicates that the doped PAn

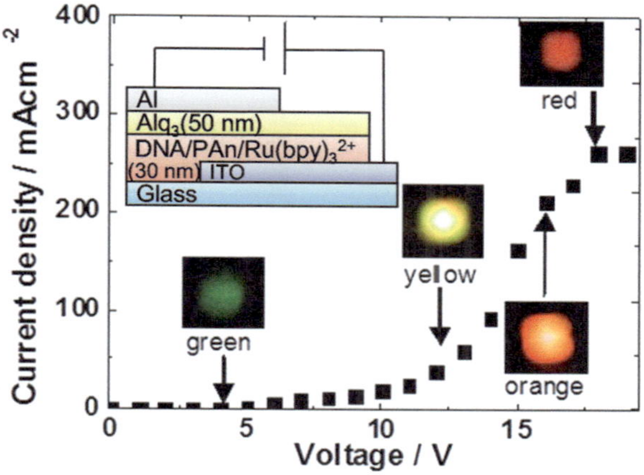


**Fig. 12.27** I-V characteristics and photographs of OLED (the *inset* shows a schematic of the OLED and digital camera images of the emission). This figure was reused from Nakamura et al. [50]. Copyright 2010, American Institute of Physics

interacted with the DNA phosphate group in the aqueous solution. Further, a CD signal corresponding to the absorption band of PAn was observed at a wavelength of around 600–800 nm, suggesting formation of the DNA/PAn complex. The absorption peak wavelength was different from that of the above-mentioned photopolymerized PAn/DNA complex (Figs. 12.19 and 12.20). This is possibly due to difference in the interaction between DNA and PAn.

We subsequently synthesized the DNA/PAn/$Ru(bpy)_3^{2+}$ complex by adding 5 mmol/L of $Ru(bpy)_3Cl_2$ aqueous solution to 50 mmol/L of DNA/PAn aqueous solution. The interaction between $Ru(bpy)_3^{2+}$ and the DNA/PAn complex was also studied by measuring their emission spectra. The luminescence spectra of $Ru(bpy)_3^{2+}$ varied with DNA concentration in the DNA/PAn/$Ru(bpy)_3^{2+}$ complex. The emission intensity of $Ru(bpy)_3^{2+}$ increased in conjunction with increasing DNA concentration. This enhancement in emission intensity indicates formation of the DNA/PAn/Ru $(bpy)_3^{2+}$ complex, because the complexation of DNA and $Ru(bpy)_3^{2+}$ suppressed the vibrational deactivation of the excited states of $Ru(bpy)_3^{2+}$ [48]. The detailed DNA complex structure and properties will be published in another paper in the near future.

We then fabricated the electroluminescent devices consisting of the DNA/PAn/ $Ru(bpy)_3^{2+}$ complex as follows. A thin layer (thickness: 30–40 nm) of the DNA/PAn/$Ru(bpy)_3^{2+}$ complex was obtained directly on the sufficiently washed ITO glass electrode by spin coating (1,000 rpm, 30 s), and the layer was dried at 50 °C for 12 h in vacuo. Subsequently, a thin layer of $Alq_3$ (thickness: 50 nm) was formed by vacuum deposition under reduced pressure ($<4.0 \times 10^{-4}$ Pa). For the top electrode, an active diode area of 0.09 $cm^2$ was formed by sequentially depositing an aluminum layer (thickness: 110 nm) in vacuo through a shadow mask on top of the $Alq_3$ thin layer without breaking the vacuum. For spectral measurements of EL, the device was then encapsulated in a glass cover and sealed with UV-cured epoxy glue.

The current density-voltage characteristics of the ITO/(DNA/PAn/$Ru(bpy)_3^{2+}$)/ $Alq_3$/Al OLED are shown in Fig. 12.27. The thickness of the DNA/PAn/$Ru(bpy)_3^{2+}$ and $Alq_3$ layers were 30 and 50 nm, respectively. The current density increased

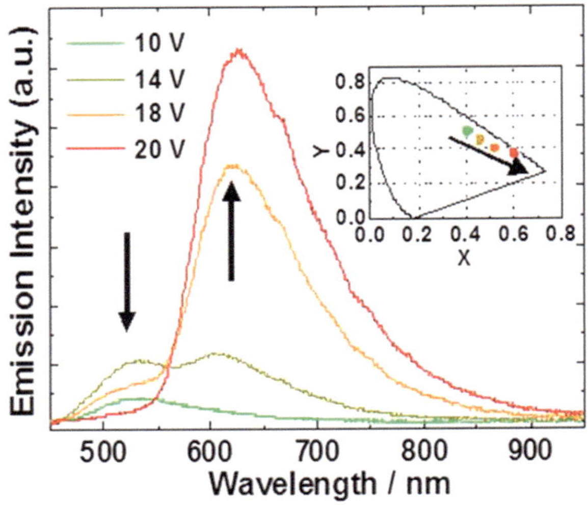


**Fig. 12.28** EL spectra of OLED at various bias voltages (the *inset* shows changes in the CIE emission coordinates of the OLED for various bias voltages). This figure was reused from Nakamura et al. [50]. Copyright 2010, American Institute of Physics

abruptly above a bias voltage of 5 V, and at 18 V, a current density of 300 mA/cm$^2$ was obtained. The current above 5 V obeyed Child's law for SCLC. Upon reflection of the I-V characteristics, green luminescence was observed from the device above a bias voltage of 5 V. Interestingly, the emission color of the device changed with increasing applied voltage, from green to yellow (14 V) to orange (16 V) and finally to red (18 V), as shown in the inset photo of Fig. 12.27. The maximum luminance of the OLED was 10 cd/m$^2$, which was lower than that of the previously reported DNA-based OLED [43, 44].

These results are of the result of insufficient amounts of PAn in the DNA complex compared with previous studies [31, 43, 44]. In a previous report, the DNA/PAn/$Ru(bpy)_3^{2+}$ complex was prepared by photopolymerization of the aniline dimer in the presence of DNA and $Ru(bpy)_3^{2+}$. Additionally, concentration of the PAn monomer was equal to that of the DNA phosphate group. On the other hand, the [DNA]/[PAn] ratio was 10/1 in this study. The present method facilitates a very easy way to introduce PAn into DNA, but it is difficult to increase the concentration of PAn in DNA. As the result, low concentration of PAn is responsible for luminance decrease.

To confirm multicolor emission from the device, we measured its EL spectra at various voltages (Fig. 12.28). At a bias voltage of 10 V, a green emission band was observed around 540 nm. This band was found to be similar to that observed in the PL spectrum of the $Alq_3$ layer [49], suggesting that the green emission originated from this layer. In addition, a red emission band with a maximum at 610 nm appeared above a bias voltage of 14 V. This band was found to be almost identical to that observed in the PL spectrum of the $Ru(bpy)_3^{2+}$ solution, indicating that the red emission originates from the $Ru(bpy)_3^{2+}$ layer. The height of the green emission band of $Alq_3$ decreased gradually when the applied voltage exceeded 14 V, and it decreased to almost zero at a bias voltage of 18 V. In contrast, the red emission band

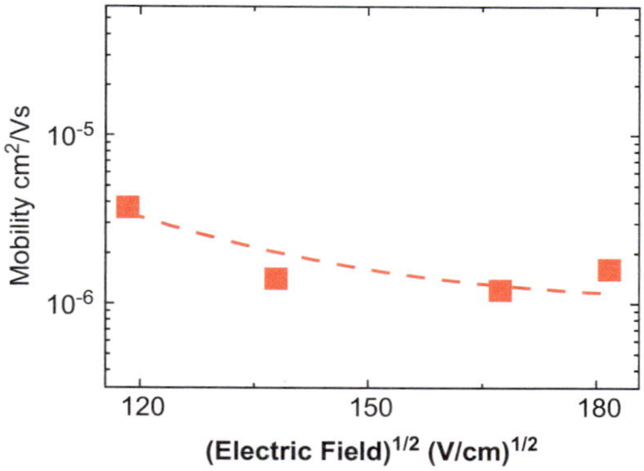


**Fig. 12.29** Applied-electric-field dependence of hole mobility for a 1:5 mixture layer of the DNA/PAn complex and native DNA (reproduced from Nakamura et al. [50]. Copyright 2010, American Institute of Physics)

of $Ru(bpy)_3^{2+}$ (610 nm) increased when the applied voltage exceeded 14 V. The inset in Fig. 12.28 shows changes in the collisional ionization equilibrium (CIE) emission coordinates of the device at various applied voltages; the coordinates changed from (0.39, 0.51) at 10 V to (0.61, 0.37) at 18 V. In this manner, the applied voltage dependence of the $Alq_3$ emission intensity was found to be different from that of $Ru(bpy)_3^{2+}$, and multicolor emission from one device was confirmed.

To discuss the multicolor emission mechanism, OLEDs with configurations different from the one mentioned above were used [50]. The current density-voltage characteristics of an OLED modified by increasing the thickness of the $Alq_3$ layer (130 nm, previous: 50 nm) was measured. A comparison with I-V characteristics of the original OLED structure (50 nm) revealed that green emission occurred from the $Alq_3$ layer at a bias voltage of 10 V, and its intensity increased above 10 V. However, no red emission was observed from the $Ru(bpy)_3^{2+}$ layer in the entire range of applied voltages. This result suggests that recombination of holes and electrons occurred in the $Alq_3$ layer, because the thickness increase of the $Alq_3$ layer caused degradation of its electron transporting ability. We also analyzed the current–voltage characteristics of another modified OLED, in which the PAn component (hole transport layer) was removed from the DNA complex layer. As a result, only red emission from $Ru(bpy)_3^{2+}$ was observed, and its intensity increased above a bias voltage of 11 V. The removal of PAn resulted in a decrease in hole mobility; therefore, carrier recombination occurred only in the DNA/$Ru(bpy)_3^{2+}$ layer.

Next, carrier mobility of the DNA complex was measured by the time of flight (TOF) technique. Fig. 12.29 shows the dependence of hole mobility on the applied electric field for a 1:5 mixture of the DNA/PAn complex and native DNA. The hole mobility of this mixture layer was found to be $1.0 \times 10^{-6}$–$4.0 \times 10^{-6}$ $cm^2/Vs$, and it showed negative field dependence. The negative-electric-field dependence of hole mobility in semiconductive or molecularly doped polymers has been previously reported [51, 52]. This negative dependence has been attributed to the diffusion of charge carriers against the direction of the applied electric field as the result of a large positional disorder. In the case of our device, the negative dependence was a result of the one-dimensional structure of the DNA/PAn complex.

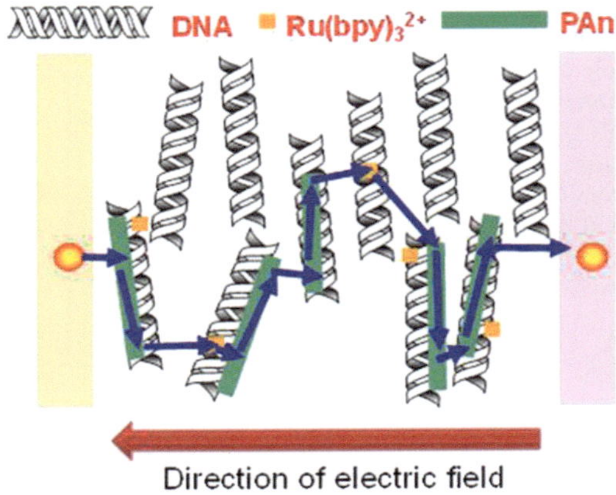


**Fig. 12.30** Suggested mechanism of charge transporting in DNA/PAn/$Ru(bpy)_3^{2+}$

In the spin-coated DNA complex layer, the main chains of the DNA complex were expected to lie on the substrate, resulting in the formation of detour-routes perpendicular to the electric field (Fig. 12.30).

In contrast, the electron mobility of the $Alq_3$ layer is known to show positive field dependence over a wide range of electric fields [53]. Based on these differences in the electric-field dependence of carriers, the recombination distribution of the OLED studied herein can be explained as follows. At low voltage, since hole mobility of the DNA complex layer was superior to electron mobility of the $Alq_3$ layer, recombination occurred mainly in the $Alq_3$ layer. As the voltage was increased, the electron mobility of the $Alq_3$ layer increased rapidly with the electric field, and the electrons could move through the interface and recombine with the holes in the DNA complex layer. This resulted in a continuous shift of the recombination region from the $Alq_3$ layer to the DNA/PAn/$Ru(bpy)_3^{2+}$ complex layer, as studied theoretically in previous literature [54].

Thus, the multicolor emission mechanism of the OLED studied herein can be summarized as follows. At a bias voltage of 5–10 V, green emission was observed to occur from the $Alq_3$ layer because carrier recombination occurred in this layer. As the applied voltage increased, the carrier recombination region shifted from the $Alq_3$ layer to the DNA/PAn/$Ru(bpy)_3^{2+}$ complex layer, and recombination occurred in both the $Alq_3$ and DNA/PAn/$Ru(bpy)_3^{2+}$ complex layers. As a result, both green emission from $Alq_3$ and red emission from $Ru(bpy)_3^{2+}$ were observed. Finally, only red emission from $Ru(bpy)_3^{2+}$ was observed above applied voltages of 18 V.

### *12.3.5 Summary*

A PAn/DNA complex was successfully prepared by the template photopolymerization of dimeric aniline (PPD) via photocatalytic reaction of $Ru(bpy)_3^{2+}$ in the presence of DNA. PPD, $Ru(bpy)_3^{2+}$, and DNA were first dissolved in the aqueous solution to analyze the structure of the resulting complex and to carry out photopolymerization. PPD and $Ru(bpy)_3^{2+}$ were bound to duplex DNA regularly in the solution through the electrostatic and/or hydrophobic interaction with DNA.

Photopolymerization was carried out by illuminating the solution containing DNA/PPD/$Ru(bpy)_3^{2+}$ with visible light. The photopolymerization reaction occurred even in solutions at pH 3.0–6.0, owing to the specific local, "lower-pH" environment provided by DNA. The absorption and CD spectra indicate that the PAn/DNA complex has an ordered structure in which PAn is associated with DNA and winds around the duplex DNA, just like a triple helix. It was revealed that DNA works as the template for photopolymerization of PPD with $Ru(bpy)_3^{2+}$ to prepare a highly ordered, novel PAn/DNA complex. The complex contains the photocatalyst, $Ru(bpy)_3^{2+}$, even after purification. $Ru(bpy)_3^{2+}$ also works as emitting material.

An Al/DNA/PAn containing $Ru(bpy)_3^{2+}$/ITO-based OLED cell was fabricated to analyze EL properties. It was revealed that the novel, processable, and water-soluble PAn/DNA complex works as material for a $Ru(bpy)_3^{2+}$ complex-based red-emitting diode with a fast turn-on response. Further, we demonstrated green-to-red color tunable emission from an OLED consisting of a DNA/PAn/$Ru(bpy)_3^{2+}$ complex (hole transport layer) and $Alq_3$ (electron transport layer). The DNA/PAn/$Ru(bpy)_3^{2+}$ complex was prepared with a simple complexation procedure between DNA and PAn, which is different from the template photopolymerization method. The emission colors of the $Alq_3$-stacked OLED could be tuned by varying the applied voltage. The color tunability of this emission was principally a result of shifts in the carrier recombination region that are attributable to increasing applied voltage.

Work is in progress to improve the luminance and the stability of the OLED. We expect that the results of our research will contribute to the development of organic electronic devices, such as BiOLED, using biomaterials characterized by DNA-hybrid materials.

**Acknowledgments** These works were partially supported by a Grants-in-Aid for Scientific Research, MEXT (No. 17067004, 20350085 and 24350090) Japan. The authors would like to thank Chiba University Global COE program "Advanced School for Organic Electronics" for supporting our research.

## References

1. J. Watson, F. Crick, Nature **171**, 737 (1953)
2. J. Watson, *The Double Helix: A Personal Account of the Discover of the Structure of DNA* (Touchstone, New York, 1968)
3. T.B. Singh, N.S. Sariciftci, J.G. Grote, Adv. Polym. Sci. **223**, 189 (2010)
4. C. Yumusak, T.B. Singh, N.S. Sariciftci, J.G. Grote, Appl. Phys. Lett. **95**, 263304 (2009)
5. E.M. Heckman, J.A. Hagen, P.P. Yaney, F.K. Hopkins, Appl. Phys. Lett. **87**, 211115 (2005)
6. A.J. Steckl, Nat. Photonics **1**, 3 (2007)
7. Y.-W. Kwon, C.H. Lee, D.-H. Choi, J.-I. Jin, J. Mater. Chem. **19**, 1353 (2009)
8. Y. Kawabe, L. Wang, S. Horinouchi, N. Ogata, Adv. Mater. **12**, 1281 (2000)

9. E. Braun, Y. Eichen, U. Sivan, G. Ben-Yoseph, Nature **391**, 775 (1998)
10. J.G. Grote, F.K. Hopkins, J. Appl. Phys. **100**, 024514 (2006)
11. T. Yukimoto, S. Uemura, T. Kamata, K. Nakamura, N. Kobayashi, J. Mater. Chem. **21**, 15575 (2011)
12. C.A. Mirkin, R.L. Letsinger, R.C. Mucic, J.J. Storhoff, Nature **382**, 607 (1996)
13. P.F. Baude, D.A. Ender, M.A. Haase, T.W. Kelley, D.V. Muyres, S.D. Theiss, Appl. Phys. Lett. **82**, 3964 (2003)
14. Z.-T. Zhu, J.T. Mason, R. Dieckmann, G.G. Malliaras, Appl. Phys. Lett. **81**, 4643 (2002)
15. L. Zhou, A. Wanga, S. Chu Wu, J. Sun, S. Park, T.N. Jackson, Appl. Phys. Lett. **88**, 083502 (2006)
16. M. Kawasaki, S. Imazeki, S. Hirota, T. Arai, T. Shiba, M. Ando, Y. Natsume, T. Minakata, S. Uemura, T. Kamata, J. SID **16**, 161 (2008)
17. L. Zhou, S. Park, B. Bai, J. Sun, S.-C. Wu, T.N. Jackson, S. Nelson, D. Freeman, Y. Hong, IEEE Electron Device Lett. **26**, 640 (2005)
18. M. Mizukami, N. Hirohata, T. Iseki, K. Ohtawara, T. Tada, S. Yagyu, T. Abe, T. Suzuki, Y. Fujisaki, Y. Inoue, S. Tokito, T. Kurita, IEEE Electron Device Lett. **27**, 249 (2006)
19. R.C.G. Naber, C. Tanase, P.W.M. Blom, G.H. Gelinck, A.W. Marsman, F.J. Touwslager, S. Setayesh, D.M. de Leeuw, Nat. Mater. **4**, 243 (2005)
20. G. Zhu, Z. Zeng, L. Zhang, X. Yan, Appl. Phys. Lett. **89**, 102905 (2005)
21. M. Hasegawaa, N. Kobayashia, S. Uemura, T. Kamata, Synth. Met. **159**, 961 (2009)
22. K. Tanaka, Y. Okahata, J. Am. Chem. Soc. **118**, 10679 (1996)
23. L. Wang, J. Yoshida, N. Ogata, Chem. Mater. **13**, 1273 (2001)
24. A.J. Steckl, H. Spaeth, H. You, E. Gomez, J. Grote, Opt. Photonics News **22**, 34 (2011)
25. A.-L. Deman, M. Erouel, D. Lallemand, M. Phaner-Goutorbe, P. Lang, J. Tardy, J. Non Cryst. Solids **354**, 1598 (2008)
26. C.J. Murphy, M.R. Arkin, Y. Jenkins, N.D. Ghatlia, S.H. Bossmann, N.J. Turro, J.K. Barton, Science **262**, 1025 (1993)
27. M.R. Arkin, E.D.A. Stemp, R.E. Holmlin, J.K. Barton, A. Hormann, E.J.S. Olson, P.F. Barbara, Science **273**, 475 (1996)
28. H.W. Fink, S. Schonenberger, Nature **398**, 407 (1999)
29. A.Y. Kasumov, M. Kociak, S. Gueron, B. Reulet, V.T. Volkov, D.V. Klinov, H. Bouchiat, Science **291**, 280 (2001)
30. D. Porath, A. Bezryadin, S. De Vries, C. Dekker, Nature **403**, 635 (2000)
31. S. Uemura, T. Shimakawa, K. Kusabuka, T. Nakahira, N. Kobayashi, J. Mater. Chem. **11**, 267 (2001)
32. R. Nagarajan, W. Liu, J. Kumar, S.K. Tripathy, F.F. Bruno, L.A. Samuelson, Macromolecules **34**, 3921 (2001)
33. A. Rakitin, P. Aich, C. Papadopoulos, Y. Kobzar, A.S. Vedeneev, J.S. Lee, J.M. Xu, Phys. Rev. Lett. **86**, 3670 (2001)
34. K. Teshima, K. Yamada, N. Kobayashi, R. Hirohashi, Chem. Commun. **829** (1996)
35. K. Teshima, S. Uemura, N. Kobayashi, R. Hirohashi, Macromolecules **31**, 6783 (1998)
36. S. Uemura, K. Teshima, S. Tokuda, N. Kobayashi, R. Hirohashi, Synth. Met. **101**, 701 (1999)
37. Y. Kim, N. Kobayashi, K. Teshima, R. Hirohashi, Synth. Met. **101**, 699 (1999)
38. Y. Kim, K. Teshima, N. Kobayashi, Electrochim. Acta **45**, 1549 (2000)
39. W. Liu, A.L. Cholli, R. Nagarajan, J. Kumar, S. Tripathy, F.F. Samuelson, J. Am. Chem. Soc. **121**, 11345 (1999)
40. S. Uemura, M. Yoshie, N. Kobayashi, T. Nakahira, Polym. J. **32**, 987 (2000)
41. S. Uemura, S. Tokuda, T. Nakahira, N. Kobayashi, Synth. Met. **119**, 89 (2001)
42. Y. Furukawa, F. Ueda, Y. Hyodo, I. Harada, T. Nakajima, T. Kawagoe, Macromolecules **21**, 1297 (1988)
43. N. Kobayashi, S. Uemura, K. Kusabuka, T. Nakahira, H. Takahashi, J. Mater. Chem. **11**, 1766 (2001)
44. N. Kobayashi, M. Hashimoto, K. Kusabuka, Elecrtochemistry **72**, 440 (2004)

45. M. Berggren, O. Inganas, G. Gustafsson, J. Rasmusson, M.R. Andersson, T. Hjertberg, O. Wennerstrom, Nature **372**, 444 (1994)
46. M. Hamaguchi, K. Yoshino, Appl. Phys. Lett. **69**, 143 (1996)
47. F.L. Lu, F. Wudl, M. Nowak, A.J. Heeger, J. Am. Chem. Soc. **108**, 8311 (1986)
48. X.H. Zou, B.H. Ye, H. Li, Q.L. Zhang, H. Chao, J.G. Liu, L.N. Ji, X.Y. Li, J. Biol. Inorg. Chem. **6**, 143 (2001)
49. Y.F. Xu, H.J. Zhang, H.Y. Li, S.N. Bao, P. He, Appl. Surf. Sci. **252**, 2328 (2006)
50. K. Nakamura, T. Ishikawa, D. Nishioka, T. Ushikubo, N. Kobayashi, Appl. Phys. Lett. **97**, 193301 (2010)
51. A.J. Mozer, N.S. Sariciftci, Chem. Phys. Lett. **389**, 438 (2004)
52. S.R. Mohan, M.P. Joshi, M.P. Singh, Chem. Phys. Lett. **470**, 279 (2009)
53. H. Park, D.S. Shin, H.S. Yu, H.B. Chae, Appl. Phys. Lett. **90**, 202103/1 (2007)
54. C.H. Chen, H.F. Meng, Appl. Phys. Lett. **86**, 201102/1 (2005)

# Part III
# Theoretical Study

# Chapter 13
# Theory of Photoelectron Spectroscopy

**Takashi Fujikawa and Kaori Niki**

## 13.1 What Is Special in Theory of Photoelectron Emission from Organic Solids?

Detailed analyses of photoemission spectra from inorganic solids provide useful information on electronic and geometric structures. In particular angle-resolved photoemission spectroscopy (ARPES) from valence levels gives us useful and important information on band dispersion for various crystalline systems [1]. Most of their valence bands are well described in terms of one-electron orbitals.

In contrast organic solids have strongly anisotropic characters showing specific behaviors of one- and two-dimensional solids. In some organic solids a giant Kohn anomaly is found in the dispersion relation of longitudinal acoustic phonons already at room temperature. Electron–phonon interaction have considerable influence on their properties revealed by UPS analyses [2]. So far most of ARPES spectra have been analyzed on the basis of one-electron tight-binding model, where weak inter-molecular interaction is taken into account by use of resonant (or transfer) integrals. Of course this theoretical framework cannot describe those inter-molecular interaction caused by dynamic polarization and electron–phonon interaction. We thus may have a question: Do ARPES analyses simply give us band dispersion $\varepsilon_n(\mathbf{q})$ (n; band index, $\mathbf{q}$; crystal momentum in a Brillouin zone) even for such weakly interacting organic molecular solids? As discussed in Sect. 13.3.1, from ARPES analyses we obtain information on $\varepsilon_\lambda + \Sigma_\lambda$ $(\lambda = (n, \mathbf{q}))$, where $\Sigma_\lambda$ is the electron self-energy influenced by electron-electron and electron–phonon interaction.

T. Fujikawa (✉) • K. Niki
Graduate School of Advanced Integration Science, Chiba University, Chiba, Japan
e-mail: tfujikawa@faculty.chiba-u.jp



H. Ishii et al. (eds.), *Electronic Processes in Organic Electronics*,
Springer Series in Materials Science 209, DOI 10.1007/978-4-431-55206-2_13

Electron–phonon interaction has usually been studied in the framework of adiabatic (Born–Oppenheimer) approximation. In this framework the electron–phonon interaction is related to vibronic effects [3]. It is not so obvious, however, to relate that approximation to the widely used interaction Hamiltonian

$$H_{ep} = \sum_{\mathbf{q},G} M(\mathbf{q}+\mathbf{G})\rho(\mathbf{q}+\mathbf{G})(a_\lambda + a^\dagger_{-\lambda}),$$

where $\rho(\mathbf{q})$ is the Fourier component of the electron density, $M(\mathbf{q})$ is given by use of the Fourier component of the unscreened electron-atom interaction potential $v_{ei}$ and phonon polarization vector $\mathbf{e}(\lambda)$

$$M(\mathbf{q}) = -i\frac{\mathbf{q}\cdot\mathbf{e}(\lambda)}{\sqrt{2NM\omega_\lambda}}v_{ei}(\mathbf{q}).$$

In the above equation, the reciprocal lattice vector $\mathbf{G}$ is used, which implies the crystal lattice periodicity. Quantum field theory provides us with natural language to discuss the electron–phonon interaction, which is incorporated in the electron self-energy $\Sigma(\omega)$ in addition to the electron-electron interaction. The above simple model can have sound theoretical basis referring to the field theoretical framework.

In this short review, we discuss some specific problems in theoretical analyses of UPS and XPS spectra excited from organic solids.

## 13.2 Basic Theory

A sophisticated many-body one-step quantum theory has been developed on the basis of many-body scattering theory as proposed by Hedin et al. [4–6]. This theoretical approach is practically useful to describe the photoemission processes in terms of the damping photoelectron wave function under the influence of the optical potential.

First principle formal photoemission theories based on Keldysh Green's functions were first developed by Caroli et al. [7], and further refined by Almbladh [8]. These theories give formally exact perturbation expansion of the photoemission intensity within nonrelativistic theoretical framework, however, no practical formulas to analyze photoemission spectra. Great advantages of this theoretical framework are feasibility of temperature effects [7, 9], radiation field effects [10, 11], and also relativistic effects [12]. Here brief description on both of the theories are given for later discussion.

### 13.2.1 Photoemission Intensities Derived from Many-Body Scattering Theory

The basic formula to describe the photoemission intensity measuring photoelectrons with momentum $\mathbf{k}$ and energy $\varepsilon_k(= k^2 2)$ is given in the golden rule approximation for the excitation from the ground target state $|\Psi_0>$

$$I(\mathbf{k}) = 2\pi \sum_n |<\Psi^-_{n\mathbf{k}}|H_{ep}|\Psi_0>|^2 \delta(E_0 + \omega - E^*_n - \varepsilon_k) \tag{13.1}$$

where $H_{ep}$ is the electron–photon interaction operator written in terms of electron field operators $\psi^\dagger(x)$ and $\psi(x)$,

$$\begin{aligned} H_{ep} &= \int dx \psi^\dagger(x) \Delta(x) \psi(x), \\ x &= (\mathbf{r}, \sigma). \end{aligned} \tag{13.2}$$

The photoelectron state $|\Psi^-_{n\mathbf{k}}>$ asymptotically approaches to the direct product of the hole state $|n^*>$ and the photoelectron state $|\phi_{\mathbf{k}}{}^->$, $|n^*>|\phi^-_{\mathbf{k}}>= c^\dagger_{\mathbf{k}}|n^*>$ far from the target. The photoelectron wave function $|\phi^-_{\mathbf{k}}>$ should satisfy "out-going" boundary condition. For the practical purpose, we choose $|\phi^-_{\mathbf{k}}>$ as a solution of one-electron equation

$$[T_e + <n^*|V_{es}|n^*>]|\phi_{\mathbf{k}}{}^-> = \varepsilon_k|\phi_{\mathbf{k}}{}^-> \tag{13.3}$$

where $T_e$ is the one-electron kinetic energy operator of electrons, and $V_{es}$ is the photoelectron-solid interaction. Except the threshold excitation, the state $|\Psi^-_{n\mathbf{k}}>$ is related to $|n^*>|\phi^-_{\mathbf{k}}>$ as [5]

$$\begin{aligned} |\Psi_{n\mathbf{k}}{}^-> &= |n^*>|\phi_{\mathbf{k}}{}^-> + \frac{1}{E - H - i\eta} V_n |n^*>|\phi^-_{\mathbf{k}}>, \\ E &= E^*_n + \varepsilon_k, (\eta \to +0), \end{aligned} \tag{13.4}$$

where $V_n$ is the fluctuation potential

$$V_n = V_{es} - <n^*|V_{es}|n^*> \tag{13.5}$$

for the target state $|n^*>$ which is responsible for inelastic scatterings inside solids. The second term of Eq. (13.4) describes the interaction between the photoelectrons and the target in the hole state $|n^*>$.

The amplitude in Eq. (13.1) is thus rewritten in terms of Dyson orbital $g_n(x) =<n^*|\psi(x)|\Psi_0>$, and the damping photoelectron wave function $\psi^-_{\mathbf{k}}$ under the influence of the non-Hermitian optical potentials [4]

$$
\begin{aligned}
< \Psi_{n\mathbf{k}}{}^{-}|H_{ep}|\Psi_0 > = < \psi_{\mathbf{k}}^{-}|\Delta|g_n > + \sum_{m(\neq n)} < \psi_{\mathbf{k}}^{-}| < n^*|V_{es}|m^* \\
> g(\varepsilon_k - E_m^* + E_n^*)\Delta|g_m > + \cdots . \qquad (13.6)
\end{aligned}
$$

In the above equation, $g(\varepsilon)$ is the damping one-electron Green's function with the optical potential $\Sigma_m(E)$, which includes important many-body effects responsible for the photoelectron wave damping. The Hermitian part also has important role for photoelectron elastic scatterings. It satisfies an important relation [13]

$$
\Sigma_m(E) = \Sigma_0(E - E_m + E_0). \qquad (13.7)
$$

Owing to this relation it is sufficient to calculate the optical potential for the ground state with and without hole: We can obtain the optical potential even for some excited states by shifting the energy. Dyson orbital $g_n(x)$ is written in terms of, for example, Hartree-Fock basis $\{\phi_i(x)\}$ and the associated annihilation operators $\{c_i\}$,

$$
\begin{aligned}
g_n(x) &= \sum_i \phi_i(x) S_i^n, \\
S_i^n &= < n^*|c_i|0 > .
\end{aligned} \qquad (13.8)
$$

The amplitude $S_i^n$ is the intrinsic (shake-up) amplitude. In the second term $< n^* \mid V_{es} \mid m^* >$ in Eq. (13.6) is the extrinsic loss amplitude during the propagation of photoelectrons, and the Green's function $g(\varepsilon_k - E_m^* + E_n^*)$ describes the damping photoelectron propagation before the loss. The above theoretical approach is direct and transparent, whereas the extension to the photoemission from finite temperature systems is rather hard. For those purposes Keldysh Green's function approach is appropriate, even though the formulation is quite complicated.

### *13.2.2 Photoemission Intensities Derived from Keldysh Green's Functions*

In quantum mechanics one-electron current density $\mathbf{j}$ is given in terms of one-electron wave function $\psi$ and momentum operator $\mathbf{p} = -i\nabla$

$$
\begin{aligned}
\mathbf{j}(\mathbf{r},t) &= \frac{1}{2}\Big[\psi^*(\mathbf{r},t)\mathbf{p}\psi(\mathbf{r},t) + \big(\mathbf{p}\psi(\mathbf{r},t).^*\psi(\mathbf{r},t). \\
&= \frac{i}{2}\Big(\frac{\partial}{\partial \mathbf{r}'} - \frac{\partial}{\partial \mathbf{r}}.\psi^*(\mathbf{r},t)\psi(\mathbf{r}',t)|_{\mathbf{r}'} = \mathbf{r}
\end{aligned} \qquad (13.9)
$$

where we take $\mathbf{r} = \mathbf{r}'$ after the differential operation.

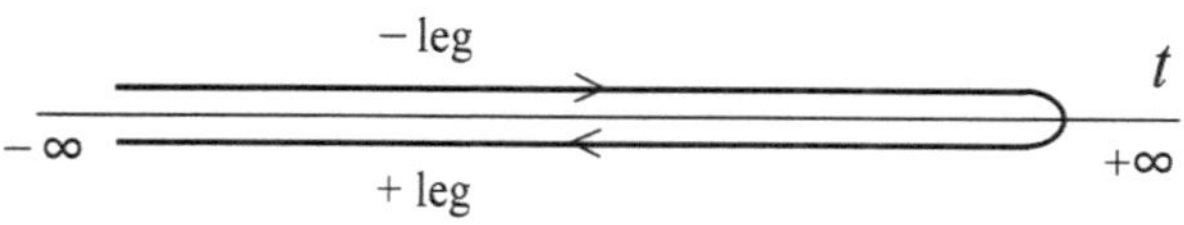


**Fig. 13.1** Keldysh contour

In the field theory to describe the photoelectron current from many-electron systems, the total photoelectron current $\mathbf{j}(\mathbf{r},t)$ induced by the X-ray field is given at a detection position $\mathbf{r}$ and spin $\sigma$ ($x=(\mathbf{r},\sigma)$) and time $t$ referring to Eq. (13.9)

$$\begin{aligned}\mathbf{j}(\mathbf{r},t) &= \frac{i}{2}\Big(\frac{\partial}{\partial \mathbf{r}'}-\frac{\partial}{\partial \mathbf{r}}\cdot < \psi^{\dagger}(x,t)\psi(x',t) > \Big|_{x'=x} \\ &= \frac{1}{2}\Big(\frac{\partial}{\partial \mathbf{r}'}-\frac{\partial}{\partial \mathbf{r}}\cdot g^{<}(x't,xt)\Big|_{x'=x}, \qquad (13.10)\\ ig^{<}(x't',xt) &= - < \psi^{\dagger}(xt)\psi(x't') > . \qquad (13.11)\end{aligned}$$

The basic difficulty in the nonequilibrium field theory is to keep track of which operators are negatively (+) or positively (−) time ordered, i.e., on what contour they lie (see Fig. 13.1). One way to keep track of this is to artificially distinguish the external perturbation $V^{(-)}$ which takes the system forward in time from $V^{(+)}$ which takes the system backward in time. To obtain physically meaningful results we must set $V^{(-)}(t)=V^{(+)}(t)=V(t)$ at an appropriate point in the calculation [14]. We now consider the time-dependent electron–photon interaction $M$ instead of $H_{ep}$ defined by Eq. (13.2)

$$M(x,t)=\psi^{\dagger}(x)\Delta(x)\psi(x)e^{-i\omega t}. \qquad (13.12)$$

The above time-dependent perturbation forces us to use the Keldysh Green's function approach to the calculation of the lesser Green's function $g^{<}$ in Eq. (13.10) because we study the time-dependent (nonequilibrium) processes. We look for a direct photoelectron current applying alternating field as shown by Eq. (13.12). We thus deal with quadratic response theory because in linear response the output electron current oscillates with the same frequency $\omega$. We thus have to calculate $g^{<}$ at least to second order in the X-ray field $M$ [7]. The simplest second order term contributing to the current $\mathbf{j}$ in regard to $M$ is given by

$$\int_c d1d2[G(xt,1)M(1)G(1,2)M^{\dagger}(2)G(2,xt)]^{<}.$$

where $1=(x_1,t_1)$ and the integral along the closed path (Keldysh path) $c$ from $-\infty$ to $\infty$ and back to $-\infty$ $\int_c dt$ is used. The Keldysh Green's function $G(1,2)$ is defined by use of path ordering operator on the contour $c$ shown in Fig. 13.1

$$iG(1,2) = < T_c[\psi(1)\psi^{\dagger}(2)] > .$$

For the practical calculations, we should change the integral $\int_c dt$ to the conventional integral $\int_{-\infty}^{\infty} dt$. For that purpose it is useful to apply the Langreth rule [15] which yields

$$\int_{-\infty}^{\infty} d1d2[g^<(xt,1)M(1)g^a(1,2)M^\dagger(2)g^a(2,xt)$$
$$+g^r(xt,1)M(1)g^<(1,2)M^\dagger(2)g^a(2,xt)+g^r(xt,1)M(1)g^r(1,2)M^\dagger(2)g^<(2,xt). \tag{13.13}$$

where $g^r$, $g^a$ are the retarded and advanced Green's functions. Now the integral in Eq. (13.13) is taken from $-\infty$ to $\infty$, and then the Fourier transform of Eq. (13.13), $\int_{-\infty}^{\infty} dt[\cdots]exp(i\varepsilon t)$ yields a simple formula

$$g^<(\varepsilon)\Delta g^a(\varepsilon-\omega)\Delta^* g^a(\varepsilon)+g^r(\varepsilon)\Delta g^<(\varepsilon-\omega)\Delta^* g^a(\varepsilon)+\cdots,$$

where the integrals over spin-coordinates are not shown for simplicity.

The dressed one-electron Green's functions $g^<$, $g^r$, $g^a$ have spectral representations

$$g^<(x,x';\varepsilon)=2\pi i\sum_n g_n(x)g_n^*(x')\delta(\varepsilon-\varepsilon_n), \tag{13.14}$$

$$(\varepsilon_n=E_0-E_n^*)$$

$$g^r(x,x';\varepsilon)=\sum_q\frac{f_q(x)f_q^*(x')}{\varepsilon-\varepsilon_q+i\eta}+\sum_n\frac{g_n(x)g_n^*(x')}{\varepsilon-\varepsilon_n+i\eta}, \tag{13.15}$$

$$g^a(x,x';\varepsilon)=\sum_q\frac{f_q(x)f_q^*(x')}{\varepsilon-\varepsilon_q-i\eta}+\sum_n\frac{g_n(x)g_n^*(x')}{\varepsilon-\varepsilon_n-i\eta}. \tag{13.16}$$

We have defined the particle Dyson orbitals $f_q$'s,

$$f_q(x)=<0|\psi(x)|q,N+1>,\quad \varepsilon_q=E_q(N+1)-E_0. \tag{13.17}$$

The Dyson orbitals $f_q$ and $g_n$ are dressed one-electron functions which include complicated many-body effects. They are reduced to unoccupied and occupied orbitals in the Hartree-Fock approximation. From a bound state $|q,N+1>$ the quasiparticle state $f_q(x)$ decays exponentially outside the target which has no contribution to the photoelectron current given by Eq. (13.10). In contrast the photoelectron quasiparticle state

$$f_{\mathbf{p}}^-(x)=<0|\psi(x)|\Psi_{0\mathbf{p}}^-(N+1)>$$

satisfies the outgoing (−) boundary condition and approaches to the plane wave at the detection point far from the target. In solids it propagates under the influence of the non-Hermitian optical potential [16]. We thus obtain a basic formula for the photoelectron current density measuring photoelectrons with momentum **k** and kinetic energy $\varepsilon_k$ from the second term of Eq. (13.13).

$$\begin{aligned}\mathbf{j}(\mathbf{k},\varepsilon_k) &= -\frac{i\mathbf{k}}{(2\pi)^2} < f_{\mathbf{k}}^- |\Delta g^<(\varepsilon_k-\omega)\Delta^*| f_{\mathbf{k}}^- > \\ &= \frac{\mathbf{k}}{2\pi}\sum_n | < f_{\mathbf{k}}^- |\Delta| g_n > |^2 \delta(\varepsilon_k-\varepsilon_n-\omega).\end{aligned} \tag{13.18}$$

The final expression is obtained by use of the spectral representation (13.14) of $g^<$, which describes the photoemission from the dressed one-electron state $g_n$ to the dressed photoelectron state $f_{\mathbf{k}}^-$. From the above equation, the photoemission intensity $I(\mathbf{k})$ is given by

$$\begin{aligned}I(\mathbf{k}) &= \mathrm{Im} < f_{\mathbf{k}}^- |\Delta g^<(\varepsilon_k-\omega)\Delta^*| f_{\mathbf{k}}^- > \\ &= 2\pi\sum_n | < f_{\mathbf{k}}^- |\Delta| g_n > |^2 \delta(\varepsilon_k-\varepsilon_n-\omega).\end{aligned} \tag{13.19}$$

This is quite similar to the one-electron expression where $f_{\mathbf{k}}^-$ and $g_n$ are replaced by simple one-electron functions.

In order to discuss loss and resonant effects, we have to take higher-order terms in regard to the electron-electron interaction [9, 10]. The amplitude $< f_{\mathbf{k}}^- |\Delta| g_n >$ corresponds to $< \psi_{\mathbf{k}}^- |\Delta| g_n >$ in Eq. (13.6). In order to discuss the extrinsic loss effects described by the second term of Eq. (13.6), we have to consider the higher order terms [9]. From now on we use the present theoretical framework in order to discuss the phonon effects on photoemission spectra.

### 13.2.3 Phonon Effects: Debye–Waller Factor, Electron–Phonon Interaction

Phonon effects on photoemission spectra have been observed both in UPS and XPS spectra. High-energy XPS spectra probe more accurately bulk rather than surface electronic structures, but some questions are raised in regard to the possibility of band mapping. Osterwalder et al. report the angular distribution from Al valence band excited at $\omega = 1{,}254$ eV which presents strong maxima at main crystallographic directions resembling closely to the angular distribution of photoelectrons from the $2s$ core level [17].

For the study of phonon effects on XPS and UPS spectra, Debye–Waller (DW) and Franck–Condon (FC) factors have extensively been discussed from theoretical sides. On the other hand systematic first principle theory has rarely been proposed to study the influence of electron–phonon interaction on photoemission spectra. Recent progress in theoretical treatment of these phonon effects are found in some literatures [18] which are built on the basis of Keldysh Green's function theory [9] and the Baym–Hedin approach to the electron–phonon interaction [19, 20].

In order to study interference effects such as X-ray diffraction, extended X-ray absorption fine structure (EXAFS) and photoemission spectra, the Debye–Waller factors play an important role, which destroy the interference and suppress the structures associated with the interference. Later detailed discussion will be given on high-energy photoemission from extended levels. There the two center terms describes the interference processes of photoelectron waves excited from different atomic sites, which include the factor like

$$\exp[i\mathbf{Q} \cdot (\mathbf{R}_\alpha - \mathbf{R}_\beta)], \quad \mathbf{Q} = \mathbf{k} - \mathbf{K}$$

where $\mathbf{k}$ and $\mathbf{K}$ are wave vectors of the photoelectrons and the incident X-rays, $\mathbf{R}_\alpha$ and $\mathbf{R}_\beta$ position vectors of atoms $\alpha$ and $\beta$. They are written as

$$\mathbf{R}_\alpha = \mathbf{R}_\alpha^0 + \mathbf{u}_\alpha$$

where $\mathbf{R}_\alpha^0$ and $\mathbf{u}_\alpha$ are the equilibrium position and the displacement of the $\alpha$th atom. Within the harmonic vibration approximation, we can apply the Mermin's theorem [21] because the displacements $\mathbf{u}_\alpha$ and $\mathbf{u}_\beta$ are both linear in phonon operators. We thus obtain the thermal average

$$< \exp[i\mathbf{Q} \cdot (\mathbf{u}_\alpha - \mathbf{u}_\beta)] >= \exp(-Q^2\Delta_{\alpha\beta}^2 2), \Delta_{\alpha\beta}^2 =< [\widehat{\mathbf{Q}} \cdot (\mathbf{u}_\alpha - \mathbf{u}_\beta)]^2 > . \quad (13.20)$$

This damping factors are quite important in high-energy region, whereas not so important in the UPS spectra.

As shown below (Sect. 13.3.1) the photoemission intensity formula includes the retarded Green's function and the photoelectron function $f_{\mathbf{k}}^-$ ; the former is influenced by the electron self-energy $\Sigma^r$. In low energy region the electron–phonon interaction has finite contribution to $\Sigma^r$ in addition to electron-electron interaction. The Baym–Hedin theory allows us to write the total screened Coulomb interaction $W$ [19, 20]

$$W(1,2) = W_e(1,2) + W_{ph}(1,2), \quad 1 = (x_1, t_1). \quad (13.21)$$

The electron screened Coulomb interaction $W_e$ is given in terms of bare Coulomb interaction $v$, inverse dielectric function $\varepsilon_e^{-1}$ and irreducible polarization $P_e$ due to electron-electron interaction:

$$W_e = \varepsilon_e^{-1} v = (1 - vP_e)^{-1} v.$$

The phonon part of the screened Coulomb interaction $W_{ph}$ is given by use $W_e$ and the phonon Green's function $D$

$$W_{ph} = W_e D W_e.$$

Corresponding to the sum $W = W_e + W_{ph}$, the electron self-energy $\Sigma$ and electron Green's function $G_e$ can be written

$$\begin{aligned} \Sigma &= \Sigma_e + \Sigma_{ph}, \\ G_e &= G_0 + G_0 \Sigma_e G_e. \end{aligned} \tag{13.22}$$

For example, in the lowest order approximation, *GW* approximation, they are explicitly given

$$\begin{aligned} \Sigma_e(1,2) &\approx -ip_1 G_e(1,2) W_e(2,1^+), \\ \Sigma_{ph}(1,2) &\approx -ip_1 G_e(1,2) W_{ph}(2,1^+), \end{aligned} \tag{13.23}$$

where $p_1 = 1$ when $t_1$ is on the + leg and $-$ 1 on the $-$ leg (see Fig. 13.1). These formulas can be applied to metals and also insulating systems (polaron). Experiments dealing with electron–phonon coupling at metal surface are reviewed by Kröger [22]: Kohn anomalies, Peierls distortions, superconductivity are discussed referring to ARPES analyses.

## 13.3 Photoemission from Valence Levels

For the photoemission from organic solids, we have paid much attention to the excitation from extended valence levels. The derivation is quite straightforward and simple in the many-body scattering theory, however the Keldysh Green's function approach has some great advantages to handle the phonon effects in the photoemission processes as discussed below. UPS and XPS spectra provide quite different information even though we discuss the valence excitations from the same levels. Arai et al. have briefly discussed the electron–phonon interaction on X-ray absorption and UPS spectra based on Keldysh theory [23].

### *13.3.1 Angle-Resolved Photoelectron Spectroscopy Excited with Low Energy Photons*

Recent experimental results from angle-resolved UPS spectra have suggested the several specific phonon modes may be of relevance to pairling in high-temperature superconductors. A simple theoretical model has been proposed for the analyses of their ARPES to study electron–phonon interactions [24]. A sophisticated theory is proposed on the basis of Keldysh theory [25]. We can rewrite the first expression of Eq. (13.19) in terms of the retarded Green's function $g^r$,

$$I(\mathbf{k}) \propto -2f(\varepsilon_k - \omega)\mathrm{Im} < f_{\mathbf{k}}^{-}|\Delta g^r(\varepsilon_k - \omega)\Delta^*|f_{\mathbf{k}}^{-} > \tag{13.24}$$

where $f(\omega)$ is the Fermi distribution function. In the Keldysh theory, the retarded function $g^r(\omega)$ satisfies a closed Dyson equation in energy space together with the retarded electron self-energy $\Sigma^r(\omega)$,

$$g^r(\omega) = g_0^r(\omega) + g_0^r(\omega)\Sigma^r(\omega)g^r(\omega). \tag{13.25}$$

Keeping diagonal elements of the electron self-energy $\Sigma^r(\omega)$ in regard to band states $\lambda = (j, \mathbf{q})$ where $j$ is the band index, $\mathbf{q}$ is the crystal momentum, we obtain a convenient formula for the UPS analyses excited from normal states [25]

$$I(\mathbf{k}) \propto f(\varepsilon_k - \omega)\sum_{\lambda}\frac{| < f_{\mathbf{k}}^{-}|\Delta|\phi_\lambda > |^2\Gamma_\lambda}{(\varepsilon_k - \omega - \varepsilon_\lambda - \Delta_\lambda)^2 + \Gamma_\lambda^2}, \tag{13.26}$$

Here we have defined $\Delta_\lambda$ and $\Gamma_\lambda \quad (> 0)$ by

$$< \lambda|\Sigma^r(\varepsilon_k - \omega)|\lambda >= \Delta_\lambda - i\Gamma_\lambda. \tag{13.27}$$

From ARPES analyses we obtain information on $\varepsilon_\lambda + \Delta_\lambda$ and $\Gamma_\lambda$ as suggested from Eq. (13.26). In case of photoemission from near Fermi level, the electron–phonon mediated electron self-energy $\Sigma_{ph}^r$ plays an important role: Direct information on the electron–phonon interaction can be provided by UPS analyses: They have observed prominent band bending near Fermi level caused by the electron–phonon interaction [24].

In order to calculate the amplitude $< f_{\mathbf{k}}^{-}|\Delta|\phi_\lambda >$, the initial one-electron state $\phi_\lambda$ is assumed to be written as linear combination of atomic orbitals (tight-binding approximation):

$$\phi_\lambda(\mathbf{r}) = \sum_{\alpha} c_{\lambda\alpha}\chi_\alpha(\mathbf{r} - \mathbf{R}_\alpha)$$

where $\chi_\alpha$ is $\alpha$th atomic orbital centered on the site $\mathbf{R}_\alpha$, $c_{\lambda\alpha}$ is the expansion coefficient. For the practical calculations, one can use Bloch function for $f_{\mathbf{k}}^{-}$ [26]. We rather prefer to using real space expression to extend the applicability even to disordered systems.

When the potential is given as the sum of nonoverlapping atomic potentials, $V = \sum_{\alpha} v_\alpha$, we have an expression for the total $T$-matrix expanded in terms of the site $T$-matrix $t_\alpha$ and damping free propagator $g_0$

$$g_0(\varepsilon) = \frac{1}{\varepsilon - T_e + i\Gamma}, \tag{13.28}$$

$$
\begin{aligned}
T &= V + Vg_0V + Vg_0Vg_0V + \ldots \\
&= \sum_\alpha (v_\alpha + v_\alpha g_0 v_\alpha + \ldots) \\
&+ \sum_{\alpha\neq\beta} (v_\beta + v_\beta g_0 v_\beta + \ldots) g_0 (v_\alpha + v_\alpha g_0 v_\alpha + \ldots) + \ldots \\
&= \sum_\alpha t_\alpha + \sum_{\beta\neq\alpha} t_\beta g_0 t_\alpha + \ldots
\end{aligned}
\tag{13.29}
$$

In Eq. (13.28) $\Gamma$ is the imaginary part of the optical potential and is usually approximated by a positive constant. Here we have defined the site $T$-matrix $t_\alpha$ at site $\alpha$

$$
t_\alpha = v_\alpha + v_\alpha g_0 v_\alpha + \ldots = v_\alpha + v_\alpha g_0 t_\alpha .
$$

The photoemission amplitude $M_\lambda(\mathbf{k})$ is thus given by

$$
\begin{aligned}
M_\lambda(\mathbf{k}) = &< f_\mathbf{k}{}^-|\Delta|\phi_\lambda >= \sum_\alpha c_{\lambda\alpha} < f_\mathbf{k}^-|\Delta|\chi_\alpha >= \sum_\alpha c_{\lambda\alpha}\Big[ < \phi_\mathbf{k}^{\alpha-}|\Delta|\chi_\alpha \\
& > + \sum_{\beta(\neq\alpha)} < \phi_\mathbf{k}^0|t_\beta g_0 t_\alpha \Delta|\chi_\alpha > + \sum_{\gamma\neq\beta(\neq\alpha)} \\
& < \phi_\mathbf{k}^0|t_\gamma g_0 t_\beta g_0 t_\alpha \Delta|\chi_\alpha > + \ldots.
\end{aligned}
\tag{13.30}
$$

We have used the photoelectron wave function $\phi_\mathbf{k}^{\alpha-}$ only for the atomic potential $v_\alpha$, which is related to the plane wave $\phi_\mathbf{k}^0$

$$
< \phi_\mathbf{k}^{\alpha-}| =< \phi_\mathbf{k}^0|(1 + t_\alpha g_0).
$$

The first term in the large parenthesis of Eq. (13.30) describes the direct photoemission amplitude without suffering elastic scatterings from neighboring atoms, the second term describes the single elastic scatterings from surrounding atoms, and so on. In the UPS region the electron scatterings from surrounding atoms are strong enough, so that infinite sum over the multiple scatterings is inevitable. For simplicity muffin-tin potentials will be used, but the extension to the full potential is straightforward [27]. For the linear polarization parallel to $z$-axis, the renormalized multiple scattering formula for the amplitude $M_\lambda(\mathbf{k})$ is now given by [28]

$$
M_\lambda(\mathbf{k}) = \sum_{\alpha\beta} c_{\lambda\alpha} \exp(-i\mathbf{k}\cdot\mathbf{R}_\beta) \sum_{LL'} Y_{L'}(\hat{\mathbf{k}}) \big[(1-X)^{-1}\big]^{\beta\alpha}_{L'L} M_{LL_\alpha},
\tag{13.31}
$$

$$
X^{\alpha\alpha'}_{LL'} = t_l^\alpha(k) G_{LL'}(k\mathbf{R}_\alpha - k\mathbf{R}_{\alpha'})(1 - \delta^{\alpha\alpha'}).
\tag{13.32}
$$

where the angular momentum representation of the site $T$-matrix $t_\alpha$ is given by use of the phase shifts $\delta_l^\alpha$ at the site $\alpha$ with orbital angular momentum $l$

$$
t_l^\alpha(k) = -\frac{\exp[2i\delta_l^\alpha(k)] - 1}{2ik}.
$$

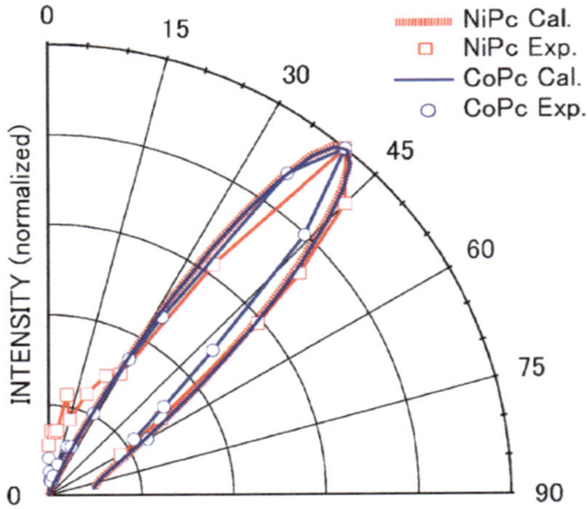


**Fig. 13.2** Theoretical and observed photoemission angular distribution from HOMO's of NiPc and CoPc [29]. (Copyright permission)

The phase shifts $\delta_l^\alpha$'s are sensitive to the electronic structure on the atom $\alpha$ in particular in low energy region, and the propagator $G_{LL'}(k\mathbf{R}_\alpha - k\mathbf{R}_\beta)$ reflects the atomic arrangement. The square matrix $X$ is labeled by a set of atomic sites $(\alpha, \beta, \ldots)$ and angular momentum $L = (l, m)$ whose matrix dimension is $N(l_{max} + 1)^2$ for the cluster of $N$ atoms and maximum angular momentum $l_{max}$. In UPS region $l_{max}$ is only 3–5, so that the full multiple scattering calculations are practical even for quite large clusters. The full multiple scattering is taken into account by use of the inverse matrix $(1 - X)^{-1} = 1 + X + X^2 + X^3 + \ldots$. In the electric dipole approximation, the photoexcitation matrix element $M_{LL_\alpha}$ excited by linearly polarized light parallel to the $z$-axis is given by

$$\begin{aligned} M_{LL_\alpha} &= \sqrt{\frac{2}{\pi}} i^{-l} \exp(i\overset{\alpha}{\underset{l}{\delta}}) \rho(l)_\alpha G(L_\alpha 10|L), \\ \rho(l)_\alpha &= \int R_l(kr) R_{l_\alpha}(r) r^3 dr, \end{aligned} \tag{13.33}$$

where $R_l(kr)$ and $R_{l_\alpha}(r)$ describe the radial part of the photoelectron wave function and composite occupied atomic orbital $\chi_\alpha(\mathbf{r}) = R_{l_\alpha}(r) Y_{L_\alpha}(\widehat{\mathbf{r}})$. The Gaunt's integral $G(LL'|L'')$ is defined and calculated by use of Clebsch–Gordan coefficients,

$$\begin{aligned} G(LL'|L'') &= \int \overset{*}{\underset{L''}{Y}}(\widehat{\mathbf{r}}) Y_{L'}(\widehat{\mathbf{r}}) Y_L(\widehat{\mathbf{r}}) d\widehat{\mathbf{r}} \\ &= \sqrt{\frac{(2l+1)(2l'+1)}{4\pi(2l''+1)}} < l0l'0|l'' >< LL'|L'' > . \end{aligned}$$

Shang et al. have applied the above formulas to the angular dependence of UPS spectra excited from the highest occupied molecular orbitals (HOMO) of NiPc and CoPc (Pc = phthalocyanine) [29]. The HOMO's ($a_{1u}$) are in-plane orbitals dominated by carbon $2p$ atomic orbitals in Pc ligands. Figure 13.2 shows the calculated angular distribution of photoelectrons excited from the HOMO's of NiPc and CoPc compared with experimental angle resolved UPS spectra. All of the data shown here have been normalized at the maximum. As we can see the maximum intensity locates at around 40° for both of the calculated and experimental results of NiPc and CoPc. The observed and calculated angular distribution are in good agreement except for satellites locating in the range $0° < \theta < 30°$. This result demonstrates the success of the multiple scattering calculations for these complicated systems. The Debye–Waller factor which should have influence on the angular distribution can be neglected in UPS spectra as shown by Eq. (13.20).

### 13.3.2 High-Energy Photoelectron Emission

In these days high-energy XPS measurements have been employed to study bulk electronic structures [30, 31]. High-energy XPS spectra provide new information, however, they raise some questions about recoil effects and breakdown of the electric dipole approximation. The former phenomena will be discussed later. The latter problem has been studied in photoemission from free atoms such as Ne, Ar and Kr from experimental [32, 33] and theoretical sides [34]. A general discussion is given for condensed systems [35].

In principle the multiple scattering formula (13.31) can be applied to high-energy XPS analyses. We however, use an alternative approach because of the two reasons. First in high-energy region, $l_{max}$ amounts to about 50. For a cluster composed of 100 atoms, the dimension of the multiple scattering matrix $(1-X)^{-1}$ is about 250,000. This matrix inversion is too time consuming. Second reason is the strong suppression of the interference due to the Debye–Waller factor. As shown by Eq. (13.20) it is quite small in the high-energy region. Furthermore the ejected photoelectrons are weakly scattered from surrounding atoms in particular from light elements as H, C, N and O. We thus have a reasonable approximation to pick up only a few terms in the expansion shown in Eq. (13.30). The lowest order approximation

$$M_\lambda(\mathbf{k}) \approx \sum_\alpha c_{\lambda\alpha} < \phi_{\mathbf{k}}^{\alpha-} |\Delta| \chi_\alpha >= \sum_\alpha c_{\lambda\alpha} \exp(i\mathbf{Q}\cdot\mathbf{R}_\alpha) \sum_L Y_L(\widehat{\mathbf{k}}) M_{LL_\alpha} \quad (13.34)$$

is sometimes called “the independent atom approximation”, which gives a formula to describe the high-energy photoemission intensity

$$I_\lambda(\mathbf{k}) \propto |M_\lambda(\mathbf{k})|^2 \approx I_\lambda^1(\mathbf{k}) + I_\lambda^2(\mathbf{k}). \quad (13.35)$$

In the above formula, $I_\lambda^1(\mathbf{k})$ and $I_\lambda^2(\mathbf{k})$ are the one- and two-center terms: The former describes the photoemission processes excited from the same atomic sites, whereas the latter describes the interference excited from the different sites because the initial states are delocalized over the systems. The explicit formulas are shown by [35]

$$I^1_\lambda(\mathbf{k}) = \mathrm{Re}\left[\sum_{\alpha\alpha'} c^*_{\lambda\alpha} c_{\lambda\alpha'} \sum_{LL'} Y^*_L(\widehat{\mathbf{k}}) Y_{L'}(\widehat{\mathbf{k}}) M^*_{LL_\alpha} M_{L'L_{\alpha'}}\right], \tag{13.36}$$

$$I^2_\lambda(\mathbf{k}) = 2\mathrm{Re}\Big[\sum_{\alpha\neq\beta} c^*_{\lambda\alpha} c_{\lambda\beta} \exp[i\mathbf{Q}\cdot(\mathbf{R}_\alpha - \mathbf{R}_\beta)] \times \sum_{LL'} Y^*_L(\widehat{\mathbf{k}}) Y_{L'}(\widehat{\mathbf{k}}) M^*_{LL_\alpha} M_{L'L_\beta}\Big]. \tag{13.37}$$

The atomic orbitals $\chi_\alpha$ and $\chi_{\alpha'}$ are on the same atomic site $\alpha$ in Eq. (13.36), whereas $\chi_\alpha$ and $\chi_\beta$ are on the different sites. The one-center term is not influenced by the nuclear vibrations, whereas the two-center term is strongly influenced by the vibrations, which gives rise to the Debye–Waller factors. With aid of the procedure discussed in Sect. 13.2.3, the two-center term $I^2_\lambda(\mathbf{k})$ should be replaced by

$$I^2_\lambda(\mathbf{k}) = 2\mathrm{Re}\Big[\sum_{\alpha\neq\beta} c^*_{\lambda\alpha} c_{\lambda\beta} \exp[i\mathbf{Q}\cdot(\mathbf{R}^0_\alpha - \mathbf{R}^0_\beta)]\exp[-Q^2\Delta^2_{\alpha\beta}2] \times \sum_{LL'} Y^*_L(\widehat{\mathbf{k}}) Y_{L'}(\widehat{\mathbf{k}}) M^*_{LL_\alpha} M_{L'L_\beta}\Big]. \tag{13.38}$$

where $\mathbf{R}^0_\alpha$ is now the equilibrium position of the $\alpha$th atom. The thermal factor $\Delta^2_{\alpha\beta}$ is the same as those used in the EXAFS analyses [see Eq. (13.20)]. In the high-energy region the thermal damping factor $\exp[-Q^2\Delta^2_{\alpha\beta}2]$ is typically small enough, which allows us to neglect the two-center term. The one-center term is well approximated by the well-known Gelius formula [35, 36]

$$I^1_\lambda(\mathbf{k}) = \sum_\alpha |c_{\lambda\alpha}|^2 \sigma_\alpha, \quad \sigma_\alpha = \frac{1}{\sqrt{4\pi}} \sum_L |M_{LL_\alpha}|^2, \tag{13.39}$$

where $\sigma_\alpha$ is proportional to the photoionization cross section of the $\alpha$th atomic orbital. The Gelius formula provides us with useful information about the electronic structure. The ARPES analyses in UPS region give us information on $\varepsilon_\lambda + \Delta_\lambda$ and $\Gamma_\lambda$ as functions of crystal momentum $\mathbf{q}$. On the other hand XPS analyses based on the Gelius formula (13.39) give us the information about the net charge $|c_{\lambda\alpha}|^2$ on the $\alpha$th atomic orbital in $\lambda$th crystal orbital ($j$th band with crystal momentum $\mathbf{q}$). The turnover from the ARPES region to the X-ray photoelectron diffraction (XPD) region is expected with increase of photon energy [25].

The one-center term is not influenced by the Debye–Waller factor, but the recoil effects have some finite contribution. In Eq. (13.34) we notice that $\mathbf{R}_\alpha = \mathbf{R}^0_\alpha + \mathbf{u}_\alpha$. We should carefully treat $\mathbf{u}_\alpha$ for the thermal average; $\mathbf{u}_\alpha$ is q-number. This extension allows us to discuss recoil effects in high-energy photoemission from deep core levels [37–39] and from extended levels [39]. In the approximation where the Franck–Condon effect is neglected, the Gelius formula is changed to include the recoil effects which give rise to the temperature independent recoil shift $\delta\varepsilon_\alpha$ and the temperature dependent broadening $F_\alpha$; both are dependent on the recoil atom $\alpha$. It is explicitly shown

$$I_\lambda^1() = \sum_{p\subset\alpha}\sigma_p|c_{\lambda p}|^2 g_\alpha(-\varepsilon_\lambda - \Delta_\lambda - \delta\varepsilon_\alpha), \tag{13.40}$$

$$g_\alpha(\varepsilon) = \sqrt{\frac{2\pi}{F_\alpha}}\exp\Big(-\frac{\varepsilon^2}{2F_\alpha}\Big). \tag{13.41}$$

$$\delta\varepsilon_\alpha = -Q^2 2M_\alpha. \tag{13.42}$$

The recoil shift of $\alpha$th atom $\delta\varepsilon_\alpha$ is of course dependent on the mass $M_\alpha$: Heavy atoms are not influenced by the recoil. We should note that isotope effects can be incorporates with aid of the recoil effects in photoemission spectra.

Suga et al. have measured valence and core photoemission spectra from $V_3Si$ and observed prominent recoil shifts [40]. There the formula (13.40) is successfully applied. Organic solids are composed of mainly light elements like C and H. Thus we can expect recoil effects even in soft X-ray region. Shang et al have applied the formula (13.40) to valence band photoemission from some organic molecular solids to study the recoil effects [41, 42]. XPS spectra for the valence bands are calculated for abb-trifluorostylene [42]. Figure 13.3 shows the calculated XPS spectra with and without the recoil effects for soft and hard X-rays, which are normalized to the maximum height. Prominent recoil shifts are observed for hard X-rays. Recoil spectra enhances the light composite atoms, which gives further check for the net population on those atoms [40]. Furthermore with aid of the recoil shifts, for example, contribution by 1*s* of H and D to each band could be separated out in high-resolution soft-Xray region XPS spectra.

## 13.4 Concluding Remarks

Two different theoretical approaches are shown in this article, many-body scattering theory and quantum field theory by use of nonequilibrium Green's functions. Both have their merits and demerits, and yield quite similar theoretical formula to describe the photoemission processes. In order to discuss phonon effects at finite temperature the latter approaches have some advantages: We can naturally include polaron effects on UPS spectra. It is important to notice that we measure electron self-energy $<\lambda|\Sigma|\lambda>$ in addition to the band energy spectra $\varepsilon_\lambda$ from ARPES analyses. Some experimental works have succeeded in measuring $\Delta_\lambda = \mathrm{Re} <\lambda|\Sigma|\lambda>$ caused by the electron–phonon interaction near Fermi levels [22, 24].

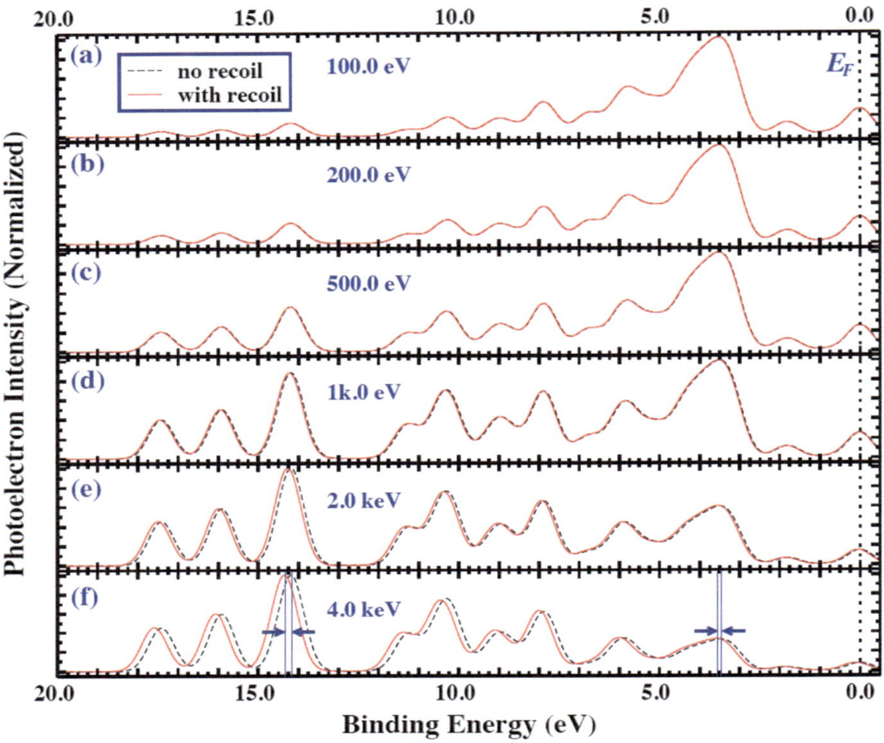


**Fig. 13.3** Calculated XPS spectra from valence band of the abbtrifluorostylene excited by X-ray from 100 eV to 4.0 KeV [42]. (Copyright permission)

## References


1. S. Hüfner, *Photoemission Spectroscopies and Applications*, 3rd edn. (Springer, New York, 2003)
2. N. Ueno, S. Kera, Prog. Surf. Sci. **83**, 490 (2008)
3. I.B. Bersuker, *The Jahn-Teller Effect* (Cambridge University Press, Cambridge, 2006)
4. W. Bardyszewski, L. Hedin, Physica Scripta **32**, 439 (1985)
5. L. Hedin, J. Michiels, J. Inglesfield, Phys. Rev. B **58**, 15565 (1998)
6. L. Hedin, in *Solid State Photoemission and Related Methods*, ed. by W.A. Schattke, M.A. Van Hove (Wiley-VCH, Weinheim, 2003), pp. 116–140
7. C. Caroli, D. Leder-Rozenblatt, B. Roulet, D. Saint-James, Phys. Rev. B **8**, 4552 (1973)
8. C.-O. Almbladh, Physica Scripta **32**, 341 (1985)
9. T. Fujikawa, H. Arai, J. Electron Spect. Relat. Phenom. **123**, 19 (2002)
10. T. Fujikawa, H. Arai, Chem. Phys. Lett. **368**, 147 (2003)
11. T. Fujikawa, H. Arai, J. Electron Spect. Relat. Phenom. **149**, 61 (2005)
12. T. Fujikawa, H. Arai, J. Electron Spect. Relat. Phenom. **136**, 85 (2004)
13. T. Fujikawa, L. Hedin, Phys. Rev. B **40**, 11507 (1989)
14. D.F. DuBois, *Lecture in Theoretical Physics*, ed. by W.E. Brittin (Gordon and Breach, New York, 1967), pp. 469–619

15. D.C. Langreth, *Linear and Nonlinear Transport in Solids*, ed. by J. Devreese, V.E. van Doren (Plenum, New York, 1976), 3 p
16. H. Arai, T. Fujikawa, Phys. Rev. B **72**, 075102 (2005)
17. J. Osterwalder, T. Greber, S. Hüfner, L. Schlapbach, Phys. Rev. Lett. **64**, 2683 (1990)
18. T. Fujikawa, H. Arai, *XAFS13*, ed. by E. Hedman, P. Pianetta. AIP-Conference Proceedings, vol. CP882, p. 75 (American Institute of Physics, Melville, 2007)
19. G. Baym, Ann. Phys. **14**, 1 (1962)
20. L. Hedin, S. Lundqvist, *Solid State Physics*, vol. 23, ed. by F. Seitz, D. Turnbull, H. Ehrenreich (Academic, New York, 1969), p. 1
21. N.D. Mermin, J. Math. Phys. **7**, 1038 (1966)
22. J. Kröger, Rep. Prog. Phys. **69**, 899 (2006)
23. H. Arai, N. Ueno, T. Fujikawa, *XAFS13*, ed. by E. Hedman, P. Pianetta. AIP Conference Proceedings, vol. CP882, p. 108 (American Institute of Physics, Melville, 2007)
24. T. Cuk, D.H. Lu, X.-J. Zhou, X.-J. Shen, T.P. Devereaux, N. Nagaosa, Phys. Status Solidi (b) **242**, 11 (2005)
25. T. Fujikawa, H. Arai, J. Electron Spect. Relat. Phenom. **174**, 85 (2009)
26. J. Igarashi, K. Hirai, Phys. Rev. B **54**, 17820 (1994)
27. K. Hatada, K. Hayakawa, M. Benfatto, C.R. Natoli, J. Phys. Condens. Matter **22**, 185501 (2010)
28. T. Fujikawa, J. Phys. Soc. Jpn. **50**, 1321 (1981)
29. M. Shang, M. Nagaosa, S. Nagamatsu, S. Hosoumi, S. Kera, T. Fujikawa, N. Ueno, J. Electron Spect. Relat. Phenom. **184**, 261 (2011)
30. A. Sekiyama, S. Suga, Physica B **312–313**, 634 (2002)
31. A. Sekiyama, S. Suga, J. Electron Spect. Relat. Phenom. **137–140**, 681 (2004)
32. M.O. Krause, Phys. Rev. **177**, 151 (1969)
33. O. Hemmers, R. Guillemin, D.W. Lindle, Rad. Phys. Chem. **70**, 123 (2004)
34. J.W. Cooper, Phys. Rev. A **47**, 184 (1993)
35. T. Fujikawa, R. Suzuki, H. Arai, H. Shinotsuka, L. Kövér, J. Electron Spect. Relat. Phenom. **159**, 14 (2007)
36. U. Gelius, K. Siegbahn, J. Chem. Soc. Faraday Discussion **54**, 257 (1972)
37. R. Suzuki, H. Arai, H. Shinotsuka, T. Fujikawa, e-J. Surf. Sci. Nanotech. **3**, 373 (2005)
38. T. Fujikawa, R. Suzuki, L. Kövér, J. Electron Spect. Relat. Phenom. **151**, 170 (2006)
39. T. Fujikawa, H. Arai, R. Suzuki, H. Shinotsuka, L. Kövér, N. Ueno, J. Electron Spect. Relat. Phenom. **162**, 146 (2008)
40. S. Suga, S. Itoda, A. Sekiyama, H. Fujiwara, S. Komori, S. Imada, M. Yabashi, K. Tamasaka, A. Higashiyama, T. Ishikawa, M. Shang, T. Fujikawa, Phys. Rev. **B86**, 035146 (2012)
41. M. Shang, T. Fujikawa, N. Ueno, e-J. Surf. Sci. Nanotech. **10**, 128 (2012)
42. M. Shang, T. Fujikawa, N. Ueno, Anal. Chem. **85**, 3739 (2013)

# Chapter 14
# Theory of Metal-Atom Diffusion in Organic Systems

**Yoko Tomita and Takashi Nakayama**

## 14.1 Introduction

Organic semiconductors are promising materials for future optical and electronic devices, such as light emitting device (LED), solar cells, and field effect transistor (FET) due to their fascinating characters like light-mass, softness, and flexible design of electronic structures by molecular replacement. During ordinary gas or liquid phase growth of organic solids using coupling reactions, various impurity atoms, such as N, O, Pd, Pt, and Ni atoms, are often incorporated into semiconductor films from source and catalytic materials (For example, [1]). On the other hand, in device structures, metal atoms in electrodes, such as Al and Au atoms, easily diffuse from electrode interfaces into semiconductor films [2, 3]. These impurity atoms are believed to move easily in semiconductors under heating and voltage applications, which changes semiconducting properties and finally promotes the degradation of device performance [4]. In case of inorganic semiconductors like Si, the behavior of atomic impurities is well understood [5]. However, in cases of organic systems, our knowledge is still limited. In this article, based on our recent theoretical calculations [6–8], we explain basic properties of metal-atom thermal diffusion in $\pi$-conjugated organic systems. There exist millions of organic semiconducting systems and each system has unique crystal structure and fascinating physical properties. Here, we focus on common features among these systems because various impurity atoms show similar diffusion properties, as shown below,

Y. Tomita (✉)
Graduate School of Pure and Applied Sciences, University of Tsukuba, 1-1-1 Tennoudai, Tsukuba, Ibaraki 305-8571, Japan

Japan Science and Technology Agency, CREST, 5 Sanbancho, Chiyoda, Tokyo 102-0077, Japan
e-mail: cosmic-eye@phys8.s.chiba-u.ac.jp

T. Nakayama
Department of Physics, Chiba University, 1-33 Yayoi Inage, Chiba 263-8522, Japan



H. Ishii et al. (eds.), *Electronic Processes in Organic Electronics*,
Springer Series in Materials Science 209, DOI 10.1007/978-4-431-55206-2_14

even in quite different carbonic systems; in graphene sheets (two-dimensional metal), polyacetylene bundles (one-dimensional small-gap semiconductor), and in oligoacene molecular solids (naphthalene and pentacene, zero-dimensional semiconductors). We hope that the present results help us to develop reliable organic devices.

## 14.2 Atomic Impurity States in Organic Solid

Before considering the metal-atom diffusion, we briefly explain electronic structures of metal atoms in organic systems [7, 9]. Figure 14.1a shows the schematic energy levels of various atomic impurities around the band gap of the model organic solid [9, 10], which were calculated by the first-principles calculation based on the density functional theory. The model solid is made of naphthalene molecules with similar thin-film crystal structure to pentacene film [7], as shown in Fig. 14.1b. Impurity atoms are assumed incorporated into the model solid as interstitial atoms, especially around the edge of naphthalene molecule, which is basically different from the cases of inorganic semiconductors with substitutional impurities. This difference occurs because most of organic systems are made of strong covalent carbon skeletons and condense by a weak van der Waals interaction, while inorganic systems condense by a strong covalent bonding.

As shown in Fig. 14.1a, the impurity states are roughly classified into four groups, A to D. (A): Cation metal atoms having small electronegativity, such as Al, In, and Mg, bond to a host molecule by an ionic-like interaction and produce donor-like states. This is because the HOMO (highest occupied molecular orbital) states of metal atoms, e.g., Al-p states, appear above the LUMO (lowest unoccupied molecular orbital) states of host molecules, i.e., C $\pi^*$ states, and the electron transfer occurs, as shown in Fig. 14.2a, from a metal atom to host molecules, thus metal atoms being partially ionized. (B): d-orbital metal atoms, such as Au, Pt, Ni, Ti, and W, produce s and d-orbital deep-level states around the center of HOMO–LUMO gap of molecules and bound to host molecules with a hybridization-induced weak covalent-like bond. Figure 14.2b shows the electron density near Au, where the weak covalent bond can be seen.

(C): Anion atoms such as N and P produce covalent bonds with molecules. However, since the number of valence electrons is odd, there appear localized half-occupied acceptor-like states within the HOMO–LUMO gap. (D): Anion atoms having large negativity, such as O and S, produce strong covalent bonds with molecules. Reflecting large negativity, bonding states are located far below the HOMO state, thus there appearing no impurity states within the HOMO–LUMO gap.

It is interesting to note that group-A and C atoms work as donors and acceptors, quite contrary to the case of inorganic Si. Such difference originates from the location of impurity atoms; substitutional vs. interstitial, and the large electronegativity of carbon atom.

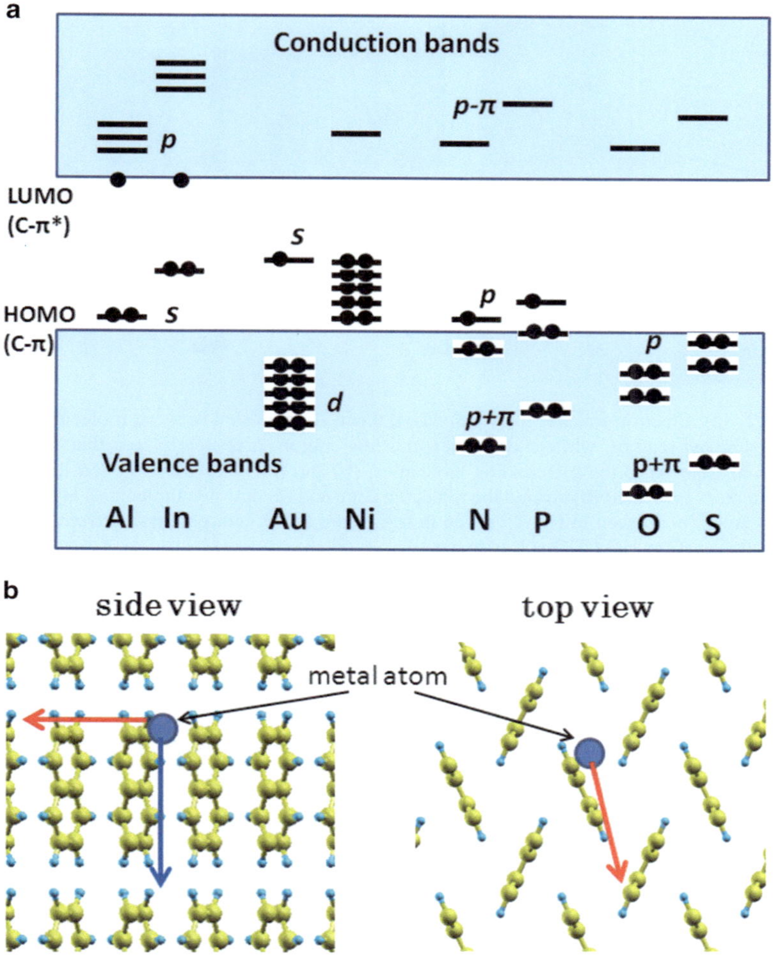


**Fig. 14.1** (**a**) Schematic picture of energy-level positions of various doped impurity atoms around the HOMO–LUMO gap of model oligoacene solid, obtained by the first-principles calculations. Main orbital characters of energy levels are shown in *italic*, while *solid dots* indicate the electron occupation, e.g., p-orbital electrons of Al and In are transferred to the LUMO states of host molecule. (**b**) *Side* and *top* views of oligoacene model solid used in the present first-principles calculations. *Yellow* (large) and *blue* (small) *balls* indicate carbon and hydrogen atoms, respectively, while *blue solid ball* (largest) is a doped metal atom. The diffusion paths along and perpendicular to molecular axis are schematically indicated by *blue* and *red arrows*, respectively

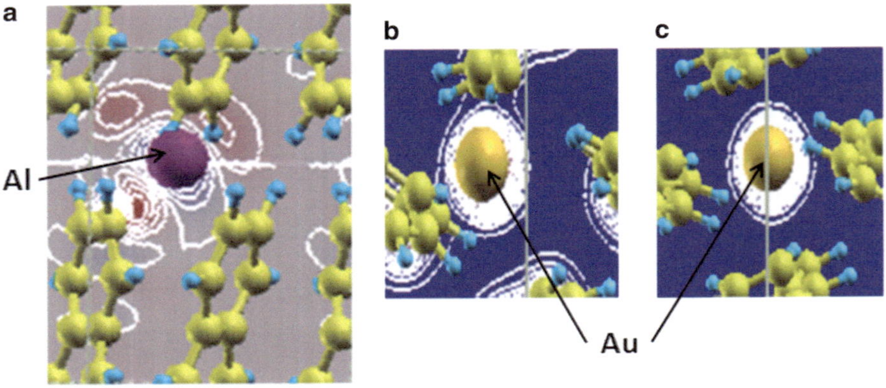


**Fig. 14.2** (**a**): Electron-transfer map (*side view*) when Al is located between molecules. Electron increases in *red* regions, while it decreases in a *blue* region. It is clearly seen that electrons are transferred from Al to the surrounding molecules. (**b**) and (**c**): Total electron density (*top view*) when Au atom is located (**b**) around the molecule edge and (**c**) between molecules. Hybridization-induced Au–C bond seen in (**b**) is broken in (**c**). Reproduced with permission from Tomita and Nakayama [7] (Copyright © 2012, Elsevier)

## 14.3 Adiabatic Potential and Atom Diffusion

In this section, we explain the theoretical background of adiabatic diffusion potential [11]. We consider the system with $N_n$ nuclei and $N_e$ electrons. Atomic nucleus, A, has the mass $M_A$, the charge $Z_A$, and the coordinate $\mathbf{R}_A$, while electrons have the coordinates, $\mathbf{r}_1, \ldots$. In this case, the Hamiltonian is written as follows;

$$H = -\sum_{A}\frac{\hbar^2}{2M_A}\nabla_A^2 - \sum_{i}\frac{\hbar^2}{2m}\nabla_i^2 - \sum_{i,A}\frac{Z_A e^2}{|\mathbf{r}_i - \mathbf{R}_A|} + \sum_{i<j}\frac{e^2}{|\mathbf{r}_i - \mathbf{r}_j|} + \sum_{A<B}\frac{Z_A Z_B e^2}{|\mathbf{R}_A - \mathbf{R}_B|}. \quad (14.1)$$

Assuming that all nuclei are stationary, the Hamiltonian of electron system is simply written as

$$H_e = -\sum_{i}\frac{\hbar^2}{2m}\nabla_i^2 - \sum_{i,A}\frac{Z_A e^2}{|\mathbf{r}_i - \mathbf{R}_A|} + \sum_{i<j}\frac{e^2}{|\mathbf{r}_i - \mathbf{r}_j|} + \sum_{A<B}\frac{Z_A Z_B e^2}{|\mathbf{R}_A - \mathbf{R}_B|}. \quad (14.2)$$

So the Schrödinger equation of the electron system can be written as follows;

$$H_e\Phi_\mu(\mathbf{R_A},\cdots,\mathbf{r}_1,\cdots) = W_\mu(\mathbf{R_A},\cdots)\Phi_\mu(\mathbf{R_A},\cdots,\mathbf{r}_1,\cdots). \quad (14.3)$$

It should be noticed that both the eigenvalue and the eigen wavefunction have the quantum number $\mu$, and depend on $\mathbf{R_A},\cdots$ as parameters. Since the electron wave-functions $\Phi_{\mu}$ produce the complete set, the total wave function of nuclei and electrons is generally written as

$$\Psi(\mathbf{R_A},\cdots,\mathbf{r}_1,\cdots)=\sum_{\mu}F_{\mu}(\mathbf{R_A},\cdots)\Phi_{\mu}(\mathbf{R_A},\cdots,\mathbf{r}_1,\cdots), \tag{14.4}$$

where $F_{\mu}(\mathbf{R_A},\cdots)$ corresponds to the expansion coefficient. Inserting this equation into Schrödinger equation, $H\Psi=E\Psi$, and adopting the Born–Oppenheimer adiabatic approximation: $|\nabla_A F_{\mu}|\gg|\nabla_A \Phi_{\mu}|$, we obtain the equation of $F_{\mu}$ for each $\mu$ independently as

$$\left[-\sum_{A}\frac{\hbar^2}{2M_A}\nabla_A^2+W_{\mu}(\mathbf{R_A},\cdots)\right]F_{\mu}(\mathbf{R_A},\cdots)=EF_{\mu}(\mathbf{R_A},\cdots). \tag{14.5}$$

The above approximation is often justified because the mass of an atomic nucleus is $10^4$–$10^5$ times heavier than that of an electron and thus the wavefunction of electron is more extended than that of nucleus. (In case of light nucleus like H and He or in case when more than two adiabatic potentials approach each other, this approximation does not work and we have to consider non-adiabatic (diabatic) processes.) Since $W_{\mu}(\mathbf{R_A},\cdots)$ acts as some effective potential for the movement of nuclei in Eq. (14.5) and this potential reflects a single electronic state $\mu$ corresponding to a single electron occupation with a constant entropy, $W_{\mu}(\mathbf{R_A},\cdots)$ is called an adiabatic potential. Moreover, when we consider the nuclear motion around room temperature, the ground state ($\mu=g$) is sufficient, which is the situation in the following sections. (Even in case of metallic systems having continuous electronic states, $\mu$s, one can often define an effective single adiabatic potential.)

In case of uniform solids, $W_{\mu}(\mathbf{R_A},\cdots)$ gives the coherent motion of nuclei, i.e., the phonon. In case of diffusion motion of impurity atom, on the other hand, one can obtain the adiabatic diffusion potential of impurity atom $W_g(\mathbf{R})$ as a function of only a coordinate of impurity atom, $\mathbf{R}$, by calculating the total energy of electronic ground state with relaxing all the positions of atoms around the impurity atom. Figure 14.3 shows the schematic illustration of the adiabatic potential for the diffusion process of an impurity atom from $\mathbf{R}_0$ to $\mathbf{R}_2$, where an impurity atom is denoted by a solid circle in a lower panel. The diffusion path with minimum potential barriers is searched by various numerical methods.

The adiabatic potential for an impurity atom is difficult to be observed directly in experiments. However, an impurity atom, which is weakly bounded to other host atoms, can move its position in the potential by receiving the thermal random forces caused by phonons etc. This movement is detected experimentally as the atom diffusion in solids. The diffusion is normally divided by a series of atom movements and each movement is described by the transition probability between stable atom positions as the Arrhenius formula [12], so far as the concentration of impurities is low. Next, we derive this formula.

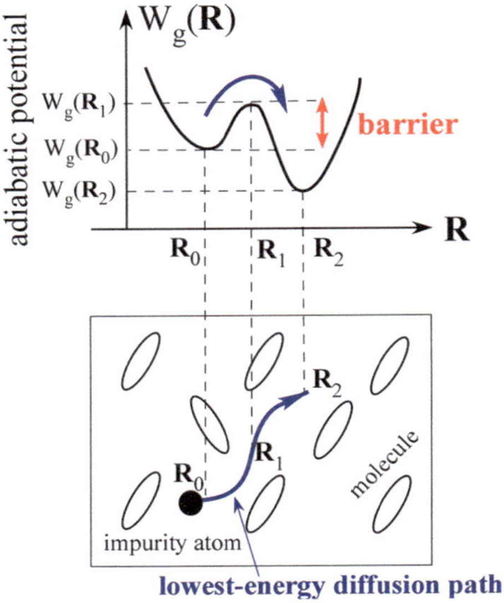


**Fig. 14.3** *Upper panel*: Schematic illustration of adiabatic potential for metal-atom diffusion from a minimum point $R_0$ to another minimum point $R_2$. The potential has a barrier at the saddle point $R_1$. *Lower panel*: Lowest energy diffusion path of metal atom in crystal. *Solid circle* and *open ellipses* indicate metal atom and host molecules, respectively

We consider the atom diffusion between stable positions, from $\mathbf{R}_0$ to $\mathbf{R}_2$, over the potential barrier as shown in Fig. 14.3 [11]. Since the potential has local minima around $\mathbf{R}_0$ and $\mathbf{R}_2$ and thus the existence probabilities around these points are large in the thermal equilibrium, the impurity atom stays for a certain amount of time around these points. On the other hand, since the de Broglie wave length of an atom (nucleus) is extremely small compared to the typical length of adiabatic potential variation, we can replace the quantum equation of impurity-atom motion in Eq. (14.5) as the classical equation. Moreover, we can assume that the diffusion occurs in one-dimensional direction along the leading lowest-energy path over a saddle point of potential. (In case of the diffusion in three-dimensional systems, there always exist saddle points between nearest potential-minimum points.)

In thermal equilibrium, the number of atoms, f (P, $\mathbf{R}$)dPdR, which are located within the volume dR near the position $\mathbf{R}$ and have the momentum P ~ P + dP, is given by the Boltzmann distribution as

$$f\,(P,\ \mathbf{R}) = \alpha \exp\left[-\beta\left\{\left(\frac{P^2}{2M}\right) + W(\mathbf{R})\right\}\right], \tag{14.6}$$

where α represents the normalization constant, M the mass of an impurity atom, and $\beta = 1/k_BT$. When we define dN/dt as the number of atoms that are moving from the valley point $\mathbf{R}_0$ to the other one $\mathbf{R}_2$ over the potential barrier at $\mathbf{R}_1$, this dN/dt is given by summing up impurity atoms with positive momentum as

$$\left(\frac{dN}{dt}\right) = \int_0^\infty dP\left(\frac{P}{M}\right) f(P,\ \mathbf{R}_1) = \frac{\alpha}{\beta}\exp(-\beta W(\mathbf{R}_1)). \tag{14.7}$$

On the other hand, assuming that the adiabatic potential $W_g(\mathbf{R})$ could be approximately rewritten around the valley $\mathbf{R}_0$ with one-dimensional variable R as $W(\mathbf{R}) = W(R_0) + \frac{M\omega^2}{2}(R - R_0)^2$, the total number of impurity atoms $N_0$ that is located around the valley $\mathbf{R}_0$ can be written as

$$N_0 = \int_{-\infty}^{\infty}\int_{R_0-d}^{R_0+d} dPdR\, f\,(P,\ R_1) \simeq \frac{2\pi\alpha}{\beta\omega}\exp(-\beta W(R_0)) \tag{14.8}$$

where d is the width of the valley, $\mathbf{R}_0$. Since the integrand in this equation rapidly reaches to zero when the position R leaves away from the center of the valley $R_0$, d was replaced by the infinity, $\infty$, to obtain the last equation.

Eliminating $\alpha$ with the use of Eqs. (14.7) and (14.8), we obtain the Arrhenius formula;

$$\frac{1}{N_0}\left(\frac{dN}{dt}\right) \simeq \frac{\omega}{2\pi}\exp[-\beta\{W(\mathbf{R}_1) - W(\mathbf{R}_0)\}] \tag{14.9}$$

This Arrhenius formula can physically understood as follows; the impurity atom in the valley $\mathbf{R}_0$ attacks the barrier with the frequency, $\omega/2\pi$, per unit of time, and it gets over the barrier with the probability of Boltzmann factor related to the barrier height, $W(\mathbf{R}_1)$–$W(\mathbf{R}_0)$.

When one writes the transition probability per unit time in Eq. (14.9) as k and the distance between $\mathbf{R}_0$ and $\mathbf{R}_2$ as L, the diffusion constant D between $\mathbf{R}_0$ and $\mathbf{R}_2$ is given by the Einstein relation as $D \simeq kL^2$. In this way, the behavior of atom diffusion can be characterized by the diffusion constant, which can be measured in experiments, and is theoretically understood by studying the variation of adiabatic potentials.

## 14.4 Diffusion between/on Graphene Sheets

Then, we consider the metal-atom diffusion between/on graphene sheets [6]. Figure 14.4a, b show the calculated diffusion potentials for Al and Au, respectively, while atom positions are displayed in Fig. 14.4c. We first consider the case of AA stacking of graphene sheets and the sheet separation of c = 7.52Å. It is seen that Al atoms are stable at the center of hexagonal carbon ring and prefer to diffuse across the C–C bonds as shown in Fig. 14.4c with a potential barrier around 0.4 eV. On the other hand, Au atom stabilizes on the carbon ring, i.e., at the A and E points, and prefers to move along the C–C bonds as shown in Fig. 14.4c with a small barrier around 0.05 eV. Similar results are obtained, for example, when we consider K and Pt metal atoms instead of Al and Au, respectively.

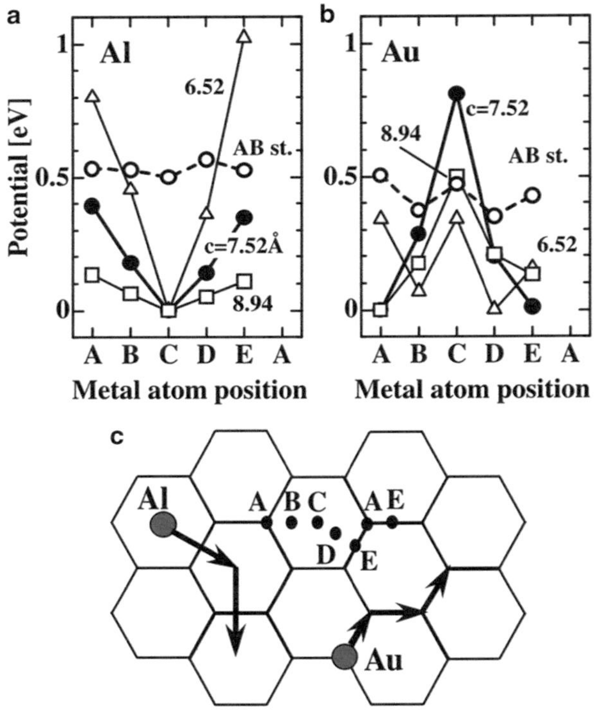


**Fig. 14.4** Calculated diffusion potentials between graphene sheets for (**a**) Al and (**b**) Au atoms, for various interlayer distances, c, and stackings, AA (*solid lines*) and AB (*dashed line*). The energies of the most stable points are taken to be zero. (**c**) Atom positions between graphene sheets (*dots*) and diffusion paths of Al and Au atoms (*arrows*). Reproduced with permission from Tomita and Nakayama [6] (Copyright (2010) The Japan Society of Applied Physics)

It is well known that, when alkali-metal atoms like K and Li are intercalated into graphite, the stacking changes from AB to AA and the metal atoms are located at the center of hexagonal carbon ring and produce the superstructures along the graphene layer such as $2 \times 2$ for $C_8K$ [13]. On the other hand, when Pt atoms are deposited on the graphite surface, they are located stable on the carbon atoms [8, 14]. The stable positions of Al and Au discussed above are in good agreement with these observations of K and Pt, respectively.

The reason why the stable position and diffusion behavior are quite different between Al and Au is clarified by examining the electronic structures. As explained in Sect. 14.2, since Al p-orbital states are located above the Fermi energy of graphene, i.e., the Dirac point, electron transfer occurs from Al atom to $\pi^*$-states of graphene sheets. Figure 14.5a shows the change of electron density, which is calculated as the difference of electron density between Al + graphene system and the sum of Al and graphene systems. It is clearly seen that the electron density decreases around Al atom, while it increases around the inner boundary of hexagonal carbon ring where $\pi^*$-states are located. This electron transfer produces the ionic coupling between Al and graphene sheets and thus stabilizes Al atom at the center of carbon ring. When Al atom is located on the carbon bond, the electron transfer from Al to graphene sheets decreases, indicating that the diffusion barrier is created by the breakdown of such ionic coupling.

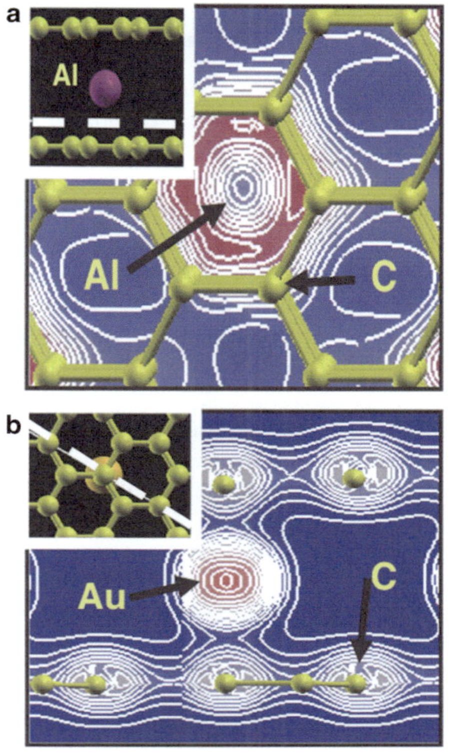


**Fig. 14.5** (**a**) Electron density change when Al atom is located at the center of hexagonal carbon ring. Density increases/decreases in *red*/*blue* regions. (**b**) Total valence electron distribution around Au between graphene sheets. *Dashed lines* in both *insets* show the positions of displaying planes. Reproduced with permission from Tomita and Nakayama [6] (Copyright (2010) The Japan Society of Applied Physics)

In case of Au, little electron transfer occurs between Au and graphene sheets because Au has larger electronegativity than Al, thus the Fermi energy being pinned to the original Dirac point. However, due to the hybridization between Au s + d and C $\pi$-bonding orbital states, a weak covalent-like bonding emerges, which is confirmed in the increase of electron density between Au and C atoms as shown in Fig. 14.5b. This is the reason why the Au atom is stable on the carbon ring and indicates that the breakdown of such bonding induces the potential barrier.

Next, we consider how the diffusion changes when the interlayer distance and the stacking of graphene sheets changes, which is also shown in Fig. 14.4a, b. When the interlayer distance increases to $c = 8.94$Å, corresponding to isolated graphene sheet, since metal atoms are bounded to a single graphene sheet, the barrier decreases to about half of the case of $c = 7.52$Å, around 0.2 eV for Al. When the distance decreases to $c = 6.52$Å, the barrier increases to around 0.8 eV for Al, while the stable position changes into inner points of carbon ring for Au. These changes occur due to the elastic energy loss by the approach of metal atoms to carbon atoms. On the other hand, when the graphene sheets become AB stacking (dashed lines), i.e., the staggered stacking, the diffusion potentials become almost flat with barriers around 0.1 eV. This is because metal atoms are bounded to graphene sheets on both sides, and thus the potential variation between the center and boundary of the carbon ring is averaged.

## 14.5 Diffusion in Polyacetylene Bundles

In this section, we consider the diffusion near the polyacetylene chain [6]. Figure 14.6a shows the calculated diffusion potentials of Al and Au, while Fig. 14.6b schematically displays their diffusion paths. Note that, although the Peierls transition is broken near metal atoms, A and C atom positions are not equivalent because metal atoms prefer to move on one side of zigzag chain and the other chain in the unit cell is located near the C point. It is seen that Al atom diffuses across the C–C bonds with a barrier around 0.60 eV, while Au atom moves along the C–C bonds with a barrier around 0.25 eV. We also found by the electronic-structure analysis that the origins of diffusion potentials are basically the same as the cases of graphene sheets.

Comparing the results in Sects. 14.4 and 14.5, we can conclude that the metal-atom diffusion near the $\pi$-conjugated C–C bonds has the common feature, not depending on the geometry of carbon skeletons such as ring or chain, but on the electronegativity of metal atoms.

## 14.6 Diffusion in Oligoacene Model Solid

Finally, we consider the metal-atom diffusion in the oligoacene model solid shown in Fig. 14.1b [7]. There exist two directions of the metal-atom diffusion; the diffusion along the molecule axis and the diffusion perpendicular to the molecule axis. Figure 14.7a shows the calculated diffusion potentials for Al and Au atoms as a function of atom positions along the molecular axis, i.e., the c-axis direction. Here, the energy of the most stable position, which is realized at the molecule edge, is set to zero and the c-axis coordinate of atom position is displayed in Fig. 14.7b.

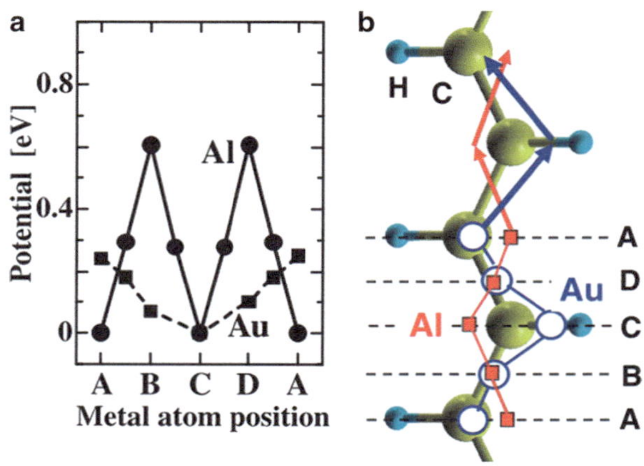


**Fig. 14.6** (**a**) Calculated diffusion potentials near polyacetylene chain for Al and Au atoms. (**b**) Schematic view of atom positions and diffusion paths of Al and Au. Reproduced with permission from Tomita and Nakayama [6] (Copyright (2010) The Japan Society of Applied Physics)

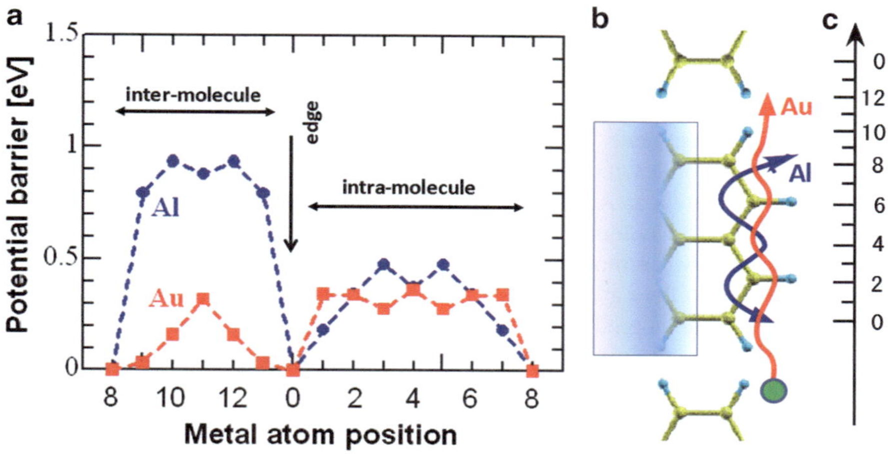


**Fig. 14.7** (**a**) Calculated adiabatic potentials for Al and Au atoms along the molecule axis, c, as a function of the metal-atom position. The c-axis coordinate of metal-atom position is displayed in (**b**). (**b**) Schematic illustration of diffusion paths for Al and Au atoms. The closest molecule is tilted to the displaying plane and is located in the left half side shown by a *gradated square*, which right edge corresponds to the edge of the closest molecule. Reproduced with permission from Tomita and Nakayama [7] (Copyright © 2012, Elsevier)

First, we consider the diffusion within the molecule; the intra-molecule diffusion. As seen in Fig. 14.7a, Al atom has the minimum energy at the molecule edge and the local minimum with higher energy around the center of molecule. On the other hand, Au atom has the almost flat potential height within the molecule and has the minimum energy only at the molecule edge. These features are explained by considering the diffusion paths, which is schematically shown in Fig. 14.7b. Both metal atoms avoid the center line of molecule and move along the right edge line of the molecule. This occurs because the wing of neighboring molecule approaches to the center line of the molecule and the space for metal atoms to diffuse remarkably reduces. It is also interesting to note that Al atom diffuses across the C–C bond in the right edge of molecule, while Au atom diffuses along the C–C bond in the same edge. Such difference comes from the difference of bonding character between Al and Au; the ionic bonding for Al and the covalent bonding for Au, similar to cases of graphene-sheet and polyacetylene-bundle systems.

Second, we explain why metal atoms prefer to locate at the molecule edge. The Al and Au atoms are located with smaller heights from the molecular plane, around 0.16 and 0.19 Å, respectively, at the molecule edges, compared to those at other inter-molecule positions, around 0.20 and 0.23 Å. This occurs because the atomic radius of H (0.4 Å) is smaller than that of C (0.77 Å) and thus the metal atom can approach to the molecule without receiving the repulsion from the molecule atoms. Reflecting such difference of atomic height, the attractive interaction between metal atom and molecule becomes largest around the molecule edge.

Third, we consider the diffusion between molecules; i.e., the inter-molecule diffusion. It is seen in Fig. 14.7a that Al has a large potential barrier around 0.9 eV, while Au has a barrier of about 0.3 eV. The origins of such barriers between molecules are again basically the same as those observed in cases of graphene sheets and polyacetylene bundles. Namely, even when Al is located between molecules, Al supplies its valence electrons to the surrounding molecules and realizes the ionization as $Al^{+}$, as shown in Fig. 14.2a. Since this ionization induces the ionic attractive Coulomb interaction between Al and molecules, the potential becomes large between molecules where the distance between Al and molecule is large, compared to the molecule edge. In case of Au, on the other hand, the Au-C weak covalent bond is produced as shown in Fig. 14.2b when Au is located around the molecule edge. However, when Au moves between molecules, such bond is broken as shown in Fig. 14.2c. In this way, the breaking of Au-C bond is the origin of barrier between molecules. Reflecting the difference of character between ionic and covalent bondings, the barrier of Al is larger than that of Au.

Next, we consider the metal-atom diffusion perpendicular to the molecular axis. Figure 14.8a shows the calculated adiabatic potentials for Al and Au atoms as a function of metal-atom position, while Fig. 14.8b schematically shows their diffusion paths viewed from the top c direction. Both metal atoms, Al and Au, show potential barriers around 0.6 eV at the positions, 2 and 6. On the other hand, the potentials have lower energies at 1 and 4 points because these points have large inter-molecule space to locate metal atoms and possess little repulsive strain energy from neighboring molecules. It is noted that Al and Au have almost the same potential variation, indicating that the repulsive interaction with neighboring molecules manifests the diffusion perpendicular to molecule axis in the present solid.

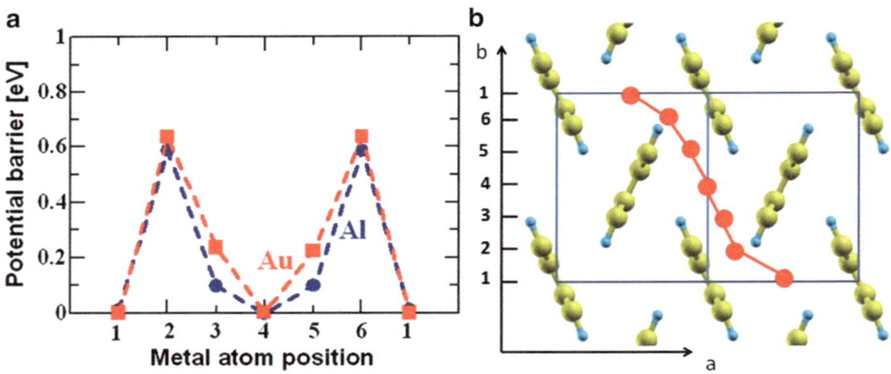


**Fig. 14.8** (**a**): Calculated adiabatic potentials for Al and Au atoms, as a function of metal-atom position perpendicular to the molecular axis. Atomic positions, 1 to 6, are displayed in (**b**). (**b**): Schematic top view of diffusion path for Al and Au atoms. Both atoms have almost the same path. *Dots* indicate the most stable positions at each b-axis coordinate. Molecules are displayed at positions when there exist no metal atoms. Reproduced with permission from Tomita and Nakayama [7] (Copyright © 2012, Elsevier)

Comparing the potential barrier between intra and inter-molecule diffusions, we can conclude that the inter-molecule diffusion becomes the limiting process for Al, thus the diffusion being sensitive to the inter-molecule distance and expected to change largely under some pressures. In case of Au, on the other hand, the barrier is small and both the intra and inter-molecule barrier contributes to the diffusion equally. Considering both the results of diffusions along and perpendicular to the molecule axis, moreover, it is expected around the room temperature [6, 7, 11] that Au atom prefers to diffuse along the molecule axis and shows the large anisotropy, while Al atom prefers to locate around the molecule edges and is difficult to diffuse along any directions.

Finally, we shortly comment on the magnitude of diffusion coefficient. By using calculated potential barriers and simply applying the Ahrenius formulas derived in Sect. 14.3, the diffusion coefficients of Al and Au atoms in the present various organic systems are respectively estimated ranging as $10^{-17}$–$10^{-14}$ and $10^{-12}$–$10^{-8}$ $cm^2/s$ at room temperature. The former value is comparable to recently observed values, for example, for fast non-reacted diffusion of Al in PTCDA [15].

## 14.7 Concluding Remarks

In this article, we explained the fundamental properties of metal-atom diffusion in characteristic carbonic systems, i.e., graphene sheets, polyacetylene bundles, and oligoacene model solid, and demonstrated that most of $\pi$-conjugated organic molecular systems have common features for metal-atom diffusion, which are quite different from inorganic systems. Namely, the small negativity atoms such as Al partially supply electrons to C–C $\pi^*$-antibonding states, are bounded to carbon systems with the ionic interaction, and diffuse across the C–C bonds with large barriers around 0.4–0.8 eV. On the other hand, the large negativity atoms like Au produce weak covalent-like bonds with $\pi$-orbitals of carbon, and prefer to move along the C–C bonds with small barriers less than 0.2 eV. The potential barrier is made of two factors; one is the breaking of such atomic bonding between metal atom and molecule. The other is the elastic repulsion caused by the condensed molecule configuration. In addition, we have shown that the diffused Al atoms become donors, while Au atoms produce deep levels that can capture electron and hole carriers (carrier traps).

One of the subjects that should be next studied is how the diffusion promotes the distribution and the clustering of impurity atoms. This is because such properties often induce leakage currents and charged defects in organic devices. To elucidate these properties, one has first to study the interaction between impurity metal atoms. Figure 14.9 shows the interaction energy between Al (Au) atoms on graphene sheet as a function of inter-atomic distance (T. Park and T. Nakayama, unpublished; [16]). Al atoms have a long-range repulsive interaction caused by the ionization of Al on carbonic supports, which is the origin of superstructures observed when low-negativity metal atoms are adsorbed with low coverage on graphene-like

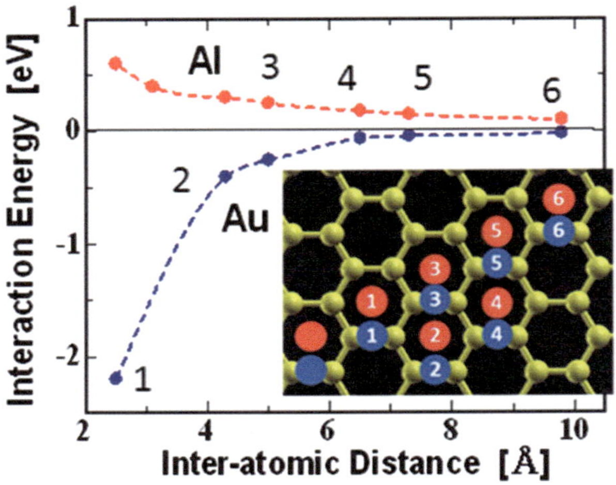


**Fig. 14.9** Interaction energies between two Al atoms (*red*) on graphene sheet as a function of inter-atomic distance, and those between two Au atoms (*blue*). *Inset* shows the adsorption positions of Al (*red*) and Au (*blue*) atoms. One atom is located at the left-below point without the number, while the other atom is located at the numbered point

systems. On the other hand, Au atoms have a short-range attractive interaction, which appears due to the metallic bonding between neighboring Au atoms. From this result, one can expect that Au atoms prefer to produce a cluster in carbonic systems, while Al atoms not. These fundamental studies are indispensable to develop reliable organic devices.

**Acknowledgements** This work is partially supported by Grants-in-Aid for Scientific Research from the Ministry of Education, Culture, Sports, Science and Technology, Japan, and the Global COE Program of Chiba University. We also acknowledge the supercomputing center of the Institute for Solid State Physics for the use of facilities.

## References

1. K. Tamao, K. Sumitani, M. Kumada, J. Am. Chem. Soc. **94**, 4374 (1972)
2. K. Suemori, M. Yokoyama, H. Hiramoto, J. Appl. Phys. **99**, 036109 (2006)
3. T. Sawabw, K. Okamura, T. Sueyoshi, T. Miyamoto, K. Kudo, N. Ueno, M. Nakamura, Appl. Phys. A **95**, 225 (2009)
4. S.T. Lee, Z.Q. Gao, L.S. Hung, Appl. Phys. Lett. **75**, 1404 (1999)
5. S.M. Sze, *Physics of Semiconductor Devices* (Wiley, New York, 1981)
6. Y. Tomita, T. Nakayama, Appl. Phys. Express **3**, 091601 (2010)
7. Y. Tomita, T. Nakayama, Org. Electron. **13**, 1487 (2012)
8. T. Park, Y. Tomita, T. Nakayama, Surf. Sci. **621**, 7 (2014)
9. Y. Tomita, T. Nakayama, in *31st International Conference on the Physics of Semiconductors*, Zurich Switzerland (2012), Tu-133
10. Y. Tomita, T. Nakayama, in *2012 International Conference on Solid State Devices and Materials*, Kyoto Japan (2012), M-4-4
11. Y. Tomita, First-principles Study of Metal-atom Diffusion in Organic Semiconductor Systems, Ph.D. thesis, Chiba University, (2011)
12. P. Hanggi, P. Talkner, M. Borkovec, Rev. Mod. Phys. **62**, 251 (1990)

13. A. Grüneis, C. Attaccalite, A. Rubio, D. Vyalikh, S.L. Molodtsov, J. Fink, R. Follath, W. Eberhardt, B. Büchner, T. Pichler, Phys. Rev. B **79**, 205106 (2009)
14. T. Kondo, K. Izumi, K. Watahiki, Y. Iwasaki, T. Suzuki, J. Nakamura, J. Phys. Chem. C **112**, 15607 (2008)
15. H. Yoshida, N. Sato, Appl. Phys. Lett. **91**, 141915 (2007)
16. T. Park, Adsorption behavior of metal atoms on carbonic substrates; interactions and Pt catalysis (in Japanese), Ph.D. thesis, Chiba University (2014)

# Chapter 15
# Numerical Approach to Charge Transport Problems on Organic Molecular Crystals

**Hiroyuki Ishii**

## 15.1 Problems in Charge Transport of Organic Semiconductors

Organic semiconductors have the potential to be used in future electronic devices requiring structural flexibility and large-area coverage that can be fabricated by low-cost printing processes. Charge mobilities in organic materials were first measured by the time-of-flight technique by Kepler [1] and LeBlanc [2]. Friedman investigated the electrical transport properties of organic crystals using the Boltzmann-equation treatment of narrow-band limit in the case of small polaron band motion [3]. Sumi discussed the change from the band-type mobility of large polarons to the hopping-type of small polarons, using the Kubo formula with the adiabatic treatment of lattice vibrations in the single-site approximation [4].

In the field of organic electronics, Kudo et al. reported the first field-effect mobilities of merocyanine dye films in 1984 [5]. Then, Koezuka et al. fabricated the first actual field-effect transistors (FETs) made of polythiophene in 1987 [6]. A major industrial breakthrough occurred in the application to electroluminescent devices: Tang and VanSlyke reported the first electroluminescent device based on a $\pi$-conjugated molecular material [7]. After that, typical applications spread to light-emitting diodes (LEDs), [8, 9], solar cells, [10–12] and high-mobility FETs [13–18].

As an example, we summarize some common issues concerning these devices, referring to the structure of FETs shown in Fig. 15.1. The energy level alignment at the metal-organic semiconductor interface is a crucial issue concerning unknown

H. Ishii (✉)
JST-PRESTO, University of Tsukuba, 1-1-1 Tennodai, Tsukuba, Ibaraki 305-8573, Japan

Division of Applied Physics, Faculty of Pure and Applied Sciences, University of Tsukuba, 1-1-1 Tennodai, Tsukuba, Ibaraki 305-8573, Japan
e-mail: ishii@bk.tsukuba.ac.jp



H. Ishii et al. (eds.), *Electronic Processes in Organic Electronics*,
Springer Series in Materials Science 209, DOI 10.1007/978-4-431-55206-2_15

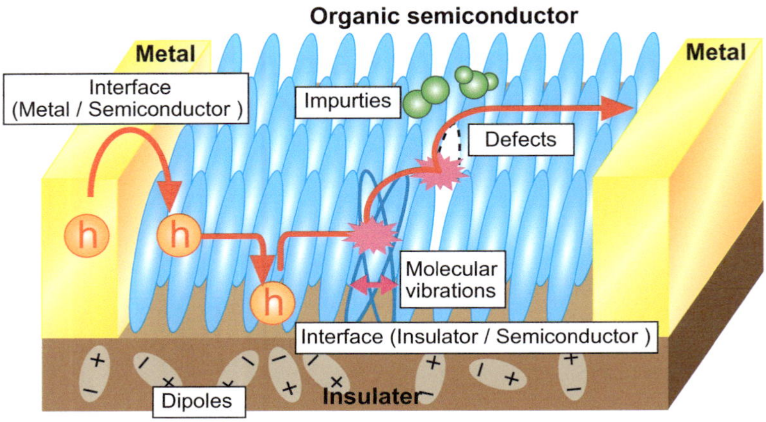


**Fig. 15.1** Schematic image summarizing various transport problems of organic electronic devices

charge injection processes and the origin of contact resistance. In addition, a very high gate voltage of 10–100 eV is required to inject charges into organic semiconductors. After charge injection, the charges drift in the semiconductors under the influence of the static electric field generated by the bias voltage. The charge carriers are frequently scattered by dynamical disorder induced by molecular vibrations and by static disorder such as lattice defects and charged impurities. Furthermore, since the organic semiconductor is formed on the dielectric film for any type of FET, it is believed that the surface structural roughness or the local dipoles in the film affect the transport properties. To create organic devices with higher performance, we must systematically understand the carrier scattering mechanisms from the atomistic viewpoint. In this chapter, we focus on the intrinsic transport property of organic semiconductors and investigate it by a numerical approach.

Unlike inorganic materials such as silicon semiconductors, organic materials are formed with very weak van der Waals interactions between molecules. For typical organic semiconductors, the intermolecular bonding energies, called transfer integrals, are small, in the range of $10 - 10^2$ meV [19, 20], which is comparable to the depth of carrier-trap potentials due to the static structural disorder, impurities, and grain boundaries [21]. Thus, the charge carriers are strongly scattered and localized in a molecule, creating a quasi-particle state called a polaron as a result of the strong interaction with intramolecular vibrations of the local lattice environment. The polaronic state should affect the transport properties of organic materials because the binding energy of a polaron (reorganization energy) is the same in the order of magnitude as transfer integrals [19] and thermal excitation energy $k_BT \simeq 25$ meV at room temperature. In fact, the incoherent molecule-to-molecule hopping of a polaron, which is a slow-moving electron coupled with lattice distortion, has been observed in experiments as a thermally activated behavior of temperature-dependent mobility, where the carrier mobility is generally below 0.1 cm$^2$/Vs at around room temperature [14]. In such a low-mobility regime, the charge-transport properties have been investigated theoretically using the following Marcus equation [22, 23], adopting polaron effects described by the Holstein model [24].

$$\mu = \frac{2\pi}{\hbar}\frac{ea^2}{k_B T}\frac{\gamma^2}{\sqrt{4\pi\lambda k_B T}}\exp\left(-\frac{(\lambda 4) - \gamma}{k_B T}\right) \tag{15.1}$$

Equation (15.1) describes the strong coupling between the intramolecular vibration and the electron. Here, $\hbar$, $a$, $T$, $\gamma$, and $\lambda$ represent the Planck's constant, the distance between molecules, temperature, transfer energy, and reorganization energy, respectively.

### *15.1.1 Recent Experimental Studies*

The recent rapid progress in technology enables us to fabricate single-crystal organic thin films and to construct flexible thin-film transistor (TFT) devices with high carrier mobility of up to $\sim$40 cm$^2$/Vs [25], which exceeds the mobility of amorphous silicon [14]. This indicates that structural disorder is almost removed from the organic TFTs. The temperature dependence of mobility is considerably different from those of strongly disordered materials. In recent organic single crystals, the mobility decreases with increasing temperature according to $\mu \propto T^{-n}$. Such power-law dependence is a typical characteristic of coherent band transport due to delocalized carriers and originates from the coupling between carriers and the intermolecular vibrations, i.e., the lattice phonons, in the scattering processes. In fact, Ueno et al. detected the clear electronic-band dispersion by angle-resolved photoemission spectroscopy (ARPES) [26–28]. Moreover, recent experimental measurements of Hall effects on organic TFTs [18, 29, 30] have provided us with evidence of the possibility of coherent charge transport in single-crystal organic materials.

On the other hand, a difficult problem arises in the coherent band-transport picture. That is, the estimated mean free path is comparable to or shorter than the distance between adjacent molecules [31], which implies a breakdown of coherent band transport. Actually, pentacene single-crystal transistors exhibit sharp electron spin resonance (ESR) signals, which indicates carrier localization in about 10 molecules [32, 33]. The temperature dependence of the ESR linewidth implies the motional narrowing effect with increasing temperature. This feature is highly consistent with the thermally activated multiple trap-and-release transport with an activation energy on the order of 10 meV [34], but the origin of trap potentials is still unknown. To clarify the characteristics of trap potentials, Kalb et al. estimated the trap states of pentacene from the transport characteristics of FETs [21]. Ohashi et al. measured the potential fluctuation in the channel of pentacene thin-film transistors by atomic-force-microscope potentiometry [35]. Recently, Yogev et al. reported on the direct determination of the hole density of states in amorphous organic films by Kelvin probe force microscopy with high lateral resolution and high energy resolution [36].

## 15.1.2 Recent Theoretical Studies

Recent theoretical studies have been aimed at the above-mentioned transport problems on single-crystal organic materials with high mobility, employing quantum mechanical theory from an atomistic viewpoint.

Troisi and Orlandi solved the time-dependent Schrödinger equation of electrons coupled with the classical equation of molecular motion taking account of the polaron effects by means of the Peierls model, which describes the strong coupling between the intermolecular vibration and an electron [37, 38]. They showed that large thermal fluctuations of molecular motion are sufficient to destroy the translational symmetry of the molecular lattice, and the carrier state alternates between delocalized bandlike and strongly localized. Fratini and Ciuchi also employed the Peierls model and found the simultaneous presence of band carriers and incoherent localized states [39]. Ciuchi, Fratini, and Mayou analyzed the optical conductivity observed in experiments using the Kubo formula with the Peierls model and found an incipient electron localization caused by large dynamical lattice disorder [40]. Böhlin, Linares, and Stafstöm investigated the polaron dynamics in a molecular lattice subjected to an electric field with a finite magnitude within the Peierls model and found a change in the transport process from an adiabatic polaron drift process to a combination of sequences of adiabatic drift and nonadiabatic hopping events [41]. Brédas suggested that the thermal bandwidth narrowing of organic semiconductors observed in ARPES experiments [42] is caused by the reduction of transfer energies induced by the thermal expansion of the lattice [43].

Although these approaches focus on the intermolecular coupling in the Peierls (large polaron) model, the intramolecular coupling described by the Holstein (small polaron) model cannot be ignored when accounting for the transport properties of single-crystal organic materials because the Holstein-type polaron has a large binding energy and describes the charge transport in the localized hopping regime owing to static disorder, which inevitably exists in molecular crystals [21]. Hannewald and Ortmann have analytically evaluated mobility by the method of canonical transformation of the Holstein model [44–47]. They explained the bandwidth-narrowing effect within the polaron picture, and studied the contributions of both coherent and incoherent scattering events to the transport properties. However, they assumed the transport to be bandlike, i.e., that the phase coherence is maintained and that scattering by phonons is infrequent. In general, this assumption is valid at temperature much lower than room temperature.

## 15.2 Fundamental Theory of Transport Phenomena in Crystals

Understanding the unknown electron transport properties in organic molecular crystals is the key not only to the progress of basic material science but also to realizing flexible and printed device applications such as organic transistors [13, 25, 48], organic LEDs [8], and organic solar cells [10]. We start this section with some basic concepts of electron transport phenomena from the viewpoint of solid-state physics. When a bias voltage $V$ is applied to an organic thin film, we can observe the electric current $I$ between source and drain electrodes. It is well known that the conductance $G$ ($\equiv I/V$) of a thin film is directly proportional to its width $W$ and inversely proportional to its length $L$:

$$G = \sigma \frac{W}{L}, \tag{15.2}$$

where the conductivity $\sigma$ is a material property of the thin film. Although the conductivity is a "macroscopic" physical quantity, its value is determined by the "microscopic" characteristics of the material. It is very important to investigate the relationships of macroscopic transport properties with the intermolecular electronic coupling, molecular vibration, and molecular packing structure, because the molecular design of novel organic materials is essential for obtaining organic devices with improved performance.

### *15.2.1 Electronic States of Molecular Crystals*

#### 15.2.1.1 Tight-Binding Approximation

The electrons in an isolated single molecule form localized eigenstates known as molecular orbitals with discrete energy levels. They naturally retain this strong localization when the molecule participates in the formation of molecular crystals, because the molecular crystals are formed with very weak van der Waals interactions between molecules. Therefore, an obvious approach to describe the crystal electron is a linear combination of molecular orbitals. The valence band states of crystal $|\Psi\rangle$ can be described using the highest occupied molecular orbitals (HOMOs) $|\Phi_n\rangle$ of the $n$th single molecule as follows:

$$|\Psi\rangle = \sum_{n=-\infty}^{+\infty} C_n |\Phi_n\rangle, \tag{15.3}$$

where $C_n$ represents the probability amplitude. As an example, we show the HOMO of an isolated pentacene molecule in Fig. 15.2a. The crystal electrons must be

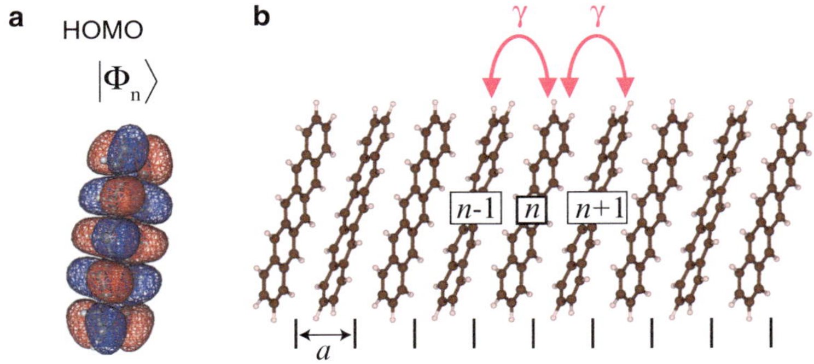


**Fig. 15.2** (**a**) HOMO of single pentacene molecule. (**b**) Simple model of one-dimensional molecular crystal with intermolecular distance $a$. The transfer energy between the nearest neighbor HOMOs is represented by $\gamma$

described as quantum particles, and the calculation of electronic states requires that we solve the time-independent Schrödinger equation for the molecular crystal, $\hat{H}|\Psi\rangle = E|\Psi\rangle$, where the crystal Hamiltonian is represented by $\hat{H}$. We assume that the molecular orbital basis set $\{|\Phi_n\rangle\}$ satisfies the conditions of a complete orthonormal system, i.e., $\langle\Phi_n|\Phi_m\rangle = \delta_{nm}$ and $\sum_n |\Phi_n\rangle\langle\Phi_n| = 1$. Using these conditions, the Schrödinger equation is rewritten as $\sum_n \hat{H}|\Phi_n\rangle C_n = E\sum_n |\Phi_n\rangle C_n$. Then, applying $\langle\Phi_m|$ to this equation, we obtain the matrix form of the Schrödinger equation:

$$\sum_n \langle\Phi_m|\hat{H}|\Phi_n\rangle C_n = EC_m. \tag{15.4}$$

Here, the diagonal elements of the Hamiltonian matrix $\langle\Phi_n|\hat{H}|\Phi_n\rangle$ are approximately equal to the HOMO level of the $n$th isolated single molecule, while the off-diagonal elements $\langle\Phi_m|\hat{H}|\Phi_n\rangle$ represent the transfer energies $\gamma_{nm}$ between the HOMO of the $n$th molecule and that of the $m$th molecule and directly affect the transport properties of the material.

As a simple example, we consider a one-dimensional molecular crystal of pentacene with intermolecular distance $a$ and the HOMO overlap included only up to nearest neighbor molecules, i.e., $\langle\Phi_{n\pm 1}|\hat{H}|\Phi_n\rangle = \gamma$ and the others are zero, since the molecular orbital is localized mainly in a single molecule. A schematic picture of the model is shown in Fig. 15.2b. We obtain the following Schrödinger equation with a sparse Hamiltonian matrix,

$$\begin{pmatrix} \ddots & \ddots & \ddots & \ddots & \ddots \\ \ddots & 0 & \gamma & 0 & \ddots \\ \ddots & \gamma & 0 & \gamma & \ddots \\ \ddots & 0 & \gamma & 0 & \ddots \\ \ddots & \ddots & \ddots & \ddots & \ddots \end{pmatrix} \begin{pmatrix} \vdots \\ C_{n-1} \\ C_n \\ C_{n+1} \\ \vdots \end{pmatrix} = E \begin{pmatrix} \vdots \\ C_{n-1} \\ C_n \\ C_{n+1} \\ \vdots \end{pmatrix}. \tag{15.5}$$

We can evaluate the eigenstates $\{C_n\}$ and their eigenenergies $E$ by the numerical diagonalization of the Hamiltonian matrix. The use of the molecular basis set has advantages in real-space representation, for instance, the local electron density on the $n$th molecule is easily expressed by $|C_n|^2$.

#### 15.2.1.2 Electronic Band Structure, Carrier Velocity and Effective Mass

In contrast to the real-space representation, the reciprocal-space representation is also useful for understanding the electron state in a crystal with periodic potentials. Mathematically, the extended electron states $|\Phi_k\rangle$ with wave vector $k$ can be transformed from the localized molecular orbitals $|\Phi_n\rangle$ by the Fourier transform

$$|\Phi_k\rangle = C_k \sum_{n=-\infty}^{+\infty} e^{ikR_n} |\Phi_n\rangle, \tag{15.6}$$

where $R_n(\equiv na)$ is the position vector of the $n$th molecule and $C_k$ is the normalization constant resulting in $\langle\Phi_k|\Phi_k\rangle = 1$. The eigenenergy of the extended state with wave vector $k$ corresponds to the electronic band structure, as shown below.

$$\begin{aligned} E(k) &\equiv \langle\Phi_k|\hat{H}|\Phi_k\rangle \\ &= C_k \sum_n e^{ikR_n} \langle\Phi_k|\hat{H}|\Phi_n\rangle \\ &= C_k \sum_n \gamma\left(e^{ik(R_{n-1}+a)}\langle\Phi_k|\Phi_{n-1}\rangle + e^{ik(R_{n+1}-a)}\langle\Phi_k|\Phi_{n+1}\rangle\right) \\ &= \gamma\left(e^{ika}\langle\Phi_k|\Phi_k\rangle + e^{ik(-a)}\langle\Phi_k|\Phi_k\rangle\right) \\ &= 2\gamma\cos(ka) \end{aligned} \tag{15.7}$$

The velocity of a crystal electron is given by the group velocity defined as

$$v(k) \equiv \frac{1}{\hbar}\frac{dE(k)}{dk}. \tag{15.8}$$

A crystal electron with charge $q$ in the presence of an external electric field $\mathcal{E}$ gains additional energy $dE$ in an infinitesimal time $dt$, i.e., $dE = q\mathcal{E} \cdot v(k)dt$. Then using Eq. (15.8), we obtain $(dE/dk)dk = q\mathcal{E}(dE/\hbar dk)dt$, resulting in

$$\frac{dk}{dt} = \frac{q\mathcal{E}}{\hbar}. \tag{15.9}$$

From Eqs. (15.8) and (15.9), it follows that the rate of change of the group velocity is

$$\begin{aligned} \frac{dv(k)}{dt} &= \frac{1}{\hbar}\frac{d^2E(k)}{dkdt}, \\ &= \frac{1}{\hbar}\frac{d^2E(k)}{dk^2}\frac{dk}{dt}, \\ &= \frac{1}{\hbar^2}\frac{d^2E(k)}{dk^2}q\mathcal{E}. \end{aligned} \tag{15.10}$$

This equation is analogous to the classical equation of motion $dv/dt = q\mathcal{E}/m$ for a point charge $q$ in an electric field $\mathcal{E}$ if the mass $m$ is replaced by the effective mass $m^*$ defined as

$$\frac{1}{m^*} = \frac{1}{\hbar^2}\frac{d^2E(k)}{dk^2}. \tag{15.11}$$

The effective mass is simply given by the curvature of the electronic band $E(k)$.

In the case of the valence band of the one-dimensional molecular crystal discussed above, the group velocity and effective mass are given as $v(k) = -2\gamma a \sin(ka)/\hbar$ and $m^* = -\hbar^2/2\gamma a^2 \cos(ka)$. To increase the group velocity and reduce the effective mass, we must synthesize molecular crystals with higher transfer energies $\gamma$ and longer lattice constants $a$.

### 15.2.2 Electric Conductivity, Mobility and Mean Free Path of Crystals

In about 1900, Drude described conductivity using the assumption of an ideal electron gas in a metal. For an ideal electron gas in an external electric field $\mathcal{E}$, the classical equation of motion for electrons with velocity $v$ is described as

$$m\frac{dv}{dt} = -\frac{m}{\tau_m}v + q\mathcal{E}, \tag{15.12}$$

where the friction term is given by the first term on the right-hand side and $\tau_m$ has the meaning of a momentum relaxation time. For a steady-state current, we have $v = (q\tau_m/m) \cdot \mathcal{E}$. The electron mobility characterizes how quickly an electron can move through a material in an applied electric field $\mathcal{E}$ and is defined as

$$\mu \equiv \frac{v}{\mathcal{E}} = \frac{q\tau_m}{m}, \tag{15.13}$$

which is always specified in units of cm$^2$/Vs. The current density $J$ along an external electric field is given by the product of the charge density $qn$ and its velocity $v$, i.e., $J = qnv = (q^2 n\tau_m/m)\mathcal{E}$; therefore, the conductivity is written as

$$\sigma = \frac{J}{\mathcal{E}} = \frac{q^2 n\tau_m}{m}. \tag{15.14}$$

For FETs using a gate insulator with capacitance $C$, we obtain the relationships among the conductivity, mobility, and gate voltage from Eqs. (15.13) and (15.14) as

$$\sigma = nq\mu = \mu C(V_G - V_{th}), \tag{15.15}$$

since the carrier density is evaluated by $nq = C(V_G - V_{th})$, where the applied gate voltage and threshold voltage are represented by $V_G$ and $V_{th}$, respectively. In experimental observations, the mobility is obtained from the rate of increase of $\sigma$ with increasing $V_G$. Improving the current-amplification factor of FETs requires higher-mobility materials, because the ratio of output current to bias voltage is proportional to the conductivity.

The mean free path $\ell_m$ is one of the fundamental quantities in understanding the charge transport mechanism and is defined as the distance that an electron travels before its initial momentum is destroyed.

$$\ell_m = v \cdot \tau_m \tag{15.16}$$

It varies widely from one material to another and is also strongly affected by temperature and the existence of static disorder. When the mean free path becomes smaller than the lattice constant $a$ in a strongly disordered system, the concept of the mean free path breaks down. In this case, the carriers are localized spatially in the strong disorder potential, thus the localization length $L_\xi$ becomes the meaningful quantity instead of the mean free path $\ell_m$.

### 15.2.3 Kubo Formula Based on Quantum Theory

In the previous subsection, we extracted the conductivity within the classical theory. However, to evaluate the transport properties from the atomistic viewpoint, we should obtain the conductivity $\sigma$ by adopting quantum-mechanical theory using the Kubo formula [49–52],

$$\sigma = \frac{q^2}{\Omega} \int_{-\infty}^{+\infty} dE \left( -\frac{df(E)}{dE} \right) \int_{0}^{+\infty} dt Tr\left[ \delta(E - \widehat{H}_e) \widehat{v}_x(t) \widehat{v}_x(0) \right], \tag{15.17}$$

where the volume of material is represented by $\Omega$, the Fermi distribution function is $f(E)$, the velocity operator along $x$ axis of charge is written as $\widehat{v}_x(t) \equiv \widehat{U}^\dagger(t) \widehat{v}_x \widehat{U}(t)$ in the Heisenberg picture, and $\widehat{U}(t)$ is the time-evolution operator. Since the velocity-correlation function $\widehat{v}_x(t)\widehat{v}_x(0)$ satisfies the following relation,

$$\int_{0}^{+\infty} dt \langle \widehat{v}_x(t) \widehat{v}_x(0) \rangle = \lim_{t \to +\infty} \frac{1}{2t} \int_{0}^{t} \int_{0}^{t} ds ds' \langle \widehat{v}_x(s) \widehat{v}_x(s') \rangle \tag{15.18}$$

$$= \lim_{t \to +\infty} \frac{1}{2t} \langle (\widehat{x}(t) - \widehat{x}(0))(\widehat{x}(t) - \widehat{x}(0)) \rangle, \tag{15.19}$$

we can rewrite the expression of conductivity as follows,

$$\sigma = \lim_{t \to +\infty} \int_{-\infty}^{+\infty} dE \quad \left( -\frac{df(E)}{dE} \right) q^2 Tr\left[ \frac{\delta(E - \widehat{H}_e)}{\Omega} \cdot \frac{\{\widehat{x}(t) - \widehat{x}(0)\}^2}{t} \right]. \tag{15.20}$$

Here the position operator of charge is written as $\widehat{x}$.

From Eq. (15.15), the mobility of a carrier with elementary charge $q$ is written as the conductivity given by Eq. (15.20) divided by the charge density, $\mu = \sigma/qn$. The charge density is defined as $qn = q \int d\,E\, f(E)\nu(E)$, where $\nu(E)$ is the density of states (DOS) given by $Tr[\delta(E - \widehat{H}_e)/\Omega]$.

The diffusion coefficient $D$ is evaluated as the long-time limit of the time-dependent diffusion coefficient,

$$\begin{aligned} D &\equiv \lim_{t \to +\infty} D(t) \\ &= \lim_{t \to +\infty} \frac{\langle \{\widehat{x}(t) - \widehat{x}(0)\}^2 \rangle}{t}, \end{aligned} \tag{15.21}$$

where $\langle \cdots \rangle \equiv \int dEf(E)Tr[\delta(E - \widehat{H}_e) \cdots ] \int dEf(E)Tr[\delta(E - \widehat{H}_e)]$. When we approximate the Fermi distribution function as $f(E) \simeq \exp{-E/k_B T}$, since we consider the low carrier states of the semiconductor in the present study, we can extract the well-known Einstein relation as follows [53, 54],

$$
\begin{aligned}
D &\simeq \lim_{t\to+\infty} \frac{1}{t} \frac{\int dE e^{\frac{-E}{k_B T}} Tr[\delta(E-\widehat{H}_e)\{\widehat{x}(t)-\widehat{x}(0)\}^2]}{\int dE e^{\frac{-E}{k_B T}} Tr[\delta(E-\widehat{H}_e)]}, \\
&\simeq k_B T \frac{\lim_{t\to+\infty} \int dE \frac{1}{k_B T} e^{\frac{-E}{k_B T}} q^2 Tr\left[\frac{\delta(E-\hat{H}_e)}{\Omega}\frac{\{\widehat{x}(t)-\widehat{x}(0)\}^2}{t}\right]}{q^2 \int dE e^{\frac{-E}{k_B T}} Tr\left[\frac{\delta(E-\hat{H}_e)}{\Omega}\right]}, \\
&\simeq k_B T \frac{\sigma}{q^2 n}, \\
&\simeq \frac{k_B T}{q}\mu.
\end{aligned} \tag{15.22}
$$

The definitions of the DOS $\nu$ and the diffusion coefficient $D$ are applied to Eq. (15.20); then we obtain another well-known equation, $\sigma \simeq q^2 \nu D$.

To clarify the transport properties in detail, the mean free path $\ell_m$ and momentum relaxation time $\tau_m$ are thoroughly discussed. Because the diffusion coefficient satisfies $D = v^2 \tau_m$ in the diffusive band-transport limit, the mean free path and relaxation time can be obtained using the diffusion coefficient as $\ell_m = D/v$ and $\tau_m = D/v^2$, respectively, where $v$ is the carrier velocity defined by $\lim_{t\to 0} \sqrt{D(t)/t}$ [51, 52].

## 15.2.4 Inter- and Intramolecular Vibration Effects

To clarify the charge transport properties of organic semiconductors from an atomistic viewpoint, we adopt a semiclassical approach for molecular vibrations (phonons), assuming that the nuclear motion of the lattice is treated classically, whereas the charge carrier dynamics is evaluated purely on the basis of the quantum mechanical approach. Note that this assumption has been frequently adopted in studies of conductive polymers [55, 56]. We employ the mixed Holstein–Peierls model for the carriers interacting tightly with both the intramolecular distortions and intermolecular vibrations.

### 15.2.4.1 Holstein Model

The Holstein model [24] is based on local electron–phonon coupling, which is purely intramolecular, i.e., it acts at a single molecule ionized by the charge of the carrier. Figure 15.3a shows the potential energy surfaces of the neutral state and charged state of a molecule as a function of the intramolecular displacement $\Delta u$. When a neutral molecule is charged by a carrier, the total energy increases from

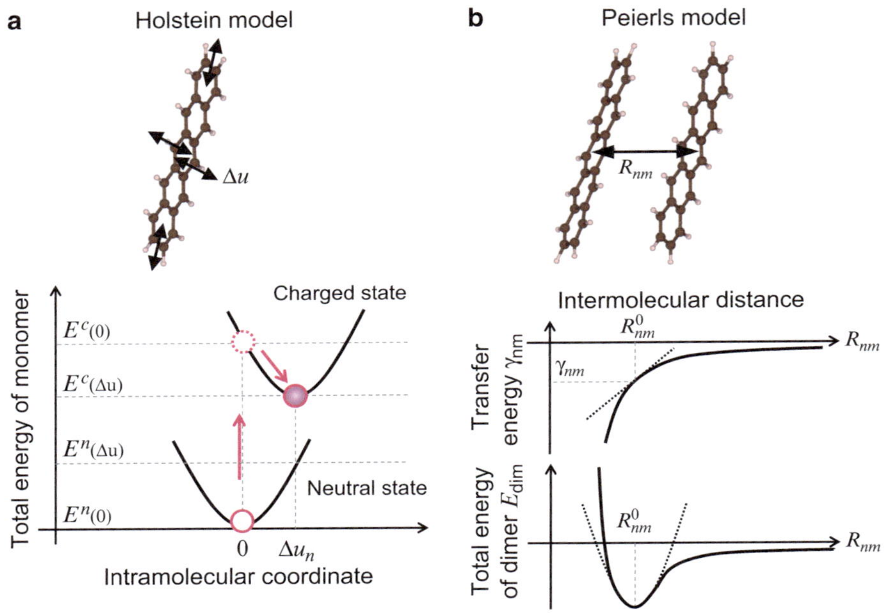


**Fig. 15.3** (**a**) (*Upper*) Schematic picture of intramolecular displacement $\Delta\, u$ of a single molecule. (*Lower*) Potential energy surface of neutral state and charged state of a molecule as a function of $\Delta\, u$. (**b**) (*Upper*) Intermolecular distance $R_{nm}$ between $n$th and $m$th molecules. (*Lower*) Schematic picture of the transfer energy $\gamma_{rm}$ and total energy of a dimer $E_{dim}$ as a function of intermolecular distance $R_{nm}$. The bond length at the equilibrium position is represented by $R^0_{nm}$

$E^n(0)$ to $E^c(0)$. Then the charged molecule takes the lowest total energy $E^c(\Delta\, u)$ with the intramolecular deformation $\Delta\, u$. The reorganization energy $\lambda$ corresponds to the energy difference between $E^c(0)$ and $E^c(\Delta\, u)$. When we employ the harmonic-potential approximation for the intramolecular deformation energy, the Hamiltonian of the Holstein model is given by

$$\hat{H}_H = \sum_n \{\tilde{\varepsilon}_n(\Delta u_n)\}\hat{c}_n^\dagger \hat{c}_n + \sum_n \frac{1}{2} K_H \{\Delta u_n\}^2, \tag{15.23}$$

where the elastic constant of intramolecular deformation is represented by $K_H$ and $\tilde{\varepsilon}_n$ is the on-site energy of the $n$th molecule defined as a function of the intramolecular deformation $\Delta\, u$,

$$\tilde{\varepsilon}_n(\Delta u_n) = \varepsilon_n + \alpha_H \cdot \Delta u_n. \tag{15.24}$$

The Holstein-type electron–phonon coupling constant is represented by $\alpha_H$. The operators, $\hat{c}_n$ and $\hat{c}_n^\dagger$, are annihilation and creation operators of a hole (electron) at the HOMO (the lowest unoccupied molecular orbital (LUMO)) with on-site energy $\varepsilon_n$.

#### 15.2.4.2 Peierls Model

The intermolecular vibration influences the time dependence of transfer energies between adjacent molecules. For organic semiconductors, even small molecular displacements induced by the intermolecular vibrations lead to significant changes in the transfer energies. The resulting nonlocal coupling leads to the Peierls model, such as the Su-Schrieffer-Heeger model [55, 56]. We show schematic pictures of the Peierls model in Fig. 15.3b. When the change in bond length between the $n$th and $m$th molecules is written as $\Delta R_{nm}$, the change in transfer energy is well expressed by $\alpha_P \cdot \Delta R_{nm}$ around the equilibrium position, where the Peierls-type electron–phonon coupling constant is denoted by $\alpha_P$. Therefore, the Hamiltonian of the Peierls model can be written as

$$\hat{H}_P = \sum_{n,m} \tilde{\gamma}_{nm} \left(\Delta R_{nm}(t)\right)\left(\hat{c}_n^{\dagger}\hat{c}_m + \hat{c}_m^{\dagger}\hat{c}_n\right) + \sum_{n} \frac{1}{2} M \left\{\frac{d\Delta \mathbf{R}_n(t)}{dt}\right\}^2 + \sum_{n,m} \frac{1}{2} K_P \{\Delta R_{nm}(t)\}^2, \quad (15.25)$$

where the transfer energy $\tilde{\gamma}_{nm}$ coupled with the intermolecular vibrations is defined by

$$\tilde{\gamma}_{nm}(\Delta R_{nm}) = \gamma_{nm} + \alpha_P \cdot \Delta R_{nm}. \quad (15.26)$$

Here, $\gamma_{nm}$ is the bare transfer energy between the $n$th and $m$th molecules. The second and third terms of Eq. (15.25) correspond to the kinetic energy of a molecule with mass $M$ and the potential energy of intermolecular vibrations, respectively. The displacement of the center of mass of the $n$th molecule is written as $\Delta \mathbf{R}_n$. The elastic constant $K_P$ between adjacent molecules can be evaluated from the change in the total energy of a dimer, $E_{dim}$, as a function of $R_{nm}$, i.e., as shown in Fig. 15.3b, and $K_P$ is obtained from the fitting to the analytic expression $E_{dim} = (1/2)K_P \Delta R_{nm}{}^2$ around the equilibrium position.

#### 15.2.4.3 Holstein–Peierls Model

As mentioned above, the Holstein model or Peierls model has been employed so far in theoretical studies of the transport properties of organic semiconductors. However, it seems that the charge carriers in organic semiconductors are strongly coupled with both the intra- and intermolecular vibrations. Therefore, for organic semiconductors, we should employ the mixed Holstein–Peierls model [44, 45]. The Hamiltonian of the mixed Holstein–Peierls model is obtained as the sum of those of Holstein model and Peierls model, that is to say, $\hat{H}_{tot} = \widehat{H}_H + \widehat{H}_P$. The electron part of the total Hamiltonian is written as

$$\hat{H}_e = \sum_{n,m} \{\tilde{\gamma}_{nm}(\Delta R_{nm})\}\left(\hat{c}_n^{\dagger}\hat{c}_m + \hat{c}_m^{\dagger}\hat{c}_n\right) + \sum_n \{\tilde{\varepsilon}_n(\Delta u_n)\}\hat{c}_n^{\dagger}\hat{c}_n, \tag{15.27}$$

where the transfer energy $\tilde{\gamma}_{nm}$ and orbital energy $\tilde{\varepsilon}_n$ are coupled with the intermolecular vibrations and the intramolecular deformation. For the intramolecular deformations and intermolecular vibrations, we obtain the following Hamiltonian:

$$\hat{H}_v = \sum_n \frac{1}{2} M \left\{ \frac{d\Delta \mathbf{R}_n(t)}{dt} \right\}^2 + \sum_{n,m} \frac{1}{2} K_P \{\Delta R_{nm}(t)\}^2 + \sum_n \frac{1}{2} K_H \{\Delta u_n(t)\}^2. \tag{15.28}$$

We can write the total Hamiltonian as $\hat{H}_{tot} \equiv \widehat{H}_e + \widehat{H}_v$. Using this Hamiltonian, we can investigate the transport properties of charge coupled with both the intramolecular deformations and intermolecular vibrations of organic molecular crystals.

## 15.3 Methodology of Charge Transport Calculations for Organic Semiconductors

To investigate the transport properties at around room temperature, we have developed a new methodology based on the time-dependent wave-packet diffusion (TD-WPD) method that can be used with the mixed Holstein–Peierls model [57]. In our approach, we can evaluate the carrier mobilities, mean free paths, and diffusion coefficients, including both the Holstein-type and Peierls-type polaronic states, on the same footing, with the use of the Kubo formula based on the quantum-mechanical calculations of electron wave-packet dynamics combined with a classical molecular-dynamics simulation.

### *15.3.1 TD-WPD Methodology for Organic Semiconductors*

In the TD-WPD method, the electron motion in a material with thermally excited lattice vibrations is described directly by combining the quantum-mechanical time-evolution calculation of an electron wave packet with a classical molecular lattice dynamics simulation [51, 52]. In our previous studies, we applied the TD-WPD method to carbon nanotubes formed by covalent bonding [58, 59]. The molecules in organic materials, on the other hand, are coupled to each other by weak van der Waals interactions, making these materials much more flexible than covalently bonded materials. The carrier motion in organic materials is strongly affected by the lattice distortions that form the so-called polaron state, that is, a quasi-particle state coupled with electron and molecular vibrations. Therefore the charge-carrier transport must be described using models different from those for covalently bonded

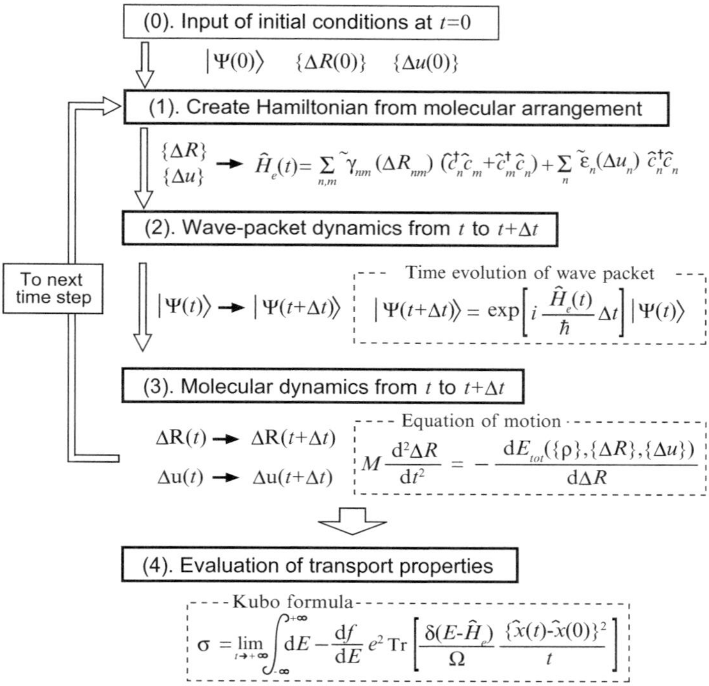


**Fig. 15.4** Flowchart of numerical computation according to the TD-WPD methodology in the case of the Holstein–Peierls model

semiconductors. We newly introduce the polaron effect into the TD-WPD methodology [57].

We show the schematic flowchart for the mixed Holstein–Peierls model in Fig. 15.4.

(0) First of all, we set the initial conditions at $t = 0$ for the electron wave packet $|\Psi(t = 0)\rangle$, intermolecular configurations $\{\Delta\mathbf{R}_n(t = 0)\}$, and intramolecular configurations $\{\Delta u_n(t = 0)\}$.

(1) Then, we determine the Hamiltonian $\hat{H}_e(t = 0)$ for the given inter- and intramolecular configurations.

(2) The time-evolution operator of an electron defined by $\exp\{-i\hat{H}_e(t = 0)\Delta t/\hbar\}$ is created from the Hamiltonian, and we evaluate the time evolution of the wave packet from $t = 0$ to $t = \Delta t$.

(3) Then, we obtain the inter- and intramolecular configurations at the next time step, $\{R_n(t = \Delta t)\}$ and $\{\Delta u_n(t = \Delta t)\}$, using the canonical equation of motion, including the effect of not only the thermally excited molecular vibration but also the polaronic lattice distortions.

(1)' After returning to process (1), we extract the Hamiltonian at the next time step, $\hat{H}_e(t = \Delta t)$, from the renewed inter- and intramolecular configurations.

(4) Repeating the above processes, we take into account the complicated trajectory of electron motion in its induced time-dependent lattice distortion, and evaluate the conductivity $\sigma$ using Eq. (15.29).

#### 15.3.1.1 Conductivity of Electron Coupled with Molecular Vibrations

To evaluate the electron conductivity $\sigma$ of materials with volume $\Omega$ from the time evolution of the initial electron wave packet $|\Psi_n(0)\rangle$, we employ the time-dependent form of the Kubo formula [49–52],

$$\sigma = \lim_{t\to+\infty} \int_{-\infty}^{+\infty} dE \left(-\frac{df(E)}{dE}\right) \frac{q^2}{\Omega} \times \frac{N}{N_w} \sum_{n=1}^{N_w} \frac{\langle \Psi_n(0)|\delta(E-\widehat{H_e})\{\widehat{x}(t)-\widehat{x}(0)\}^2|\Psi_n(0)\rangle}{t}, \quad (15.29)$$

where the Fermi distribution function is represented by $f(E)$. $N$ and $N_w$ represent the number of molecules and the number of initial wave packets, respectively. The position operator of charge is written as $\widehat{x}(t) \equiv \widehat{U}^\dagger(t)\widehat{x}\widehat{U}(t)$ in the Heisenberg picture. The time-evolution operator is defined as $\widehat{U}(t = N_t\Delta t) \equiv \Pi_{n=0}^{N_t-1}\exp\{i\hat{H}_e(n\Delta t)\Delta t/\hbar\}$. We can evaluate the electron wave packet at time $t$ by solving the time-dependent Schrödinger equation $|\Psi(t)\rangle = \widehat{U}(t)|\Psi(0)\rangle$. The dynamical change in electronic states owing to the inter- and intramolecular vibrations is included in the time-dependent electron Hamiltonian $\hat{H}_e(t)$.

#### 15.3.1.2 Motion of Molecules Coupled with Electrons

Using real-time classical molecular dynamics simulations, we describe the dynamical lattice distortions due to both the thermal fluctuation and the reorganization upon ionization by charge carriers.

When we employ the generalized coordinate system $\{\mathbf{Q}\}$ to describe the lattice distortion, the equation of motion for the $n$th site with mass $M$ is derived from the canonical equation $M \times d^2\mathbf{Q}_n/dt^2 = -\partial E_{tot}/\partial \mathbf{Q}_n$, where the total energy $E_{tot}$ is defined as the sum of the electron energy and molecular vibration energy. Here, for instance, $\mathbf{Q}_n$ represents the displacement of the $n$th molecular position $\boldsymbol{\Delta}\mathbf{R}_n$ and the intramolecular distortion $\boldsymbol{\Delta}\mathbf{u}_n$. In general, the total energy is divided into four terms, $E_{tot} = E_e(\{\rho\}) + E_v^{pot}(\{\mathbf{Q}\}) + E_v^{kin}(\{\dot{\mathbf{Q}}\}) + E_{ev}(\{\rho \cdot \mathbf{Q}\})$, where $E_e$, $E_v^{pot}$, $E_v^{kin}$, and $E_{ev}$ represent the electron energy, the elastic potential energy, the kinetic energy of molecular vibration, and the electron–phonon coupling energy, respectively.

Here, the electron density matrix is written as $\{\rho\}$. We obtain the following equation of motion for the lattice vibration and polaronic distortion:

$$M\frac{d^2\mathbf{Q}_n}{dt^2} = -\frac{\partial E_v^{pot}(\{\mathbf{Q}\})}{\partial \mathbf{Q}_n} - \frac{\partial E_{ev}(\{\rho\cdot\mathbf{Q}\})}{\partial \mathbf{Q}_n}, \tag{15.30}$$

where the first term on the right-hand side corresponds to the elastic force and the second term induces the lattice distortion due to polaron formation. The temperature $T$ is fixed by normalizing the kinetic energy of lattice vibration at each time step under the condition of $\sum_{n=1}^{N} M\dot{\mathbf{Q}}_n^2/2 = dNk_BT/2$ in $d$-dimensional systems. Solutions of the coupled Eqs. (15.29) and (15.30) allow us to describe the motion of an electron that is strongly coupled with both the thermal fluctuating lattice and the distortion due to polaron formation.

The total Hamiltonian of the mixed Holstein–Peierls model is given by $\hat{H}_{tot}(t) \equiv \widehat{H}_e(t) + \widehat{H}_v(t)$, and we can obtain the total energy as

$$\begin{aligned} E_{tot} &= \int dE \quad f(E-E_F)Tr\left[\delta(E-\widehat{H}_e)\hat{H}_{tot}\right] \\ &= \sum_{n,m}\tilde{\gamma}_{nm}(\Delta R_{nm})\cdot\{\rho_{nm}(t)+\rho_{mn}(t)\} + \sum_n \tilde{\varepsilon}_n(\Delta u_n)\cdot\rho_{nn}(t) \\ &\quad + \sum_n \frac{1}{2}M\left\{\frac{d\Delta\mathbf{R}_n(t)}{dt}\right\}^2 + \sum_{n,m}\frac{1}{2}K_P\{\Delta R_{nm}(t)\}^2 \\ &\quad + \sum_n \frac{1}{2}K_H\{\Delta u_n(t)\}^2, \end{aligned} \tag{15.31}$$

where $\rho_{nm} \equiv \int dEf(E-E_F)Tr[\delta(E-\widehat{H}_e)\hat{c}_n\hat{c}_m^\dagger]$ is the density matrix element.

- Canonical equation of motion for intermolecular vibrations Electronic intermolecular interactions in molecular crystals with transfer integrals $\gamma \simeq 100\,\mathrm{meV}$ should give rise to the formation of mesoscopically extended Bloch electrons on a time scale of $\tau_e = \hbar/\gamma \simeq 10$ fs. Intermolecular vibrations on a slower time scale of $\tau_{inter} > 100$ fs lead to dynamic fluctuations of the transfer integrals [60]. Replacing $\mathbf{Q}_n$ by $\Delta\mathbf{R}_n$ in Eq. (15.30), we can derive the equation of motion for the intermolecular vibrations as $M \times d^2\Delta\mathbf{R}_n/dt^2 = -\partial E_{tot}/\partial\Delta\mathbf{R}_n$.
- Canonical equation of motion for intramolecular vibrations The intramolecular vibration occurs on a time scale comparable to that for charge delocalization $\tau_{intra} \sim \tau_e$ and should result in Holstein-type polaronic relaxation. Therefore, for the intramolecular distortions, we determine $\Delta\ u_n$ by the variational principle for the minimum total energy, $\partial\ E_{tot}/\partial\ \Delta\ u_n = 0$, resulting in $\Delta u_n(t) = (\alpha_H/K_H)\cdot\{\rho_{nn}(t)/\sum_m \rho_{mm}(t)\}$.

### 15.3.1.3 Parameters of Organic Semiconductors

Since van der Waals force originates from a temporary fluctuating dipoles of a molecule, a rigorous treatment of van der Waals interaction within a simple tight-binding approximation is very difficult. Therefore, we employ a semi-empirical approach for the van der Waals interactions. When we extract the physical quantities of molecular semiconductors such as transfer energy and elastic constant between molecules, we employ the DFT-D approach, which consists in adding a semi-empirical dispersion potential to the conventional Kohn–Sham DFT energy [61]. The obtained quantities are in good agreement with some experimental observations [62]. Although various two-dimensional organic thin films can be treated with the present approach, we consider, for simplicity, a one-dimensional stack of 1,600 pentacene molecules here. This is justified since the transport properties of organic semiconductors are strongly anisotropic in real space.

We obtain the bare transfer integrals between adjacent molecules, $\gamma = -75\,\mathrm{meV}$, by fitting the bandwidth of the three-dimensional pentacene crystal computed by DFT calculations (see Hummer and Ambrosch-Draxl [63] for details) with the bandwidth $4\gamma$ of the one-dimensional model used here. The Peierls-type electron–phonon coupling is estimated as $\alpha_P \equiv d\ \gamma/d\ \Delta\ R$ at the equilibrium molecular position, where the bond-length-dependent $\gamma$ is calculated by the dimer-splitting approach [20, 64] at the DFT-D/B3LYP-D3/6-31G(d) level. For these quantities, we use $\alpha_P = 104\,\mathrm{meV/Å}$ [65]. The Peierls-type elastic constant can be estimated as $K_P = 2.19\,\mathrm{eV/Å^2}$ by fitting the computed bond-length-dependent total energy $U(\Delta\ R)$ with the parabolic form $U = K_P(\Delta\ R)^2/2$ around the equilibrium molecular position [62]. These results are in good agreement with previously reported values ($\alpha_P = 86 \sim 174\,\mathrm{meV/Å}$ and $K_P = 1.50\,\mathrm{eV/Å^2}$) [37, 38]. The intermolecular vibration has a continuous phonon-band structure, $\hbar\omega_P(q) = \hbar\sqrt{2K_P(1 - \cos qa)/M}$ with the bandwidth of 3.8 meV, where the bond length $a = 5.16$ Å and the mass of a pentacene molecule $M$ is $4.151 \times 10^{-24}$ kg. This phonon dispersion is in good agreement with an acoustic phonon branch of oligoacene crystals [66]. By coupling the wave packet dynamics with the classical molecular dynamics, we can use various phonon modes from $q = 0$ to $\pm\pi/a$, which are thermally excited at a finite temperature, to study the realistic phonon-scattering effects on the charge-transport properties [51, 52].

The Holstein-type electron–phonon coupling and the elastic constant of $\alpha_H = -93$ and $K_H = 92\,\mathrm{meV}$, respectively, are taken from Lei and Shimoi [67]. These quantities are also estimated using the DFT calculations at the B3LYP/6-31G(d) level [68]. Here, we employ the dimensionless-effective-scalar coordinate $\Delta\ u_n$ as the intramolecular distortion. We confirmed that when the effective coordinate $\Delta\ u$ is equal to 1, the relaxation energy defined as $K_H \cdot (\Delta u)^2/2$ and the binding energy of the Holstein-type polaron are in good agreement with those obtained in other theoretical studies [20, 67].

For the numerical computation of the time evolution of wave packets combined with the molecular dynamics within the TD-WPD methodology, we employ a time

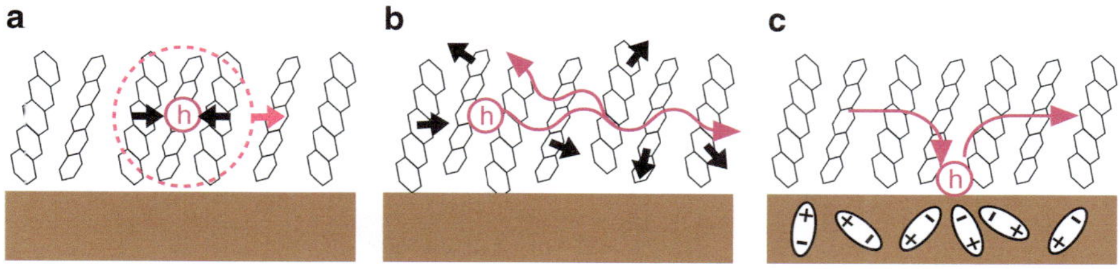


**Fig. 15.5** Schematic images of some major scattering factors in organic semiconductors. (**a**) Creation of polaron, which is a quasi-particle state composed of a charge and its accompanying lattice deformations. (**b**) Dynamical disorder on electronic states induced by thermally excited molecular motion (lattice phonons). (**c**) Static disorder originating from structural disorder, impurities, grain boundaries, and random dipoles in the substrate

step $\Delta t$ of $2 \times h/(1\,\text{eV}) \simeq 8.272\,\text{fs}$ for stacked pentacene molecules. We evaluate Eq. (15.29) using 256 initial conditions for wave packets of localized random phase states [69], distributions of static disorder, and Maxwell velocity distributions of molecules.

## 15.4 Transport Properties of Carrier Coupled with Inter- and Intramolecular Vibrations of Organic Semiconductors

For typical organic semiconductors, the transfer energies of carriers are small in the range of $10\text{–}10^2\,\text{meV}$, which is comparable to the depth of carrier-trap potentials due to static disorder, such as impurities and dipoles in the substrate [21, 70]. Furthermore, the magnitude of transfer energies is similar to both the dynamical disorder induced by thermal fluctuations of molecules and the polaron binding energies. It is important for us to understand the carrier transport mechanism in competition among the thermal fluctuations, the polaron formation, and the static disorder, as shown in Fig. 15.5. Therefore, we take these into account and investigate the temperature dependence of transport properties of organic semiconductors.

### *15.4.1 Polaron-Formation Effect on Electronic States*

First, we investigate the electron–phonon coupling effect on the polaron formation of single-crystal organic semiconductors without both thermal fluctuation and static disorder. The polaron state is achieved by self-consistent calculations to minimize the total energy $E_{tot}$ with respect to the intra- and intermolecular displacements, i.e., $\Delta u_n$ and $\Delta \mathbf{R}_n$, respectively, under the condition of keeping the length of the molecular lattice fixed [41, 55, 56].

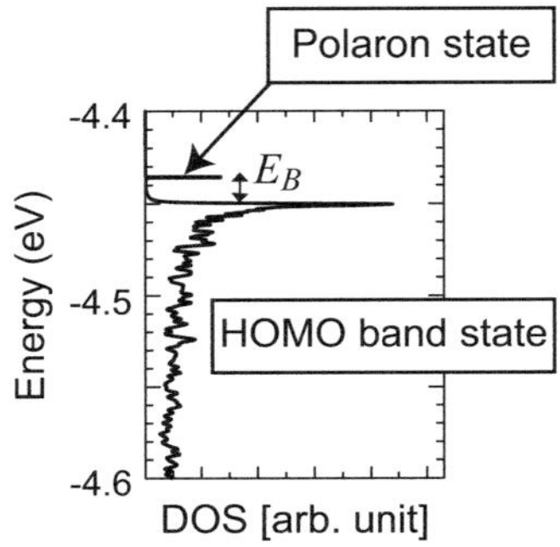


**Fig. 15.6** Density of states (*DOS*) for the HOMO band of one-dimensional pentacene crystal. A discrete energy level of the polaron state is created above the HOMO band, as indicated by the *arrow*. The binding energy $E_B$ is defined as the energy difference between the polaron level and the top of the HOMO band. (Reproduced, with permission, from Ishii et al. [57])

**Table 15.1** Polaron binding energies calculated by the Holstein, Peierls, and Holstein–Peierls models for pentacene single crystal. (Reproduced, with permission, from Ishii et al. [57])

| | Holstein | Peierls | Holstein–Peierls |
|---|---|---|---|
| Binding energy $E_B$ (meV) | 7.5 | 1.3 | 14.3 |

#### 15.4.1.1 Binding Energy of Polaron State

The obtained spatial distribution of the polaron state is localized in the range of 50 Å ( $\sim$ 10 molecules). We show the energy spectrum of the DOS of the HOMO bands in Fig. 15.6. The spectrum shows van Hove singularities since we employ the one-dimensional model. We can see a discrete polaron state in the energy spectrum of the DOS located above the HOMO band. The binding energy $E_B$ is evaluated to be 14.3 meV, which is of the same order as the transfer energy $\gamma$ and thermal excitation energy $k_BT$ at around room temperature. This implies that the Holstein–Peierls-type electron–phonon coupling affects the carrier transport properties through polaron formation.

In Table 15.1, we present the calculated polaron binding energies for three different models. One is the mixed Holstein–Peierls model discussed above, and the others are the Peierls model where $\alpha_P \neq 0$ and $\alpha_H = 0$, and the Holstein model where $\alpha_P = 0$ and $\alpha_H \neq 0$. We observe significant differences among these models. We see that the Holstein–Peierls model gives rise to a much larger binding energy than the Holstein and Peierls models. These calculated results indicate that we must take into account both the Holstein-type and Peierls-type electron–phonon couplings simultaneously when we study the transport properties of organic semiconductors.

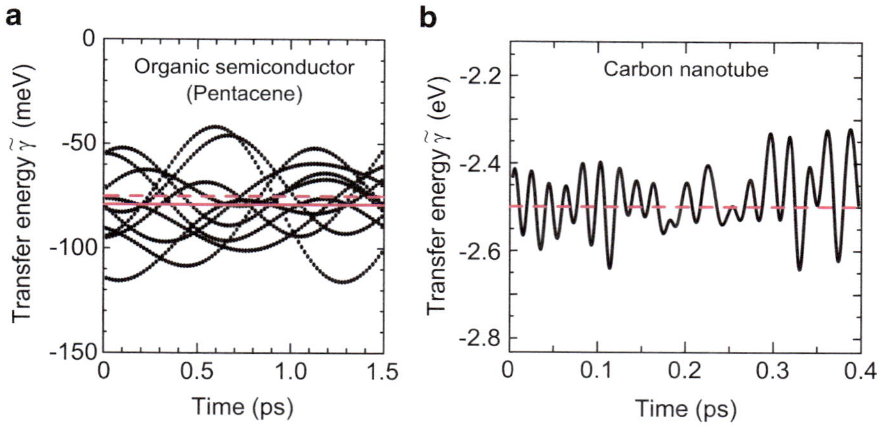


**Fig. 15.7** (**a**) Some time-dependent transfer energies induced by the intermolecular vibrations at $T = 300$ K. The transfer energy without electron–phonon coupling ($\alpha_H = \alpha_P = 0$) is presented by the *red broken line*, and the maximum value of transfer integrals enhanced by polaronic distortion is shown by the *solid red straight line*. (**b**) For a comparison, the time-dependent transfer energy between $\pi$-orbitals of neighboring carbon atoms in a carbon nanotube at $T = 300$ K is shown

#### 15.4.1.2 Thermal-Fluctuation and Polaron-Formation Effects on Transfer Energy

Next, we investigate the electron–phonon coupling effects on the transfer energies, because the charge transport properties strongly depend on the transfer energies. In the above discussion, we obtained the polaron state by self-consistent calculations to minimize the total energy $E_{tot}$ with respect to $\Delta\ u_n$ and $\Delta \mathbf{R}_n$. Here, we consider the low-temperature limit ($T \to 0$) to exclude the thermal fluctuation effects in the molecular dynamics simulations. The transfer energies are enhanced from − 75 to −79 meV by the shrinkage of the molecular bond length, resulting in the formation of the polaron state.

Since the organic semiconductor devices are operated at room temperature, it is essential to consider how the thermal fluctuations of molecular motion, as well as the polaron formation, affect the transfer energies. Figure 15.7a shows the time-dependent transfer energies for several bonds at 300 K. For comparison, the transfer energy without any distortion, $\Delta R_{nm} = 0$, is shown by the red broken line. The thermal lattice vibrations give rise to a large amount of dynamic disorder in the transfer energies, whereas the polaron formation slightly contributes to the change in transfer energy. The amplitude of the thermally fluctuating transfer energies reaches about 80 meV, which is comparable to the magnitude of the long-time-averaged transfer energies. This large dynamic disorder in the transfer energy is one of the significant characteristics of organic semiconductors. For comparison with covalently bonded materials, we show the time-dependent transfer energies for

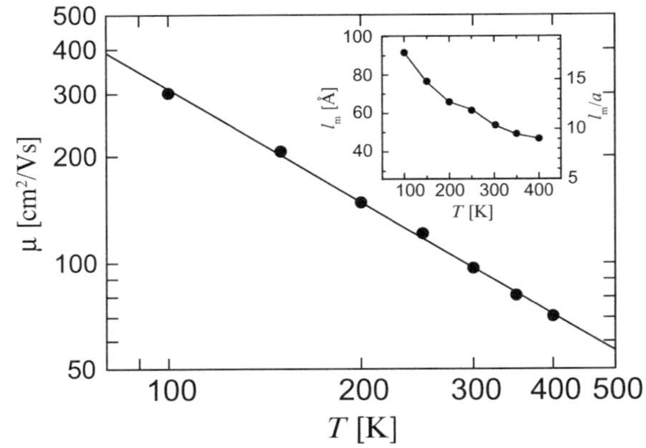


**Fig. 15.8** Intrinsic carrier mobility $\mu$ as a function of temperature $T$ in the case of no static disorder. $\mu$ decreases monotonically with increasing $T$ and reaches about $100\,\mathrm{cm}^2/\mathrm{Vs}$ at around room temperature. (*Inset*) Mean free path $\ell_m$ as a function of temperature $T$. The ratio of $\ell_m$ to the lattice constant $a$ is also shown along the right axis. (Reproduced, with permission, from Ishii et al. [57])

carbon nanotubes in Fig. 15.7b, where the transfer energy is defined as the coupling between $\pi$-orbitals of carbon atoms. We see that the ratio of the amplitude of fluctuation to the long-time averaged transfer energy is very small compared with that of organic semiconductors.

The dynamic disorder of transfer energies in the organic semiconductors is comparable to the thermal excitation energy $k_B T$, and the Holstein–Peierls-type polaron binding energy $E_B$. These calculated results indicate that the charge-transport properties of organic semiconductors are determined within a subtle balance among the energies of the thermal fluctuation of molecular motion, thermal excitation, and polaron formation.

### 15.4.2 *Inter- and Intramolecular-Vibration Effects on Intrinsic Charge-Transport Properties*

We study the intrinsic charge-transport properties of single-crystal organic semiconductors, excluding static disorder. This provides the possible maximum value of mobility. The main plot in Fig. 15.8 shows the calculated carrier mobility $\mu$ as a function of temperature $T$. The mobility decreases monotonically with increasing $T$ in accordance with the power-law dependence, which is apparent evidence of bandlike transport [71]. Similar power-law temperature dependence of mobility has been reported in recent theoretical works [37–39]. The present calculated results show that the intrinsic mobility reaches approximately $100\,\mathrm{cm}^2/\mathrm{Vs}$ at around room temperature and increases to $300\,\mathrm{cm}^2/\mathrm{Vs}$ with decreasing temperature to 100 K.

The mean free path $\ell_m$ is important in understanding the transport mechanism. If $\ell_m$ is equivalent to or shorter than the lattice constant $a$ (distance between adjacent molecules), $\ell_m < a$, then the carrier assumes the hopping transport, whereas bandlike transport is assumed when $\ell_m$ is much longer than $a$, $\ell_m \gg a$. In the inset of Fig. 15.8, we show the mean free path $\ell_m$ and its $T$-dependence. The carrier scattering is included in the changes in the time-dependent transfer energies, and is induced by the thermal fluctuation of molecular motion through Peierls-type electron–phonon coupling. The mean free path decreases monotonically as the temperature increases as $\ell_m = v\tau \propto v/(\Delta R_{nm})^2 \propto v/(k_B T)$ [39]. For reference, the ratio of $\ell_m$ to $a$ is plotted along the right axis. We see that the mean free path is 10 times longer than the lattice constant around at around room temperature, supporting the possibility of bandlike behavior in intrinsic transport when static disorder is absent.

## 15.4.3 *Static Disorder Effects on Charge Transport*

Next, we consider the effect of static disorder, which inevitably exists in molecular crystals in forms such as impurities and dipoles in the substrate [21, 70]. We introduce the Anderson-type disorder potentials $W_n$, which modulate the on-site orbital energies randomly within the energy width $[-W/2, +W/2]$ [58, 59], then investigate how the transport properties are affected by static disorder. Several experimental evaluations show that the potential depth of static disorder in high-mobility organic FETs is about 50 meV [21]. Thus we change $W$ from 50 to 200 meV. We note that this static disorder is comparable to the HOMO bandwidth of the present organic semiconductors. In fact, the ratio $W/\gamma$ ranges from 0.67 to 2.67. Therefore, the charge-transport properties are expected to be strongly disturbed by the effects of the presence of static disorder in the various temperature regimes.

### 15.4.3.1 Temperature Dependence of Carrier Mobility and Mean Free Path

Figure 15.9a shows a logarithmic plot of the carrier mobility $\mu$ as a function of the temperature $T$ for several strengths of static disorder, $W = 50$, 100, and 200 meV. The existence of static disorder decreases the carrier mobility significantly from 100 (no static disorder) to 2 cm$^2$/Vs ($W = 200$ meV) at around 300 K, and more importantly, the temperature dependence of mobility is changed completely.

In the case of $W = 50$ meV, the magnitude of $\mu$ becomes larger than 50 cm$^2$/Vs at around room temperature, and $d\mu/dT$ takes negative values in the entire temperature regime studied here, indicating that $\mu$ increases monotonically with decreasing $T$. The $T$-dependence of $\mu$ above 200 K is close to the power-law dependence seen in Fig. 15.9a. On the other hand, the slope of the increase in $\mu$ becomes gentle in the

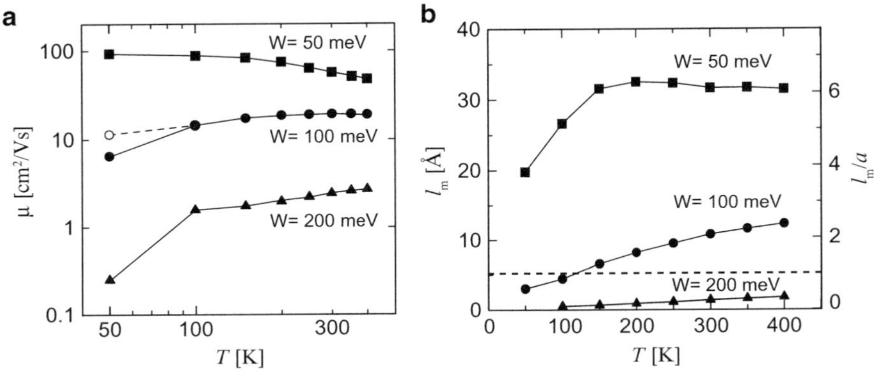


**Fig. 15.9** (**a**) Logarithmic plot of carrier mobility $\mu$ as a function of temperature $T$ for static disorder where $W = 50$, 100, and 200 meV. As $W$ increases, the $T$-dependence of $\mu$ at around room temperature changes from nearly power-law dependence to thermally activated behavior. (**b**) Temperature dependence of the mean free path $\ell_m$ for $W = 50$, 100, and 200 meV. The ratios of $\ell_m$ to $a$ are also shown along the right axis. (Reproduced, with permission, from Ishii et al. [57])

low-$T$ regime since scattering by static disorder independent of temperature becomes dominant in carrier motion. Figure 15.9b, in which the mean free path $\ell_m$ is presented as a function of $T$, shows $\ell_m$ to be $4-6$ times longer than the lattice constant $a$ in the case of $W = 50$ meV.

When the disorder $W$ increases to 100 meV, $d\mu/dT$ changes its sign below 300 K, as shown in Fig. 15.9a, that is to say, $\mu$ changes its behavior from bandlike in the high-$T$ regime to thermally activated in the low-$T$ regime. In this situation, we expect the carrier to be localized by the static disorder, which induces intramolecular distortions, resulting in low-mobility polaron formation. From the comparison with the calculated mobility taking only Peierls-type coupling into account, we find that the presence of Holstein-type coupling, which describes the interaction between a hole and the intramolecular distortion, decreases the mobility from 11.4 to 6.4 $cm^2/Vs$ at 50 K and from 15.0 to 14.5 $cm^2/Vs$ at 100 K, as shown by white circles in Fig. 15.9a. At temperatures exceeding 150 K, the mobility reduction due to polaron formation becomes negligible, and the mobility reaches its maximum value of 19 $cm^2/Vs$ at 300 K. Then $\mu$ decreases again with increasing $T$. Note that such $T$-dependence of $\mu$ has been observed in several recent experiments using organic TFTs [13, 48]. Let us investigate the $T$-dependence of $\ell_m$. From Fig. 15.9b, we can see that $\ell_m$ is comparable to the lattice constant $a$, which is in good agreement with experimental observations [31]. The polaronic localization state is destroyed with increasing $T$ because thermal excitation energy exceeding the polaron binding energy is given to the carrier.

When the static disorder becomes more significant and is increased to $W = 200$ meV, the magnitude of $\mu$ takes a much lower value of about 1 $cm^2/Vs$ in comparison with those for $W = 50$ and 100 meV. The sign of $d\mu/dT$ becomes positive in the whole temperature regime, as shown in Fig. 15.9a. In this situation, because

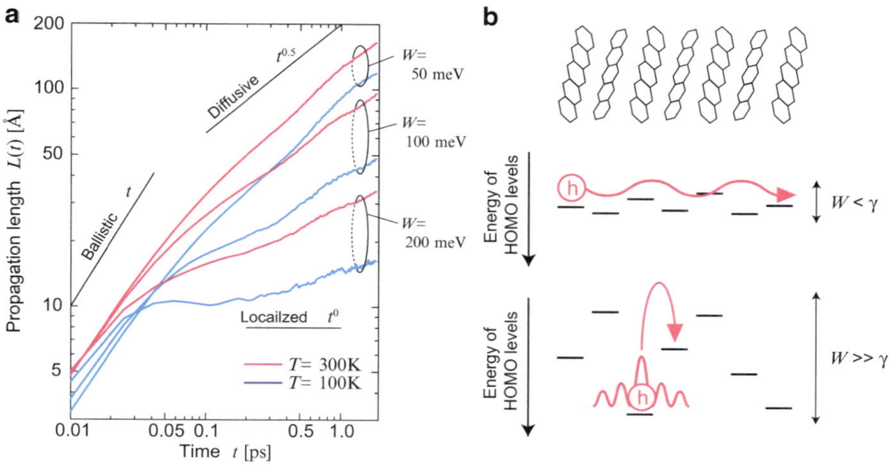


**Fig. 15.10** (**a**) Propagation length, $L(t) \equiv \sqrt{D(t) \times t}$, as a function of time for several values of $T$ and $W$. We plot the ideal time-dependent behavior of $L$ in the limit of the ballistic ($L \propto t$), diffusive ($L \propto t^{0.5}$), and localization (time-independent) characteristics. (**b**) Schematic diagrams of carrier transport for $W \leq \gamma$ and $W \gg \gamma$. (Reproduced, with permission, from Ishii et al. [57])

the carrier is trapped tightly by the static disorder, the calculated $\ell_m$ in Fig. 15.9b is less than the lattice constant $a$. The evaluated mean free path of the carrier is shorter than the molecule-to-molecule distance, which renders the concept of bandlike transport meaningless. The carrier transport properties are very close to typical thermally activated behaviors observed for organic TFTs with low quality [14].

### 15.4.3.2 Time-Scale of Electron Propagation in Organic Semiconductors

Finally, we study the time dependence of carrier motion in organic semiconductors. Figure 15.10a shows the propagation length of carriers as a function of time, defined as $L(t) \equiv \sqrt{D(t) \times t}$. We also plot the ideal time-dependent behaviors of $L(t)$ by thin solid lines in the limit of the ballistic ($L \propto t$), diffusive ($L \propto \sqrt{t}$), and localized (time-independent) characteristics. When the time exceeds $t \simeq 30$ fs, the carrier begins to be scattered both by the electron–phonon coupling and by the static disorder, and its time dependence deviates gradually from that of ballistic behavior. The localized nature of carriers is enhanced in the presence of static disorder. We find that the estimated localization length $L_\xi$ is about 10 Å in the case of $W = 200$ meV. Note that the crystal structure of organic molecular systems is considered to be quasi-one-dimensional owing to its highly anisotropic character. For a purely one-dimensional model, the localization effect acts more efficiently; therefore, the above estimation should be an upper bound of the localization length. As $T$ increases or $W$ decreases, the carrier begins to move and its motion changes from thermally activated hopping transport due to the localized carrier to diffusive transport, as shown in Fig. 15.10b.

## 15.5 Summary

We introduced some recent topics addressed in studies of the charge transport of organic semiconductors and presented the fundamental transport theories from an atomistic viewpoint. Then we presented our theoretical study on charge transport using the time-dependent wave-packet diffusion (TD-WPD) method, taking the polaron effects into account. By this method, we carried out quantum-mechanical time-evolution calculations of wave packets and classical molecular dynamics simulations simultaneously, including the inter- and intramolecular vibrations and static disorder on an equal footing without any perturbative treatment. Using this method, we studied the carrier mobilities and mean free paths of single-crystal pentacene organic semiconductors, focusing on their temperature dependencies in competition with the thermal fluctuation of molecular motion, polaron formation, and static disorder.

When the static disorder is absent, the intrinsic carrier mobility follows the power-law temperature dependence. This is mainly due to the transfer-energy modulation originating from the thermal fluctuation of molecular motion, whereas the polaron formation slightly contributes to the transport properties. As the static disorder increases, the carrier becomes localized and coupled with the intramolecular distortions, forming the polaron state. Therefore, the magnitude of mobility is dramatically decreased and its temperature dependence changes from bandlike power-law dependence to thermally activated hopping transport behavior. These calculated results indicate that competition among the thermal fluctuation, polaron formation, and static disorder provides important clues to understanding the transport mechanism of realistic organic semiconductors.

### *15.5.1 Forthcoming Challenges in Theoretical Studies*

As far as we know, the theoretical studies on the charge transport of organic semiconductors are divided into two approaches. One is the semiclassical approach, including our present study, where the thermal fluctuations of molecular motion determine the carrier scattering time and induce decreases in mobility with increasing temperature in pure organic molecular crystals without static disorder. Here, polaron formation slightly contributes to the temperature dependence of mobility.

In the full quantum approach, on the other hand, the concept of the polaron is obtained from the canonical transformed Holstein–Peierls Hamiltonian. Unlike in the semiclassical approach, the origin of the decreases in mobility with increasing temperature can be understood as the polaron-band-narrowing effect. It is assumed that the scattering time of the polaron is determined by the static disorder, because the interaction between polarons and molecular vibrations is not taken into account in this approach.

The two approaches mentioned above give similar power-law temperature dependencies of intrinsic mobility at around room temperature, but the physical origins are considerably different from each other. Theoretical challenges regarding this issue still remain, preventing deeper understanding of the intrinsic charge transport in pure organic semiconductors. Furthermore, one of the main scientific challenges is to clarify the microscopic origin of carrier traps that are inevitably present in realistic organic semiconductors. The inclusion of trap potentials is crucial for understanding and improving the device performance of organic semiconductors. In our present study, we introduced Anderson-type disorder potentials and treated the fluctuating energy width $W$ as a parameter. To enable a detailed understanding of the origin of trap potentials on a microscopic scale, further experimental and theoretical investigations are required.

**Acknowledgements** I would like to thank the 21C0E program of Chiba University for the financial and educational support to my student life. The knowledge, which I obtained from activities of this program, introduced me to the study of organic semiconductors recently. I also would like to thank Prof. Nobuhiko Kobayashi of University of Tsukuba and Dr. Kenji Hirose of NEC for valuable comments and suggestions. This work was supported by the Grant-in-Aid for Young Scientists B [24760024] from the Japan Society for the Promotion of Science.

## References

1. R.G. Kepler, Phys. Rev. **119**, 1226 (1960)
2. O.H. LeBlanc, J. Chem. Phys. **33**, 626 (1960)
3. L. Friedman, Phys. Rev. **133**, A1668 (1964)
4. H. Sumi, J. Phys. Soc. Jpn. **33**, 327 (1972)
5. K. Kudo, M. Yamashina, T. Moriizumi, Jpn. J. Appl. Phys. **23**, 130 (1984)
6. H. Koezuka, A. Tsumura, T. Ando, Synth. Met. **18**, 699 (1987)
7. C.W. Tang, S.A. VanSlyke, Appl. Phys. Lett. **51**, 913 (1987)
8. J.H. Burroughes, D.D.C. Bradley, A.R. Brown, R.N. Marks, K. Mackay, R.H. Friend, P.L. Burns, A.B. Holmes, Nature **347**, 539 (1990)
9. Y. Ohmori, M. Uchida, K. Muro, K. Yoshino, Jpn. J. Appl. Phys. **30**, L1941 (1991)
10. C.W. Tang, Appl. Phys. Lett. **48**, 183 (1986)
11. J. Xue, S. Uchida, B.P. Rand, S.R. Forrest, Appl. Phys. Lett. **85**, 5757 (2004)
12. J.Y. Kim, K. Lee, N.E. Coates, D. Moses, T.-Q. Nguyen, M. Dante, A.J. Heeger, Science **317**, 222 (2007)
13. S.F. Nelson, Y.-Y. Lin, D.J. Gundlach, T.N. Jackson, Appl. Phys. Lett. **72**, 1854 (1998)
14. C.D. Dimitrakopoulos, P.R.L. Malenfant, Adv. Mater. **14**, 99 (2002)
15. V. Podzorov, V.M. Pudalov, M.E. Gershenson, Appl. Phys. Lett. **82**, 1739 (2003)
16. M. Kiguchi, M. Nakayama, K. Fujiwara, K. Ueno, T. Shimada, K. Saiki, Jpn. J. Appl. Phys. **42**, L1408 (2003)
17. O.D. Jurchescu, J. Baas, T.T.M. Palstra, Appl. Phys. Lett. **84**, 3061 (2004)
18. J. Takeya, K. Tsukagoshi, Y. Aoyagi, T. Takenobu, Y. Iwasa, Jpn. J. Appl. Phys. **44**, L1393 (2005)
19. J.-L. Brédas, D. Beljonne, V. Coropceanu, J. Cornil, Chem. Rev. **104**, 4971 (2004)
20. V. Coropceanu, J. Cornil, D.A. da Silva Filho, Y. Oliver, R. Silbey, J.-L. Brédas, Chem. Rev. **107**, 926 (2007)
21. W.L. Kalb, B. Batlogg, Phys. Rev. B **81**, 035327 (2010)

22. R.A. Marcus, Rev. Mod. Phys. **65**, 599 (1993)
23. Z. Bao, J. Locklin, *Organic Field-Effect Transistors: Optical Science and Engineering Series* (CRC Press, London, New York, 2007)
24. T. Holstein, Ann. Phys. **8**, 325 (1959); Ann. Phys. **8**, 343 (1959)
25. J. Takeya, M. Yamagishi, Y. Tominari, R. Hirahara, Y. Nakazawa, T. Nishikawa, T. Kawase, T. Shimoda, S. Ogawa, Appl. Phys. Lett. **90**, 102120 (2007)
26. H. Yamane, S. Nagamatsu, H. Fukagawa, S. Kera, R. Friedlein, K.K. Okudaira, N. Ueno, Phys. Rev. B **72**, 153412 (2005)
27. H. Kakuta, T. Hirahara, I Matsuda, T. Nagao, S. Hasegawa, N. Ueno, K. Sakamoto, Phys. Rev. Lett. **98**, 247601 (2007)
28. S.-I. Machida, Y. Nakayama, S. Duhm, Q. Xin, A. Funakoshi, N. Ogawa, S. Kera, N. Ueno, H. Ishii, Phys. Rev. Lett. **104**, 156401 (2010)
29. V. Podzorov, E. Menard, J.A. Rogers, M.E. Gershenson, Phys. Rev. Lett. **95**, 226601 (2005)
30. M. Yamagishi, J. Soeda, T. Uemura, Y. Okada, Y. Takatsuki, T. Nishikawa, Y. Nakazawa, I. Doi, K. Takimiya, J. Takeya, Phys. Rev. B **81**, 161306(R) (2010)
31. J. Takeya, C. Goldmann, S. Haas, K.P. Pernstich, B. Ketterer, B. Batlogg, J. Appl. Phys. **94**, 5800 (2003)
32. K. Marumoto, S.-I. Kuroda, T. Takenobu, Y. Iwasa, Phys. Rev. Lett. **97**, 256603 (2006)
33. A.S. Mishchenko, H. Matsui, T. Hasegawa, Phys. Rev. B **85**, 085211 (2012)
34. H. Matsui, T. Hasegawa, Y. Tokura, M. Hiraoka, T. Yamada, Phys. Rev. Lett. **100**, 126601 (2008)
35. N. Ohashi, H. Tomii, R. Matsubara, M. Sakai, K. Kudo, M. Nakamura, Appl. Phys. Lett. **91**, 162105 (2007)
36. S. Yogev, E. Halpern, R. Matsubara, M. Nakamura, Y. Rosenwaks, Phys. Rev. B **84**, 165124 (2011)
37. A. Troisi, G. Orlandi, Phys. Rev. Lett. **96**, 086601 (2006)
38. A. Troisi, G. Orlandi, J. Phys. Chem. A **110**, 4065 (2006)
39. S. Fratini, S. Ciuchi, Phys. Rev. Lett. **103**, 266601 (2009)
40. S. Ciuchi, S. Fratini, D. Mayou, Phys. Rev. B **83**, 081202(R) (2011)
41. J. Böhlin, M. Linares, S. Stafström, Phys. Rev. B **83**, 085209 (2011)
42. N. Koch, A. Vollmer, I. Salzmann, B. Nickel, H. Weiss, J.P. Rabe, Phys. Rev. Lett. **96**, 156803 (2006)
43. Y. Li, V. Coropceanu, J.-L. Brédas, J. Phys. Chem. Lett. **3**, 3325 (2012)
44. K. Hannewald, V.M. Stojanović, J.M.T. Schellekens, P.A. Bobbert, G. Kresse, J. Hafner, Phys. Rev. B **69**, 075211 (2004)
45. K. Hannewald, P.A. Bobbert, Appl. Phys. Lett. **85**, 1535 (2004)
46. F. Ortmann, F. Bechstedt, K. Hannewald, Phys. Rev. B **79**, 235206 (2009)
47. F. Ortmann, S. Roche, Phys. Rev. B **84**, 180302(R) (2011)
48. V. Podzorov, E. Menard, A. Borissov, V. Kiryukhin, J.A. Rogers, M.E. Gershenson, Phys. Rev. Lett. **93**, 086602 (2004)
49. S. Roche, D. Mayou, Phys. Rev. Lett. **79**, 2518 (1997)
50. S. Roche, J. Jiang, F. Triozon, R. Saito, Phys. Rev. Lett. **95**, 076803 (2005)
51. H. Ishii, N. Kobayashi, K. Hirose, Appl. Phys. Express **1**, 123002 (2008)
52. H. Ishii, N. Kobayashi, K. Hirose, Phys. Rev. B **82**, 085435 (2010)
53. S.M. Sze, *Semiconductor Devices: Physics and Technology*, 2nd edn. (Wiley, Hoboken, New Jersey, 2001)
54. D.K. Ferry, *Semiconductor Transport* (Taylor & Francis, London, 2000)
55. W.P. Su, J.R. Schrieffer, A.J. Heeger, Phys. Rev. Lett. **42**, 1698 (1979)
56. W.P. Su, J.R. Schrieffer, A.J. Heeger, Phys. Rev. B **22**, 2099 (1980)
57. H. Ishii, K. Honma, N. Kobayashi, K. Hirose, Phys. Rev. B **85**, 245206 (2012)
58. H. Ishii, S. Roche, N. Kobayashi, K. Hirose, Phys. Rev. Lett. **104**, 116801 (2010)
59. H. Ishii, N. Kobayashi, K. Hirose, Appl. Phys. Express **3**, 095102 (2010)

60. J.-F. Chang, T. Sakanoue, Y. Olivier, T. Uemura, M.-B. Dufourg-Madec, S.G. Yeates, J. Cornil, J. Takeya, A. Troisi, H. Sirringhaus, Phys. Rev. Lett. **107**, 066601 (2011)
61. S. Grimme, J. Comp. Chem. **25**, 1463 (2004)
62. T. Uemura, T. Fukami, H. Ishii, N. Kobayashi, K. Hirose, J. Takeya (in preparation)
63. K. Hummer, C. Ambrosch-Draxl, Phys. Rev. B **72**, 205205 (2005)
64. J.L. Brédas, J.P. Calbert, D.A. da Silva Filho, J. Cornil, Proc. Natl. Acad. Sci. USA **99**, 5804 (2002)
65. M.W. Schmidt, K.K. Baldridge, J.A. Boatz, S.T. Elbert, M.S. Gordon, J.H. Jensen, S. Koseki, N. Matsunaga, K.A. Nguyen, S.J. Su, T.L. Windus, M. Dupuis, J.A. Montgomery Jr., We use the GAMESS program at DFT-D/B3LYP-D3/6–31G(d) level. J. Comput. Chem. **14**, 1347 (1993)
66. I. Natkaniec, E.L. Bokhenkov, B. Dorner, J. Kalus, G.A. Mackenzie, G.S. Pawley, U. Schmelzer, E.F. Sheka, J. Phys. C Solid State Phys. **13**, 4265 (1980)
67. J. Lei, Y. Shimoi, J. Phys. Soc. Jpn. **80**, 034702 (2011)
68. M.J. Frisch, G.W. Trucks, H.B. Schlegel, G.E. Scuseria, M.A. Robb, J.R. Cheeseman, J.A. Montgomery Jr., T. Vreven, T. Kudin, J.C. Burant, J.M. Millam, S.S. Iyengar, J. Tomasi, V. Barone, B. Mennucci, M. Cossi, G. Scalmani, N. Rega, G.A. Petersson, H. Nakatsuji, M. Hada, M. Ehara, K. Toyota, R. Fukuda, J. Hasegawa, M. Ishida, T. Nakajima, Y. Honda, O. Kitao, H. Nakai, M. Klene, X. Li, J.E. Knox, H.P. Hratchian, J.B. Cross, C. Adamo, J. Jaramillo, R. Gomperts, R.E. Stratmann, O. Yazyev, A.J. Austin, R. Cammi, C. Pomelli, J.W. Ochterski, P.Y. Ayala, K. Morokuma, G.A. Voth, P. Salvador, J.J. Dannenberg, V.G. Zakrzewski, S. Dapprich, A.D. Daniels, M.C. Strain, O. Farkas, D.K. Malick, A.D. Rabuck, K. Raghavachari, J.B. Foresman, J.V. Ortiz, Q. Cui, A.G. Baboul, S. Clifford, J. Cioslowski, B.B. Stefanov, G. Liu, A. Liashenko, P. Piskorz, I. Komaromi, R.L. Martin, D.J. Fox, T. Keith, M.A. Al-Laham, C.Y. Peng, A. Nanayakkara, M. Challacombe, P.M.W. Gill, B. Johnson, W. Chen, M.W. Wong, C. Gonzalez, J.A. Pople, *We Use the Gaussian03 Program at B3LYP/6-31G(d) Level: GAUSSIAN 03*, Revision B. 03 (Gaussian Inc., Pittsburgh, 2003)
69. T. Markussen, R. Rurali, M. Brandbyge, A.-P. Jauho, Phys. Rev. B **74**, 245313 (2006)
70. I.N. Hulea, S. Fratini, H. Xie, C.L. Mulder, N.N. Iossad, G. Rastelli, S. Ciuchi, A.F. Morpurgo, Nat. Mater. **5**, 982 (2006)
71. N. Karl, Synth. Met. **133**, 649 (2003)

# Part IV
# Bridging Different Fields: Challenges for the Future

# Chapter 16
# Function of Conjugated π-Electronic Carbon Walled Nanospaces Tuned by Molecular Tiling

**Toshihiko Fujimori, Fitri Khoerunnisa, Tomonori Ohba, Suzana Gotovac-Atlagic, Hideki Tanaka, and Katsumi Kaneko**

## 16.1 Introduction

Graphene has gathered an intensive interest in physical and chemical sciences [1–4]. Graphene can be regarded as a giant molecule of polyaromatic hydrocarbons (PAH), in which the number ratio of hydrogen against carbon is zero. Then graphene is an ideally π-conjugated molecular system, being one of the popular research targets in chemical science. The most striking point of graphene from interface chemistry is that all carbon atoms are faced to the front and rear surfaces. If we roll the graphene sheet, single-wall carbon nanotube (SWCNT) is obtained. All carbon atoms in SWCNT are faced both two surfaces as well as graphene. However, both surfaces of SWCNT can be distinguished from each other according to the sign of the nanoscale curvature. Therefore, all carbon atoms of SWCNT are

T. Fujimori • K. Kaneko (✉)
Center for Energy and Environmental Science, Shinshu University,
4-17-1 Wakasato, Nagano, 380-8553, Japan
e-mail: kkaneko@shinshu-u.ac.jp

F. Khoerunnisa
Department of Chemistry, Indonesia University of Education, Bandung 40154, Indonesia

T. Ohba
Department of Chemistry, Graduate School of Science, Chiba University, 1-33 Yayoi-cho, Inage-ku, Chiba 263-8522, Japan

S. Gotovac-Atlagic
Public Health Institute, Jovana Ducica 1, 78000, Banja Luka, Bosnia and Herzegovina

H. Tanaka
Department of Chemical Engineering, Kyoto University, Katsura, Nishikyo,
Kyoto 615-8510, Japan

 
H. Ishii et al. (eds.), *Electronic Processes in Organic Electronics*,
Springer Series in Materials Science 209, DOI 10.1007/978-4-431-55206-2_16

on the surfaces which can offer different surface function according to the nanoscale curvature sign. Thus, SWCNT is an intriguing solid, which should be named "an interfacial solid" or "a surface solid" [5]. One of the important subjects in chemical science is catalysis. The catalyst requests high efficiency and selectivity for a chemical reaction and hereby the surface area of the catalyst should be as large as possible for the high efficiency in addition to a nanoscale structure. Then SWCNT being "a surface solid" is the highest potential for catalyst application. The geometrical surface area of SWCNT and graphene is 2,630 $m^2/g$ which guarantees the high efficiency for chemical function [6]; SWCNT has a nanoscale tubular space which can lead to an excellent selectivity for chemical functions. However, the π-conjugated surface is highly inert, although recent study on the reactivity of the π-conjugated carbon surface has shown a specific reactivity [7]. Then, generally speaking, the graphene surface itself is not promising as catalyst. It has been indicated recently that the edge carbon and defects in the graphene structure exhibit unique chemical reactivities [8–10]. The π-conjugated carbon structure consists of $sp^2$ carbon atoms, whereas $sp^3$ carbon atoms contribute to the edge and defect structures. There should be a mixed valence state of $sp^2$ and $sp^3$ carbon atoms around the edges and defects. SWCNT and the related carbons can be converted into the excellent catalysts by introduction of the edge carbons and defect carbons. From these characteristics it is clear that the SWCNT should be studied from the view point of chemistry for contributing to the sustainable technology.

Adsorption itself has a key role in concentration, storage, and separation of substances, being inevitable to construct clean energy and energy-and space-saving technologies. SWCNT has provided a model adsorbent system to develop adsorption science. Figure 16.1 shows the interaction potential profiles of SWCNT with a $CH_4$ molecule [11]. Here the tube diameter of SWCNT is 3.2 nm. The interaction potential of the internal wall of SWCNT having a negative curvature is deeper than that of the external one having a positive curvature, by 2.5 kJ/mol. The bisurface nature of SWCNT of different nanoscale curvatures for molecules is quantitatively described by their interaction potential profile [12, 13]. We can modify the chemical function of SWCNT using the bundle structure; SWCNTs form the bundle structure of hexagonal symmetry which gives the X-ray diffraction in the low diffraction angle range [14, 15]. The SWCNT bundle has three adsorption sites of the interstitial space, the internal tube spaces, and the groove surfaces. M, Ti, Di, and G denote these different sites, as shown in Fig. 16.2a. Here M is the monolayer site on the internal tube surface and molecules can occupy the internal tube space in addition to the monolayer site with increase of the gas pressure. If the SWCNT is bold, two kinds of interstitial sites of Ti and Di are available for molecular adsorption. In the case of fine SWCNTs, molecules cannot access the Di site, but only the Ti one; even Ti sites are not available for molecular adsorption for finer SWCNTs. The groove surface is faced to the external surface and is surrounded by

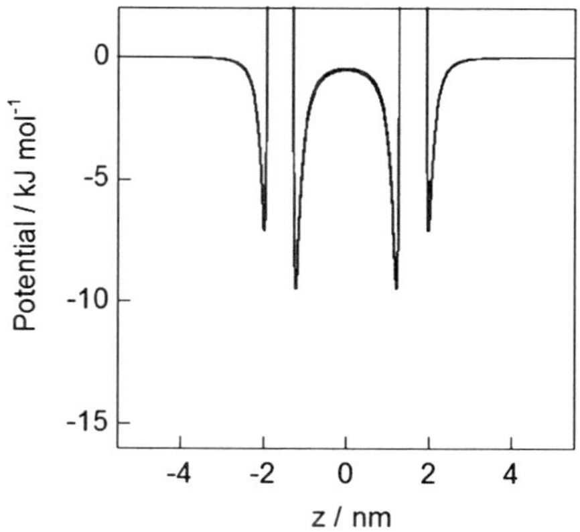


**Fig. 16.1** The interaction potential profile of SWCNT (diameter-3.2 nm) with a $CH_4$ molecule

two tubes, which is denoted G. The external surface of the bundle is expressed by the shadowed region in Fig. 16.2a. Figure 16.2b shows the interaction potential profiles for an $H_2$ molecule. Here the SWCNTs are considerably bold (3.2 nm in the diameter) and the intertube distance is 0.70 nm. The interstitial space at Di has the deepest interaction potential well and the M site at the internal tube surface gives a relatively deep potential well. In this geometry, molecules are preferentially adsorbed in the interstitial spaces around the Di sites. This interaction potential profile varies sensitively with the tube diameter of SWCNT. If the tube diameter is 1 nm, Di site has no attractive interaction potential. On the contrary, Ti sites having the deepest interaction potential well can accept small molecules such as $H_2$, because they are surrounded by three contacting SWCNTs even though these three surfaces have the positive curvature. However, the interstitial space volume (pore volume) is very small compared with other two adsorption sites. Interestingly, even the groove site has a considerably deep interaction potential well because of overlapping of the interaction potentials from two tubes. As the groove sites have the best accessibility for molecules, except for the external surface, their application to beneficial phenomena such as catalysis is promising.

There are several tuning routes of the adsorption sites of SWCNT bundles such as control of the tube diameter and bundle size, intercalation in the interstitial sites, and partial embracing of the internal tube spaces. Here molecular tiling with PAH molecules using liquid phase adsorption is a powerful method for the intercalation and embracing the internal tube space. Also the intercalation and embracing routes with PAH molecules can tune the electronic properties of the SWCNT through charge transfer interaction. In this chapter, unique and important functions of tubular carbon nanospaces surrounded by highly π-conjugated walls are described and the useful examples of the tuning routes with molecular tiling using PAH molecules are shown.

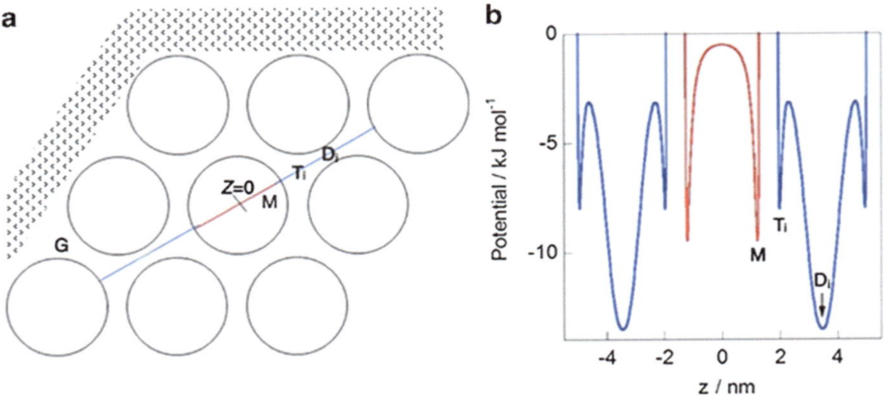


**Fig. 16.2** The SWCNT bundle model (**a**) and the interaction potential profile of SWCNT with an $H_2$ molecule (**b**) along the axis illustrated in Fig. 16.2 a with a *solid line*. Here the tube diameter and intertube distance are 3.2 and 0.70 nm, respectively. The external surface area of the bundle is illustrated with a shadowed region

## 16.2 Single- and Double-Wall Nanocarbons

There are SWCNT and single wall nanohorn (SWCNH) in single wall nanocarbons. SWCNT is produced by arc discharge method and laser ablation method from graphite in the presence of metal catalysts such as Fe, Ni, and Co. Also SWCNT is synthesized by the chemical vapor deposition (CVD) with the metal catalysts [16]. Generally speaking, the discharge and laser ablation methods produce highly crystalline SWCNT. On the other hand, the CVD method produces less-crystalline SWCNT. When the catalyst metal nanoparticles are embedded in the substrate, highly pure SWCNTs without bundle formation are obtained by the CVD method [17]. The crystallinity of defective SWCNT can be improved by the heat-treatment above 2,000 K in an inert gas. The metallic catalysts must be removed to elucidate the intrinsic properties and functions of SWCNT. The perfect purification is not necessarily easy. The post heating treatment after the dissolution of metallic catalyst with mixed acid is efficient [16, 18, 19]. However, the purification procedure often gives rise to defective structures. Figure 16.3 shows transmission electron microscopic (TEM) images of SWCNTs by laser ablation (SWCNT-LA) and CVD (SWCNT-CVD). SWCNT-LA has the bundle structure of an ordered hexagonal symmetry.

The internal tube space of SWCNT is closed by the caps at the ends. The caps can be removed by selective oxidation treatment and thereby the internal tube spaces are available for molecular adsorption and reaction. On the other hand, SWCNT-CVD has almost no bundle structure and each SWCNT has many defects. This SWCNT-CVD is unique and other SWCNT prepared by CVD method has a less ordered bundle structure. This difference can be clearly observed in the

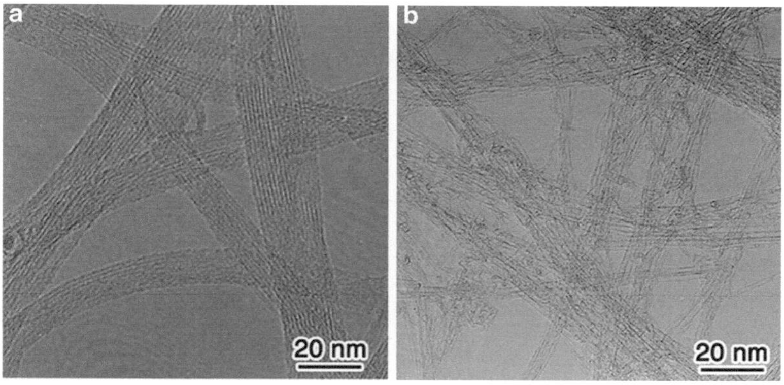


**Fig. 16.3** TEM images of (**a**) SWCNT prepared by laser ablation (SWCNT-LA) and (**b**) SWCNT prepared by CVD (SWCNT-CVD)

**Fig. 16.4** XRD patterns of SWCNT prepared by laser ablation (SWCNT-LA) and SWCNT prepared by CVD (SWCNT-CVD)

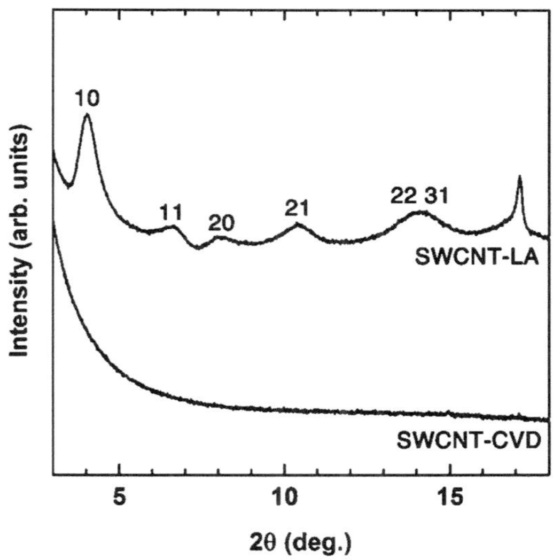


X-ray diffraction patterns. Figure 16.4 shows the XRD patterns of both SWCNTs using synchrotron X-ray source ($\lambda = 0.100$ nm). SWCNT-LA has clear peaks due to the ordered bundle structure, whereas SWCNT-CVD has no peak. Here the observed Bragg peaks of SWCNT-LA are assigned to the 10, 11, 20, 21, 30, 22, and 31 reflections, respectively, of the super lattices of the ordered bundle structure, which are influenced sensitively by the above-mentioned tuning treatment. XRD is an effective method to understand the modified structure of the SWCNT bundles with the molecular tiling method.

Single-wall carbon nanohorn (SWCNH) is one of single-wall nanocarbons. SWCNH is produced by $CO_2$ laser ablation of graphite under Ar atmosphere [20]. SWCNH has an assembly structure of horn-shaped particles whose tube

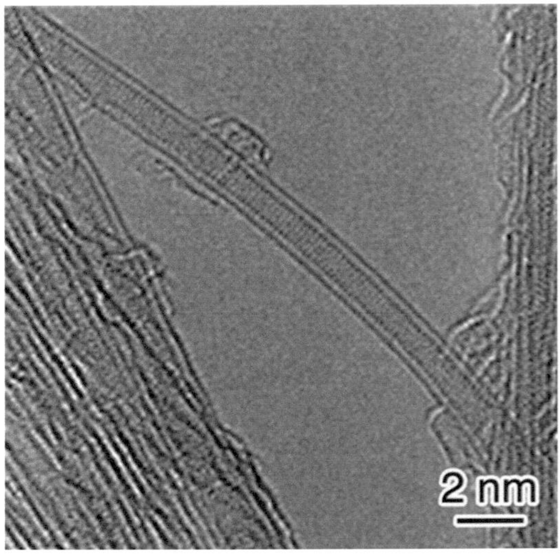


**Fig. 16.5** TEM image of DWCNT prepared by CVD technique

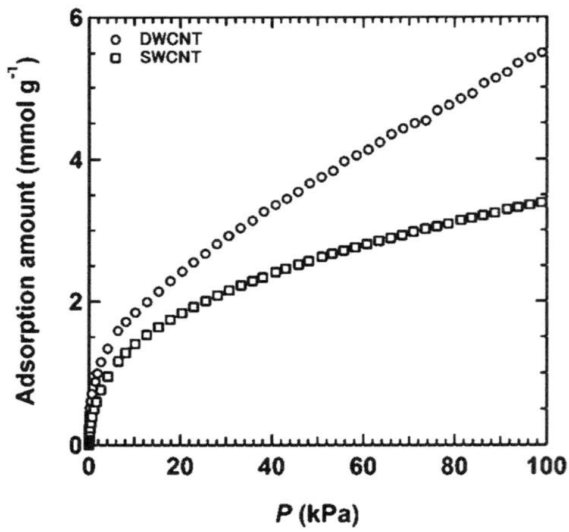


**Fig. 16.6** $H_2$ adsorption isotherms of DWCNT and SWCNT at 77 K

diameter is in the range of 2–5 nm. As SWCNH is more defective than SWCNT, detailed feature of SWCNH is described in the next section: Sect. 16.3.

Double wall carbon nanotube (DWCNT) has a bundle structure similar to SWCNT [21]. Figure 16.5 shows TEM image of DWCNT prepared by CVD technique. We can see well-grown double wall structure and well-ordered bundle structure, although the DWCNT is prepared by the CVD method. In this case, we can observe considerably sharp X-ray diffraction peaks coming from the ordered bundles of the hexagonal symmetry [22]. The interstitial pores of DWCNT have deeper interaction potential wells even for $H_2$ than those of SWCNT, because the double-wall structure has the deeper interaction potential well than the single-wall one. Figure 16.6 shows the $H_2$ adsorption isotherms of DWCNT and SWCNT at

77 K. The adsorption amount of $H_2$ on DWCNT is larger than that on SWCNT even in terms of adsorption amount per unit weight of nanotube samples, because the interstitial pores of this DWCNT have an attractive interaction potential large enough to adsorb $H_2$ molecules. As the critical temperature of $H_2$ is 35 K, the above result explicitly indicates the importance of the enhanced interaction of DWCNT for $H_2$.

## 16.3 Morphological Defect Structure

SWCNT consists of rolled graphene and cap structures, as described above. The graphene has a carbon hexagon network, being planar due to a perfect aromaticity. The addition of pentagon or heptagon in the flat graphene brings about positively and negatively curved structures, respectively, as shown in Fig. 16.7. These pentagons and heptagons play an essential role in a dynamical change of carbon nanotube structure. The hemispherical cap part of SWCNT has several pentagons. The pentagon and heptagon can supply a quasi-free electron and hole, respectively, and thereby these defects are strongly associated with electronic properties of SWCNT. Figure 16.8 shows a high resolution TEM (HR-TEM) image of SWCNHs.

SWCNH consists of tube and tip parts; the tube diameter distributes from 2 to 5 nm. SWCNH has many defects such as pentagons and heptagons [23, 24]. The electrical conductivity responses of SWCNH on $CO_2$ and $O_2$ adsorption indicate an n-type semiconductor of SWCNH; adsorption of $CO_2$ as an electron donor increases the electrical conductivity, while adsorption of $O_2$ as an electron acceptor decreases it. Generally speaking, carbon materials including carbon nanotube exhibit p-type semiconductivity and thereby the electronic property of SWCNH is quite unique, being ascribed to predominant presence of pentagons. These morphological defects are slightly unstable compared with the hexagon structure; oxidation of single-wall nanocarbons begins from these morphological defects,

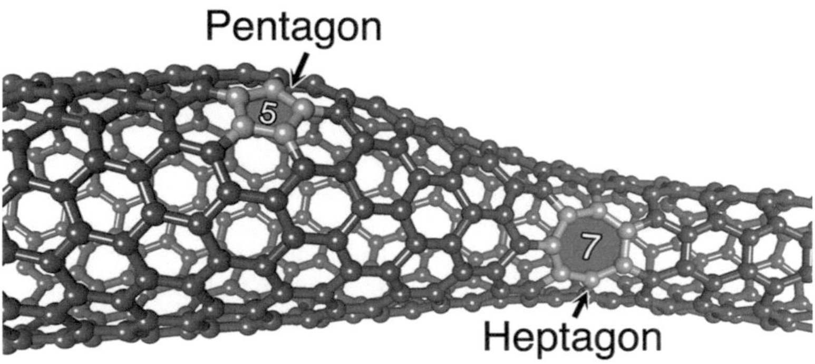


**Fig. 16.7** Pentagnal (positive curvature) and heptagonal (negative curvature) carbon rings incorporated in the hexagonal lattice of a SWCNT

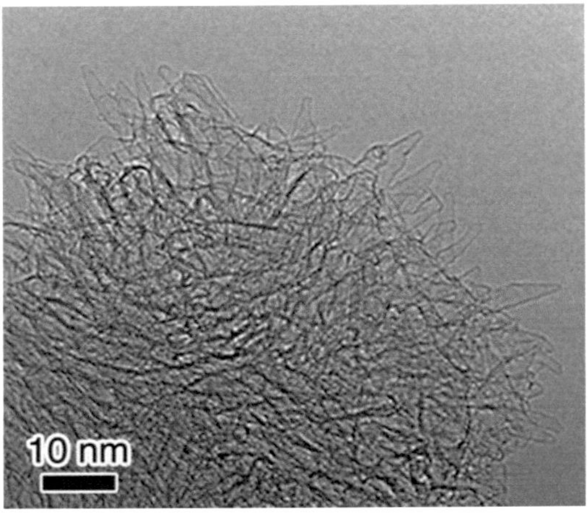


**Fig. 16.8** A TEM image of SWCNHs

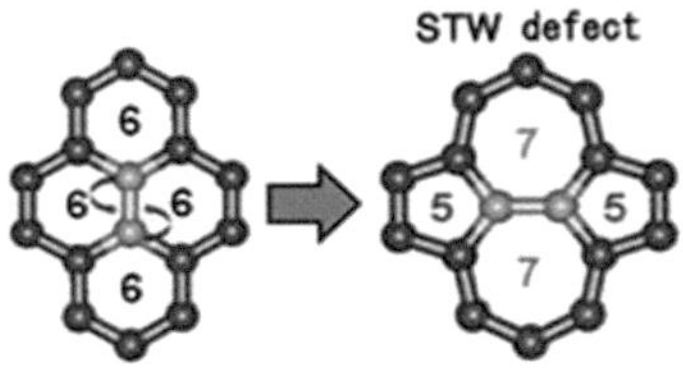


**Fig. 16.9** A schematic of formation of Stone-Thrower-Wales (STW) defect from carbon hexagon network

inducing the selective removal of the cap structures with the oxidation treatment. If we need to use the internal spaces of single-wall nanocarbons, the caps must be removed by the oxidation in advance. Thus, morphological defects are essentially important in chemical reactivity of single-wall nanocarbons.

It is noteworthy to describe the importance of Stone-Thrower-Wales (STW) defects in the structural dynamics of single wall nanocarbons [23]. The formation model of the STW defect in the carbon hexagon network is illustrated in Fig. 16.9. The STW defect can maintain the planar structure of considerable stability. However, the STW defect can work as the trigger of the structure change of the graphene structure.

We need a highly sensitive characterization method of these defects to unveil the stability and dynamic nature of the single-wall nanocarbons. Suenaga et al. showed explicitly the STW structure of SWCNT with high resolution TEM [25]. However, the defects are irradiated by electrons in vacuo during the TEM observation. If we have another methodology which is available in any atmosphere and gives less damages to single wall nanocarbons than TEM, it should be applicable to study the role of the defects in the surface reactivity of single wall nanocarbons. Fujimori et al. applied the surface-enhanced Raman scattering (SERS) with the DFT calculation to characterize these defects [23, 26, 27]. Figure 16.10 shows the SERS spectrum of SWCNH together with ordinary Raman spectrum under ambient atmosphere. The SERS gives several new peaks which are not observed in the ordinary Raman spectrum. Some of these peaks can be assigned to the characteristics jelly fish vibrations which are evaluated

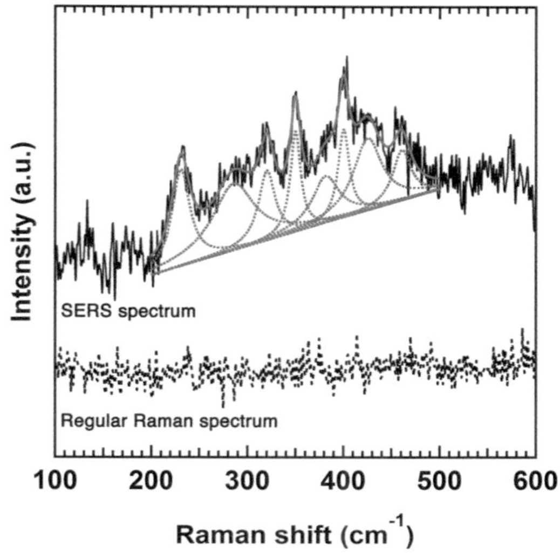


**Fig. 16.10** SERS spectrum and ordinary Raman spectrum of SWCNH. Here a broad SERS spectrum is deconvoluted into various peaks

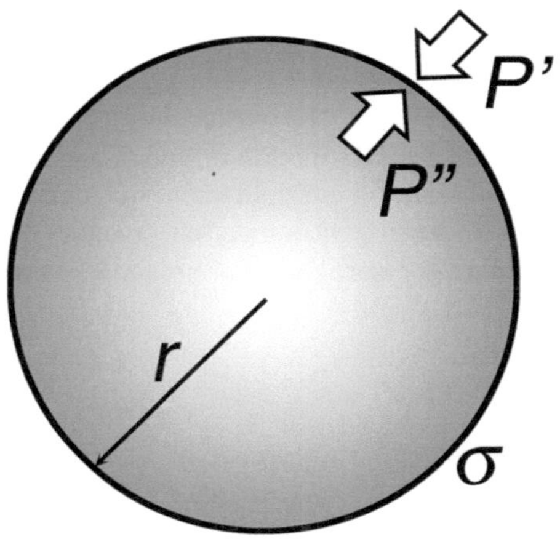


**Fig. 16.11** A schematic of the internal pressure $P''$ and the external pressure $P'$ in a spherical cavity; $\sigma$ and $r$ are the surface tension of the wall and radius of the cavity, respectively

with the DFT calculation. Therefore, the SERS is applicable to study surface process around the defects on the single-wall nanocarbons. In particular, characterization of PAH molecules adsorbed on single wall nanocarbons tuned with molecular tiling method using SERS is promising in future.

## 16.4 In-Pore Super High Pressure Effect

Thermodynamics tells the pressure difference between the concave and convex sides of the curved surface according to Laplace's formula as given by Eq. (16.1), for a simple cavity shown in Fig. 16.11 [28]. Here, the surface is assumed to be a graphene-like wall which has a spherical structure having radius of 1 nm. $P'$ and $P''$ are the mutually opposite pressures working against the external and internal surfaces of the cavity. When the surface tension, σ, of graphite ($150 \pm 30$ mJ/m$^2$) and the cavity is stable [29], the pressure difference $\triangle p$ is 0.3 GPa.

$$\Delta \mathrm{P} = \mathrm{P}'' - \mathrm{P}' = \frac{2\sigma}{\mathrm{r}} \tag{16.1}$$

Hence the internal nanoscale cavity wall should be exposed to the high compression pressure. When the adsorbed layer is formed on the internal cavity wall, the adsorbed layer may be compressed in a similar way. Also we can estimate very briefly the plausible pressure for molecules adsorbed on the pore walls using the molecule-pore interaction potential profile; the effective pressure for a molecule be evaluated from the slope of the interaction potential per the cross-sectional area of the molecule. This simple estimation also predicts the high pressure application for molecules in micropores.

Kaneko et al. showed perfect dimerization of NO in the slit-shaped micropores of activated carbon fiber [30, 31] and cylindrical micropores of zeolite [32] at 303 K with an aid of magnetic susceptibility measurements. The NO dimers are found in the liquid state of NO below boiling temperature (121 K) and then the observed NO dimerization in the pores of ACF at 303 K is unusual. We can determine the dimerization of NO molecules using the magnetic susceptibility measurement, because the NO monomer and dimer are paramagnetic and diamagnetic, respectively, giving greatly different magnetic susceptibilities.

The dimerization reaction given by Eq. (16.2) at 303 K in the bulk phase and ambient pressure does not proceed (<less than 1 mol%).

$$2\mathrm{NO(g)} \leftrightarrows \mathrm{(NO)_2(g)} \tag{16.2}$$

Hence, this fact indicates that NO molecules in the micropores are effectively compressed by high pressure. Later the complete dimerization of NO in the tube spaces of SWCNT was also evidenced by Byl et al. using the FT-IR measurement [33]. Molecular simulation supported the enhanced NO dimerization in carbon micropores [34]. Also the disproportionation reaction of the NO dimers into $NO_2$ and $N_2O$ can be accelerated in the slit-shaped carbon pores below 0.1 MPa at 303 K [35, 36], although the reaction in the bulk phase requests the highly compressed NO gas above 20 MPa [37]. Also $CO_2$ reduction of a representative high pressure electrochemical reaction proceeds in the micropores of activated carbon fiber without application of the high pressure requested in the bulk phase reaction [38]. Carbon micropores are appropriate to clarify the high pressure effect, because the π-conjugated pore walls are stable and uniform and the interaction with molecules is quite strong per unit weight. Carbon micropores have no specific interaction sites as observed in zeolite. As the cylindrical tube spaces have much deeper

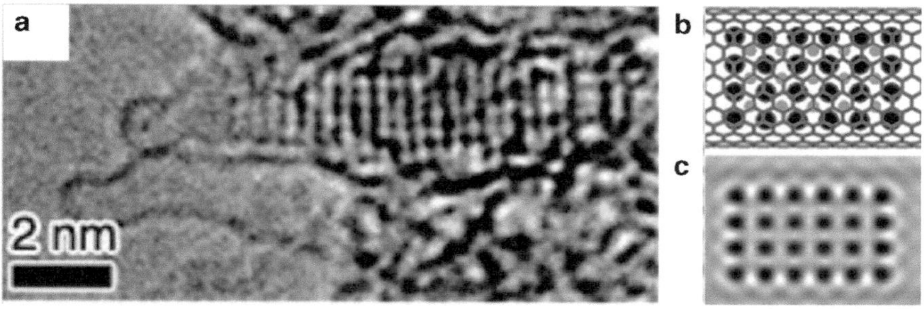


**Fig. 16.12** (**a**) A TEM image of KI inside SWCNH, (**b**) model of KI in carbon tube, and (**c**) simulated TEM image of the model KI in carbon nanotube

interaction potential well than the slit-shaped ones, a more prominent high pressure effect can be evidenced in the carbon nanotube.

Urita et al. [39] showed evidently the formation of KI of high pressure phase in the tube spaces of SWCNH even below ambient pressure using the high resolution TEM and synchrotron X-ray diffraction. The bulk KI crystals transform from ambient pressure phase of B1 structure to the high pressure phase of B2 structure at 1.9 GPa [40]. KI was deposited in the tube spaces of SWCNH at 1073 K below 0.1 MPa. Figure 16.12 shows HR-TEM images of the KI crystallite grown in the SWCNH tube spaces. Here the KI crystallite image in the tube space at the open entrance is also shown. We can see the lattice structure which is attributed to the ordered structure of iodine atoms. The structure can be assigned to the slightly distorted high pressure phase of KI. Synchrotron X-ray diffraction of KI formed in the tube spaces was measured; there were many peaks coming from the lattices of distorted high pressure phase. Accordingly, the effective compression pressure in the tube spaces of the SWCNH of 2.5 nm in the average diameter is explicitly shown to be more than 1.9 GPa.

Furthermore, more intriguing evidence on the high pressure compression effect was observed in sulfur. Well-known solid sulfur consists of eight member rings of sulfur atoms ($S_8$), being a typical insulator. However, compression of the solid sulfur with ~90 GPa dissociates the $S_8$ ring structure to form a two-dimensional atomic layer showing metallic property [41]. Also theoretical study predicts that atomically one dimensional (1D) structures of linear and zigzag forms exhibit metallic nature [42–44]. Fujimori et al. [45] succeeded to prepare a completely 1D sulfur chain of high crystallinity in the tube spaces of well-crystalline SWCNT and DWCNT. Sulfur was introduced in the tube spaces at 873 K. Figures 16.13 and 16.14 show TEM image and synchrotron X-ray diffraction patterns of 1D sulfur chains inside a SWCNT and a DWCNT. Zigzag and linear chain structures of sulfur atoms inside broader and narrower tube spaces are clearly shown. Surprisingly, X-ray diffraction patterns of these 1D sulfur chains are observed, as given in Fig. 16.14.

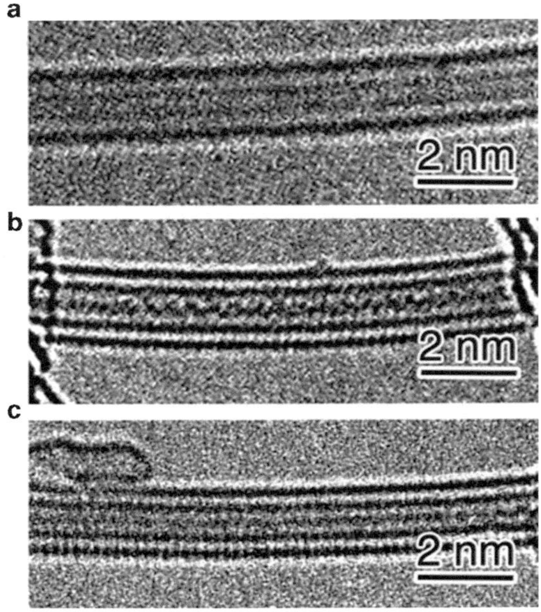


**Fig. 16.13** TEM images of sulfur chains inside (**a**) SWCNT and (**b**)–(**c**) DWCNT. (Reproduced from Fujimori et al. [44])

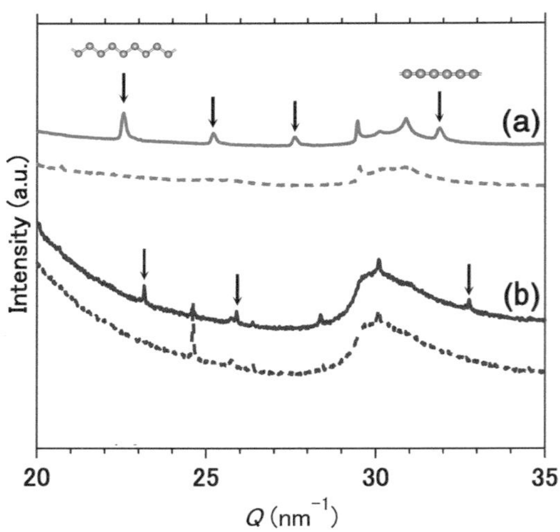


**Fig. 16.14** XRD patterns of (**a**) SWCNT and (**b**) DWCNT with and without 1D sulfur chains. *Arrows* indicate the Bragg peaks of 1D sulfur chains. *Solid* and *dotted lines* correspond to XRD patterns with and without 1D sulfur chains, respectively (Reproduced from T. Fujimori et al. [44])

The peak positions correspond to S-S and S-S-S distances in the linear and zigzag chains, respectively; the zigzag chain is formed in the wider tube space of 0.68 nm in the diameter, whereas the linear one is formed in the narrower tube space of 0.60 nm in the diameter. The X-ray peak shape analysis provides that the lengths of the sulfur chains inside SWCNT and DWCNT are 40 and 100 nm in average, respectively. The direct-current (DC) electrical conductivity of carbon nanotubes (SWCNT and

DWCNT) and sulfur-encapsulated carbon nanotubes was measured from 2 to 300 K using the buckypaper form which is prepared by filtration of the dispersed carbon nanotubes. The formation of the sulfur chains inside SWCNT and DWCNT enhanced the electrical conductivity evidently. Therefore, these atomically 1D-sulfur chains should be metallic. The dimensions of the variable hopping conduction for sulfur-encapsulated carbon nanotubes and pure carbon nanotubes are three and two, respectively, indicating that the 1D sulfur chain has an inherent high conductivity. In this case, confinement of sulfur in the carbon nanotube spaces of about 1 nm order induces the super high pressure reaction for the formation of the metallic sulfur without application of the high pressure.

The in-pore high pressure effect of carbon nanospaces comes from graphene-like walls of the highly dense carbon hexagon structure bonded by the conjugated π-electrons. Recently statistical mechanical studies on the in-pore high pressure effect have been carried out for the slit-shaped pores by Gubbins et al., indicating the importance of the horizontal pressure effect [46, 47].

## 16.5 Adsorption-Aided Tuning of Electronic and Interfacial Properties of SWCNT

As the SWCNT bundles of well-crystalline SWCNTs have three kinds of strong adsorption sites for atoms and molecules on modifier molecules, we can modify electronic properties of SWCNT through charge transfer interaction with adsorption of molecules on the different sites. The most prominent and stable tuning of the SWCNT properties should stem from the adsorption of the modifier atoms or molecules in the interstitial spaces, which guarantees the strongest interaction between the modifiers and SWCNT. This tuning can be called "intercalation" as the analogy of the intercalated graphite in which the intercalants are inserted between the stacked graphene layers, varying the electronic properties [48]. If we can encapsulate the modifier atoms or molecules only in the internal tube spaces of SWCNT, unique tuning of the electronic properties of SWCNT should be possible, because the modifier atoms or molecules can interact with the concave conjugated π-electron structure in the different way from the convex one of the external surface. Tuning using adsorption of modifier atoms or molecules on the groove sites or on the external surface is another route. Generally speaking, this tuning route is less efficient in comparison with the preceding routes.

We need to introduce an electronic interaction between the modifier molecule and SWCNT to tune the electronic properties of SWCNT. One of the key interactions is charge transfer interaction. If we choose the electron donor and acceptor modifier molecules from π-electronically conjugated molecules, a systematic tuning should be attained.

### 16.5.1 Charge Transfer Interaction

The charge transfer interaction between an aromatic modifier molecule and SWCNT can be approximately described by the Mulliken's resonance theory for a weak donor-acceptor complex [49, 50]. The weak charge transfer interaction between an electron donor molecule (D) and an electron acceptor molecule (A) can be described in terms of the following wave function, $\Psi_N$ (Eq. (16.3)).

$$\Psi_N(AD) = a\Psi_0(A,D) + b\Psi_1(A^- - D^+) \tag{16.3}$$

Here $\Psi_0$ has been termed by Mulliken the "non-bonding" wave function. This bonding corresponds to the dispersion interaction for neutral D and A molecules. The wave function $\Psi_1$ has been termed the "dative" wave function. This function expresses the complete transfer of an electron from the donor to the acceptor. Here a and b are constants. Therefore, the weak charge transfer interaction is described in terms of the resonance concept between the non-bonding contact state and electron-transfer one.

The ground state energy ($W_N$) for weak interactions is given by sum of the energy $W_0$ of the separated D and A, the non-bonding energy, and the resonance energy $X_0$. The sum of $W_0$ and the non-bonding energy expresses the energy of D and A molecules without the charge transfer interaction. The $W_N$ is approximated by the second-order perturbation theory, as given by Eq. (16.4).

$$W_N \equiv \int \Psi_N H \Psi_N d\tau \approx W_0 - \frac{(H_{01} - S_{01}W_0)^2}{(W_1 - W_0)} \tag{16.4}$$

where $W_0 = \int \Psi_0 H \Psi_0 d\tau$ and $W_1 = \int \Psi_1 H \Psi_1 d\tau$. $W_1$ is the energy of the dative structure $A^-D^+$. $H_{01} = \int \Psi_0 H \Psi_1 d\tau$, $S_{01} = \int \Psi_0 \Psi_1 d\tau$, and H is the exact Hamiltonian for the charge transfer complex. Hence the resonance energy $X_0$ depends effectively on the energy difference ($W_1$ - $W_0$) and the overlap integral $S_{01,}$ because $H_{01}$ is governed by $S_{01}$. The coefficient ratio of b/a is given by the second-order perturbation theory (Eq. (16.5)).

$$b/a \approx (H_{01} - S_{01}W_0)/(W_1 - W_0) \tag{16.5}$$

The large (b/a) indicates a greater contribution of the electron transfer state in the A· D complex. Here an electron jumps from an occupied molecular orbital in D, $\varphi_D$, to an unoccupied molecular orbital in $\varphi_A$, on the charge transfer interaction between A and D. Then $S_{01}$ can be approximately expressed by the overlap integral between $\varphi_A$ and $\varphi_D$, as given by Eq. (16.6).

$$S_{01} \equiv \sqrt{2}S_{AD}\left(1 + S_{AD}^2\right)^{1/2} \tag{16.6}$$

where $S_{AD} = \int \varphi_A \varphi_B d\tau$ As $S_{AD}$ is much smaller than 1, $S_{01}$ is proportional to $S_{AD}$. Also $H_{01}$ almost varies linearly with $S_{AD}$ for small $S_{AD}$ values. Consequently the electron donor and electron acceptor moieties tend to orientate themselves to give the maximum overlap integral $S_{AD}$. ($W_1$ - $W_0$) can be expressed by the ionization energy of the donor, $I_D$, electron affinity of the acceptor, $E_A$, the non-bonding energy of A· D contact ($G_0$) and the Coulombic attractive energy of the $A^-$,$D^+$ ions ($G_1$) as given by Eq. (16.7),

$$W_1 - W_0 = I_D - E_A - (G_1 - G_0) \tag{16.7}$$

($I_D - E_A$) is more molecular structure sensitive than ($G_1 - G_0$) and thereby both of $I_D$ and $E_A$ are key factors in charge transfer interaction. Therefore, a series of modifier molecules having different $I_D$ and $E_A$ are recommended to use for tuning the electronic properties of SWCNT. Also, the above classical charge transfer theory is effective for understanding the interaction between modifier atoms or molecules and SWCNT, although it is qualitative.

It is well-known that graphitic structured carbon materials can work as electron donor and acceptor depending on the interacting molecules and atoms. This is because graphite has a unique density of state structure; the band gap is almost zero. Therefore, the graphitic carbon can work as a donor and an acceptor, depending on the interacting atoms or molecules. On the other hand, SWCNT can be obtained as the mixture of metallic and semiconducting SWCNTs. The work function of metallic SWCNT is used for description of the electronic state, whereas the electron affinity and ionization potential can be used to describe the electronic properties of semiconducting SWCNT. Accordingly we need to use several modifier molecules having different electron affinity values to tune the electronic properties of SWCNT through the charge transfer interaction. Here we chose polycyclic aromatic hydrocarbons (PAHs) of a planar structure which can interact strongly with the SWCNT wall through the π-π charge transfer interaction [51–53].

### 16.5.2 $C_{60}$-Intercalated SWCNT

A $C_{60}$ molecule of conjugated π electrons has a spherical shape of 1 nm in the external diameter [54, 55]. Therefore, $C_{60}$ molecules are one of optimum intercalants for the SWCNT bundle with preservation of conjugated π-electron system. Intercalation of $C_{60}$ in the SWCNT (SWCNT-LA) bundle can be easily attained by ultrasonication of a $C_{60}$ toluene solution in the presence of SWCNT [56]. The TEM image is taken on the cross-section of the bundle as shown in Fig. 16.15. The considerably ordered bundle structure is still preserved even after the intercalation treatment with sonication. We can observe local structures of tetragonal symmetry in addition to the distorted hexagonal structure. The interlayer distances of the bundle of hexagonal and tetragonal superlattices of the $C_{60}$ –intercalated SWCNT bundle are 2.03 and

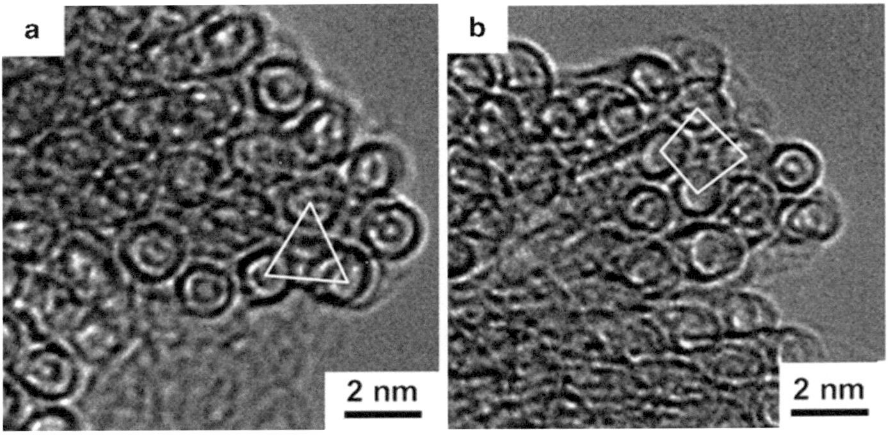


**Fig. 16.15** TEM images of distorted (**a**) hexagonal and (**b**) tetragonal bundle structures having $C_{60}$

**Fig. 16.16** XRD patterns of SWCNT with and without doped $C_{60}$

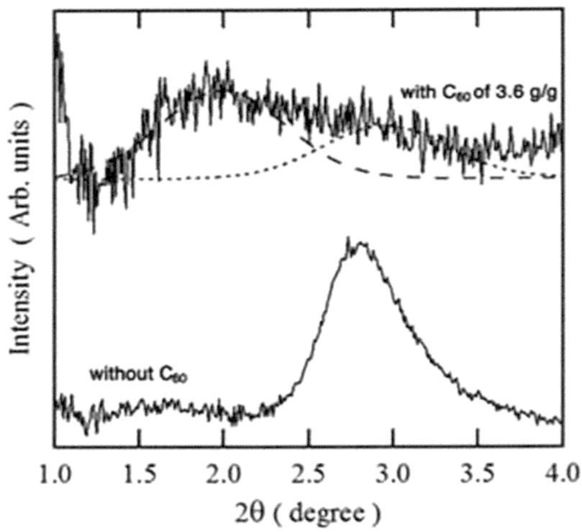


1.92 nm, respectively, from the geometrical evaluation, being close to the interlayer distances of the distorted hexagonal and tetragonal lattices observed by TEM (Fig. 16.15). Figure 16.16 shows X-ray diffraction patterns of SWCNT and $C_{60}$ –intercalated SWCNT. The SWCNT bundle has a clear peak due to the superlattice of the bundle of the hexagonal symmetry at $2\theta = 2.81°$ (interlayer distance = 1.44 nm). After intercalation, a very broad peak appears around $2\theta = 2.0°$, corresponding to the interlayer distance of 2.0 nm, which is close to the value observed by TEM.

Consequently, X-ray diffraction gives an evidence on the intercalation formation between SWCNT bundle and $C_{60}$ molecules. Also Raman spectroscopic study shown in Fig. 16.17 supports the intercalation of $C_{60}$ molecules. The Raman spectrum of the $C_{60}$-intercalated SWCNT has no strong peak of $C_{60}$ vibration.

Fig. 16.17 Raman spectra of $C_{60}$-intercalated SWCNT (doped amount = 3.6 g/g), SWCNT, and $C_{60}$

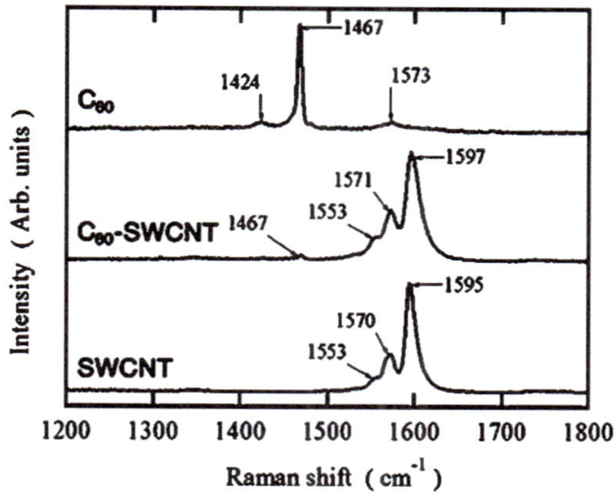


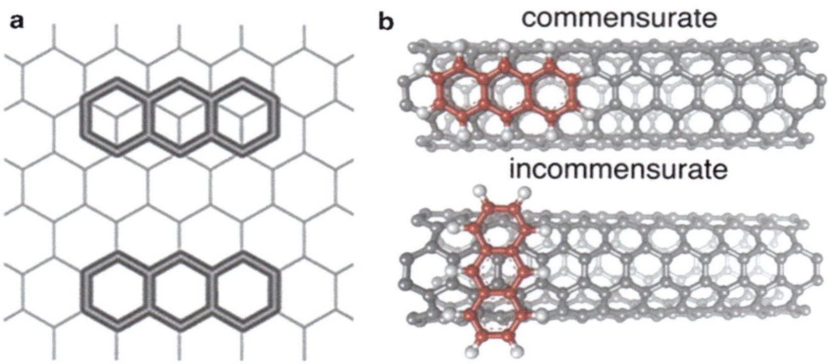


**Fig. 16.18** Possible commensurate / incommensurate configurations of anthracene molecules on graphene (**a**) and SWCNT (**b**)

The information from Raman spectroscopy is limited to the surface layers of the samples and thereby $C_{60}$ molecules are inside the SWCNT bundles. However, the G-band shift on the intercalation of $C_{60}$ is tiny, suggesting no intensive charge transfer interaction between $C_{60}$ and SWCNT.

### *16.5.3 Polycyclic Aromatic Hydrocarbon Intercalated SWCNT*

When PAH molecules are adsorbed on graphene, the adsorbed PAH structure is commensurate or incommensurate to the graphene structure [52, 57], as shown in Fig. 16.18a. However, SWCNT has a nanoscale curved surface and so that the SWCNT surface cannot offer the best contact for PAH molecules. In particular, a

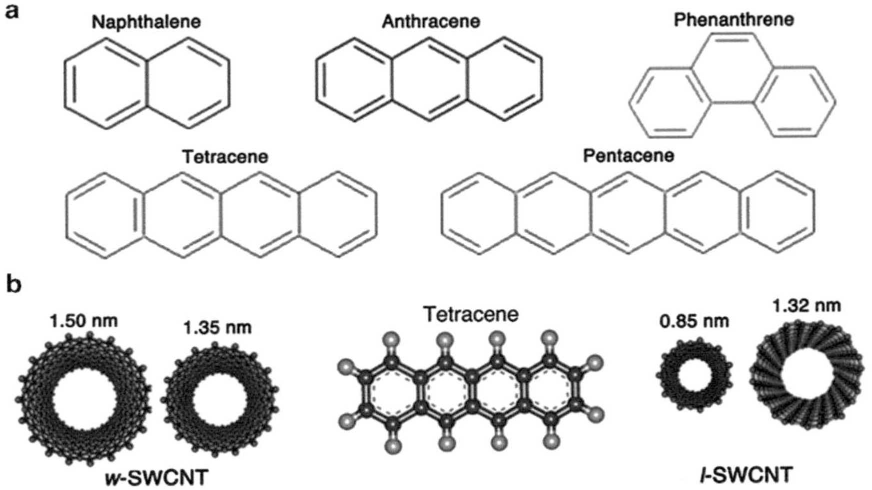


**Fig. 16.19** Chemical structures of PAH molecules (**a**) and relative size of a tetracene molecule to SWCNT of different tube diameters (**b**)

considerably stable commensurate structure having the best π–π contact is very limited in the case of fine SWCNT. Figure 16.18b shows possible commensurate structures of an anthracene molecule on the external surface of SWCNT. Anthracene molecules prefer the molecular orientation parallel to the tube axis than perpendicular to the tube axis. The commensurate structure depends on the PAH molecular structure, the coverage of the surface with PAH molecules, and the chirality of the SWCNT. If SWCNT has a highly defective surface, the π-π interaction is not necessarily predominant and thereby the adsorption of PAH also depends on the surface roughness of SWCNT. Gotovac et al. measured adsorption isotherms of naphthalene, anthracene, phenanthrene, tetracene, and pentacene on well-crystalline and less-crystalline SWCNTs with the liquid phase adsorption [58–60]. Figure 16.19 shows the chemical structure of these PAH molecules and the relative dimension of SWCNTs against tetracene. The well-crystalline SWCNT (*w*-SWCNT) is produced by laser ablation method, while less-crystalline SWCNT (*l*- SWCNT) is HiPco SWCNT purified with $HNO_3$ and HCl. *w* -and *l*-SWCNTs have the surface areas of 380 and 545 $m^2/g$, respectively. Here HiPco SWCNT is a popular SWCNT, being produced with high-pressure CO ("HiPco" comes from **hi**gh-**p**ressure **CO**). Consequently, most of SWCNT samples have caps on both ends. The tube diameters from the RBM bands are 1.35 and 1.50 nm for *w*-SWCNT and 0.85 and 1.32 nm for *l*-SWCNT. Accordingly almost PAH molecules cannot be adsorbed in the internal tube spaces. The crystallinity difference of two SWCNT samples is evident through the G/D band ratio of Raman spectra (*w*-SWCNT : 55,: *l*-SWCNT 12).

Tetracene and pentacene are more preferentially adsorbed on *w*-SWCNT, whereas phenanthrene is more adsorbed on *l*-SWCNT. Here the molecular width of both tetracene and pentacene in the short axis is 0.74 nm, while that of

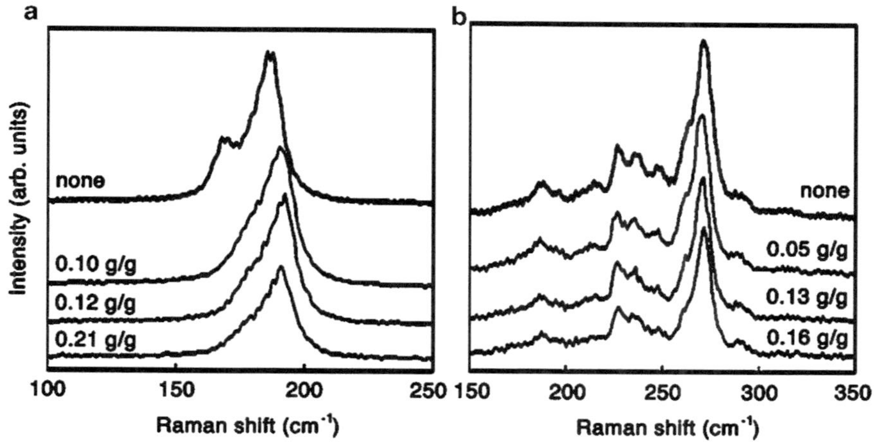


**Fig. 16.20** RBM spectra of w-SWCNT (**a**) and *l*-SWCNT (**b**) having different amounts of adsorbed tetracene

phenanthrene is 0.81 nm. Also the RBM of Raman spectrum of *w*-SWCNT induces a clear high frequency shift on adsorption of tetracene, while adsorption of tetracene does not change the RBM of *l*-SWCNT as shown in Fig. 16.20. Consequently tetracene interacts strongly with *w*-SWCNT surface through π-π interaction. However, tetracene does not interact with the conjugated π-electrons of *l*-SWCNT, but with local surface functional groups. Then a linear PAH molecule should prefer the commensurate structure in which the molecular long axis is parallel to the SWCNT tube axis, as shown in Fig. 16.18b.

Naphthalene with the molecular length being shorter than anthracene and tetracene can more strongly interact with *w*-SWCNT on liquid phase adsorption, although naphthalene crystals have the extremely low sublimation temperature and the high vapor pressure at ambient temperature. The naphthalene-adsorption treated w-SWCNT shows very interesting property. Here the *w*-SWCNT was filtrated after adsorption of naphthalene in the liquid phase to obtain the buckypaper form. The naphthalene-treated SWCNT was evacuated by the ultrahigh vacuum system to study the electronic structure with ultraviolet photoelectron spectroscopy (UPS) and meta-stable atom electron spectroscopy. The spectroscopic study clarified that naphthalene molecules are not on the external surface of the SWCNT bundle and there is a new peak in the density of state near the Fermi level. Accordingly, the naphthalene-adsorbed SWCNT should be metallic, which stems from the strong interaction of naphthalene molecules with SWCNT. It was ascertained that even pentacene molecules were not stably adsorbed on the external surface of the SWCNT bundle. Also an intensive evacuation at 673 K in ultrahigh vacuum is required to desorb naphthalene completely, indicating physical interaction of naphthalene in the interstitial spaces of the SWCNT bundle. The direct evidence on the intercalation of naphthalene was obtained by the appearance of the new peak at the intertubular spacing of 2.20 nm which increased from 1.43 nm of the intertubular

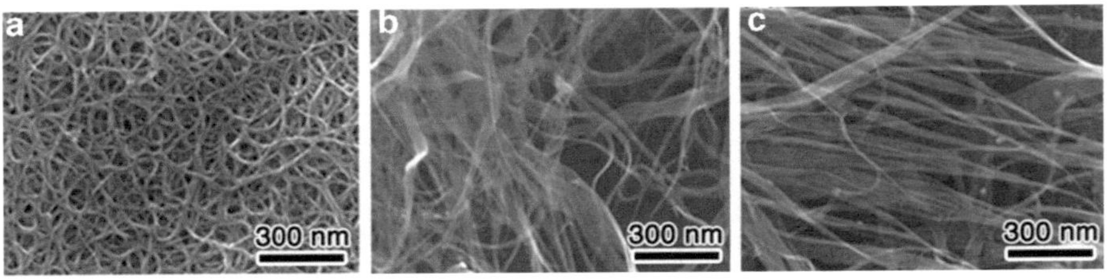


**Fig. 16.21** SEM images of non-treated SWCNT (**a**), DHN-adsorption treated SWCNT (**b**) and DNN-adsorption treated SWCNT (**c**)

spacing of non-adsorbed SWCNT bundles; the difference of 0.77 nm corresponds to the bilayer thickness of the naphthalene molecule. Accordingly, naphthalene molecules are adsorbed on each SWCNT of the bundle form in the monolayer. Then, the naphthalene-SWCNT interaction is strong enough to induce metallicity [60]. Although we have no evidence on the intercalation of anthracene, tetracene, and pentacene molecules in the SWCNT bundle, these PAH molecules also may form the intercalation structure with SWCNT.

### *16.5.4 Naphthalene-Derivatives Intercalated SWCNT*

As naphthalene molecules interact intensively with SWCNT, systematic study on tuning electronic properties of SWCNT using naphthalene derivatives is necessary. We used naphthalene derivatives (NDs) of which moieties are hydroxyl ($-OH$), amine ($-NH_2$), methyl ($-CH_3$) and nitro ($-NO_2$) of different electron donor and acceptor properties; the chemical formulas of the NDs are 1,5-dyhydroxynaptalene ($C_{10}H_8O_2$; DHN), 1,5-diaminonaphthalene ($C_{10}H_{10}O_2$; DAN), 1,5 dimethyl naphthalene ($C_{12}H_{12}$; DMN), and 1,5-dinitronaphalene ($C_{10}H_6N_2O_4$; DNN) [61]. The intercalation treatment is carried out with liquid phase adsorption of the ND molecules at 298 K under sonication. The surface coverage of SWCNT with ND molecules is regulated to be $0.30 \pm 0.05$. The ND-adsorption treatment of purified SWCNT having caps brings about unraveling of the SWCNT bundle, as shown in the field emission scanning electron microscope (SEM) images in Fig. 16.21. The SWCNT bundles have typical entangled structure. However the adsorption treatment with DHN and DNN gives thinner bundles, showing partial debundling. The other ND-adsorption treatments also give similar SEM images. As the SWCNT bundles have a well-ordered two dimensional hexagonal lattice structure, we can observe the 10, 11, 20, and 21 diffraction peaks of the hexagonal lattice at 2.72°, 4.40°, 7.06°, and 9.56°, respectively, as shown in Fig. 16.22. The ND-adsorption treatment induces a lower angle shift, depression, and broadening of the 10 peak. In particular, naphthalene-adsorption reduces markedly the 10 diffraction peak. These results indicate the partial collapse of the ordered bundle structure and insertion of ND molecules in the SWCNT bundles, although we cannot succeed to show it

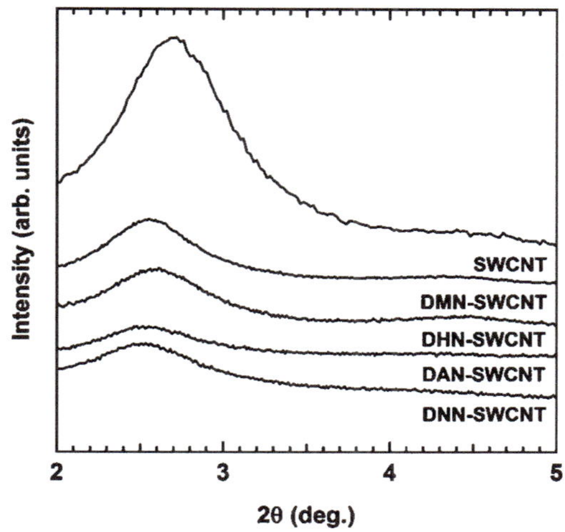


**Fig. 16.22** The X-ray diffraction patterns of SWCNT and ND-adsorbed SWCNT: SWCNT, DMN-SWCNT, DHN-SWCNT, DAN-SWCNT, and DNN-SWCNT

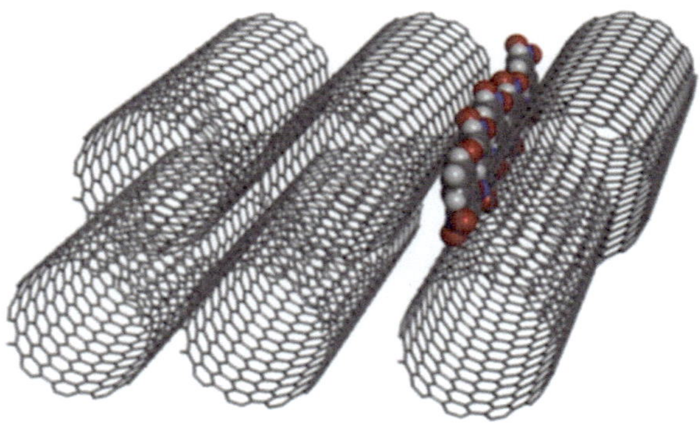

**Fig. 16.23** The plausible configuration of DNN molecules in SWCNT bundle from molecular simulation

directly with TEM observation yet. The lower angle shift of the 10 peak corresponds to the inter-SWCNT layer spacing of about 0.8 nm, being larger than the short molecular-width of the ND molecules. Then, ND molecules can occupy the expanded interstitial spaces, which blocks the interstitial pores. The porosity measurement using $N_2$ adsorption at 77 K supports this mechanism; the micropore volume decreases from 0.14 mL/g for SWCNT to 0.06 mL/g for DNN-treated SWCNT. Preliminary molecular simulation suggests the plausible configuration of DNN molecules in SWCNT bundle, as shown in Fig. 16.23. DNN molecules can occupy the interstitial space of two SWCNTs in the parallel arrangement to the SWCNT walls. Thus, DNN molecules can interact strongly with SWCNTs to tune the electronic properties of SWCNT. The DC electrical conductivity is enhanced by the ND-adsorption treatment, as shown in Fig. 16.24. The conductivity at room temperature increases with elevation of the compression pressure of the SWCNT disk samples in vacuo. The enhancement of the electrical conductivity indicates the charge transfer interaction between the naphthalene derivative molecules and

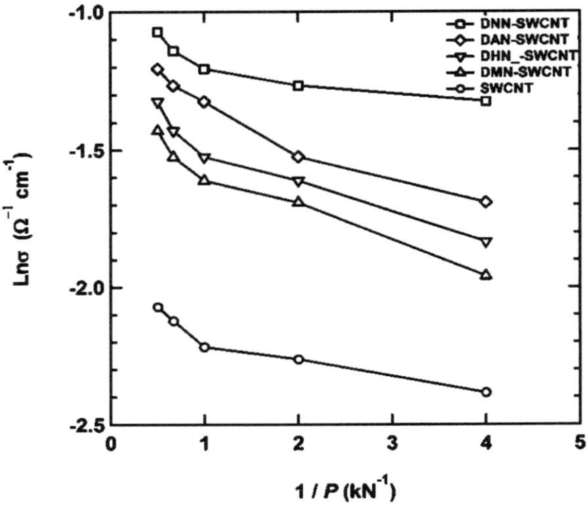


**Fig. 16.24** Compression pressure dependence of electrical conductivity of naphthalene derivative-adsorbed SWCNTs at room temperature in vacuo

SWCNT; the charge transfer should be associated with the debundling behavior observed by SEM. DNN-adsorption is especially efficient for the enhancement. This electrical conductivity increase should stem from the charge transfer interaction between ND molecules and SWCNT [62].

## *16.5.5 TCNQ-Adsorbed SWCNT*

7,7,8,8-tetracyanoquinodimethane (TCNQ) and 2,3-dichloro-5,6-dicyano-p-benzoquinone (DDQ) are representative electron acceptor molecules. TCNQ and DDQ were adsorbed on SWCNT at 298 K with the liquid phase adsorption. A charge transfer between those electron acceptor molecules and SWCNT was evidenced by a higher frequency shift of the Raman G band, agreeing with preceding study [63]. DC electrical conductivity of SWCNT decreased to less than a half value of the SWCNT without adsorption of these molecules for both of TCNQ and DDQ at 298 K, although the adsorption treatment of SWCNT with TCNQ and DDQ is expected to induce the electrical conductivity increase [64]. There is a possibility that TCNQ and DDQ molecules are too large to form the stable intercalation structure with the stable SWCNT bundle. Consequently, the adsorption treatment of SWCNT with TCNQ and DDQ may unravel the ordered bundle structure irregularly, increasing the contact resistance between SWCNTs.

## *16.5.6 Methylene Blue-Encapsulated SWCNT*

When we use open-SWCNT without caps, modifier molecules can occupy the internal tube spaces. Even in this case, the modifier molecules can work as the intercalants in the SWCNT bundle. When the size of the modifier molecule is

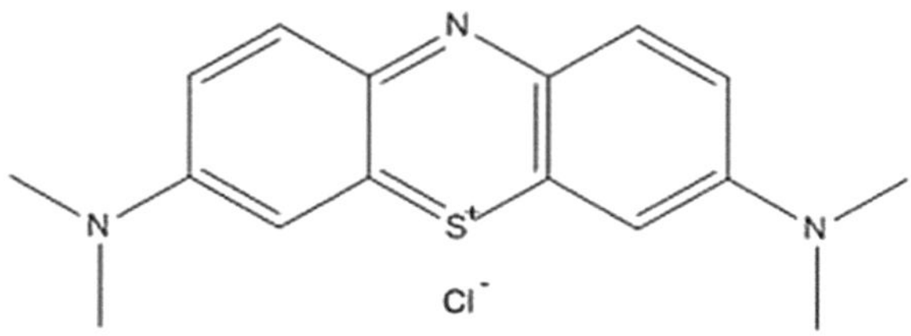


**Fig. 16.25** Molecular structure of methylene blue ($C_{16}H_{18}N_3SCl$)

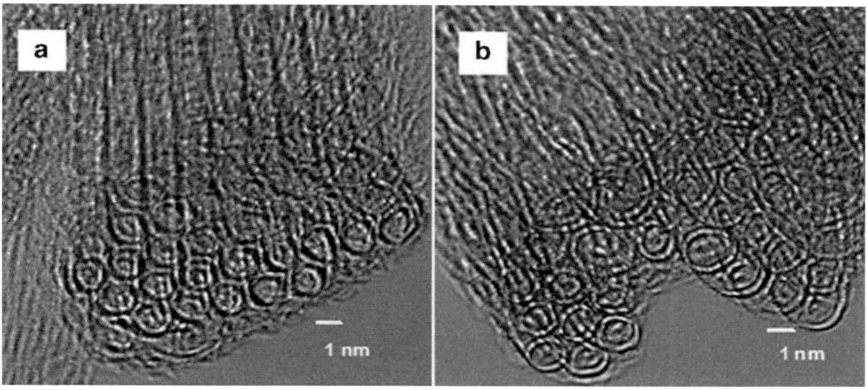


**Fig. 16.26** TEM images of SWCNT (**a**) and MB-adsorbed SWCNT (**b**)

smaller than the internal tube diameter, but larger than the interstitial space size, the modifier molecules favor the internal tube spaces without marked distortion of the super lattice structure of the SWCNT bundle. Both of TEM observation and X-ray diffraction studies can provide an evidence on the encapsulation of modifier molecules in the internal tube spaces.

Methylene blue (MB) being a kind of basic dyes with the planar structure (1.432 nm × 0.665 nm × 0.407 nm) belongs to the phenothiazine compounds, as shown in Fig. 16.25. The molecular structure of MB is close to polycyclic aromatic hydrocarbon. Entrapment of the dye molecules in the internal tube spaces can donate new optical properties to SWCNT. We encapsulated MB molecules using liquid phase adsorption technique, although the encapsulation of dye molecules is carried out with the desorption in the gas phase [65, 66]. The saturated adsorption amount of MB is 110 mg/g, being 0.18 of the functional filling in the internal tube spaces; the MB adsorption amount can be expressed by $(SWCNT)_{0.996}(MB)_{0.004}$. Then, the content of MB molecules is just in the impurity level. Surprisingly, the encapsulation of small amount of MB molecules tunes the electronic property of SWCNT, as described later [15].

Figure 16.26 shows TEM images of SWCNT and MB-adsorbed SWCNT. Both of SWCNT and MB-adsorbed SWCNT have a highly ordered and uniform bundle

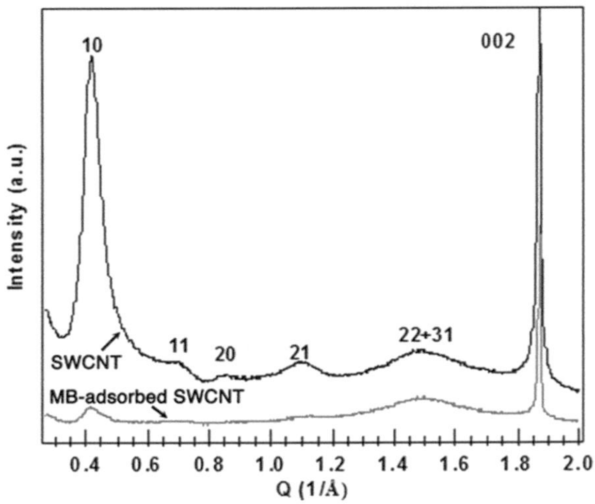


**Fig. 16.27** XRD patterns of SWCNT and MB-adsorbed SWCNT

structure and the intertube distance does not vary on adsorption of MB molecules. The TEM observation cannot clearly indicate both the intercalation of MB molecules in the bundles and encapsulation in the tube spaces. The X-ray diffraction supports the encapsulation. The peaks from the super lattice structure of the bundles almost disappear, as shown in Fig. 16.27, although the ordered bundle structure is preserved even after adsorption treatment according to the TEM observation. If MB molecules are randomly distributed in the internal tube spaces, SWCNTs cannot form the regular lattice structure, giving rise to decrease in the X-ray diffraction peaks of the super lattices of the bundles.

The MB adsorption induces the notable shifts and hyperchromic effect in both metallic and semiconducting optical absorption transitions of SWCNT due to a perturbation in the extended $\pi$-network of nanotubes, supporting modification of electronic structure of SWCNT by the charge transfer interaction. MB adsorption increases the electrical conductivity of SWCNT, indicating the charge transfer interaction between SWCNT and MB molecules; the dc electrical conductivity increases from $4.6 \times 10^{-3}$ to $9.7 \times 10^{-3}$ $\mathrm{Scm}^{-1}$ for the 0.18 filling ratio of pore spaces with MB. As major carriers of SWCNT are holes, the dc electrical conductivity-increase suggests the charge transfer from SWCNT to MB molecules. Then, the SWCNT walls should be positively charged. The effect of the positively charged walls can be supported by $CO_2$ adsorption. This is because adsorption of $CO_2$ having a large quadrupole moment is sensitive to the surface electrical field coming from the pore wall charges. The encapsulation of MB clearly increases the $CO_2$ adsorption capacity at 283 K by more than 10 %, indicating the presence of an enhanced interaction between $CO_2$ molecules and the charged pore walls.

**Conclusions**

This article shows the importance of weak electronic interaction between adsorbed PAH molecules and carbon nanotube walls of highly conjugated π electron network in tuning the physical and chemical properties of carbon nanotubes. As SWCNT and DWCNT have efficient pore spaces which induce distinguished in-pore high pressure effect, SWCNT and DWCNT tuned by the molecular tiling with PAH molecules should induce unusually efficient functions for nanomaterials and separation technologies. Also we can produce easily highly conductive, transparent, and flexible films using the naphthalene derivative-tiled SWCNT because of high electrical conductivity and dispersibility, which can be applied to various electronic technologies. Furthermore, this molecular tiling method can contribute to develop dispersion technology which accelerates to produce light and strong materials due to hybridization between the dispersed carbon nanotube and polymer materials. We need to elucidate and design interfacial functions of carbon nanotubes for construction of better sustainable society.

**Acknowlegements** We thank the great support for T.F., F.K., and G.A. from Global Center of Excellent (G-COE) and for T.O. and H.T. from Center of Excellence Programs of MEXT at Chiba University and for F. K. from scholarship by Graduate School of Science, Chiba University. K.K. and T.F. were supported by Exotic Nanocarbons, Japan Regional Innovation Strategy Program by the Excellence, JST. The Grant-in-Aid for Scientific Research (S) (No. 15101003 (2004–2008)) and Grant-in-Aid for Scientific Research (A) (No. 21241026 (2009–2011) and No. 24241038 (2012-)) supported these researches. The synchrotron radiation experiments were performed at SPring-8 with the approval of Japan Synchrotron Radiation Research Institute (JASRI) as Nanotechnology Support Project of the Ministry of Education, Culture, Sports, Science and Technology.

## References

1. H.E. Chan (ed.), *Graphene and Graphite Materials* (Nova Science Pub Inc, New York, 2010)
2. A. Jorio, M.S. Dresselhaus, G. Dresselhaus (eds.), *Carbon Nanotubes: Advanced Topics in the synthesis, Structure, Properties and Applications* (Springer, Berlin, 2008)
3. C.N.R. Rao, A.K. Sood (eds.), *Graphene Synthesis, Properties, and Phenomena* (Wiley-VCH, Weinheim, 2013)
4. D.M. Guldi, N. Martín (eds.), *Carbon Nanotubes and Related Structures Synthesis, Characterization, Functionalization, and Applications* (Wiley-VCH, Weinheim, 2010)
5. K. Kaneko, T. Itoh, T. Fujimori, Collective interactions of molecules with an interfacial solid. Chem. Lett. **41**, 466–475 (2012)
6. K. Kaneko, C. Ishii, M. Ruike, H. Kuwabara, Origin of superhigh surface area and microcrystalline graphitic structures of activated carbons. Carbon **30**, 1075–1088 (1992)
7. L.R. Radovic, B. Bockrath, On the chemical nature of graphene edges: Origin of stability and potential for magnetism in carbon materials. J. Am. Chem. Soc. **127**, 5917–5927 (2005)

8. M. Asai, T. Ohba, T. Iwanaga, H. Kanoh, M. Endo, J. Campos-Delgado, M. Terrones, K. Nakai, K. Kaneko, Marked adsorption irreversibility of graphitic nanoribbons for $CO_2$ and $H_2O$. J. Am. Chem. Soc. **133**, 14880–14883 (2011)
9. M. Fujita, K. Wakabayashi, K. Nakada, K. Kusakabe, Peculiar localized state at zigzag graphite edge. J. Phys. Soc. Jpn. **65**, 1920–1923 (1996)
10. A. Chuvilin, E. Bichoutskaia, M.C. Gimenez-Lopez, T.W. Chamberlain, G.A. Rance, N. Kuganathan, J. Biskupek, U. Kaiser, A.N. Khlobystov, Self-assembly of a sulphur-terminated graphene nanoribbon within a single-walled carbon nanotube. Nat. Mater. **10**, 687–692 (2011)
11. H. Tanaka, M. El-Merraoui, W.A. Steele, K. Kaneko, Methane adsorption on single-walled carbon nanotube: a density functional theory model. Chem. Phys. Lett. **352**, 334–341 (2002)
12. T. Ohba, T. Matsumura, K. Hata, M. Yumura, S. Iijima, H. Kanoh, K. Kaneko, Nanoscale curvature effect on ordering of $N_2$ molecules adsorbed on single wall carbon nanotube. J. Phys. Chem. C **111**, 15660–15664 (2007)
13. M. Yamamoto, T. Itoh, H. Sakamoto, T. Fujimori, K. Urita, Y. Hattori, T. Ohba, H. Kagita, H. Kanoh, S. Niimura, K. Hata, K. Takeuchi, M. Endo, F. Rodrígues-Reinoso, K. Kaneko, Effect of nanoscale curvature sign and bundle structure on supercritical $H_2$ and $CH_4$ adsorptivity of single wall carbon nanotube. Adsorption **17**, 643–651 (2011)
14. S. Utsumi, K. Kaneko, in *Carbon Nanotubes –From Research to Applications*, ed. by S. Bianco (InTech-Open Access Company, Rijeka, 2011), pp. 37–54
15. F. Khoerunnisa, T. Fujimori, T. Itoh, K. Urita, T. Hayashi, H. Kanoh, T. Ohba, S.Y. Hong, Y.C. Choi, S.J. Santosa, M. Endo, K. Kaneko, Enhanced $CO_2$ adsorptivity of partially charged single walled carbon nanotubes by methylene blue encapsulation. J. Phys. Chem. C **116**, 11216–11222 (2012)
16. P.J.F. Harris, *Carbon Nanotube Science: Synthesis, Properties and Applications* (Cambridge University Press, Cambridge, 2011)
17. K. Hata, D.N. Futaba, K. Mizuno, T. Namai, M. Yumura, S. Iijima, Water-assisted highly efficient synthesis of impurity-free single-walled carbon nanotubes. Science **306**, 1362–1364 (2004)
18. C.-M. Yang, K. Kaneko, M. Yudasaka, S. Iijima, Effect of purification on pore structure of HiPco single wall carbon nanotube aggregates. Nano Lett. **2**, 385–388 (2002)
19. D. Noguchi, H. Tanaka, T. Fujimori, T. Kagita, Y. Hattori, H. Honda, K. Urita, S. Utsumi, Z.-M. Wang, T. Ohba, H. Kanoh, K. Hata, K. Kaneko, Selective $D_2$ adsorption enhanced by quantum sieving effect on entangled single-wall carbon nanotubes. J. Phys. Condens. Matter **22**, 334207–334221 (2010)
20. S. Utsumi, T. Ohba, H. Tanaka, K. Kaneko, in *Novel Carbon Absorbent*, ed. by J.M.D. Tascón (Elsevier, Amsterdam, 2012), pp. 401–433
21. M. Endo, H. Muramatsu, T. Hayashi, Y.A. Kim, M. Terrones, M.S. Dresselhaus, 'Buckypaper' from coaxial nanotubes. Nature **433**, 476 (2005)
22. J. Miyamoto, Y. Hattori, D. Noguchi, H. Tanaka, T. Ohba, S. Utsumi, H. Kanoh, Y.A. Kim, H. Muramatsu, T. Hayashi, M. Endo, K. Kaneko, Efficient $H_2$ adsorption by nanopores of high-purity double-walled carbon nanotube. J. Am. Chem. Soc. **128**, 12636–12637 (2006)
23. T. Fujimori, L.R. Radovic, A.B. Silva-Tapia, M. Endo, K. Kaneko, Structural importance of Stone-Thrower-Wales defects in rolled and flat graphenes from surface-enhanced Raman scattering. Carbon **50**, 3274–3279 (2012)
24. S. Berber, Y.K. Kwon, D. Tománek, Electronic and structural properties of carbon nanohorns. Phys. Rev. B. **62**, R2291–R2294 (2000)
25. K. Suenaga, H. Wakabayashi, M. Koshino, Y. Sato, K. Urita, S. Iijima, Imaging active topological defects in carbon nanotubes. Nat. Nanotechnol. **2**, 358–360 (2007)
26. T. Fujimori, K. Urita, T. Ohba, H. Kanoh, K. Kaneko, Evidence of dynamic pentagon-heptagon pairs in single-wall carbon nanotubes using surface-enhanced Raman scattering. J. Am. Chem. Soc. **132**, 6764–6767 (2010)

27. T. Fujimori, K. Urita, D. Tománek, T. Ohba, I. Moriguchi, M. Endo, K. Kaneko, Selective probe of the morphology and local vibrations at carbon nanoasperities. J. Chem. Phys. **136**, 064505 (2012)
28. R. Defay, I. Prigogine, A. Bellemans, D.H. Everett, *Surface Tension and Adsorption* (Wiley, New York, 1966), p. 7
29. P. Baumli, G. Kaptay, Wettability of carbon surfaces by pure molten alkali chlorides and their penetration into a porous graphite substrate. Mater. Sci. Eng. A **495**, 192–196 (2008)
30. K. Kaneko, N. Fukuzaki, S. Ozeki, The concentrated NO dimer in micropores above room temperature. J. Chem. Phys. **87**, 776–777 (1987)
31. K. Kaneko, N. Fukuzaki, K. Kakei, T. Suzuki, S. Ozeki, Enhancement of nitric oxide dimerization by micropore fields of activated carbon fibers. Langmuir **5**, 960–965 (1989)
32. K. Kaneko, A. Kobayashi, A. Matsumoto, Y. Hotta, T. Suzuki, S. Ozeki, Enhanced intermolecular interaction of NO in cylindrical micropores. Chem. Phys. Lett. **163**, 61–64 (1989)
33. O. Byl, P. Kondratyuk, J.T. Yates Jr., Adsorption and dimerization of NO inside single-walled carbon nanotubes – an infrared spectroscopic study. J. Phys. Chem. B **107**, 4277–4279 (2003)
34. C.H. Turner, J.K. Johnson, K.E. Gubbins, Effect of confinement on chemical reaction equilibria: The reactions $2NO \Leftrightarrow (NO)_2$ and $N_2 + 3H_2 \Leftrightarrow 2NH_3$ in carbon micropores. J. Chem. Phys. **114**, 1851 (2001)
35. J. Imai, M. Souma, S. Ozeki, T. Suzuki, K. Kaneko, Reaction of dimerized $NO_x$ ($x = 1$ or 2) with $SO_2$ in a restricted slit-shaped micropore space. J. Phys. Chem. **95**, 9955–9960 (1991)
36. S.F. Agnew, B.I. Swanson, L.H. Jones, R.L. Mills, Disproportionation of nitric oxide at high pressure. J. Phys. Chem. **89**, 1678–1682 (1985)
37. T. Yamamoto, D.A. Tryk, K. Hashimoto, A. Fujishima, M. Okawa, Electrochemical reduction of $CO_2$ in the micropores of activated carbon fibers. J. Electrochem. Soc. **147**, 3393–3400 (2000)
38. K. Urita, Y. Shiga, T. Fujimori, T. Iiyama, Y. Hattori, H. Kanoh, T. Ohba, H. Tanaka, M. Yudasaka, S. Iijima, I. Moriguchi, F. Okino, M. Endo, K. Kaneko, Confinement in carbon nanospace-induced production of KI nanocrystals of high-pressure phase. J. Am. Chem. Soc. **133**, 10344–10347 (2011)
39. K. Asaumi, T. Mori, High-pressure optical absorption and x-ray-diffraction studies in RbI and KI approaching the metallization transition. Phys. Rev. B. **28**, 3529–3533 (1983)
40. R. Steudel, B. Eckert, Solid sulfur allotropes. Top. Curr. Chem. **230**, 1–79 (2003)
41. M. Springborg, R.O. Jones, Energy surfaces of polymeric sulfur: structure and electronic properties. Phys. Rev. Lett. **57**, 1145–1148 (1986)
42. M. Springborg, R.O. Jones, Sulfur and selenium helices: structure and electronic properties. J. Chem. Phys. **88**, 2652–2658 (1988)
43. S. Okano, D. Tománek, Effect of electron and hole doping on the structure of C, Si, and S nanowires. Phys. Rev. B. **75**, 195409 (2007)
44. T. Fujimori, A. Morelos-Gómez, Z. Zhu, H. Muramatsu, R. Futamura, K. Urita, M. Terrones, T. Hayashi, M. Endo, S.Y. Hong, Y.C. Choi, D. Tománek, K. Kaneko, Conducting linear chains of sulphur inside carbon nanotubes. Nat. Commun. **4**, 2162 (2013)
45. Y. Long, J.C. Palmer, B. Coasne, M. Śliwinska-Bartkowiak, K.E. Gubbins, Pressure enhancement in carbon nanopores: a major confinement effect. Phys. Chem. Chem. Phys. **13**, 17163–17170 (2011)
46. Y. Long, J.C. Palmer, B. Coasne, M. Śliwinska-Bartkowiak, K.E. Gubbins, Microporous Mesopor ous Mater. **154**, 19–23 (2012)
47. M.S. Dresselhaus, G. Dresselhaus, P.C. Eklund, *Science of Fullerenes and Carbon Nanotubes: Their Properties and Applications* (Academic, New York, 1996), pp. 37–39
48. R.S. Mulliken, Structures of complexes formed by halogen molecules with aromatic and with oxygenated solvents. J. Am. Chem. Soc. **72**, 600–608 (1950)
49. R. Foster, *Organic Charge-Transfer Complexes* (Academic Press, New York, 1969). Chap. 2

50. J. Zhao, J.P. Lu, J. Han, C.-K. Yang, Noncovalent functionalization of carbon nanotubes by aromatic organic molecules. Appl. Phys. Lett. **82**, 3746–3748 (2003)
51. F. Tournus, S. Latil, M.I. Heggie, J.-C. Charlier, π-stacking interaction between carbon nanotubes and organic molecules. Phys. Rev. B **72**, 075431 (2005)
52. R.G.A. Veiga, R.H. Miwa, Ab initio study of TCNQ-doped carbon nanotubes. Phys. Rev. B **73**, 245422 (2006)
53. A. Goel, J.B. Howard, J.B. Vander Sande, Size analysis of single fullerene molecules by electron microscopy. Carbon **42**, 1907–1915 (2004)
54. P.W. Fowler, D.E. Manolopoulos, *An Atlas of Fullerenes* (Dover, New York, 1995), pp. 43–67
55. M. Arai, S. Utsumi, M. Kanamaru, K. Urita, N. Yoshizawa, D. Noguchi, K. Nishiyama, Y. Hattori, F. Okino, T. Ohba, H. Tanaka, H. Kanoh, K. Kaneko, Enhanced hydrogen adsorptivity of single-wall carbon nanotube bundles by one-step $C_{60}$-pillaring method. Nano Lett. **9**, 3694–3698 (2009)
56. J. Lu, S. Nagase, X. Zhang, D. Wang, M. Ni, Y. Maeda, T. Wakahara, T. Nakahodo, T. Tsuchiya, T. Akasaka, Z. Gao, D. Yu, H. Ye, W.N. Mei, Y. Zhou, Selective interaction of large or charge-transfer aromatic molecules with metallic single-wall carbon nanotubes: critical role of the molecular size and orientation. J. Am. Chem. Soc. **128**, 5114–5118 (2006)
57. S. Gotovac, Y. Hattori, D. Noguchi, J. Miyamoto, M. Kanamaru, S. Utsumi, H. Kanoh, K. Kaneko, Phenanthrene adsorption from solution on single wall carbon nanotubes. J. Phys. Chem. B **110**, 16219–16224 (2006)
58. S. Gotovac, L. Song, H. Kanoh, K. Kaneko, Assembly structure control of single wall carbon nanotubes with liquid phase naphthalene adsorption. Colloid Surf ace A Physicochem Eng Aspect **300**, 117–121 (2006)
59. S. Gotovac, H. Honda, Y. Hattori, K. Takahashi, H. Kanoh, K. Kaneko, Effect of nanoscale curvature of single wall carbon nanotubes on adsorption of polycyclic aromatic hydrocarbons. Nano Lett. **7**, 583–587 (2007)
60. S. Gotovac-Atlagic, T. Hosokai, T. Ohba, N. Ueno, H. Kanoh, K. Kaneko, Pseudometallization of single wall carbon nanotube bundles with intercalation of naphthalene. Phys. Rev. B **82**, 075136 (2010)
61. F. Khoerunnisa, D. Minami, T. Fujimori, S.Y. Hong, Y.C. Choi, H. Sakamoto, M. Endo, K. Kaneko, Enhanced $CO_2$ adsorptivity of SWCNT by polycyclic aromatic hydrocarbon intercalation. Adsorption **20**, 301–309 (2014)
62. F. Khoerunnisa, A. Morelos-Gomez, H. Tanaka, T. Fujimori, D. Minami, R. Kubobat, T. Hayashi, S. Y. Hong, Y. C. Choi, M. Miyahara, M. Terrones, M. Endo, K. Kaneko, Metal-semiconductor transition like behavior of naphthalene-doped single wall carbon nanotube bundles. Faraday Dis. **173**, 1–12 (2014)
63. A.M. Rao, P.C. Eklund, S. Bandow, A. Thess, R.E. Smalley, Evidence for charge transfer in doped carbon nanotube bundles from Raman scattering. Nature **388**, 257–259 (1997)
64. S. Gotovac-Atlagic, H. Kanoh, K. Kaneko, in preparation
65. Y. Yan, M. Zhang, K. Gong, L. Su, Z. Guo, L. Mao, Adsorption of methylene blue dye onto carbon nanotubes: a route to an electrochemically functional nanostructure and its layer-by-layer assembled nanocomposite. Chem. Mater. **17**, 3457–3463 (2005)
66. K. Yanagi, K. Iakoubovskii, H. Matsui, H. Matsuzaki, H. Okamoto, Y. Miyata, Y. Maniwa, S. Kazaoui, N. Minami, H. Kataura, Photosensitive function of encapsulated dye in carbon nanotubes. J. Am. Chem. Soc. **129**, 1294992–1294997 (2007)

# Chapter 17
# Understanding of Unique Thermal Phase Behavior of Room Temperature Ionic Liquids: 1-Butyl-3-Methylimdiazolium Hexafluorophosphate as a Great Example

**Takatsugu Endo and Keiko Nishikawa**

## 17.1 Introduction

### *17.1.1 What are Room Temperature Ionic Liquids?*

Room temperature ionic liquids (RTILs), or sometimes simply called ionic liquids, are salts that are liquid around room temperature. The definition is a little controversy, but the melting point below 373 K seems to be widely accepted [1]. Since the definition is on the physical property but molecular structure, every kind of ions and their combinations can be RTILs. However, there are some preferable forms for ions in a salt to achieve low melting point such as imidazolium, pyridinium, pyrrolidinium, piperidinium and ammonium as cations, and halides, tetrafluoroborate, hexafluorophosphate and bis(trifluoromethylsulfonyl)imide as anions. Since cations for RTILs are generally organic ions, it is feasible to modify their structure and subsequent physical properties. Besides, there are a vast number of combinations of cations and anions to form RTILs. With this property, RTILs deserve "tunable" or "tailor-made" solvents.

Because they are composed of ions solely, which are totally different from ordinary molecular liquids or electrolyte solution, they show some outstanding properties as liquids. Practically RTILs do not show vapor pressure and flammability. They have high thermal/chemical/electrochemical stability. Some of them show unique solubility, that is, do not mix with water even though they are salts. These properties let many people realize that RTILs are potentially useful

T. Endo
University of California, Davis, One Shield Avenue, Davis, CA 95616, USA
e-mail: tkendo@ucdavis.edu

K. Nishikawa (✉)
Graduate School of Advanced Integration Science, Chiba University,
1-33 Yayoi-cho, Inage, Chiba 263-8522, Japan
e-mail: k.nishikawa@faculty.chiba-u.jp



H. Ishii et al. (eds.), *Electronic Processes in Organic Electronics*,
Springer Series in Materials Science 209, DOI 10.1007/978-4-431-55206-2_17

for electrolytes, green solvents, lubricants, extraction solvents and so forth [2–6]. RTILs even cover electronics field [7–9]. Nowadays, many researchers consider RTILs as "solvents for the future".

### *17.1.2 Importance of Understanding of Thermal Phase Behavior of RTILs*

Here we focus on thermal phase behavior of RTILs, i.e., melting, freezing, solid-solid phase transition and glass transition for fundamentally understanding of this type of material. Melting point is probably the most important property for RTILs because the definition depends on it. Why RTILs have melting point so low would be the most basic question for them. Qualitatively, it has been answered based on large size of ions, asymmetric structure, flexibility in functional groups, existence of a variety of stable conformers [10–12]. These factors decrease or increase the lattice energy or entropy in the liquid state, respectively, which results in lowering melting point. However, it would be still open to debate quantitatively. To answer the question, it will be required to investigate melting and freezing phenomena as well as ionic structure and dynamics and their change through thermal phase changes in detail. Besides, RTILs sometimes show unique thermal phase behavior, i.e., premelting phenomenon, wide temperature range of supercooled liquid state and subsequently tendency to form glassy state, large thermal hysteresis and complex solid–solid phase transitions. These characteristic features are worth investigating. To reveal the thermal phase behavior is significantly important for some applications, such as solid electrolytes. Some RTILs possess high ion conductivity even in the solid state [13–16]. This would be strongly related with the ion structure and dynamics in such states.

### *17.1.3 1-Butyl-3-Methylimidazolium Hexafluorophosphate*

1-Butyl-3-methylimidazolium hexafluorophosphate ([$C_4$mim]$PF_6$, Fig. 17.1) is a RTIL of interest here. The melting point is ca. 285 K [17], which well falls within the definition. The material is simply prepared with two steps [17], and is one of the most representative RTILs. However, despite of the relatively simple chemical structure, it shows a complicated thermal phase behavior. Two groups reported the thermal phase behavior of [$C_4$mim]$PF_6$ with differential scanning calorimetry (DSC) at different scanning rates, 10 and 5 K $min^{-1}$, in 2003 and 2005 [18, 19]. Their results are similar. Namely, nothing happened in cooling process from the liquid to glassy state (ca. 196 K). When temperature increased from the glassy state, the crystallization occurred (cold crystallization) in the range of temperatures 257 to 233 K. Then, the RTIL melted through one

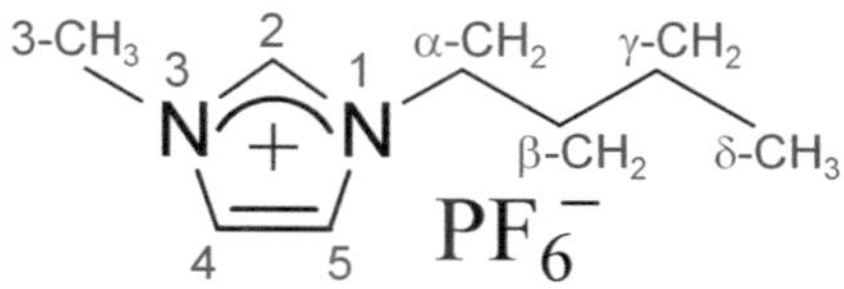


**Fig. 17.1** Chemical structure of $[C_4mim]PF_6$ and numbering system

endothermic peak (a solid-solid phase change). Choudhury et al. also observed another exothermic peak (ca. 280 K) right before melting [19], which is consistent with the data obtained by Troncoso et al. in 2006 [20]. Triolo et al. performed DSC and X-ray diffraction measurements and found two crystalline phases [21], but the DSC traces looked different from the others, which suggested there were more than two crystalline phases. All the data previously reported were consistent with the fact that there were at least two different crystals existing in $[C_4mim]PF_6$, nevertheless, the DSC traces looked differently each other more or less. At that time, only one crystal structure, which could be attributed to one of the crystals appeared in the DSC traces, was determined [19, 22], but there was no more structural information for the other phases in detail. Additionally, the fusion enthalpy, which is the heat absorbed during melting, showed large diversity ranging from 9.21 kJ $mol^{-1}$ to 19.91 kJ $mol^{-1}$ [18, 20, 23–25].

In the following sections, we addressed the complex thermal phase behavior of $[C_4mim]PF_6$. Our goal is to give a conclusion for this confusing thermal phase behavior and determine ion structures in each phase. We also performed nuclear magnetic resonance (NMR) spectroscopy to obtain dynamical aspect of this material.

## 17.2 Thermal Phase Behavior and Molecular Structure of $[C_4mim]PF_6$

### *17.2.1 Development of Simultaneous Measurement Apparatus*

The issues on the $[C_4mim]PF_6$ DSC measurements seem to come from the large thermal hysteresis and overlapping of several phase transition peaks. To solve these problems, it would require two criteria for calorimetric measurements. First, a spectroscopic technique must be carried out simultaneously with calorimetry. This approach enables us to observe molecular structural changes in situ. Second, slow scanning rates of calorimeter that separate overlapping peaks must be adopted. Slow scanning rates require high temperature stability of calorimetric traces. Also, high sensitivity is necessary for the apparatus because such slow rates reduce peak intensity.

For this purpose, we constructed an apparatus that enables simultaneous measurements of Raman spectroscopy and calorimetry as shown in Fig. 17.2 [26].

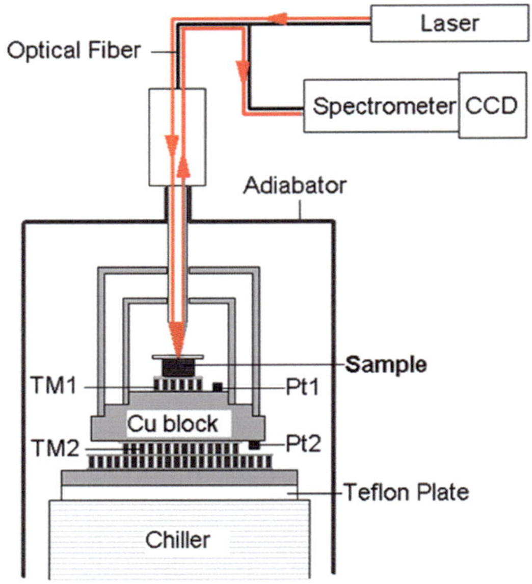


**Fig. 17.2** Schematic diagram of the calorimeter combined with the Raman spectrometer

The Raman spectrometer is a conventionally available one (HoloLab 5000, Kaiser Optical Systems). This spectrometer is equipped with an optical fiber and a GaAlAs diode laser (wavelength: 785 nm). A spectrum ranging 100–3450 $cm^{-1}$ is simultaneously measurable with resolution of 4 $cm^{-1}$ by adopting a multiplex grating backed up by a CCD camera.

A lab-built calorimetric part is a key for this apparatus. Commercially available calorimeters usually adopt a temperature sensor such as a thermopile made of multijunction thermocouples composed of chromel-alumel. On the other hand, we use a thermoelectric module (TM) as a heat flux sensor, which makes it possible to attain higher sensitivity and to downsize the calorimeter [27–29]. Heat flow from a sample is measured with TM1 (9500/018/012, Ferro Tec), which is made of 18 semiconducting thermoelectric elements of BiTe connected in series. The output voltage of the TM is about 8 mV $K^{-1}$. The calorimeter is miniaturized since TM1 is small ($6.05 \times 7.62 \times 2.64$ $mm^3$). The height and diameter of the inner cylindrical copper cap of the calorimeter are 18.8 and 21.8 mm, respectively. The sample vessel is set on TM1. The cell of the calorimeter is composed of triple covers of inner and outer cylindrical copper caps and an adiabator. These covers have a hole in the center to allow optical access. The temperature is controlled by means of heat pump TM2 (6320/157/085, Ferro Tec) and a chiller (TMA-SCUC08, Shinyei Kaisha) backed up by a PID circuit with a Pt resistance thermometer, Pt2 (EL-700-U, Teijin Engineering). The measurement can be performed in the temperature range approximately between 120 and 420 K. It is possible to carry out with very slow heating and cooling rates of a minimum value of 0.5 mK $s^{-1}$ (0.030 K $min^{-1}$). The temperature of the sample is measured with Pt1, which is placed near TM1. The temperature and the heat flux are measured by digital

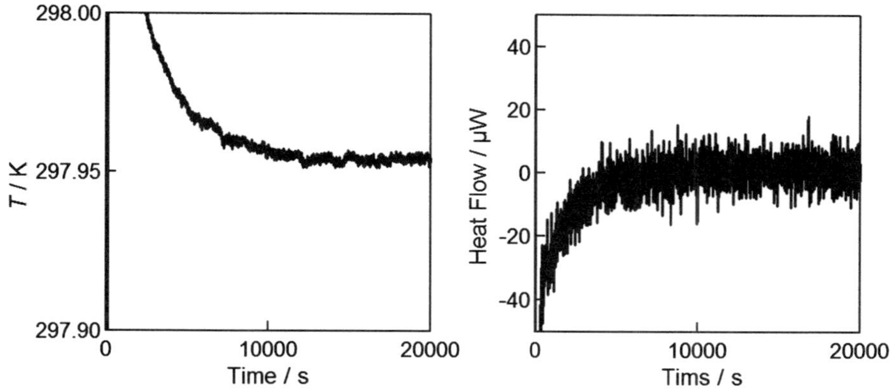


**Fig. 17.3** Temperature and baseline stabilities of the calorimeter shown in Fig. 17.2

multimeters (2000 multimeter, Keithley Instruments) and the data are stored in a computer. Original design of the calorimeter (DSC) has baseline and temperature stabilities of ±3 nW and ±0.1 mK, respectively [29]. These values would be a couple of hundred times greater than ones commercially available at that time. The calorimeter used here is a much simpler type one and even lacking a reference stage. The stabilities of the baseline and the temperature of the present calorimeter were estimated to be ±5 μW and ±1 mK, respectively (Fig. 17.3). Although these values are poor in comparison to the original one, they are still great enough to detect phase change peaks of materials. Besides, needless to say, the calorimetric measurements can be performed simultaneously with Raman spectroscopy, which is the greatest advantage of this apparatus.

### *17.2.2 Thermal Phase Behavior of [$C_4$mim]$PF_6$ Investigated with Calorimetry*

Figure 17.4 shows the calorimetric traces of [$C_4$mim]$PF_6$ using the calorimeter mentioned above. The result is similar to the previous ones [18, 19], nevertheless, the phase change peaks are well separated because of slower scanning rate, 1.2 K $min^{-1}$, with high baseline stability. The RTIL does not crystallize in cooling from the liquid state and finally form the glassy state. The cold crystallization was observed at 233 K, and the sample melts at 286 K with further heating through two endothermic peaks, at 255 and 276 K, respectively. We call these crystalline phases α, β and β′ phases in the order from lower temperature. The three phases were observed with a normal cooling and heating cycle. However, we found one more crystalline phase, called γ phase, with re-cooling and re-heating from the β phase (Fig. 17.5). When the β phase is cooled down below 200 K, no phase change occurs. With subsequent heating a very wide exothermic peak is observed around 255 K,

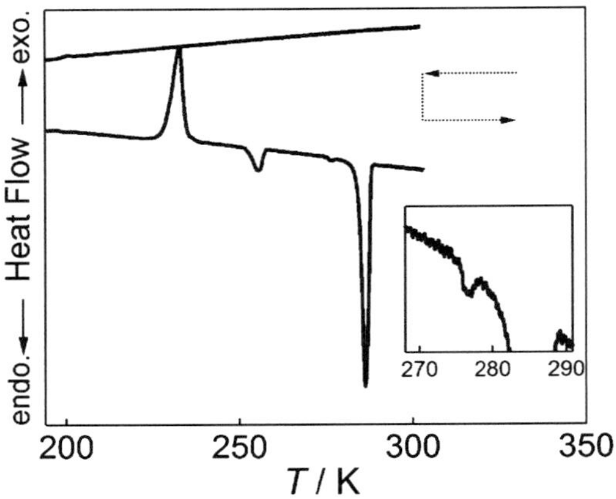


**Fig. 17.4** Calorimetric curves for $[C_4mim]PF_6$ at a scan rate of 1.2 K $min^{-1}$. The trace was initiated from the liquid state near room temperature. The *inset* is an enlargement of the tiny endothermic peak just below the melting point. Reprinted with permission from Endo and Nishikawa [30]. Copyright 2013 Elsevier

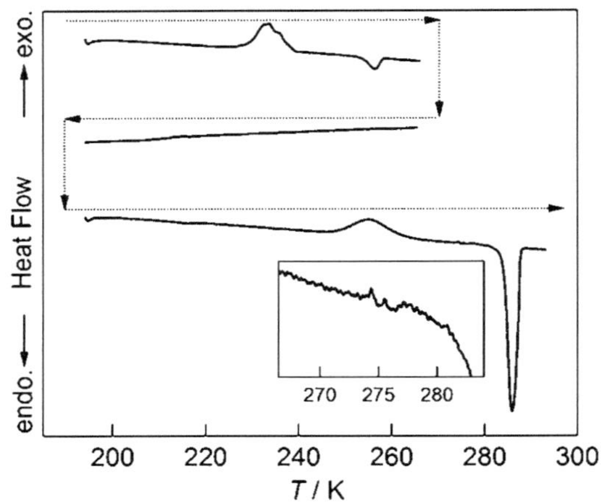


**Fig. 17.5** Calorimetric curves with different thermal histories from that shown in Fig. 17.4. The trace was initiated from the glassy state at 194 K. The *inset* corresponds to that in Fig. 17.4. The small fluctuations in the heat flow observed in the *inset* may be ascribed to the melting of a little amount of water originating from the atmospheric moisture. Reprinted with permission from Endo and Nishikawa [30]. Copyright 2013 Elsevier

which is assigned to the phase change from the β to γ phases. The γ phase melts at 285 K, which is fairly close to the melting point of the β phase. The assignments of these phases were performed simultaneous Raman spectroscopic measurements, the detail of which will be described later.

Exotherm in heating means that the phase change is not a phase transition but an irreversible change. Showing a basic thermodynamical equation would be helpful to understand these terms as;

$$G = H - TS \tag{17.1}$$

where $G$ is the Gibbs free energy, $H$ is the enthalpy, $T$ is the temperature and $S$ is the entropy. Gibbs free energy of two phases is equal at temperature where the phase transition occurs. Then, the following equation is derived from Eq. (17.1),

$$T_{trans} = \frac{\Delta_{trans}H}{\Delta_{trans}S} \tag{17.2}$$

where $T_{\mathrm{trans}}$ is the phase transition temperature, $\Delta_{\mathrm{trans}}H$ and $\Delta_{\mathrm{trans}}S$ are the difference of enthalpy and entropy, respectively. Materials usually absorb heat in the heating to get higher enthalpic and entropic state while their $G$ values are same at the phase change temperature, which is defined as phase transition. In the case observed, since the peak is exotherm, $G$ of the γ phase is thermodynamically more stable than that of the β phase at lease around the phase change temperature. The reason why the stable γ phase does not appear in the normal calorimetric measurement cycle would be explained in terms of kinetics. Although the γ phase is thermodynamically more stable than the β phase, forming that phase from α (or the melt) would be kinetically unfavored, that means, it requires higher activation energy to overcome than from the β phase. Even though such an irreversible phase change can occur at any temperature, the temperature 255 K seems to be preferred because this phase change was observed reproducibly around this temperature. This would be explained by nucleation and subsequent its growth. In some materials, preferred temperature for nucleation and its growth in a phase change is significantly different. The temperature where both phenomena effectively occur can be the phase change point.

The phase change temperature from β to γ is almost the same as that from α to β. Besides, as was already mentioned, the melting point of the β and γ phases (286 and 285 K, respectively) is fairly close. These features cause confusion and let us make a misassignment in the early paper [17] because the phase changes sometimes occur at the same time accidentally, namely, the γ phase was obtained from the α phase directly. In our first paper on the thermal phase change of $[C_4mim]PF_6$ in 2010 [17], we mistakenly assigned the β′ phase as the γ phase. Actually this confusing phase change was already observed in the previous papers [18, 19]. In the papers, the calorimetric data showed the existence of a peak that possesses both endothermic and exothermic features. However, probably because of its complex appearance, the authors did not mention about it and would not recognize the existence of two phase change peaks.

From Eq. (17.2) fusion enthalpy ($\Delta_{\mathrm{fus}}H$) and entropy ($\Delta_{\mathrm{fus}}S$) can be derived for the β and γ phases, and the values are listed in Table 17.1. While their melting points are close to identical, there is a significant difference in $\Delta_{\mathrm{fus}}H$ and $\Delta_{\mathrm{fus}}S$. On the both parameters the γ phase shows higher values, which is consistent with the fact that γ is thermodynamically more stable than β. As was already indicated above, these values for $[C_4mim]PF_6$ was reported in a wide range. To our knowledge, the reported $\Delta_{\mathrm{fus}}H$ values so far are 13.2 [23], 9.21 [18], 19.60 [24], 19.91 [20], and 12 kJ $\mathrm{mol}^{-1}$ [25]. The reason why the reported $\Delta_{\mathrm{fus}}H$ has a large diversity can be attributed to the crystal polymorphism of the RTIL. Namely, the lower and higher values are assigned to the $\Delta_{\mathrm{fus}}H$ of the β and γ phases, respectively. It should

**Table 17.1** Melting points ($T_m$), fusion enthalpies ($\Delta_{fus}H$), and entropies ($\Delta_{fus}S$) of the β and γ crystals. Values are taken from the peak top. Experiments were performed three times with different weights of the sample to estimate standard deviations

| | $T_m$/K | $\Delta_{fus}H$/kJ mol$^{-1}$ | $\Delta_{fus}S$/J K$^{-1}$ mol$^{-1}$ |
|---|---|---|---|
| β | 285.8 ± 0.7 | 13.1 ± 0.7 | 45.9 ± 2.6 |
| γ | 285.3 ± 0.7 | 22.6 ± 1.6 | 79.1 ± 5.6 |



be noted that there are also some other factors that affect $\Delta_{fus}H$, such as purity of sample, types of method and analysis to determine the value.

Finally, it would be possible to depict the thermodynamic relationship between these polymorphic crystals. Figure 17.6 is a schematic phase diagram of [$C_4$mim] $PF_6$. The phase β' is not present in the figure because the thermodynamical and structural natures are still unclear (see the following section). The Gibbs free energy of the β phase is always higher and steeper than that of the γ phase due to the higher enthalpy and entropy, respectively. The phase change from α to β is considered as a phase transition in the figure because it occurs endothermically in the heating process around the same temperature even with a different scanning rate (0.6 K min$^{-1}$, data is not shown). However, it is noted that we cannot exclude the possibility that this phase change is irreversible because the phase change from β to α was not observed in the cooling.

## *17.2.3 Determination of Ion Structure*

Since the thermal phase behavior of [$C_4$mim]$PF_6$ has been understood, next question would be how they are different, that is, difference in ion structure. Because the thermal phase behavior does not always provide the same result, we performed Raman spectroscopy with simultaneously monitoring calorimetric traces. Raman spectra of each phase are shown in Fig. 17.7. Prominent difference was observed in the range of 580–640 cm$^{-1}$ as well as 300–350 cm$^{-1}$, and the former is known to reflect the cation conformation difference of [$C_4$mim]$^+$ [32, 33]. This result suggests that observed crystalline phases except the β′ phase possess different cation conformation. Also the Raman spectra indicate that cation conformers observed in the crystalline state coexist in the liquid state because the liquid state spectrum seems to be the sum of the crystalline state spectra [32, 33]. Density functional theory (DFT) calculations (B3LYP/6-311 + G(d,p)) were carried out to predict possible cation conformations in the gas phase and their Raman spectrum with Gaussian 03 program package [34]. The calculated cation conformations with energy difference and Raman spectrum are summarized in Table 17.2 and Fig. 17.8, respectively. Ten stable conformations were observed for this cation, which agrees with the previous data with different calculation level [35]. It is known that the calculated Raman spectra in the gas phase well correspond to the observed one in both liquid and crystalline states even without scaling factor [33, 36, 37].

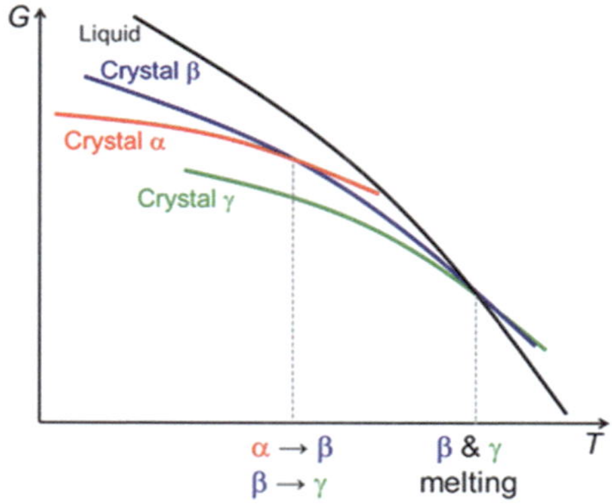


**Fig. 17.6** Schematic phase change diagram for [$C_4$mim]$PF_6$. Reprinted with permission from Endo and Nishikawa [30]. Copyright 2013 Elsevier

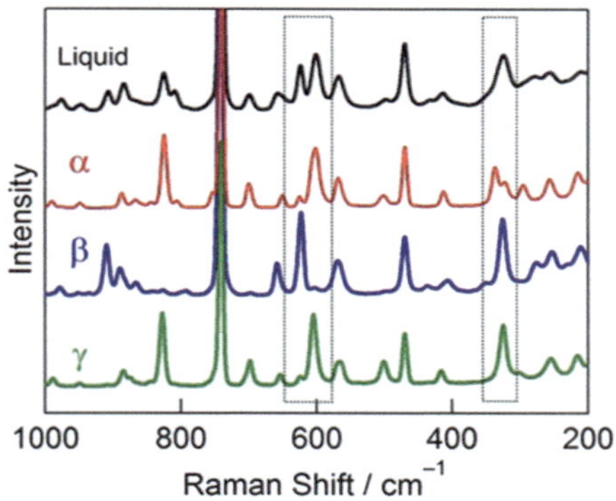


**Fig. 17.7** Raman spectra of [$C_4$mim]$PF_6$. Colors correspond to the following phases: *black*: liquid (302 K), *red*: crystal α (229 K), *blue*: β (254 K), and *green*: γ (280 K). The spectrum of the β' phase, which is almost the same as that of the β phase, is not shown here. Two peaks are observed in the Raman spectra of the α phase ranging 300–350 cm$^{-1}$. The larger peak is assigned to the GT conformer, while the smaller peak is assigned to other minor conformers. The GT conformer becomes dominant with decreasing temperature [17, 31]. Reprinted with permission from Endo and Nishikawa [30]. Copyright 2013 Elsevier

Comparing the frequency region of 580–640 cm$^{-1}$ between the observed and calculated Raman spectra, the cation conformation in crystal β could be TG, TT, TG' or pTT (see Table 17.2 for the abbreviations) because of the characteristic Raman band at 623 cm$^{-1}$. The other six conformations are the candidate for both the α and γ phases. There were many papers that determined crystal structure of RTILs including [$C_4$mim]$^+$ [10, 19, 22, 38–41], and most of them were concentrated to the three conformers, namely, G'T, TT and GT. Since the calculation results indicate that these three conformers are energetically the most stable ones (see Table 17.2), it would be reasonable to adopt only these conformers for the assignment. Namely, TT conformer is for the β phase while G'T or GT is for the α or γ phase. Taking a closer

**Table 17.2** Dihedral angle of the butyl group in the cation and energy difference (TT is set to be 0.0 kJ mol$^{-1}$) for the calculated stable $[C_4mim]^+$ conformers at the B3LYP/6-311 + G(d,p) level

| Conformer | Energy difference/kJ mol$^{-1}$ | Dihedral angle/degree C(2)-N-C(α) -C(β) | N-C(α)-C(β) -C(γ) | C(α)-C(β)-C(γ) -C(δ) |
|---|---|---|---|---|
| GG | 5.66 | −80.2 | −64.5 | −65.0 |
| GT | 2.45 | −83.1 | −65.1 | −176.6 |
| GG' | 11.67 | −62.1 | −70.4 | 77.2 |
| TG | 3.12 | −104.4 | −178.7 | −66.0 |
| TT | 0.00 | −104.3 | 179.9 | 179.8 |
| TG' | 3.13 | −104.1 | 178.1 | 65.8 |
| G'G | 10.06 | −117.0 | 70.8 | −78.6 |
| G'T | 0.67 | −106.1 | 64.6 | 177.1 |
| G'G' | 4.07 | −108.1 | 63.8 | 65.1 |
| pTT | 2.90 | −0.7 | 179.9 | −180.0 |

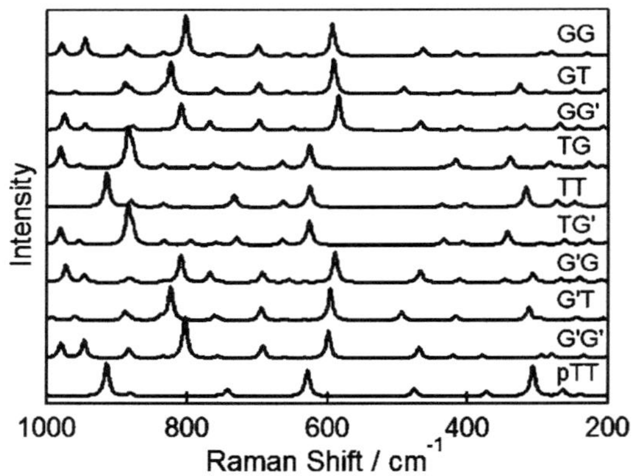


**Fig. 17.8** Calculated Raman spectra with a certain linewidth for ten stable conformers of $[C_4mim]^+$

look at the observed Raman spectral region of 300–350 cm$^{-1}$, a peak of G'T in the region is slightly higher than that of GT. Comparing to the calculated spectra in Fig. 17.8, this assesses that the former and the latter are the cation conformation for the α and γ phases, respectively. This could be consistent with the crystal structure determined by X-ray single crystal analyses. Only one crystal structure of $[C_4mim]PF_6$ was report by two different groups at that time [19, 22], the cation conformation of which is G'T. Our calorimetric measurements indicated that the γ phase was thermodynamically most stable, thereby it is possible to consider that that phase was obtained as single crystal. Figure 17.7 also shows that in the α phase the peak in the range of 300–350 cm$^{-1}$ has a shoulder peak at lower frequency side, the intensity of which decreases with decreasing temperature. This indicates that there is a minor component for the α phase, which would consist of G'T conformers, while GT would be still dominant in the phase.

Our conformational assignment for the three polymorphic crystals in $[C_4mim]PF_6$ based on Raman spectroscopy has been fully confirmed by Saouane et al. in 2013 [31]. They performed X-ray single crystal analyses for both α and β phases changing temperature or pressure, and determined their structure. It was found that there was G'T conformer in the α phase as a minor component (25 %) at 193 K as we predicted. The authors also discovered two minor components for the β phase, TG and TE', where TE' stands for a conformation between TG and TT and is not obtained in the gas phase as a stable structure.

The detailed structure of the β' phase is still unclear because there is almost no distinguishable difference in Raman spectrum between the β and β' phases. This was confirmed by the simultaneous measurement of the Raman spectrum and the calorimetry for this peak. The Raman spectrum is that of the continuous phase change of β to a phase that is partially mixed with a liquid component as the temperature increases (data is not shown). Therefore, the endothermic peak does not accompany a clear conformational change of the cation or a displacement of the relative position of the ions, both of which should significantly affect the Raman spectra of RTILs [33, 36, 42, 43]. Taking into account the fact that both the cation and anion tend to be disordered in the β crystal [31], this peak might possibly be related to further displacements of the ions.

The three main conformers of the cation, GT, TT and G'T with some minor components would coexist in the liquid state. Their stability explained as enthalpy differences can be obtained by comparing the intensities of Raman bands at different temperatures [36, 44]. If enthalpy difference trends are similar for the liquid, supercooled liquid, and glassy states, the differences can be estimated from the results shown in Fig. 17.9. Figure 17.9b and c compare the Raman spectra in the ranges of 300–350 $cm^{-1}$ and 530–670 $cm^{-1}$, respectively. Only one peak is observed in the 300–350 $cm^{-1}$ range, and it slightly shifts to lower frequency with increasing temperature. Assuming that the three conformers of GT, TT and G'T only exist in the liquid state, this peak overlaps the bands of those conformations, and the GT band is observed at higher frequency as shown in Fig. 17.7. This finding suggests that the contribution of the GT conformer is larger at lower temperature and thus has lower energy than the average of the TT and G'T conformations. As shown in Fig. 17.9c, the Raman band intensity at 624 $cm^{-1}$ attributed to the TT conformation increases with increasing temperature. This shows that the average energy of the GT and G'T conformations is lower than that of the TT conformation. These findings lead to an enthalpy order of GT < TT in these states. This result contradicts to the one in the gas phase but not in the crystalline states in low temperature region.

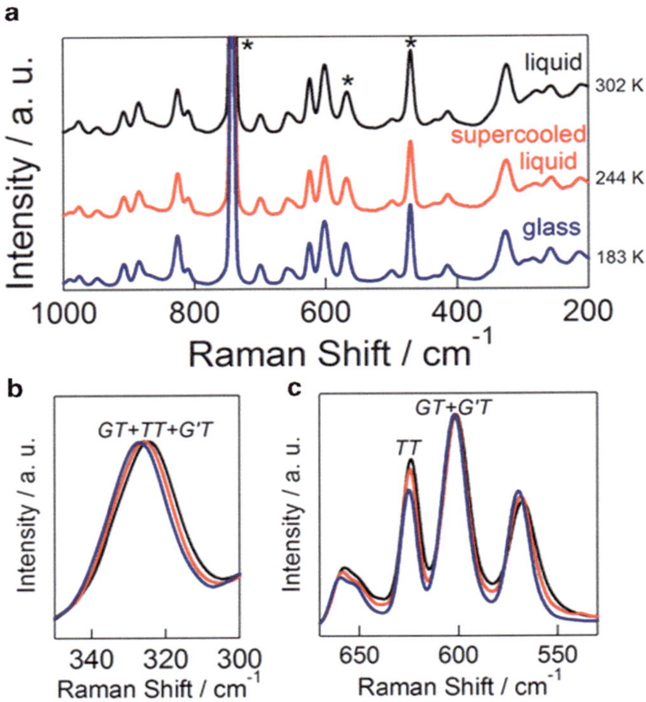


**Fig. 17.9** Raman spectra in the liquid (302 K), supercooled (244 K), and glassy states (183 K) in the range of (**a**) 200–1,000 $cm^{-1}$, (**b**) 300–350 $cm^{-1}$, and (**c**) 530–670 $cm^{-1}$. *Asterisks* indicate the anion bands. Reprinted with permission from Endo et al. [17]. Copyright 2010 American Chemical Society

## 17.3 Molecular Dynamics of $[C_4mim]PF_6$ in the Solid State

It has been demonstrated that the thermal phase change of $[C_4mim]PF_6$ is strongly linked to the conformational change of the cation. Although it was already reported about cation conformation-driven crystal polymorphism for imidazolium-based RTILs [10, 39], $[C_4mim]PF_6$ would be the most representative RITL that proves the importance of conformational flexibility in the thermal phase change. Molecular dynamics would be a key to understand this characteristic phenomenon of the RTIL more. Therefore, we carried out NMR spectroscopy, particularly second moment ($M_2$) analyses and spin–lattice relaxation time ($T_1$) measurements, to investigate both the cation ($^1$H NMR) and anion ($^{31}$P NMR) dynamics in the crystalline states. $M_2$ is a variant, and roughly corresponds to linewidth of a NMR peak. This enables us to assess the type of motion occurring in the solid state. $T_1$ is the time constant for relaxation of nucleus from excited to thermal equilibrium states. This time constant is sensitive for picosecond to nanosecond rotational dynamics. It should be noted

**Table 17.3** Theoretical $^{1}$H second moment ($M_2$) values (in Gauss$^2$) for [$C_4$mim]$PF_6$ in the crystalline states

| | Intra-cation | Inter-cation | | Total |
|---|---|---|---|---|
| | | <3.5 Å | >3.5 Å | |
| Rigid lattice | 17.7 | 2.6 | 1.7 | 22.0 |
| 3-$CH_3$ rotation | 14.1 | 2.6 | 1.7 | 18.5 |
| 3-$CH_3$ + δ-$CH_3$ rotations | 10.8 | 1.9 | 1.7 | 14.4 |
| Isotropic rotation of cation | 0.0 | 0.0 | 1.7 | 1.7 |

Reprinted with permission from Endo et al. [46]. Copyright 2012 American Chemical Society

that NMR experiments were not conducted on the β' phase because it only exist in the narrow temperature region (ca. 10 K). Also the phase is not stable and eventually turns into the γ phase.

### 17.3.1 Cation Dynamics

First, theoretical $M_2$ values of $^{1}$H NMR were calculated using the Van Vleck equation [45] and listed in Table 17.3. The structural parameters required for the calculation were taken from the crystal structure of the γ phase [19] except C-H bond lengths in the cation. Since they were underestimated in X-ray measurements, the bond lengths in the calculated gas phase structure were used. We obtained the $M_2$ value of 22.0 Gauss$^2$ for the rigid lattice structure. When the 3-$CH_3$ rotation as the fastest motion in the cation occurs [46], $M_2$ is reduced to 18.5 Gauss$^2$ by 3.5 Gauss$^2$. The rotation of the additional methyl group (δ-$CH_3$) reduces $M_2$ to 14.4 Gauss$^2$. If the rapid isotropic rotation of the cation exists in the crystal, $M_2$ is drastically reduced to 1.7 Gauss$^2$ with no contributions from intra- and intercations having a proton distance less than 3.5 Å.

Figure 17.10 shows the temperature dependent observed $M_2$ values for the α, β and γ phases. A dashed line in the figure explains the theoretical value when 3-$CH_3$ and δ-$CH_3$ rapid rotations occur (14.4 Gauss$^2$). The figure indicates three important insight into rotational dynamics of [$C_4$mim]$PF_6$ in the crystalline states. First, almost every plot is below the dashed line, which demonstrates that the both methyl groups in the [$C_4$mim]$^+$ rotate rapidly in the crystalline states. Second, the fact that all the plots are below the line verifies the existence of certain rotational motions in addition to the two methyl group rotations. This is also confirmed by the fact that all the $M_2$ curves decrease with increasing temperature. If no rotation would occur in addition to the methyl group rotations in the crystalline states, the values should be constant, i.e. temperature independence. It is reasonable to consider that the additional motions are reorientational motions of the butyl group in the cation due to its high flexibility compared to the imidazolium ring. Finally, it can be found from the figure that the $M_2$ values follow the trend crystal γ > crystal β ≈ crystal α, which means the mobility of the segmental motions follows this trend with the reverse

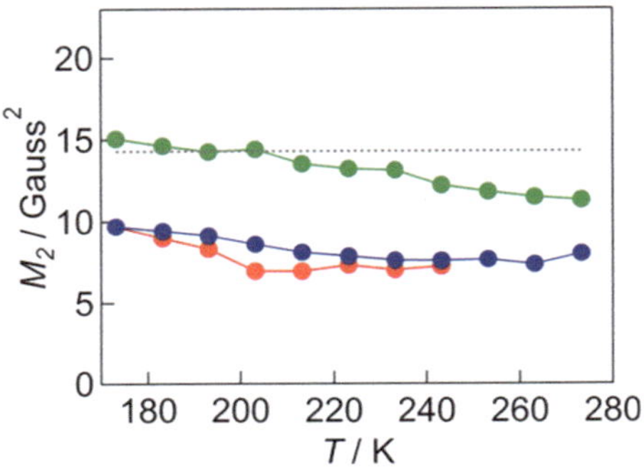


**Fig. 17.10** Temperature dependence of the second moment $M_2$ of $^1$H NMR for crystal α (*red*), crystal β (*blue*), and crystal γ (*green*). The *gray dashed line* is the theoretical value when both 3-$CH_3$ and δ-$CH_3$ rotations occur (14.4 Gauss$^2$). Reprinted with permission from Endo et al. [46]. Copyright 2012 American Chemical Society

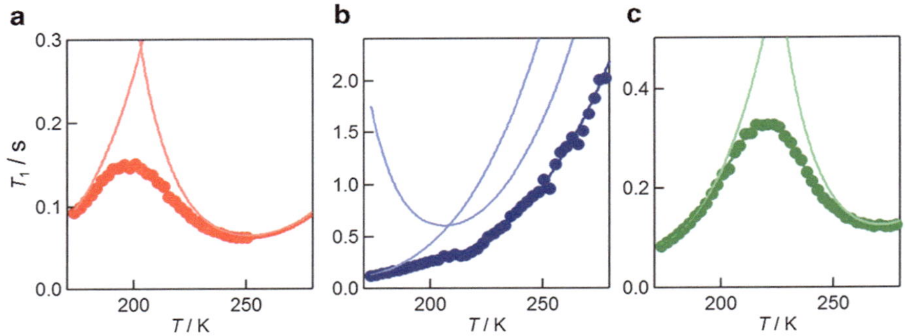


**Fig. 17.11** Theoretical fitting for the observed $T_1$ plots: (**a**) crystal α, (**b**) crystal β, and (**c**) crystal γ. The simulation of $^1$H $T_1$ data is represented by the *dark shade*, and the individual simulation components for the low and high temperature dynamical components are indicated by the *light shade*. Reprinted with permission from Endo et al. [46]. Copyright 2012 American Chemical Society

order. The latter two findings have been explained well by the later results of single crystal analyses [31]. They found the presence of minor conformations in each crystal, the population of which changes by temperature or pressure. This means that conformational changes of the cation that can reduce $M_2$ of $^1$H NMR spectra occur in the crystalline states. Crystal α and β has 25 % and 50 % minor components at 193 K, respectively, while the minor component of crystal γ is only 8 % at 263 K. This difference implies slower conformational change in γ than in α and β.

The observed $^1$H $T_1$ plots are depicted in Fig. 17.11. The figure also contains theoretical fitting curves derived from Bloembergen-Purcell-Pound theory [47] assuming that magnetic dipole–dipole interactions are dominant mechanism of the $^1$H relaxation. The fitting reveals that there are two components in the $T_1$ plots of all the crystals, which are referred to the fast and slow rotational motions.

**Table 17.4** Parameters $\Delta M_2$, $E_a$ and $\tau_0$ derived from fitting the $^1$H $T_1$ data

| | Crystal α | | Crystal β | | Crystal γ | |
|---|---|---|---|---|---|---|
| | Fast | Slow | Fast | Slow | Fast | Slow |
| $\Delta M_2$ (Gauss$^2$) | 3.3 | 3.6 | 2.8 | 0.4 | 2.8 | 1.8 |
| $E_a$ (kJ mol$^{-1}$) | 12.5 ± 1.4 | 20.1 ± 1.0 | 14.6 ± 10.9 | 16.7 ± 9.7 | 15.1 ± 0.4 | 26.8 ± 0.6 |
| $\tau_0$ (fs) | 269 | 260 | 51 | 235 | 73 | 27 |

Reprinted with permission from Endo et al. [46]. Copyright 2012 American Chemical Society

The parameters obtained from the $T_1$ fitting, $\Delta M_2$, $E_a$ and $\tau_0$, are summarized in Table 17.4. $\Delta M_2$ is a part of the second moment averaged by the considered motion. $E_a$ and $\tau_0$ are rotational activation energy and rotational correlation time at infinite temperature for the motions. Theoretical $M_2$ analyses indicated that the rotations of 3-$CH_3$ and δ-$CH_3$ decreased in $M_2$ by 3.5 Gauss$^2$ and 4.1 Gauss$^2$, respectively (see Table 17.3). The fast rotational motion for the three crystals derived from $T_1$ fittings is estimated to be ca. 3 Gauss$^2$ in $\Delta M_2$ and has similar $E_a$ values as 12.5–15.1 kJ mol$^{-1}$. Based on these findings, the fast motion can be assigned to the rotation of either 3-$CH_3$ or δ-$CH_3$. The rotational activation energy of 3-$CH_3$ and δ-$CH_3$ in the gas phase were estimated to be 1.9 kJ mol$^{-1}$ and 11.9 kJ mol$^{-1}$, respectively [46]. Therefore, the fast motion observed in the $T_1$ plots of all three crystals can be attributed to δ-$CH_3$ rotation. This assignment is reasonable, because the calculated $E_a$ for 3-$CH_3$ rotation 1.9 kJ mol$^{-1}$ is also in good agreement with the $E_a$ estimated from the neutron scattering measurement, 1.28 kJ mol$^{-1}$ [48].

There is diversity of $\Delta M_2$ and $E_a$ for the slow motion among the three crystalline phases. The relatively large $\Delta M_2$ values in crystals α and γ indicate that their slow motions would be a certain segmental motion of the butyl group, whose presence was demonstrated in the previous section as well as in Saouane et al. (2013) [31]. Their $\Delta M_2$ values are distinguishable, which indicates that they have different segmental motions. The smaller $\Delta M_2$ value in crystal γ (1.8 Gauss$^2$) than in crystal α (3.6 Gauss$^2$) indicates the slower or smaller segmental motion of the butyl group in crystal γ. This is consistent with the result of observed $M_2$ shown in Fig. 17.10 as well as X-ray single crystal analysis data [31]. The $\Delta M_2$ value in crystal β (0.4 Gauss$^2$) is small to be considered as similar segmental motions of the butyl group as observed in the crystal α or γ. The slow motion in crystal β would be attributed to either segmental motions of the butyl group that have very small rotational angle or occurs only in a small portion of the cations, or a certain rotational motion of the anion.

Rotational correlation times $\tau_c$ were derived from $^1$H $T_1$ fitting. Figure 17.12 shows temperature dependence of $\tau_c$ of the fast (i.e. δ-$CH_3$ rotation) and slow rotational motions, respectively. The $\tau_c$ values for the slow motions are ca. $10^2$ larger than that for δ-$CH_3$ rotations. The conspicuous diversity of $\tau_c$ values was observed in the slow motions of the three crystalline phases while $\tau_c$ for δ-$CH_3$ rotations depend on the crystal phase slightly. The diversity arises from the observation of the different type of motion for the slow motion. The orders of the

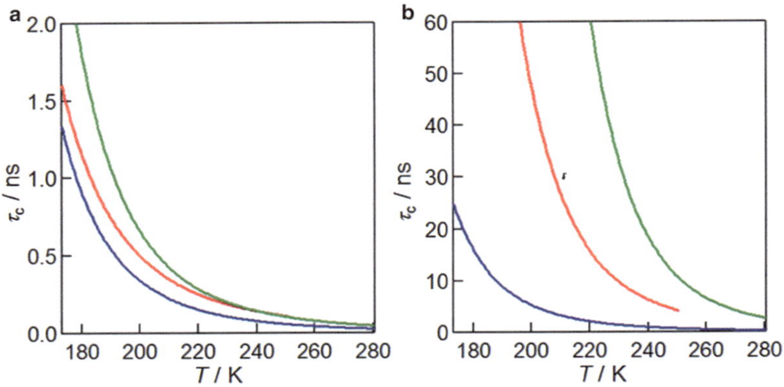


**Fig. 17.12** Rotational correlation times $\tau_c$ derived from $^1$H $T_1$ fitting for (**a**) the fast rotational motion that is assigned to δ-$CH_3$ rotation and for (**b**) the slow rotational motion. *Red*: crystal α, *blue*: crystal β, and *green*: crystal γ. Reprinted with permission from Endo et al. [46]. Copyright 2012 American Chemical Society

correlation time for the fast and slow motions are both crystal β < crystal α < crystal γ. The fact that the crystal γ shows slowest rotational dynamics coincides with the calorimetric results. In addition, the mobility of the segmental motions is likely to change with the conformation of the butyl group.

## *17.3.2 Anion Dynamics*

We also performed $^{31}$P NMR to investigate the $PF_6^-$ anion rotational dynamics in the crystalline states as well as in the glassy state with the same strategy used above [49]. The theoretical $^{31}$P $M_2$ values for the anion are listed in Table 17.5. Figure 17.13 shows temperature dependence of the experimental $^{31}$P $M_2$ values in all crystalline polymorphs obtained from the spin echo measurements. All $^{31}$P $M_2$ values are below 1.42 Gauss$^2$ suggesting that the contribution to $M_2$ from the intra-anionic $^{31}$P–$^{19}$F dipolar interactions is negligible and only small inter-ionic contributions remain. The negligible contribution from intra-anionic $^{31}$P–$^{19}$F dipolar interactions indicates that the $PF_6^-$ anion undergoes rapid isotropic rotation in all crystalline polymorphs. Rapid axial rotation of the anion with $C_3$ symmetry could also lead to such effects but would be unlikely since there are no plausible strong interactions in the crystal that would allow for preferential rotation about the $C_3$ axis. The fact that all the experimental $^{31}$P $M_2$ values are below 1.42 Gauss$^2$ and they decrease slightly with increasing temperature are consistent with the findings that there are rotational motions in the cation that exist in all three crystalline states which decrease the $^{31}$P–$^1$H dipolar interactions [46]. It is probable that the differences in the $^{31}$P $M_2$ values originate from the differences in the crystal structures of the three polymorphs that change the $^{31}$P–$^1$H dipolar interactions.

**Table 17.5** Theoretical $^{31}$P second moment $M_2$ (Gauss$^2$) for [$C_4$mim]$PF_6$ in the crystalline state

| | | Intra $^{31}$P–$^{19}$F | Inter $^{31}$P–$^{19}$F | | Inter $^{31}$P–$^{1}$H | | Inter $^{31}$P–$^{31}$P | | Total |
|---|---|---|---|---|---|---|---|---|---|
| | | | <6 Å | 6 Å< | <6 Å | 6 Å< | <6 Å | 6 Å< | |
| Rigid lattice | | 51.45 | 0.10 | 0.05 | 1.40 | 0.15 | 0.005 | 0.004 | 53.16 |
| 3-$CH_3$ and δ-$CH_3$ free rotations | | 51.45 | 0.10 | 0.05 | 1.13 | 0.15 | 0.005 | 0.004 | 52.90 |
| $PF_6^-$ rotation | $C_4$ | 25.73 | 0.08 | 0.05 | 1.13 | 0.15 | 0.005 | 0.004 | 27.16 |
| $PF_6^-$ rotation | $C_2$ | 6.43 | 0.07 | 0.05 | 1.13 | 0.15 | 0.005 | 0.004 | 7.85 |
| $PF_6^-$ rotation | $C_3$ | 0.00 | 0.07 | 0.05 | 1.13 | 0.15 | 0.005 | 0.004 | 1.42 |
| $PF_6^-$ rotation | Isotropic | 0.00 | 0.07 | 0.05 | 1.13 | 0.15 | 0.005 | 0.004 | 1.42 |

Reprinted with permission from Endo et al. [49]. Copyright 2013 American Chemical Society

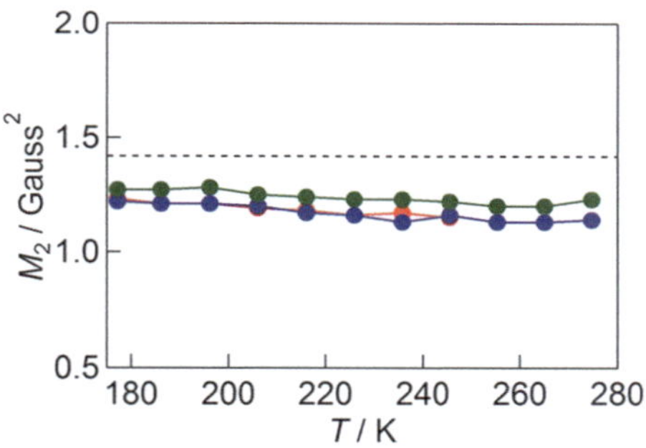


**Fig. 17.13** Temperature dependence of the second moment $M_2$ of $^{31}$P NMR spectra. *Red, blue,* and *green* symbols represent data for the [$C_4$mim]$PF_6$ crystalline polymorphs α, β, and γ, respectively. The *gray dashed line* is the theoretical limit of $M_2$ of 1.42 Gauss$^2$ for isotropic rotation of $PF_6^-$ anions (see Table 17.5). Reprinted with permission from Endo et al. [49]. Copyright 2013 American Chemical Society

Figure 17.14 shows the temperature dependence of the $^{31}$P $T_1$ for the three crystalline polymorphs and the liquid of [$C_4$mim]$PF_6$, displayed with theroretical fitting curves. The data in all cases seem to have a minimum that is located below 170 K. In addition a rather shallow minimum in $^{31}$P $T_1$ is observed for the β phase and the liquid that are located near 250 and 330 K, respectively. The dominant dynamical process would be the isotropic rotational motion of the anion. The cation dynamics that contributes $^{31}$P–$^{1}$H interaction can also exist and was attributed to $T_1$ process with the shallow minimum observed in the β phase and liquid state. The resultant fitting parameters $\tau_0$, $E_a$ and $\Delta M_2$ are listed in Table 17.6. The temperature dependence of the $\tau_c$ values for all the states are shown in Fig. 17.15.

The results presented above, when taken together, clearly demonstrate that RTILs are unique materials that enable comparison of ionic dynamics in salts

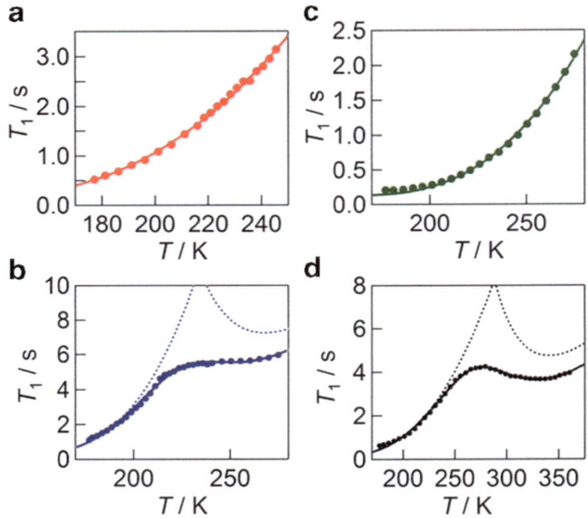


**Fig. 17.14** Theoretical fits (*solid lines*) to the measured $^{31}$P $T_1$ data (*symbols*): (**a**) α phase; (**b**) β phase; (**c**) γ phase; (**d**) the liquid. The individual simulation components in parts **b** and **d** are indicated by *dashed lines*. Reprinted with permission from Endo et al. [49]. Copyright 2013 American Chemical Society

**Table 17.6** Parameters $\Delta M_2$, $E_a$ and $\tau_0$ derived from fitting the $^{31}$P $T_1$ data

| | Crystal α | Crystal β | | Crystal γ | Liquid | |
|---|---|---|---|---|---|---|
| Interaction | P–F | P–F | P–H | P–F | P–F | P–H |
| $\Delta M_2$/ Gauss$^2$ | 51.5 (fixed) | 51.5 (fixed) | $1.0 \pm 0.0$ | 51.5 (fixed) | 51.5 (fixed) | $1.6 \pm 0.0$ |
| $E_a$/ kJ mol$^{-1}$ | $9.7 \pm 0.1$ | $14.4 \pm 0.6$ | $17.2 \pm 1.6$ | $13.9 \pm 0.2$ | $11.1 \pm 0.3$ | $20.2 \pm 0.9$ |
| $\tau_0$/fs | $92 \pm 6$ | $2 \pm 1$ | $311 \pm 241$ | $35 \pm 4$ | $38 \pm 6$ | $572 \pm 186$ |
| Motional assignment | $PF_6^-$ isotropic rotation | $PF_6^-$ isotropic rotation | Rotational motion in the cation | $PF_6^-$ isotropic rotation | $PF_6^-$ isotropic rotation | Rotational motion in the cation |

Reprinted with permission from Endo et al. [49]. Copyright 2013 American Chemical Society

between crystalline and liquid states. As can be seen in Fig. 17.15, the timescale $\tau_c$ for the rotational dynamics of the $PF_6^-$ anion in all phases is on the order of picoseconds and is the slowest in the γ phase and the fastest in the β phase. The α phase and the liquid are comparable in their behavior and are characterized by timescales intermediate between those characteristic of the β and γ phases. It was shown in our previous study [50] that the $\tau_c$ of $PF_6^-$ rotation in the supercooled [C$_4$mim]PF$_6$ liquid increases only slightly with decreasing temperature to several tens of picoseconds even below its glass transition (192 K) indicating that this dynamics is not related to the viscous slowdown and is strongly decoupled from the structural relaxation process. The cation-anion Coulombic interaction should be weaker in the liquid state compared to that in solid state due to the increase in molar volume by ~10 – 15 % [51, 52] that typically accompany melting. Therefore, one may expect faster ionic dynamics in liquid state. However, the present results show that $\tau_c$ of $PF_6^-$ rotation in the liquid state is comparable to that in the

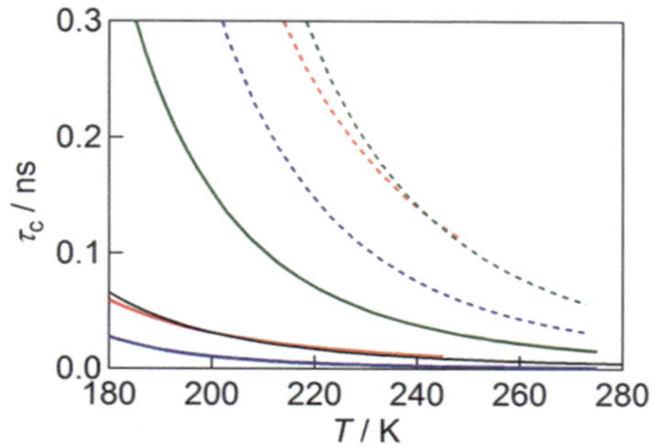


**Fig. 17.15** Rotational correlation times $\tau_c$ in $[C_4mim]PF_6$ derived from fits to the $^{31}P$ $T_1$ data. *Solid* and *dashed lines* are for rotation of $PF_6^-$ and δ-$CH_3$, respectively. *Red* crystal α, *blue* crystal β, *green* crystal γ, *black* the liquid state. The data for δ-$CH_3$ are taken from a previous report in the literature [46]. Reprinted with permission from Endo et al. [49]. Copyright 2013 American Chemical Society

α phase and is slower than the β phase. Such behavior may be related to the fact that the $PF_6^-$ rotational dynamics is governed by the local structure and/or local interactions rather than the average Columbic interaction between the cations and the anions.

It is worth comparing the anion dynamics to that of the cation in the $[C_4mim]PF_6$ crystals. Fig. 17.15 also includes the $\tau_c$ of the δ-$CH_3$ rotational motion in the cation [46]. Interestingly, the trend of the $\tau_c$ for the rotational motion is the same for both cation and anion. The $PF_6^-$ rotational motion in the three crystal phases is all faster than the δ-$CH_3$ rotation in cations. Previous results indicated that δ-$CH_3$ rotation was slower than the rotation of the methyl group attached the imidazolium ring (3-$CH_3$) but faster than the segmental motion of the butyl group [46, 48]. Hence, the order of the $\tau_c$ for the rotational motion in the $[C_4mim]PF_6$ crystals would be 3-$CH_3$ < $PF_6^-$ < δ-$CH_3$ < butyl group segmental motion. The presence of multiple timescales may suggest the presence of a dynamical hierarchy in these materials. A similar comparison between the cation and anion rotational dynamics can also be made in the case of the $[C_4mim]PF_6$ liquid for which previous studies investigated the cation dynamics [53–56]. The $\tau_c$ of $PF_6^-$ rotation in the liquid state as obtained in this study is 3.3 ps at 300 K whereas for the cation at the same temperature in the liquid state $\tau_c$ was found in previous studies to range between 700 and 900 ps for the three carbons in the imidazolium ring, 26 ps for 3-$CH_3$, 510/220/120 ps or 168/103/60 ps for the three carbons in the methylene group of the butyl chain and 26 or 13 ps for the rotation of δ-$CH_3$ [53, 55, 56]. Therefore, it is clear that all local motions in the cation have slower rotational timescale compared to that for $PF_6^-$. It should be noted that reported 3-$CH_3$ and δ-$CH_3$ rotational motions in the liquid state would not be the threefold axis rotation as observed in the crystalline state. This conclusion is also true when the anion dynamics is compared to the rotational reorientation of the whole cation as the latter is expectedly significantly slower with $\tau_c$ of ~3–4 ns near

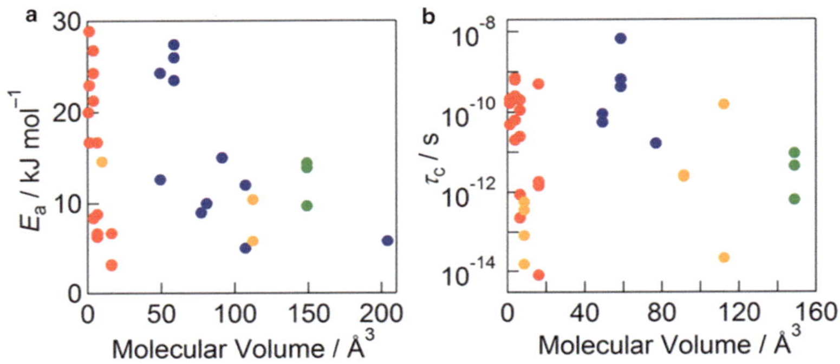


**Fig. 17.16** (**a**) $E_a$ and (**b**) $\tau_c$ at 300 K for $PF_6^-$ rotation in salts in their crystalline states. *Red*, *yellow*, *blue*, and *green* symbols represent data for salts with alkali cations, spherical nonalkali cations, nonspherical cations, and for $[C_4mim]PF_6$, respectively. See Endo et al. [49] and references therein for details. The $\tau_c$ values for some salts were obtained by minor extrapolation, as their melting points are below 300 K. Reprinted with permission from Endo et al. [49]. Copyright 2013 American Chemical Society

300 K [50, 57–59]. Previous NMR spectroscopy [60] and MD simulation [61, 62] studies indicated that in contrast with the rotational dynamics the timescale of the translational dynamics of the cation and the anion were comparable.

Finally, it is interesting to compare the timescale of the $PF_6^-$ rotational dynamics in $[C_4mim]PF_6$ to that in "ordinary" salts such as alkali and ammonium salt as significant data are available in the literature on the $PF_6^-$ rotational dynamics in such salts [49]. Figure 17.16 compare the $E_a$, $\tau_c$ (if available) versus ionic volume of the $PF_6^-$ anion in these salts. The DFT calculations carried out in this study [63] yield the ionic volumes of $[C_4mim]^+$ and $PF_6^-$ ions to be 148.8 and 71.3 Å$^3$, respectively. These values are comparable to the van der Waals volumes of these ions that were reported in the literature to be 150 and 69 Å$^3$ [64]. It is intuitively expected and was suggested in a number of previous studies that the ionic volume or radius is roughly correlated with $PF_6^-$ rotational dynamics [65–68]. This is because larger ionic volume can decrease cation-anion interaction that could result in faster $PF_6^-$ rotational dynamics. However, the data in Fig. 17.16 indicate almost no relationship between the ionic volume and the dynamical parameters for the $PF_6^-$ rotational dynamics. This observation strongly suggests that the $PF_6^-$ rotational dynamics is not significantly influenced by the averaged cation-anion interaction. It is tempting to attribute this result to the high sphericity of the octahedral $PF_6^-$ anions. It is well known that ions or molecules with spherical symmetry are expected to have small activation barriers for rotational jumps or diffusion [69–71], consistent with the activation energies of 10–15 kJ mol$^{-1}$ obtained in this study for the isotropic rotation of $PF_6^-$ anions in the $[C_4mim]PF_6$ crystals and liquid. Such low activation energies have also been observed for the rotational motion of these anions in a remarkably wide variety of salts, as shown in Fig. 17.16.

## 17.4 Summary and Perspective

Here we focused on the results of [$C_4$mim]$PF_6$, but there have been a large variety of RTILs and their series investigated, such as researches on alkyl chain length dependence and the methylation effect at the 2 position of the imidazolium ring [42, 72, 73]. Particularly on imidazolium-based RTILs that are the most well-known series probably because of tendency to have low melting point, when taken together, flexibility of an alkyl group attached to the ring plays an important role in complicated thermal phase behavior of RTILs. Namely, low melting point, easy formation of supercooling and subsequent glassy states, complex solid-solid phase change and slow phase change can be somewhat explained with the flexibility. In this sense, [$C_4$mim]$PF_6$ can be the most representative RTIL, which shows three dominant cation conformations with some minor ones during the thermal phase changes. Some pyrrolidinium-based and piperidinium-based RTILs show plastic crystallinity that significantly enhances ion transportation property in the solid state [13–16]. This is also the strong evidence that flexibility of ions drastically changes physical properties of RTILs and proves the importance of high degree of freedom in ions. It should be noted that 1,3-dimethylimidazolium hexafluorophosphate ([$C_1$mim]$PF_6$), which has no flexibility in the ions, shows two polymorphic crystals [43]. Melting point difference between polymorphic crystals of imidazolium-based RTIL is the largest in [$C_1$mim]$PF_6$, which is ca. 50 K. As can be already seen in the presence of the β' phase in [$C_4$mim]$PF_6$, there are some phase changes that do not accompany with the conformational change of ions [30, 42, 43]. Obviously, many factors, especially Columbic interactions that are characteristics of salts, should be considered for understanding the nature of RTILs in detail. However, one will not be able to underestimate the importance of flexibility in ions.

The work presented here would not directly contribute to any industrial applications such as electronic devices. However, considering the uniqueness and outstanding properties of RTILs, we believe that our fundamental researches somewhat paved the road of RTIL fields for the future.

## References

1. J.S. Wilkes, Green Chem. **4**, 73 (2002)
2. B. Kirchner (ed.), *Topics in Current Chemistry*, 1st edn. (Springer-Verlag Berlin, Berlin, 2009)
3. I. Minami, Molecules **14**, 2286 (2009)
4. A. Kokorin (ed.), *Ionic Liquids: Applications and Perspectives* (InTech, Vienna, 2011)
5. P. Wasserscheid, T. Welton, *Ionic Liquids in Synthesis* (VCH-Wiley, Weinheim, 2003)
6. H. Ohno, *Electrochemical Aspects of Ionic Liquids* (Wiley-Interscience, Hoboken, 2005)
7. M. Grätzel, J. Photochem. Photobiol. A **164**(3) (2004)
8. F. Endres, ChemPhysChem **3**, 144 (2002)
9. M. Armand, F. Endres, D.R. MacFarlane, H. Ohno, B. Scrosati, Nat. Mater. **8**, 621 (2009)
10. J.D. Holbrey, W.M. Reichert, M. Nieuwenhuyzen, S. Johnston, K.R. Seddon, R.D. Rogers, Chem. Commun. 1636 (2003)

11. W.M. Reichert, J.D. Holbrey, K.B. Vigour, T.D. Morgan, G.A. Broker, R.D. Rogers, Chem. Commun. 4767 (2006)
12. I. Krossing, J.M. Slattery, C. Daguenet, P.J. Dyson, A. Oleinikova, H. Weingärtner, J. Am. Chem. Soc. **128**, 13427 (2006)
13. J.M. Pringle, P.C. Howlett, D.R. MacFarlane, M. Forsyth, J. Mater. Chem. **20**, 2056 (2010)
14. D.R. MacFarlane, M. Forsyth, Adv. Mater. **13**, 957 (2001)
15. H.A. Every, A.G. Bishop, D.R. MacFarlane, G. Orädd, M. Forsyth, J. Mater. Chem. **11**, 3031 (2001)
16. V. Tricoli, G. Orsini, M. Anselmi, Phys. Chem. Chem. Phys. **14**, 10979 (2012)
17. T. Endo, T. Kato, K. Tozaki, K. Nishikawa, J. Phys. Chem. B **114**, 407 (2010)
18. U. Domańska, A. Marciniak, J. Chem. Eng. Data **48**, 451 (2003)
19. A.R. Choudhury, N. Winterton, A. Steiner, A.I. Cooper, K.A. Johnson, J. Am. Chem. Soc. **127**, 16792 (2005)
20. J. Troncoso, C.A. Cerdeiriña, Y.A. Sanmamed, L. Romaní, L.P.N. Rebelo, J. Chem. Eng. Data **51**, 1856 (2006)
21. A. Triolo, A. Mandanici, O. Russina, V. Rodriguez-Mora, M. Cutroni, C. Hardacre, M. Nieuwenhuyzen, H.-J. Bleif, L. Keller, M.A. Ramos, J. Phys. Chem. B **110**, 21357 (2006)
22. S.M. Dibrov, J.K. Kochi, Acta Crystallogr. Sect. C **C62**, o19 (2006)
23. D.M. Fox, W.H. Awad, J.W. Gilman, P.H. Maupin, H.C. De Long, P.C. Trulove, Green Chem. **5**, 724 (2003)
24. G.J. Kabo, A.V. Blokhin, Y.U. Paulechka, A.G. Kabo, M.P. Shymanovich, J.W. Magee, J. Chem. Eng. Data **49**, 453 (2004)
25. H. Jin, B. O'Hare, J. Dong, S. Arzhantsev, G.A. Baker, J.F. Wishart, A.J. Benesi, M. Maroncelli, J. Phys. Chem. B **112**, 81 (2008)
26. T. Endo, K. Tozaki, T. Masaki, K. Nishikawa, Jpn. J. Appl. Phys. **47**, 1775 (2008)
27. A. Kojima, C. Ishii, K. Tozaki, S. Matsuda, T. Nakayama, N. Tsuda, Y. Yoshimura, H. Iwasaki, Rev. Sci. Instrum. **68**, 2301 (1997)
28. K. Tozaki, H. Inaba, H. Hayashi, C. Quan, N. Nemoto, T. Kimura, Thermochim. Acta **397**, 155 (2003)
29. S. Wang, K. Tozaki, H. Hayashi, H. Inaba, J. Therm. Anal. Calorim. **79**, 605 (2005)
30. T. Endo, K. Nishikawa, Chem. Phys. Lett. **584**, 79 (2013)
31. S. Saouane, S.E. Norman, C. Hardacre, F.P.A. Fabbiani, Chem. Sci. **4**, 1270 (2013)
32. R. Ozawa, S. Hayashi, S. Saha, A. Kobayashi, H. Hamaguchi, Chem. Lett. **32**, 948 (2003)
33. S. Hayashi, R. Ozawa, H. Hamaguchi, Chem. Lett. **32**, 498 (2003)
34. M.J. Frisch, G.W. Trucks, H.B. Schlegel, G.E. Scuseria, M.A. Robb, J.R. Cheeseman, J.A. Montgomery Jr., T. Vreven, K.N. Kudin, J.C. Burant, J.M. Millam, S.S. Iyengar, J. Tomasi, V. Barone, B. Mennucci, M. Cossi, G. Scalmani, N. Rega, G.A. Petersson, H. Nakatsuji, M. Hada, M. Ehara, K. Toyota, R. Fukuda, J. Hasegawa, M. Ishida, T. Nakajima, Y. Honda, O. Kitao, H. Nakai, M. Klene, X. Li, J.E. Knox, H.P. Hratchian, J.B. Cross, C. Adamo, J. Jaramillo, R. Gomperts, R.E. Stratmann, O. Yazyev, A.J. Austin, R. Cammi, C. Pomelli, J.W. Ochterski, P.Y. Ayala, K. Morokuma, G.A. Voth, P. Salvador, J.J. Dannenberg, V.G. Zakrzewski, S. Dapprich, A.D. Daniels, M.C. Strain, O. Farkas, D.K. Malick, A.D. Rabuck, K. Raghavachari, J.B. Foresman, J.V Ortiz, Q. Cui, A.G. Baboul, S. Clifford, J. Cioslowski, B.B. Stefanov, G. Liu, A. Liashenko, P. Piskorz, L. Komaromi, R.L. Martin, D.J. Fox, T. Keith, M.A. Al-Laham, C.Y. Peng, A. Nanayakkara, M. Challacombe, P.M.W. Gill, B. Johnson, W. Chen, M.W. Wong, C. Gonzalez, J.A. Pople, Gaussian 03, Gaussian, Inc., Wallingford CT (2004)
35. E.A. Turner, C.C. Pye, R.D. Singer, J. Phys. Chem. A **107**, 2277 (2003)
36. Y. Umebayashi, T. Fujimori, T. Sukizaki, M. Asada, K. Fujii, R. Kanzaki, S. Ishiguro, J. Phys. Chem. A **109**, 8976 (2005)
37. R.W. Berg, Monatsh. Chem. **138**, 1045 (2007)
38. Y.U. Paulechka, G.J. Kabo, A.V. Blokhin, A.S. Shaplov, E.I. Lozinskaya, D.G. Golovanov, K.A. Lyssenko, A.A. Korlyukov, Y.S. Vygodskii, J. Phys. Chem. B **113**, 9538 (2009)

39. S. Saha, S. Hayashi, A. Kobayashi, H. Hamaguchi, Chem. Lett. **32**, 740 (2003)
40. M. Nakakoshi, M. Shiro, T. Fujimoto, T. Machinami, H. Seki, M. Tashiro, K. Nishikawa, Chem. Lett. **35**, 1400 (2006)
41. C.S. Santos, S. Rivera-R, S. Dibrov, S. Baldelli, J. Phys. Chem. C **111**, 7682 (2007)
42. T. Endo, T. Kato, K. Nishikawa, J. Phys. Chem. B **114**, 9201 (2010)
43. T. Endo, T. Morita, K. Nishikawa, Chem. Phys. Lett. **517**, 162 (2011)
44. J.C. Lassègues, J. Grondin, R. Holomb, P. Johansson, J. Raman Spectrosc. **38**, 551 (2007)
45. J.H. Van Vleck, Phys. Rev. **74**, 1168 (1948)
46. T. Endo, H. Murata, M. Imanari, N. Mizushima, H. Seki, K. Nishikawa, J. Phys. Chem. B **116**, 3780 (2012)
47. N. Bloembergen, E.M. Purcell, R.V. Pound, Phys. Rev. **73**, 679 (1948)
48. A. Triolo, O. Russina, C. Hardacre, M. Nieuwenhuyzen, M.A. Gonzalez, H. Grimm, J. Phys. Chem. B **109**, 22061 (2005)
49. T. Endo, H. Murata, M. Imanari, N. Mizushima, H. Seki, S. Sen, K. Nishikawa, J. Phys. Chem. B **117**, 326 (2013)
50. T. Endo, S. Widgeon, P. Yu, S. Sen, K. Nishikawa, Phys. Rev. B **85**, 054307/1 (2012)
51. J. Dupont, P.A.Z. Suarez, Phys. Chem. Chem. Phys. **8**, 2441 (2006)
52. J.-Z. Yang, X.-M. Lu, J.-S. Gui, W.-G. Xu, Green Chem. **6**, 541 (2004)
53. J.H. Antony, D. Mertens, A. Dölle, P. Wasserscheid, W.R. Carper, ChemPhysChem **4**, 588 (2003)
54. W.R. Carper, P.G. Wahlbeck, A. Dölle, J. Phys. Chem. A **108**, 6096 (2004)
55. J.H. Antony, D. Mertens, T. Breitenstein, A. Dölle, P. Wasserscheid, W.R. Carper, Pure Appl. Chem. **76**, 255 (2004)
56. W.R. Carper, P.G. Wahlbeck, J.H. Antony, D. Mertens, A. Dölle, P. Wasserscheid, Anal. Bioanal. Chem. **378**, 1548 (2004)
57. A. Rivera, A. Brodin, A. Pugachev, E.A. Rössler, J. Chem. Phys. **126**, 114503 (2007)
58. N. Shamim, G.B. McKenna, J. Phys. Chem. B **114**, 15742 (2010)
59. W. Fan, Q. Zhou, J. Sun, S. Zhang, J. Chem. Eng. Data **54**, 2307 (2009)
60. H. Tokuda, K. Hayamizu, K. Ishii, M.A.B.H. Susan, M. Watanabe, J. Phys. Chem. B **108**, 16593 (2004)
61. W. Zhao, F. Leroy, B. Heggen, S. Zahn, B. Kirchner, S. Balasubramanian, F. Mueller-Plathe, J. Am. Chem. Soc. **131**, 15825 (2009)
62. T.I. Morrow, E.J. Maginn, J. Phys. Chem. B **106**, 12807 (2002)
63. D.F. Parsons, B.W. Ninham, J. Phys. Chem. A **113**, 1141 (2009)
64. H. Machida, R. Taguchi, Y. Sato, R.L. Smith Jr., Fluid Phase Equilib. **281**, 127 (2009)
65. G.R. Miller, H.S. Gutowsky, J. Chem. Phys. **39**, 1983 (1963)
66. H.S. Gutowsky, S. Albert, J. Chem. Phys. **58**, 5446 (1973)
67. G. Burbach, N. Weiden, A. Weiss, Z. Naturforsch. A Phys. Sci. **47a**, 689 (1992)
68. K.J. Mallikarjunaiah, R. Damle, K.P. Ramesh, Solid State Nucl. Magn. Reson. **34**, 180 (2008)
69. R. Blinc, G. Lahajnar, J. Chem. Phys. **47**, 4146 (1967)
70. M. Makrocka-Rydzyk, S. Glowinkowski, S. Jurga, W.H. Meyer, Appl. Magn. Reson. **18**, 63 (2000)
71. R.D. Johnson, C.S. Yannoni, H.C. Dorn, J.R. Salem, D.S. Bethune, Science **255**, 1235 (1992)
72. J.D. Holbrey, K.R. Seddon, J. Chem. Soc. Dalton Trans. 2133 (1999)
73. P.A. Hunt, J. Phys. Chem. B **111**, 4844 (2007)

# Chapter 18
# Single Molecular Spintronics

**Toyo Kazu Yamada**

## 18.1 Introduction

Organic molecules have enabled the realization of cheap, flexible, and printed electronic devices, which are nowadays widely used for organic solar cells [1], organic electroluminescence devices [2], organic displays [3], and organic light-emitting diodes [4]. So far, organic molecular films with a thickness of 10–100 nm grown on a substrate have been used, and electron or hole transfer through molecular junctions has determined the electronic properties of devices. Minimization of the device size is an important challenge for near-future electronics to realize low-cost device with low power consumption and high functionalities, which is why studies on nano electronics and nano science have rapidly developed in the last decade. Fundamental studies on nano molecules of size 1–10 nm have recently been carried out and new electronic (spin) properties have been exhibited by single molecules [5–11], single-molecular chains [12], and monolayer (ML) molecular films [13, 14] on metal substrates. In particular, local density of states (LDOS) of nano molecules varies greatly owing to their unique symmetry, i.e., surface or interface effects mainly determine the nano device properties [15, 10].

Quantum spin states of single molecular magnets have been widely studied using noble metal substrates [6–9, 11, 13, 14, 16, 17]. Electronic couplings from the noble metal substrates have been considered not to affect strongly the quantum spin states of the molecules. However, we have recently found that properties of Fe-based spin-crossover molecules are affected by the electronic coupling from a Cu(001) substrate [8]. We succeeded in detecting molecular low-spin and high-spin states of the spin-crossover molecules by inserting a CuN layer between the substrate and the molecules to cut the coupling from the substrate, which means that even with noble metal electrodes, electronic or spin states of single molecules can be modified. Such

T.K. Yamada (✉)
Graduate School of Advanced Integration Science, Chiba University, 1-33, Yayoi-cho, Inage-ku, Chiba-shi 263-8522, Chiba, Japan



H. Ishii et al. (eds.), *Electronic Processes in Organic Electronics*, Springer Series in Materials Science 209, DOI 10.1007/978-4-431-55206-2_18

a decoupling layer is required to observe intrinsic molecular orbitals. MgO (2–3 ML) and NaCl (2 ML) thin films have also been used as decoupling layers [18–21].

On the other hand, strong electronic couplings between substrates and molecules can also be useful for new molecular spintronics. Strong hybridizations between 3d spin-polarized LDOS and molecular $\pi$-orbitals result in a magnetoresistance effect through a single molecule [5, 10], i.e., spin current through single molecules can be controlled by switching the magnetization directions of magnetic electrodes.

One of the most effective techniques for precisely measuring electronic spin properties of 1-nm-size molecules is tunneling spectroscopy with a scanning tunneling microscopy (STM) setup, known as scanning tunneling spectroscopy (STS). This technique measures the so-called $I(V)$ curves or $dI/dV$ curves, which are the tunneling conductivity and differential tunneling conductivity as functions of the sample bias voltage, respectively. The LDOS, minority and majority spin states, is included in the $dI/dV$ [15, 22–24].

In this work, we show experimental findings of giant magnetoresistance through a $\pi$-conjugated single phthalocyanine ($H_2Pc$) molecule by means of ultrahigh vacuum spin-polarized STM setup at 4.6 K.

## 18.2 Spin-Polarized STM

Spin-polarized scanning tunneling microscopy(STM) [25, 26] is one of the most powerful tool to resolve surface quantum spin states of magnetic thinfilms [24, 27, 28], single atoms [29], single molecules [5, 6, 10], and nanoclusters [15] by measuring sample surface spin-polarized LDOS near the Fermi energy.

Figure 18.1 shows a simple sketch of an STM setup. A sample is set in an STM stage and a bias typically $\mu$-3 V is applied. An STM tip is approached toward the sample surface until the tip detects tunneling current. Typical tip-sample distance

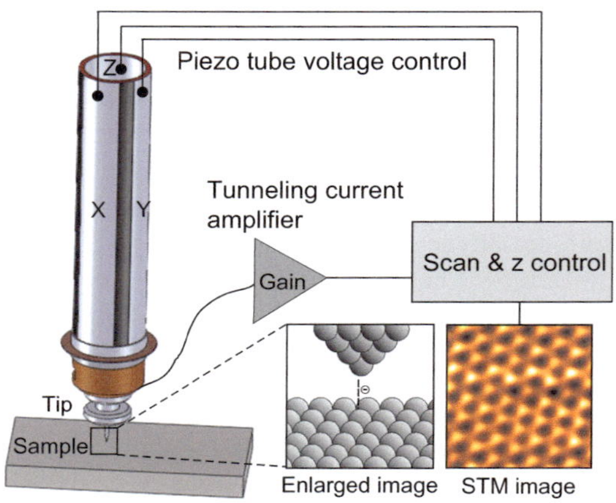


**Fig. 18.1** A scanning tunneling microscopy (STM) setup

during tunneling is 0.5–1.0 nm. Since STM detects tunneling current, usually conductive samples and tips are used, i.e. metals or semiconductors. Since the tunneling current depends exponentially on the tip-sample distance, pm scale corrugation can be detected. Since atomic corrugation of the sample surface is usually 1–50 pm, atomically resolved images can be obtained by STM.

The detected tunneling current ($10^{-12}$–$10^{-9}$ A) is $10^6 \sim 10^{10}$ times amplified and switched to a voltage signal via an I-V convertor. In a constant current mode, a surface topographic image of the sample is obtained by keeping a constant current, e.g., 100 pA. During a x-y scan, a feedback loop tries to keep the constant current by controlling the tip-sample distance with a z piezo. A map of the applied voltage to the z piezo at each pixel position on the sample surface corresponds to the STM topographic image such as Fig. 18.1 where hexagonal lattice of carbon atoms are observed.

We need a spectroscopy mode for magnetic imaging, i.e., not only the topology, but also LDOS near the Fermi energy is required. So far spin-polarized STM experiments have been performed at 0.3–300 K in ultrahigh vacuum (UHV). Spectroscopy measurements were performed by opening the feedback loop of the tunneling current, i.e., fixing the tip-sample separation ($z$). The tunneling current was measured at each pixel by varying the sample voltage to obtain $I(V)$ curves. Differential conductivity ($dI/dV$) curves were obtained by numerical differentiation of the $I(V)$ curves. We could obtain a spectroscopy image by mapping the $dI/dV$ value at each pixel position. A lock-in amplifier was also used to obtain the $dI/dV$ map.

By the way, how we could know magnetic moment vectors by measuring LDOS of nanomaterials? Electrons have not only charges but also angular momentum vectors, and generate spin up $\alpha$=(1,0) and spin down $\beta$=(0,1) states. Now we focus LDOS near the Fermi energy. In non-magnetic materials numbers of spin up and down electrons are equal, but in magnetic materials numbers of spin up and down electrons are different (Stoner model). A magnetic moment vector can be described as follows: $\vec{\mu_J}=-g_J\mu_B\vec{J}$, where $g_J$ denotes Lande's g factor, $\mu_B$ Bohr magnetron, $\vec{J}$ total angular momentum vector ($\vec{J}=\vec{L}+\vec{S}$, $\vec{L}$: orbital angular momentum vector, $\vec{S}$: spin angular momentum vector). Spin up and down electrons can stay in minority or majority LDOS ($\rho_{min}$ and $\rho_{maj}$). When the densities are different, the difference ($\rho_{maj}-\rho_{min}$) corresponds to an amplitude of a magnetic moment, e.g., bulk-Fe has a magnetic moment of 2.2 $\mu_B$. Also, an amplitude of spin polarization vector can be defined as $|\vec{P}| \equiv (\rho_{maj}-\rho_{min})/(\rho_{maj}+\rho_{min})$, i.e., for magnetic materials, spin up and down states are different near the Fermi energy ($E_F$): $\rho_{min} \neq \rho_{maj}$, generating magnetic moment. In this way, magnetic moment can be investigated by measuring the sample LDOS. But, for that we need a magnetic tip.

Preparation of the magnetic tip is one of the most important and difficult issue for success of spin-polarized STM measurements [30]. Although magnetic force microscopy, atomic force microscopy with a magnetic cantilever, uses stray field from the tip apex, such a stray field from the tip must be excluded for spin-polarized

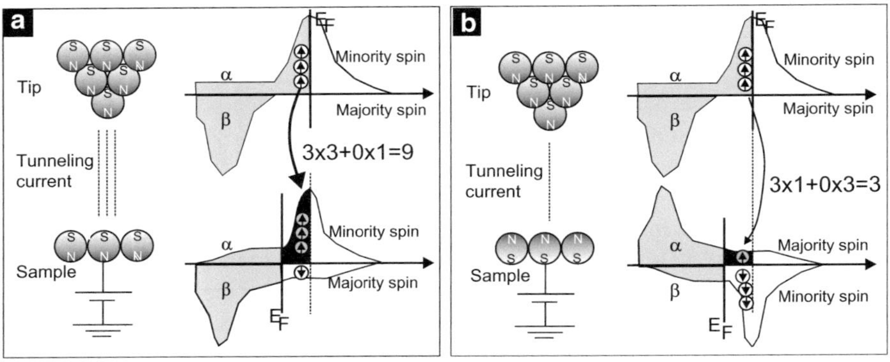


**Fig. 18.2** Simple mechanism of spin-polarized tunneling between two magnets: a magnetic tip and a magnetic sample surface. $\alpha$ and $\beta$ denote spin up and spin down. $E_F$ the Fermi energy. Here we assume a tip spin polarization of 100 %, i.e. minority spin band has spin up only. Magnetic moment vectors of the tip and the sample are (**a**) parallel and (**b**) anti-parallel. Since the sample magnetic moment vector reverses between (**a**) and (**b**), minority and majority spin bands also reverse. Electron spin transmission probability is $3 \times 3 + 0 \times 1 = 9$ for parallel and $3 \times 1 + 0 \times 3 = 3$ for anti-parallel

STM since tiny stray fields from the tip can change spin polarization vectors of the nanomagnets easily. Magnetic thinfilm coated W tips have been widely used as a spin-polarized STM tip. Tungsten (W) tips were electrochemically etched from W wires ($\phi = 0.3$ mm, purity 99.9 %) in air using KOH aq.. Subsequently, the tip was rinsed with hot water and acetone, then transferred into STM chamber. The apex of the W tip was sputtered with $Ar^+$ in UHV and annealed at temperatures of up to 2,000 K to obtain a clean apex. Then, a magnetic film was deposited on the W apex in UHV. So far, it has been reported that Fe-coated W tips have an in-plane spin-polarization of about 33 % [30] (Nagai, private communication) and Co-coated W tips have an out-of-plane anisotropy [10].

Figure 18.2 shows spin transfer between magnetic tip and sample. Figure 18.2a shows a case of parallel coupling between tip and sample magnetic moment vectors. Minority and majority spin states are shown. Now the tip has spin up ($\alpha$) in the minority states and spin down ($\beta$) in the majority states. Here we use an ideal tip with a 100 % spin polarization. In the case of parallel coupling (Fig. 18.2a), the sample has spin up ($\alpha$) in the minority states and spin down ($\beta$) in the majority states, similar to the tip.

In the case of anti-parallel coupling (Fig. 18.2b), the tip state does not change while the sample magnetic moment changes, i.e., sample minority and majority spin states reverse, thus now the sample has spin up ($\alpha$) in the majority states and spin down ($\beta$) in the minority states.

Since there is no spin flip during tunneling, spin up (down) electrons can transfer only to the spin up (down) states, i.e., spin transmission probability is $3 \times 3 + 1 \times 0 = 9$ for the parallel case and $3 \times 1 + 0 \times 3 = 3$ for the anti-parallel case.

Magnetoresistance of $(9-3)/3 = 200\,\%$ can be obtained, and the tip spin polarization is $(3-0)/(3+0) = 100\,\%$, the sample spin polarization is $(3-1)/(3+1) = 50\,\%$.

Thus, number of tunneling electrons depends on the inner product of the tip and the sample spin polarization vectors, which can be described as $I \propto I_o(1 + \vec{P_s} \cdot \vec{P_t})$ or $dI/dV \propto dI_o/dV(1 + \vec{P_s} \cdot \vec{P_t}) = dI_o/dV(1 + |\vec{P_s}||\vec{P_t}|\cos\theta)$, if $\vec{P_t} = P_t(0, 0, 1)$ and $\vec{P_s} = P_s(\sin\theta\sin\phi, \sin\theta\cos\phi, \cos\theta)$. Therefore, we can obtain the sample spin polarization by measuring *dI/dV* or normalized $(dI/dV)/T$ curves[1]:

$$
\begin{aligned}
(dI/dV)/T_P &= \rho_s^{min}\rho_t^{min} + \rho_s^{maj}\rho_t^{maj} \\
(dI/dV)/T_{AP} &= \rho_s^{min}\rho_t^{maj} + \rho_s^{maj}\rho_t^{min} \\
A_{(dI/dV)/T} &= \frac{(dI/dV)/T_P - (dI/dV)/T_{AP}}{(dI/dV)/T_P + (dI/dV)/T_{AP}} \\
&= \frac{(\rho_s^{min}\rho_t^{min} + \rho_s^{maj}\rho_t^{maj}) - (\rho_s^{min}\rho_t^{maj} + \rho_s^{maj}\rho_t^{min})}{(\rho_s^{min}\rho_t^{min} + \rho_s^{maj}\rho_t^{maj}) + (\rho_s^{min}\rho_t^{maj} + \rho_s^{maj}\rho_t^{min})} \\
&= \frac{\rho_s^{maj} - \rho_s^{min}}{\rho_s^{maj} + \rho_s^{min}} \frac{\rho_t^{maj} - \rho_t^{min}}{\rho_t^{maj} + \rho_t^{min}} \\
&= P_{sample}P_{tip},
\end{aligned}
$$

where $(dI/dV)/T_P$ and $(dI/dV)/T_{AP}$ denote normalized *dI/dV* curves by its fitted tunneling probability function $(T)$ when $\vec{P_s}$ and $\vec{P_t}$ are parallel and anti-parallel, $\rho_s^{maj}$ and $\rho_s^{min}$ sample majority and minority LDOS, $\rho_t^{maj}$ and $\rho_t^{min}$ tip majority and minority LDOS. Experimentally, $|\vec{P_t}|$ have been studied to have a (maximum) spin polarization near the Fermi energy of 40, 10, and 8 % for Fe-coated W tips [30], Cr-coated W tips (Nagai, private communication), and annealed Mn-coated W tips [31], respectively.

---

[1] During the spectroscopy mode, the feedback is off, i.e., the tip-sample separation is constant. The distance was fixed by tunneling current and the sample bias, $V_s$. Frequently contrast in the *dI/dV* is considered due to different spin polarizations, i.e., different minority or majority LDOS, at each atoms, however, *dI/dV* can be also varied by the exponential background $(T)$. When you set the bias $V_s$ at the energy position of spin polarized LDOS, the distance can be varied when the tip and the sample spin polarizations are parallel and anti-parallel, which affect *dI/dV* signals and modifying the spin contrast. The reason is that *dI/dV* is proportional to $LDOS \cdot exp(-2\kappa z)$, i.e., the change in the tip-sample distance $(z)$ varies $T \propto exp(-2\kappa z)$ and *dI/dV* varies. One good example was demonstrated on the Mn(001) surface [24], where the magnetic contrast as well as asymmetry in *dI/dV* varies drastically by using different bias $V_s$ for determining the tip-sample separation, i.e., for quantitative spin polarization measurements normalization of the spin-dependent tip-sample separation effect must be excluded, i.e., $|\vec{P_s}||\vec{P_t}| = ((dI/dV)/T)_P - (dI/dV)/T_{AP})/((dI/dV)/T_P + (dI/dV)/T_{AP})$.

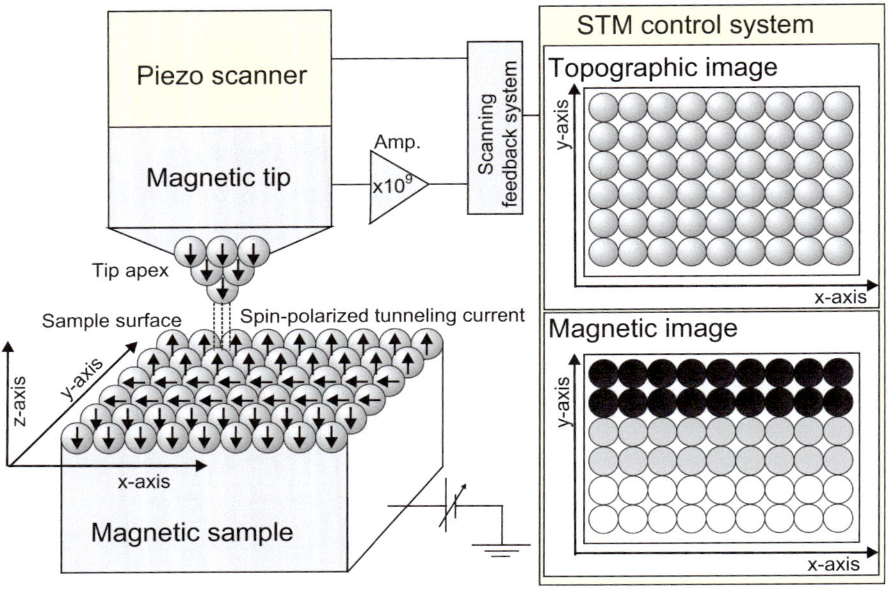


**Fig. 18.3** Magnetic imaging of spin-polarized STM. Electron spin transmission probability is proportional to an inner product of tip $(\vec{P_t})$ and sample $(\vec{P_s})$ spin polarization vectors: $dI_o/dV(1 + |\vec{P_s}||\vec{P_t}|\cos\theta)$, where $dI_o/dV$ is non-magnetic component in $dI/dV$. Magnetic image is obtained by mapping a $dI/dV$ value at each pixel position. The magnetic contrast can be proportional to $\vec{P_s} \cdot \vec{P_t}$. *Arrows* indicate spin polarization vectors of surface atoms

Figure 18.3 shows how STM magnetic imaging is practically performed. Arrows indicate directions of spin polarization vectors of surface atoms. We scan the sample surface with a magnetic tip. To obtain the magnetic image, we need to get not only STM topographic image but also a spectroscopy image simultaneously. During the scanning, at each pixel position, first the $z$ position is determined under feedback is ON, next the feedback loop is opened and $dI/dV$ curves are measured by numerical differentiation of $I(V)$ curves. This process is repeated at all pixel positions. Typically it takes 5–30 min to complete measurement. $dI/dV$ curves are obtained by numerical differentiation. Mappings of the obtained $z$ and $dI/dV$ value at each pixel position are called topographic and spectroscopy images, respectively. If there is contrast in the spectroscopy image, it may be caused by different spin distribution on the sample surface. However, cares must be required since the contrast can be caused by differences in (1) local work function (barrier height), (2) LDOS, (3) chemical spieces, and (4) spin polarization vectors. You have to exclude possibilities of (1), (2) and (3) to prove the obtained contrast in the spectroscopy image has a magnetic origin. You can also use a lockin amplifier to get $dI/dV$ signal while in this case always the feedback is closed, thus more cares are required for understanding the meaning of contrast in the spectroscopy image since the spin-dependent $z$ effect is included.[2]

Since $dI/dV = dI/dV_0(1 + \vec{P_s} \cdot \vec{P_t})$, sample areas where the coupling between tip and sample spin polarization vectors is parallel (anti-parallel) have a higher (lower) tunneling conductance. In the spectroscopy image areas with a higher and a lower conductance are mapped brighter and darker, respectively (Fig. 18.3), thus just by watching the color (so-called magnetic contrast) we can know distribution of the spin polarization amplitudes or directions on the sample surface.[3]

Spin-polarized STM can study spin polarization vectors of magnetic materials, however the studied region is limited only at surface. Due to breaking of bulk symmetry even magnets in bulk can be non-magnetic at surface. Therefore, for spin-polarized STM measurements magnetic samples with highly spin-polarized surface states have been used. So far the following samples have known to have spin-polarized surface states: e.g., Fe(110), Fe(001) [32], Mn(001) [24, 27], Co(0001), Cr(001).

To perform real spin-polarized STM measurements much knowledge of vacuum technology, surface science, and magnetism are required. We first need an atomically flat substrate. It is standard to use a surface with an atomic terrace width larger than 100 nm, i.e., one atomic step (about 0.2 nm) per 100 nm, since atomic steps have different symmetries which can quench the spin-polarized surface states [32]. Such a step effect occurs only near the step (0.5–1.0 nm). We adsorb magnetic samples such as single atoms, single molecules, nano clusters, thinfilms on the

[2] The spin polarization is defined as the difference between minority and majority spin states divided by the sum of the minority and majority spin states. Since minority and majority spin states varies in energy, spin polarization varies at each energy position. Therefore, magnetic contrast in STM images is varied at each energy position, i.e., only $I(V)$ or $dI/dV$ obtained at bias voltages near a highly spin polarized state shows contrast. Mostly highly spin polarized surface states exist near the Fermi energy. If you set bias voltage too far from the Fermi energy the contrast disappear. To exclude the spin dependent tip-sample distance effect to the $dI/dV$, it is better to choose the bias $V_s$ far from the Fermi energy (typically $|V_s| > 1.5$ V).

[3] Here we explain the relation between spin polarization $\vec{P}$ and magnetic moment vectors $\vec{\mu_J}$. Magnetic moment vector $\vec{\mu_J}$ can be described as $-g_J\mu_B\vec{J}$, while the spin polarization vector $\vec{P}$ can be defined with spin function $\gamma$ as $P \equiv <\gamma|2\vec{J}\ |\gamma>$. Now we assume the $\vec{J}$ points an arbitrary direction: $(x, y, z)$=(sin $\theta$ cos $\phi$, sin $\theta$ sin $\phi$, cos $\theta$). $\theta$ and $\phi$ denote angles from $z$ and $x$ axes, respectively. $\gamma = (\cos\frac{\theta}{2}, \sin\frac{\theta}{2}\exp(i\phi))$, i.e., if $\theta = \pi/2$ and $\phi = 0, \gamma = \frac{1}{\sqrt{2}}\alpha + \frac{1}{\sqrt{2}}\beta$, while $\alpha$=(1,0), $\beta$=(0,1). Now $\gamma$ is substituted, $P_x = 2\cos\frac{\theta}{2}\sin\frac{\theta}{2}\frac{\exp(i\phi)+\exp(-i\phi)}{2} = \sin\theta\cos\phi$, $P_y = -2i\cos\frac{\theta}{2}\sin\frac{\theta}{2}\frac{\exp(i\phi)-\exp(-i\phi)}{2} = \sin\theta\sin\phi$, $P_z = \cos^2\frac{\theta}{2} - \sin^2\frac{\theta}{2} = \cos\theta$, i.e., $\vec{P}$ points the same direction as $\vec{J}$. $\vec{P}$ is parallel to $\vec{J}$ and anti-parallel to $\vec{\mu_J}$. Since in bulk 3$d$ magnetic metals $\vec{L}$ is negligibly small, $\vec{J}$ is considered to be approximately equal to $\vec{S}$ and therefore magnetism in bulk 3$d$ metals can be explained mainly by spin angular moment vectors. However, nano magnets such as single atoms, $\vec{L}$ becomes larger and cannot be ignored. It is better to use $\vec{J}$.

atomically flat substrates. As a non-magnetic substrate, single crystal surfaces of W, Mo, and noble metals have been used.

Furthermore, spin polarization vectors of small magnets can be unstable by thermal fluctuations, which frequently decrease the Curie temperature much lower than room temperature, therefore it is better to check the Curie temperature before you perform experiment. So far 0.3, 4, 77, and 300 K STM are commercially available. For example, since 1-2 monolayers of Fe thinfilm on a W(110) substrate and Co thinfilms on Cu(111) [10] have Curie temperature lower than room temperature, spin-polarized STM experiments must be performed at 4 K by cooling with liquid He bath cryostat. It should be noted that spin-polarized STM measurements still can perform at room temperature if we use a strong exchange coupling between ferromagnet and antiferromagnet, which increase Curie temperature higher than room temperature [24, 27, 28].

## 18.3 Magnetoresistance Through a Single Molecule

Spin-polarized STM can use for developments of nanospintronic devices. Magnets (N–S poles) are nowadays used for storing information of our daily life. A harddisk drive includes more than $10^{12}$ nanomagnets, where a giant magnetoresistance (GMR) sensor detects stray field of each magnet and read alignments of N–S poles.

In this study we fabricated the smallest GMR sensor using spin-polarized STM technique with a single organic molecule. Figure 18.4 shows how the GMR sensor works in the harddisk drive. The sensor, consisting of two nanomagnets separated by a spacer film, detects conductance through this [magnet1/spacer/magnet2] junction. If the spacer is vacuum, this junction corresponds to the spin-polarized STM. Therefore the conductance through this junction is proportional to $\vec{P_1} \cdot \vec{P_2}$. When the two magnets are parallel (anti-parallel) the conductance becomes higher (lower) and then the light is ON (OFF), which effect is called magnetoresistance, i.e., we can control the conductance by switching spin polarization vectors. In Fig. 18.4a, magnet 1 is pinned and the magnet 2 is free, so the N–S pole of the magnet 2 is changed by stray field from each magnet on the harddisk. Thus, the sensor detects the variation of the conductance during scanning on the array of the nanomagnets, which correspond to binary signals (1 or 0) (reading information).

In a last decade, tunneling magnetoresistance and magnetic random access memory devices have been developed. Magnetoresistance higher than 100 % was achieved using a Fe/MgO interface [33, 34]. Such spintronic devices have developed so far mainly by inorganic materials, i.e., Si-based semiconductor and metal films. Our challenge is a use of a 1-nm-size single organic molecule for spintronics. We use a single molecule as a spacer as shown in Fig. 18.4b. Scientific interesting is whether the single organic molecule can pass spins and generate magnetoresistance effect. However, since it is hard to fabricate such a small junction with top–down techniques, we fabricate such a single molecular junction with the spin-polarized

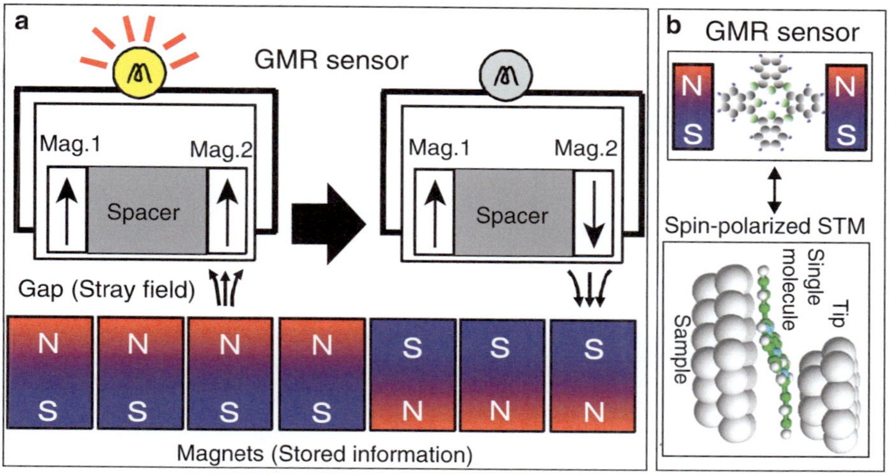


**Fig. 18.4** Magneto resistance (MR) sensor inside hard disk reads magnetic directions of nano magnetics (= stored information). The sensor consists of two nano magnets. Magnet 1 is a fixed magnet and magnet 2 is a detecting magnet. The detecting magnet follows stray field from nano magnets on the hard disk. When the magnet 1 and 2 are parallel (anti-parallel), resistance through the mag.1/spacer/mag.2 junction is low (high) and the ramp is on (off). Thus, the MR sensor can read stored information. Sketches of (**a**) a magneto resistance sensor and (**b**) a spin-polarized scanning tunneling microscopy setup. By contacting the magnetic STM tip with a single molecule deposited on a nano magnet, we can measure magneto resistance through a single molecule

STM as shown in Fig. 18.4b, where we first prepared a nanomagnet on a substrate, second deposited a single molecule on it, then gently approached the spin-polarized STM magnetic tip toward the molecule, and finally the tip contacted with the molecule. Thus, we succeeded to fabricate a [nanomagnet/single molecule/nanomagnet] junction without breaking a single molecule, and simultaneously we could measure the conductance through the single molecule. If the conductance is changed by switching the coupling from parallel to anti-parallel (or vice versa), we could prove that a single molecule can be used for magnetoresistance sensor.

All spin-polarized STM spectroscopy experiments were performed at 4.6 K in UHV. Phthalocyanine ($H_2Pc$) molecules were used as a representative of $\pi$-conjugated organic molecules. Commercial $H_2Pc$ powder (Alfa Aesar, purity 95 %) was purified by sublimation at 653 K and recrystallization at 473 K in a pressure of $10^{-3}$ Pa (yield 30 %). Nuclear magnetic resonance (NMR) and infrared spectra confirmed the absence of impurities after the purification. The clean $H_2Pc$ powder was placed in a crucible, which was set in a molecular chamber and heated to 550 K. The $H_2Pc$ powder was deposited on a substrate at 300 K by opening the shutter valve between the preparation and molecular chambers. Figure 18.5a shows an STM topographic image of $H_2Pc$ single molecules deposited on a substrate. Four benzene rings in side groups have $\pi$-orbitals, appearing as four bright spots in the STM image. An inset image in Fig. 18.5a also shows a $H_2Pc$ single molecule but with higher resolution, where we could observe atoms constructing the molecule.

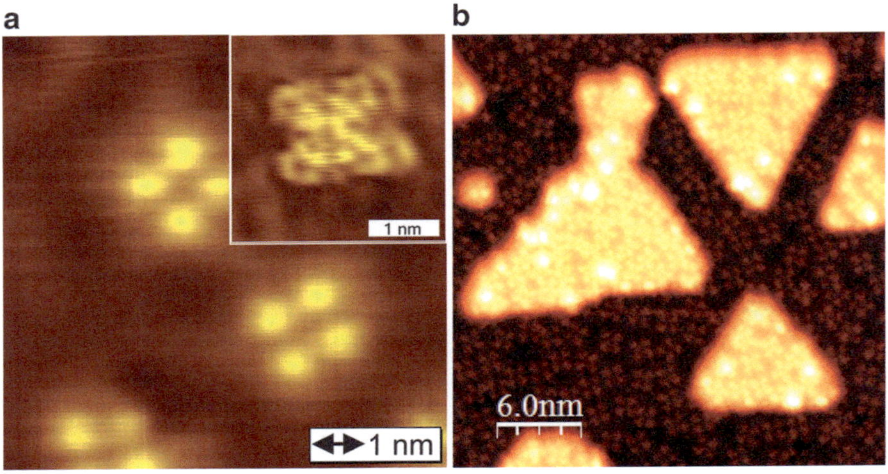


**Fig. 18.5** (**a**) STM images of phthalocyanine ($H_2Pc$) single molecules on Cu(111) substrate. An *inset* shows an atomically resolved STM image of the single $H_2Pc$ molecule. (**b**) $H_2Pc$ single molecules are deposited on a surface of 0.2 ML Co on Cu(111). *Triangulars* denote bilayer Co nano magnets

Using this single molecule, we started to fabricate the single molecular junction. Co-coated W tips were used as a ferromagnetic electrode. As another ferromagnetic electrodes we used 10–20 nm size Co bilayer nanoislands grown on a clean Cu (111), which are shown in Fig. 18.5b. The Co islands have a triangular shape because they follow a substrate fcc-Cu(111) symmetry, which have perpendicular magnetic anisotropy, and therefore N–S poles of the islands have two possibilities, i.e., upward and downward.

We deposited single molecules on the Co nanoislands in UHV at 300 K, then subsequently bring to STM at 4.6 K. The magnetoresistance measurements through the single molecule with the spin-polarized STM setup were performed by following the experimental scheme shown in Fig. 18.6a. The ferromagnetic Co/W tip first gently contacts with the molecule adsorbed on the Co island with a spin polarization vector pointing upward, and measure the conductance. Next, the same tip is withdrawn from the molecule and moved to another molecule adsorbed on the Co island with a spin polarization vector pointing downward, and measure the conductance. We repeated this process many times. If we see differences in the conductances between parallel and anti-parallel magnetic couplings of the tip and the sample, it will prove that the single molecule has a magnetoresistance effect.

However, from the normal STM topographic image in Fig. 18.5b, spin polarization directions of the nanoislands are unknown. We checked the N–S poles of each nanoisland by spin-polarized STM. Fig. 18.6b shows an STM topographic image combined with magnetic contrast in the *dI/dV* map, in which one island appears brighter and another one appears darker, indicating that the brighter and the

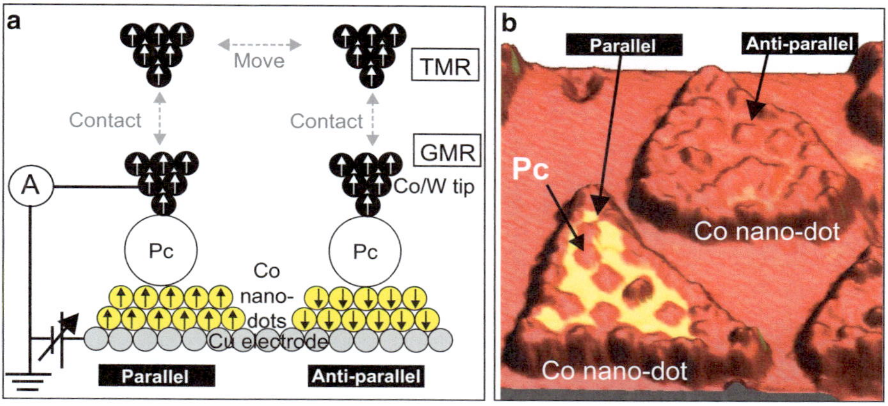


**Fig. 18.6** (**a**) Tunneling magneto-resistance and giant magneto-resistance measurements with STM setup. Here, we used bilayer Co nano-islands grown on Cu(111) substrate, in which magnetic anisotropy is perpendicular to surface. *Black* and *white arrows* indicate magnetic moments. Pc denotes a single phthalocyanine molecule. (**b**) STM magnetic imaging of single $H_2Pc$ molecules deposited on Co nano-islands with a triangular shape. Two islands are observed. One appears brighter and the other appears darker, which means that the former has parallel spin to tip spin and the latter has anti-parallel spin to tip spin

darker islands couple parallel and anti-parallel, respectively, with the ferromagnetic Co/W tip.[4]

By following Fig. 18.6a, we measured conductance as a function of the tip-sample distance. Figure 18.7a shows an example of the conductance measurement. We approached the STM tip toward the molecule from offset=0 pm, simultaneously we measured the conductance. The conductance increases exponentially by closing the tip to the surface until offset=165 pm, which is under tunneling regime. But, at 165 pm, the conductance suddenly increases almost one order of magnitude and conductance becomes almost constant. The bias during the conductance measurement was set below 10 mV to prevent phonon excitation of the phthalocyanine molecule. Since such a jump occurred only when we approached the tip at the side groups of the molecule, we considered that one of the side group contacts with the tip and the opposite side contacts with the Co nanoisland. The conductance when the tip contacted with the molecule includes two components, i.e., tunneling conductance ($G_{tun}$): electron transfer from the tip to the Co nanoisland and the conductance through the molecule ($G_{mol}$). $G_{mol}$ was obtained by subtracting $G_{tun}$ from the obtained conductance.

[4] We could obtain the tunnel magnetoresistance of the Co nanoislands by $(dI/dV_P - dI/dV_{AP})/dI/dV_{AP}$. TMR was 90 % at −300 mV, while it decreased to 5 % near the Fermi energy.

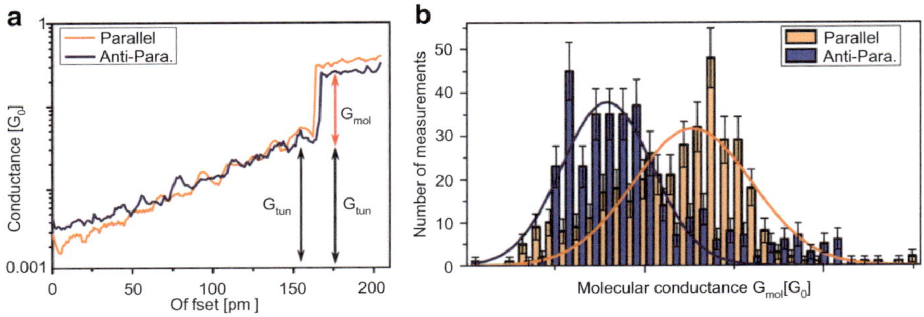


**Fig. 18.7** An STM tip contacts to $H_2$Pc molecule. $G_{tun}$ and $G_{mol}$ denote tunneling conductance and conductance through the single molecule, respectively. (**a**) Conductance measurements with an increase in the offset from the original tip position. Tip contacts the molecule around offset of 165 pm, and the conductance drastically increases. "Parallel" and "antiparallel" indicate spin configuration between tip and Co nano-islands [10]. $G_o$ denotes quantum conductance of $2e^2/h \sim 1/(12.9k\Omega)$. (**b**) Histogram of spin-dependent conductance through single $H_2$Pc molecules [10]

Figure 18.7a shows the contact conductances in the cases of the parallel (orange color) and the anti-parallel (blue color) couplings, where we could clearly see the difference in the two curves, i.e. the parallel and the anti-parallel couplings show 0.26 $G_0$ and 0.19 $G_0$, respectively (GMR ~ 37 %).[5] We repeated the conductance measurements and results are shown in Fig. 18.7b. A difficulty of conductance measurements using the spin-polarized STM is that a contacting areal slightly changes at each measurement, which varies conductance slightly. Also, in this [Co/Pc/Co] junction case, more than two molecules are required to obtain the parallel and anti-parallel couplings, i.e., we cannot measure the GMR through the same single molecule. Figure 18.7b shows a histogram of the experimentally obtained molecular conductances (parallel 381 and anti-parallel 366), where we can clearly see two peaks (peak positions are 0.253 $G_0$ and 0.158 $G_0$) which gives us GMR of about 60 % [10]. Furthermore, it should be noted that the [Co/vacuum/Co] junction show only TMR=5 % near the Fermi energy, but inserting the single molecule enhances magnetoresistance almost 10 times. Another advantage of using the single molecule is a low resistance of 52 k$\Omega$, which is good for high-frequency devices, as well as low resistance area product of 70 $\mu\Omega\mu m^2$, good for high density devices. In this study we succeeded to fabricate a single molecular junction with spin-polarized STM and experimentally obtained GMR of 60 % for [Co/Pc/Co] junction.

After this first trial, we also fabricated a single phthalocyanine junction with different magnetic materials. We deposited $H_2$Pc single molecules on a layerwise antiferromagnetic Mn(001) films [23, 24], and measured GMR by gently contacting an Fe-coated W tips [24, 30], in which system magnetic anisotropy is in-plane

[5] No clear difference was not observed using non-magnetic W tips.

(Co nanoislands and Co-coated W tips have an out-of-plane anisotropy). Markedly, negative GMR of 50 % was obtained.

Drastic changes in GMR between the [Co/Pc/Co] and the [Mn/Pc/Fe] junctions is considered to be different electronic spin hybridizations, i.e., the molecular LUMO orbital hybridizes with Co minority and Mn majority spin states. Therefore understanding of the hybridized spin states at the interface between the single molecule and the 3d magnets could be a key to develop the molecular spintronic devices. So far we have been developed a normalization technique to recover the LDOS of the nanomolecules from experimentally obtained *dI*/*dV* curves [35], and performed a systematic study how LDOS near the Fermi energy is affected by different substrates, i.e., MgO(001) ultrathin films (decoupling layers), noble metal films, and 3d magnets, which tells us a complete change of original molecular states by strong hybridization of 3d and $\pi$ orbitals, causing a new spin structure at molecule-3d metal interface, which will be the origin of the magnetoresistance effect [19].

**Acknowledgements** These experimental works had been performed with Prof. Dr. W. Wulfhekel, Dr. Y. Nahas, and Dr. S. Schmaus. We thank Dr. L. Gerhard for carefully reading our manuscript. This work was supported by Chiba University Global COE Program: Advanced School for Organic Electronics, KAKENHI(22810005,23681018), Japan Science and Technology Agency (JST)-Improvement of Research Environment for Young Researchers: Chiba University Young Research-Oriented Faculty Member Development Program in Bioscience Areas, the Deutsche Forschungsgemeinschaft (WU 349/3-1 and SPP1243), the Center for Functional Nanostructures, the French–German University, the Alexander von Humboldt foundation, and the Agence Nationale de la Recherche (ANR-06-NANO-033-01) as well as from the Yamada Science Foundation and the Asahi Glass Foundation.

# References

1. G.W. Michael, A.B. Rudine, C.C. Wamser, J. Porphyr. Phthalocyanines **14**, 759–792 (2010)
2. M. Nothaft, S. Hohla, F. Jelezko, N. Fruhauf, J. Pflaum, J. Wrachtrup, Nat. Commun. **3**, 628 (2012)
3. T.N. Jackson, IEEE J. Sel. Top. Quantum Electron. **4**, 100–104 (1998)
4. C.W. Tang, S.A. VanSlyke, C.H. Chen, J. Appl. Phys. **65**, 3610 (1989)
5. A. Bagrets, S. Schmaus, A. Jaafar, D. Kramczynski, T.K. Yamada, M. Alouani, W. Wulfhekel, F. Evers, Nano Lett. **12**, 5131 (2012)
6. J. Brede, N. Atodiresei, S. Kuck, P. Lazic, V. Caciuc, Y. Morikawa, G. Hoffmann, S. Blu gel, R. Wiesendanger, Phys. Rev. Lett. **105**, 047204 (2010)
7. T. Komeda, H. Isshiki, J. Liu, Y.F. Zhang, N. Lorente, K. Katoh, B.K. Breedlove, M. Yamashita, Nat. Commun. **2**, 217 (2010)
8. T. Miyamachi, M. Gruber, V. Davesne, M. Bowen, S. Boukari, F. Scheurer, G. Rogez, T.K. Yamada, P. Phresser, E. Beaurepaire, W. Wulfhekel, Nat. Commun. **3**, 938 (2012)
9. A. Mugarza, R. Robles, C. Krull, R. Korytar, N. Lorente, P. Gambardella, Phys. Rev. B **85**, 155437 (2012)
10. S. Schmaus, A. Bagrets, Y. Nahas, T.K. Yamada, A. Bork, F. Evers, W. Wulfhekel, Nat. Nanotechnol. **6**, 185 (2011)

11. A.F. Takacs, F. Witt, S. Schmaus, T. Balashov, M. Bowen, E. Beaurepaire, W. Wulfhekel, Phys. Rev. B **78**, 233404 (2008)
12. Y. Tanaka, P. Mishra, R. Tateishi, N.T. Cuong, H. Orita, M Otani, T. Nakayama, T. Uchihashi, K. Sakamoto, ACS Nano. **7** 1317 (2013)
13. N. Tsukahara, K. Noto, M. Ohara, S. Shiraki, N. Takagi, Y. Takata, J. Miyawaki, M. Taguchi, A. Chainani, S. Shin, M. Kawai, Phys. Rev. Lett. **102**, 167203 (2009)
14. A. Zhao, Q. Li, L. Chen, H. Xiang, W. Wang, S. Pan, B. Wang, X. Xiao, J. Yang, J.G. Hou, Q. Zhu, Science **309**, 1542 (2005)
15. L. Gerhard, T.K. Yamada, T. Balashov, A.F. Takacs, M. Daena, S. Ostanin, A. Ernst, I. Mertig, W. Wulfhekel, Nat. Nanotechnol. **5**, 792 (2010)
16. I. Chizhov, G. Scoles, A. Kahn, Langmuir **16**, 4358 (2000)
17. A. Scarfato, S.H. Chang, S. Kuck, J. Brede, G. Hoffmann, R. Wiesendaner, Surf. Sci. **602**, 677 (2008)
18. L. Gross, F. Mohn, N. Moll, P. Liljeroth, G. Meyer, Science **325** 1100 (2009)
19. S. Nakashima, Y. Yamagishi, K. Oiso, T.K. Yamada, Jpn. J. Appl. Phys. **52**, 110115 (2013)
20. S. Schintke, S. Messerli, M. Pivetta, F. Patthey, L. Libioulle, M. Stengel, A. De Vita, W. D. Schneider, Phys. Rev. Lett. **87**, 276801 (2001)
21. H.J. Shin, J. Jung, K. Motobayashi, S. Yanagisawa, Y. Morikawa, Y. Kim, M. Kawai, Nat. Mater. **9** 442 (2010)
22. V.A. Ukraintsev, Phys. Rev. B **53**, 11176 (1996)
23. T.K. Yamada, M.M.J. Bischoff, T. Mizoguchi, H. van Kempen, Surf. Sci. **516**, 179 (2002)
24. T.K. Yamada, M.M.J. Bischoff, G.M.M. Heijnen, T. Mizoguchi, H. van Kempen, Phys. Rev. Lett. **90**, 056803 (2003)
25. R. Wiesendanger, Rev. Mod. Phys. **81**, 1495 (2009).
26. T.K. Yamada, *Handbook of Magnetic Imaging*, Kyoritsu, 15–44 (2010)
27. T.K. Yamada, M.M.J. Bischoff, G.M.M. Heijnen, T. Mizoguchi, H. van Kempen, Jpn. J. Appl. Phys. **42**, 4688 (2003)
28. T.K. Yamada, M.M.J. Bischoff, T. Mizoguchi, H. van Kempen, Appl. Phys. Lett. **82**, 1437 (2003)
29. T. Balashov, T. Schuh,, A.F. Takacs, A. Ernst, S. Ostanin, J. Henk, I. Mertig, P. Bruno, T. Miyamachi, S. Suga, W. Wulfhekel, Phys. Rev. Lett. **102**, 257203 (2009)
30. T. Irisawa, T.K. Yamada, T. Mizoguchi, New J. Phys. **11**, 113031 (2009)
31. T.K. Yamada, A.L. Vazquez de parga (Submitted to Appl. Phy. Lett.)
32. M.M.J. Bischoff, T.K. Yamada, C.M. Fang, R.A. de Groot, H. van Kempen, Phys. Rev. B **68**, 045422 (2003)
33. S.S.P. Parkin et al., Nat. Mater. **3**, 862 (2004)
34. S. Yuasa et al., Nat. Mater. **3**, 868 (2004)
35. Y. Yamagishi, S. Nakashima, K. Oiso, T.K. Yamada, Nanotechnol. **24** 395704 (2013)

# Chapter 19
# Vortex Lasers Twist Materials to Form Chiral Nanostructures

**Takashige Omatsu, Nobuyuki Aoki, and Katsuhiko Miyamoto**

## 19.1 Introduction

We discovered that the angular momentum of the optical vortex can twist an irradiated sample, so as to create twisted nanostructures (which we term chiral nanostructures); and the twisted direction (chilarity) of these nanostructures can be selectively controlled by changing the sign of the angular momentum.

Such chiral nanostructures fabricated by an optical vortex will enable us to open up new materials science, such as chiral photonics and plasmonics. They might also have potential to selectively identify chiral molecules and composites on the nanoscale.

In this chapter, we review optical vortex materials processing, e.g., the formation of chiral metal nanostructure as well as chiral organic reliefs.

T. Omatsu (✉)
Graduate School of Advanced Integration Science, Chiba University, 1-33 Yayoi-cho, Inage-ku, Chiba 263-8522, Japan

CREST Japan Science and Technology Agency, Sanbancho, Chiyoda-ku
Tokyo 102-0075, Japan
e-mail: omatsu@faculty.chiba-u.jp

N. Aoki • K. Miyamoto
Graduate School of Advanced Integration Science, Chiba University, 1-33 Yayoi-cho, Inage-ku, Chiba 263-8522, Japan



H. Ishii et al. (eds.), *Electronic Processes in Organic Electronics*,
Springer Series in Materials Science 209, DOI 10.1007/978-4-431-55206-2_19

## 19.2 Optical Vortex Lasers

### *19.2.1 Optical Vortex [1–3]*

An optical vortex with an annular intensity profile carries an orbital angular momentum ($\ell\ell\hbar$) associated with a helical wavefront, owing to a phase singularity, $\ell\phi$ (where $\ell$ is an integer and $\phi$ is the azimuthal angle) in the transverse plane (Fig. 19.1).

Circularly polarized light also exhibits a spin angular momentum owing to a helical electric field. Consequently, a circularly polarized optical vortex, known as "twisted light with spin", carries a helicity, referred to as the total angular momentum ($j\hbar = \ell\hbar + s\hbar$) (Fig. 19.1), defined by the resultant vector sum of the orbital and spin angular momenta [4, 5]. The total angular momentum has been evidenced by causing acceleration (or deceleration) of the orbital motion of submicron particles in optical tweezers.

The optical vortex has been potentially utilized in various technologies, including optical tweezers [6], high-density plasma confinement [7], and super-resolution microscopes [8, 9]. For instance, in optical tweezers, an optical vortex forces submicron particles to rotate in the same direction as the orbital angular momentum. Furthermore, a technique for optical trapping of submicron particles has been demonstrated using multiple vortices with several phase singularities in the wavefront. This technique allows us to control several independent optical traps with individually specified characteristics by using one beam.

Super-resolution fluorescence microscopy, in which fluorescence depletion by stimulated emission occurs in the spatially overlapping part of a Gaussian pump and vortex erase lasers, so as to improve the spatial resolution of the fluorescence signal, has also been successfully demonstrated.

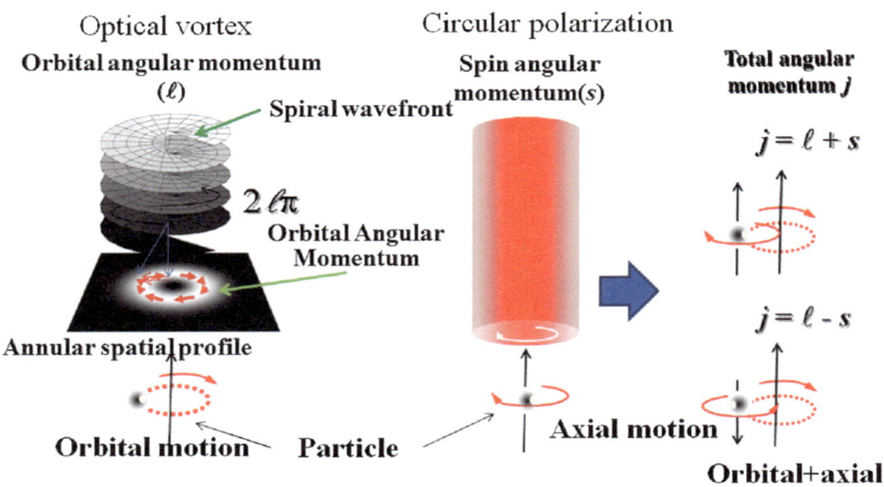


**Fig. 19.1** Optical vortex and its angular momenta

The Laguerre–Gaussian mode, which is the most conventional optical vortex, is an eigenmode of the paraxial electromagnetic equation in a cylindrical coordinate system:

$$\left(\frac{\partial^2}{\partial r^2}+\frac{1}{r}\frac{\partial}{\partial r}-\frac{1}{r^2}\frac{\partial^2}{\partial \varphi^2}-2jk\frac{\partial}{\partial z}\right)u(r,\varphi,z)=0, \qquad (19.1)$$

where $r$ is the radial coordinate, $\phi$ is the azimuthal angle, and $z$ are the propagation distance, respectively; $u(r,\phi,z)$ is the amplitude of the optical wave, and $k$ is the wave number.

The Laguerre–Gaussian beams, $u_{p,\ell}$ $(r,\phi,0)$ at $z=0$, are given by

$$u_{p,\ell}(r,\varphi,0)\propto\frac{1}{\omega_0}\left(\frac{\sqrt{2}r}{\omega_0}\right)^{|\ell|}L_p^{|\ell|}\left(\frac{2r^2}{\omega_0^2}\right)\exp\left[-\frac{r^2}{\omega_0^2}\right]\exp(-i\ell\varphi), \qquad (19.2)$$

where $L_p^{|\ell|}$ are the generalized Laguerre polynomials, $p$ is the radial index, $\ell$ is the azimuthal index, termed the topological charge, and $\omega_0$ is the spot size of the optical field, respectively.

### 19.2.2 How to Produce Optical Vortex Lasers

Several mode conversion techniques from a Gaussian beam to an optical vortex beam by utilizing additional phase elements, such as spiral phase plates and static (or dynamic) computer-generated holograms, have been proposed. In particular, a dynamic computer-generated hologram based on the use of a spatial light modulator, which is the most conventional method, enables us to create an arbitrary wavefront, such as a 2-dimentional optical vortex array [10]. However, the damage threshold and the diffraction loss of the spatial light modulator impact the power handling ability of the system.

Direct production from a laser cavity, such as a side-pumped bounce laser with a mode-size-dependent thermal lens [11, 12] is an alternative, which has been successfully demonstrated to generate a vortex output of over 20 W. However, in most vortex lasers, the wavefront helicity of the vortex output is randomly determined by the cavity alignment.

Recently, a stressed large-mode-area fiber amplifier in combination with a master laser has produced more than a 20 W vortex output in picosecond and nanosecond regimes, and it has also allowed the selective control of the wavefront helicity of the vortex outputs [13, 14].

The fiber amplifier used was a large-mode-area $Yb^{3+}$-doped double-clad fiber (a length of 4 m, a core diameter of 30 μm, and a cladding diameter of 400 μm), and

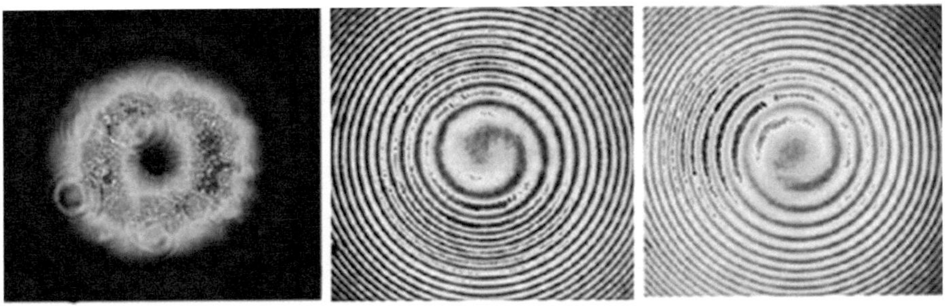

**Fig. 19.2** Spatial forms of optical vortex with $|\ell| = 1$. (*Left*) Intensity profile of the optical vortex. Spherical referenced fringes with (*middle*) clockwise and (*right*) anti-clockwise directions

it was pumped by a 975 nm fiber-coupled laser diode. The fiber was also bent into a loop with a radius of ~13 cm to prevent an undesired higher-order mode operation.

A collimated master laser output with a Gaussian profile was off-axially coupled into a large-mode-area fiber; thereby forcing two orthogonal $LP_{11}$ modes to lase at high efficiency. The fiber was appropriately stressed by two clamps, so as to provide 90° or −90° out of phase to two orthogonal $LP_{11}$ modes at the end of the fiber, and it then produced a vortex output. The sign of the spiral wavefront of the output was also controlled only by changing the stress onto the fiber.

Over 25 W of picosecond vortex output from a stressed Yb-doped fiber amplifier system in combination with a continuous-wave mode-locked Nd:$YVO_4$ master laser has been achieved, corresponding to an optical-optical efficiency of 47.9 %.

Millijoule nanosecond vortex pulses have also been generated from the fiber amplifier in combination with an actively Q-switched Nd:$YVO_4$ master laser. Spatial forms of the vortex output are summarized in Fig. 19.2.

## 19.3 Optical Vortex Laser Ablation

### *19.3.1 Nanostructure Fabrication [15, 16]*

The commercial 1.06 μm Q-switched Nd:YAG laser used had a Gaussian spatial output with a pulse duration of 30 ns; its output was converted into an optical vortex with $\ell = 1 \sim 2$ by silica spiral phase plates (SPP), azimuthally divided into 16 parts with a $n\pi/8$ phase shift (where n is an integer between 0 and 15). To control the polarization of the optical vortex (to provide spin angular momentum to the optical vortex), a quarter-wave plate was also used.

The circularly polarized optical vortex was then loosely focused by the objective lens (NA = 0.08) onto a target, a Ta plate, (complex dielectric constant of approximately $-2.54 + 10i$ at 1.06 μm) to a ϕ60 μm spot on the sample surface. The laser energy used in the present experiments was fixed to be 2 mJ.

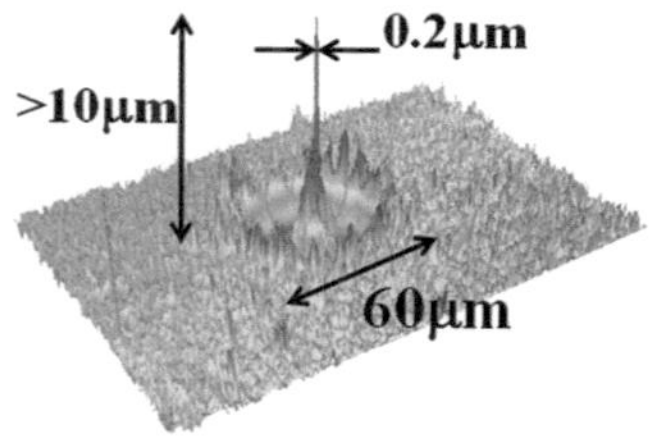


**Fig. 19.3** Laser-scanning microscope image of the metal surface ablated by optical vortex pulses

All experiments were performed at atmospheric pressure and room temperature. The morphology of the ablated target was observed using a confocal laser-scanning microscope (LSM) with a spatial resolution of 20 nm in both depth and transverse displacements.

After single-shot irradiation of an optical vortex pulse with $j = 2$, a bump appeared at the core of the optical vortex. The height (length between the top and bottom ends of the bump) was measured to be ~4.4 μm. After three $j = 2$ pulses were fired at the same position, the bump was shaped to be a needle with a height (from the target surface) of ~10 μm, and its tip diameter was measured to be less than 0.2 μm (the value was limited by the spatial resolution of LSM) (Fig. 19.3). A two-dimensional 5 × 6 nanostructure array was further fabricated by translating the target, resulting in uniformly well shaped nanostructures with an average length and tip diameter of 11 and 0.5 μm, respectively.

The target ablated by using a circularly polarized Gaussian beam only with spin angular momentum had no nanostructures, such as needles. This indicates that the formation of the nanostructure requires the collection and confinement of the melted metal to an on-axis dark core by the photon pressure.

### *19.3.2 Chiral Materials Processing [17, 18]*

A technique used to twist material, such as metal on a nanoscale to create spiral (chiral) nanostructures has not been established even by utilizing advanced chemical techniques. In particular, the laser ablation process, in which a metal is broken down into its constituent elements (melted or vaporized metal) by high-intensity laser pulses, is unsuitable for recombining the constituent elements and structuring spiral materials.

The angular momentum of the optical vortex, forcing orbital motion of submicron compositional elements (a plasma of ions and electrons, and melted matter) produced by the high intense laser pulses, might potentially provide new physical insight to laser materials processing. However, the orbital angular momentum effects on the light-matter interaction have never been studied.

As related above, the circularly polarized optical vortex carries a total angular momentum ($j$) defined as the sum of the orbital ($\ell$) and spin ($s$) angular momenta.

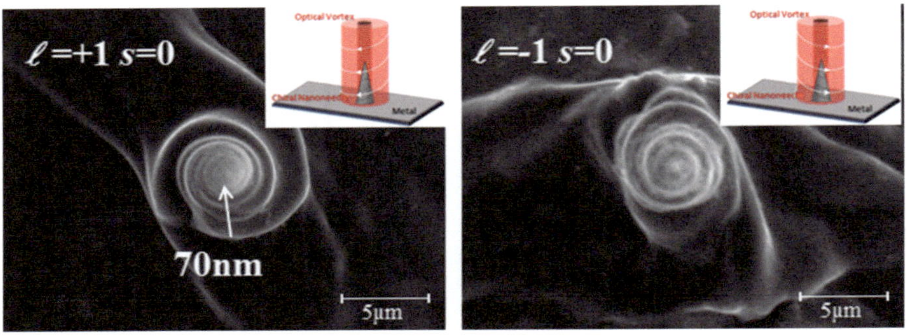


**Fig. 19.4** SEM images of a chiral nanostructure fabricated by an optical vortex with $j=-1$ (clockwise) and $j=1$ (counter-clockwise). The focusing lens had an NA of 0.08

Namely, when the sign of the orbital angular momentum, $m$, is the same as (or opposite to) that of the spin angular momentum, $s$ (1 or $-1$), the total angular momentum of the optical vortex is equal to be $\ell+1$ (or $\ell$-1).

We mention that the laser energy used in the present experiments was controlled in the range of 0.075–0.3 mJ, which is less than one-sixth that (2 mJ) used in the experiments mentioned in section 19.3.1. Above this energy level, the leading edge of the optical vortex pulse produced dense plasma (vaporized material) so as to shield the rest of the optical pulse, thereby yielding an insufficient transfer of the angular momentum to the melted material. And thus, only low energy pulse deposition enabled us to create chiral nanostructures.

The helicity of the optical vortex was reversed by inverting the spiral phase plate and the quarter wave plate. Four vortex pulses with a $\phi \sim 65$ μm annular spot onto the target were fired on the same position of the target. The morphology of the ablated target surface was observed by a scanning electron microscope (SEM) with a spatial resolution of 8 nm at 3 kV.

Figure 19.4 show a SEM image of a target surface ablated by an optical vortex with $\ell=-1$. A needle was formed at the center of the ablated zone with a smooth outline; its tip curvature and height were typically measured to be ~70 and ~10 μm, respectively. The magnified needle had a conical surface twisted azimuthally in the clockwise direction. In contrast, a twisted needle in the counter-clockwise direction was fabricated by irradiation of an optical vortex with $\ell=1$. The twisting direction (chirality) of the twisted nanostructure (termed chiral nanostructures) is selectively determined by the sign of the orbital angular momentum of the optical vortex. The spiral frequency of the nanostructure (defined as the winding number per 1 μm height) also increased as the magnitude of $\ell$ increased.

As shown in Fig. 19.5, the spiral frequency of the nanostructure as a function of $j$ of the pump laser is investigated. The spin angular momentum, $s\hbar$, can then enhance (reduce) the spiral frequency of a nanostructure when the sign of $\ell$ is the same as (opposite to) that of $s$.

We also notice that $j$ determines the spiral frequency of the chiral nanostructure instead of $\ell$. For example, linearly polarized second-order ($\ell=2$; $s=0$) and

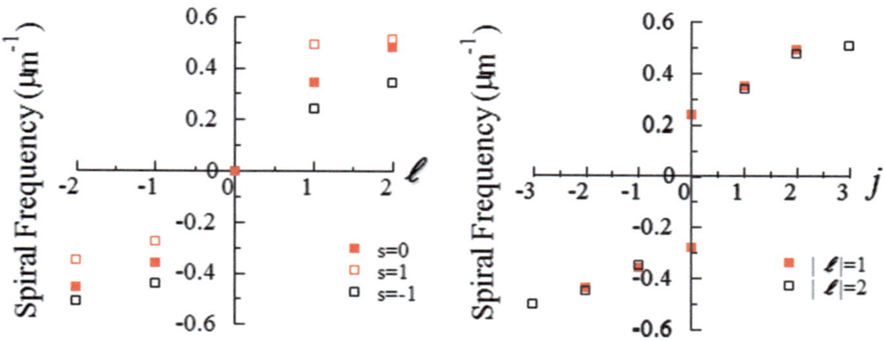


**Fig. 19.5** (*Top*) Experimental plots of spiral frequency as a function of orbital angular momentum, $\ell$, for various spin angular momenta $s$. (*Bottom*) Experimental plots of spiral frequency as a function of the orbital angular momentum, $j$, for various orbital angular momenta, $\ell$

circularly polarized first-order ($\ell = 1$; $s = 1$) optical vortices created chiral nanostructures with the same chirality and spiral frequency. Even if $j = 0$ ($\ell = 1$; $s = -1$), optical vortices also created chiral nanostructures.

Additionally, the tip curvature of the chiral nanostructure was approximately inversely proportional to the NA of the objective lens. At NA = 0.18, the tip curvature and height of the nanostructure were measured to be ~36 nm (less than 1/25th of the laser wavelength) and 7.5 μm, respectively.

The electrical properties of the nanostructure were investigated by using two 50 μm diameter tungsten probes with an internal resistance of ~1.0 Ω. One probe was contacted onto the top of the nanostructure, while the other was tightly pressed to the substrate surface so as to minimize the contact resistance.

The measured current between the nanostructure and the substrate was proportional to the supplied voltage with a resistance of ~50 mΩ, excluding the internal resistance of the probe. The Energy-dispersive X-ray (EDX) spectrum of the fabricated needle was mostly identical to that of the substrate surface. Thus, these results indicate that the fabricated needle was perfectly metallic.

### *19.3.3 Angular Momentum Effects*

The linear momentum density of the light, $\boldsymbol{P}$, given by the time-averaged real part of the Poynting vector, is written as [6, 19]

$$\boldsymbol{P} \propto \langle(\boldsymbol{E}^* \times \boldsymbol{B} + \boldsymbol{E} \times \boldsymbol{B}^*)\rangle = \left\langle i\omega_0(u^*\nabla u - u\nabla u^*) + 2\omega_0 k|u|^2\boldsymbol{z} - \omega s\frac{\partial|u|^2}{\partial r}\boldsymbol{\varphi}\right\rangle, \tag{19.3}$$

where $\omega_0$ is the frequency of the light, $s$ is the spin angular momentum (1 or $-1$) for clockwise or anti-clockwise circularly polarized light, and $\boldsymbol{z}$ and $\boldsymbol{\phi}$ are unit vectors along the $z$- and azimuthal directions, respectively. The resulting angular momentum, $\boldsymbol{M}$, of the light is given by

$$\boldsymbol{M} = \boldsymbol{r} \times \boldsymbol{P}, \tag{19.4}$$

where $\boldsymbol{r}$ is a unit vector along the radial direction.

By substituting Eqs. (19.2) and (19.3) into Eq. (19.4), the time averaged angular momentum density, $j_{\ell,s}$, of a Laguerre–Gaussian mode is obtained.

The time averaged angular momentum density, $j_{\ell,s}$, of a circularly polarized Laguerre–Gaussian mode with $p = 0$ is written by

$$j_{\ell,s} \propto \frac{\omega_0}{|\ell|!}\left(\ell - |\ell|s + s\left(\frac{\sqrt{2}r}{\omega_0}\right)^2\right)\left(\frac{\sqrt{2}r}{\omega_0}\right)^{2|\ell|} \exp\left(-\frac{2r^2}{{\omega_0}^2}\right). \tag{19.5}$$

We notice that the total angular momentum, $j\hbar$, defined as the integration of the angular momentum density, $j_{\ell,s}$, over the whole beam aperture, is given by the sum of the orbital, $\ell\hbar$, and spin, $s\hbar$, angular momenta per photon.

General relationships, written as $j_{\ell\pm1,0} = j_{\ell,\pm1}$ ($|\ell| \geq 2$) and $j_{\ell,s} = -j_{-\ell,-s}$, are established. Thus, the relationship $j_{2,0} = j_{1,1}$ is valid, which indicates that two optical vortices with $j = 2$ are degenerate so as to produce the same chiral nanostructure; $j_{1,-1}$ is also given by

$$j_{1,-1} \propto \left(2 - \left(\frac{\sqrt{2}r}{\omega_0}\right)^2\right)\left(\frac{\sqrt{2}r}{\omega_0}\right)^2 \exp\left(-\frac{2r^2}{{\omega_0}^2}\right). \tag{19.6}$$

Even though the total angular momentum spatially averaged over the whole beam aperture is zero, a nonzero total angular momentum near the dark core of the optical vortex forces the melted metal to revolve to create a chiral nanostructure. These experimental results suggest the following model. The melted metal caused by irradiation of the optical vortex is collected in the ring of the optical vortex. The melted metal then receives orbital angular momentum from the optical vortex, revolving about the axial core of the optical vortex. The melted metal also receives the optical scattering force to direct forwardly, and it is subsequently confined within the core (where it does not experience efficiently a forward scattering force) to form a chiral needle on the nanoscale.

## 19.4 Chiral Organic Materials

Surface relief formation [20–23] on the azo-polymer film, owing to mass transport based on an optical gradient force, anisotropic photo-fluidity, and cis-trans photo-isomerization, has received intense attention because it enables us to create unique optical devices, such as photonic waveguides and circuits.

The mass transport driving force induces the azo-polymer to direct from a bright fringe toward a dark fringe along the polarization of the light. Thus, linearly or circularly polarized light does not form a chiral surface relief in the azo-polymer film. Until now, there are no reports on the chiral surface relief formation in azo-polymer films. An optical vortex with angular momentum, owing to a helical wavefront, will have a potential to twist the azo-polymer film to form a chiral surface relief.

In this section, we present the first demonstration, to the best of our knowledge, of a chiral surface relief with a micro-meter scale height and depth formed in an azo-polymer film using a circularly polarized optical vortex. The azo-polymer used in this study was a Poly-Orange Tom-1(POT) with the absorption band in the wavelength range of 300–550 nm, and showed photo-isomerization behavior upon the irradiation of a green laser.

A continuous-wave 532 nm green laser output was converted to be a circularly polarized 1st order optical vortex with a total angular momentum of $j = 2$ by using a polymer spiral phase plate (RPC photonics, VPP-1c) and a quarter-wave plate. The sign of $j$ was also reversed by inverting the spiral phase plate and the quarter-wave plate. The green vortex output was focused to be a $\phi$5 μm annular spot by an objective lens (NA ~ 0.65) onto a spin-coated azo-polymer film having a thickness of ~4 μm. The exposure time and the laser intensity were then 12 s and ~4 kW/cm$^2$, respectively. The morphology of the azo-polymer surface was observed by an atomic force microscope (AFM; SHIMADZU, SPM-9700).

As shown in Fig. 19.6, a clockwise chiral surface relief with a height of 1.3 μm, a diameter of 4.5 μm, and a tip curvature of ~0.5 μm (we term this a 'conch'-shaped relief) was formed upon the irradiation of an optical vortex. When the sign of $j$ was inverted, a 'conch'-shaped relief with a counter-clockwise direction was also completed.

These experimental results can be understood by using the following model. Trans-cis photo-isomerization of the azo-polymer occurs by irradiation of an optical vortex. The cis (soft) azo-polymer then starts to revolve in a clockwise (or counter-clockwise) direction due to the total angular momentum of the circularly polarized optical vortex. The mass transport driving force also forces the azo-polymer to direct toward the dark core of the optical vortex, and to form such 'conch'-shaped structure.

The 'conch'-shaped relief in the azo-polymer with a doping of impurities, such as functional chemical composites (e.g., laser dyes) and metal (or semiconductor, magnetic) nano-particles, will be promising to produce planar chiral photonic devices, as well as high dimensional and hierarchic structures. The metal-coated 'conch'-shaped relief can also act as chiral plasmonic materials.

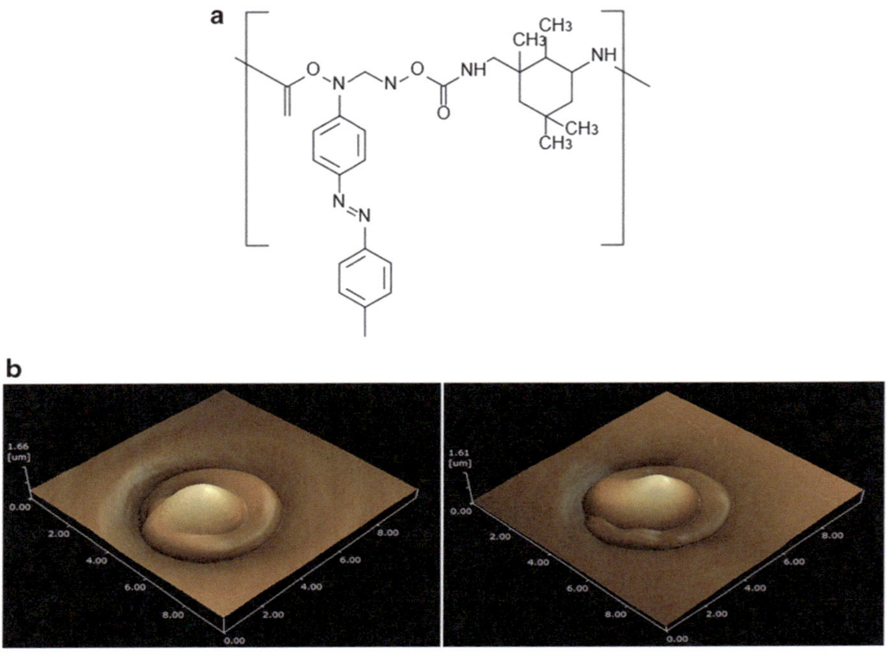


**Fig. 19.6** (**a**) Chemical structure of the azo-polymer, Poly-Orange Tom-1. (**b**) AFM images of 'conch'-shaped surface reliefs by irradiation of the optical vortex with (*left*) $j = 2$ ($\ell = 1, s = 1$) and (*right*) $j = -2$ ($\ell = -1$, $s = -1$), respectively

## Conclusion

We discovered, for the first time, that the angular momentum of the optical vortex transfers to a melted metal and a photo-isomerized polymer so as to form chiral nanostructures (i.e. nanostructures and "coch'-shaped relief). The chirality of the fabricated nanostructures can also be determined selectively by the sign of the angular momentum of the optical vortex. Further, an array of two-dimensional chiral nanostructures can also be fabricated by translating the target, and then irradiating the optical vortex onto the target.

Chiral nanostructures will open up potentially new advanced technologies, such as nanoimaging systems, energy-saving displays, and biomedical nanoelectromechanical systems. They can also be potentially applied to the selective identification of chirality in chemical reactions on plasmonic nanostructures.

**Acknowledgments** The authors acknowledge financial support of a Research Grant (CREST) from Japan Science and Technology Agency. We would also like to thank Ms. Mizuki Watabe from Chiba University and Professor R. Morita from Hokkaido University for their productive helps.

## References


1. L. Allen, M.W. Beijersbergen, R.J.C. Spreeuw, J.P. Woerdman, Orbital angular momentum of light and the transformation of Laguerre-Gaussian laser modes. Phys. Rev. A **45**, 8185–8189 (1992)
2. M. Padgett, J. Courtial, L. Allen, Optical vortices and their propagation. Phys. Today **57**, 35–40 (2004)
3. S. Franke-Arnold, L. Allen, M.J. Padgett, Advances in optical angular momentum. Laser Photon. Rev. **2**, 299–313 (2008)
4. Q. Zhang, Properties of circularly polarized vortex beams. Opt. Lett. **31**, 867–869 (2006)
5. N.B. Simpson, K. Dholakia, L. Allen, M.J. Padgett, Mechanical equivalence of spin and orbital angular momentum of light: an optical spanner. Opt. Lett. **22**, 52–54 (1997)
6. A.T. O'Neil, I. MacVicar, L. Allen, M.J. Padgett, Intrinsic and extrinsic nature of the orbital angular momentum of a light beam. Phys. Rev. Lett. **88**, 053601–1-4 (2002)
7. D.G. Grier, A revolution in optical manipulation. Nature **424**, 810–816 (2003)
8. V. Skarka, N.B. Aleksi´c, V.I. Berezhiani, Evolution of singular optical pulses towards vortex solitons and filamentation in air. Phys. Lett. A **319**, 317–324 (2003)
9. S. Bretschneider, C. Eggeling, S.W. Hell, Breaking the diffraction barrier in fluorescence microscopy by optical shelving. Phys. Rev. Lett. **98**, 218103–1-4 (2007)
10. T. Watanabe, Y. Iketaki, T. Omatsu, K. Yamamoto, M. Sakai, M. Fujii, Two-point-separation in super-resolution fluorescence microscope based on up-conversion fluorescence depletion technique. Opt. Express **11**, 3271–3276 (2003)
11. K. Ladavac, D.G. Grier, Microoptomechanical pumps assembled and driven by holographic optical vortex arrays. Opt. Express **12**, 1144–1149 (2004)
12. M. Okida, T. Omatsu, M. Itoh, T. Yatagai, Direct generation of high power Laguerre-Gaussian output from a diode-pumped Nd:$YVO_4$ 1.3 μm bounce laser. Opt. Express **15**, 7616–7622 (2007)
13. M. Okida, Y. Hayashi, T. Omatsu, J. Hamazaki, R. Morita, Characterization of 1.06 μm optical vortex laser based on a side-pumped Nd:$GdVO_4$ bounce oscillator. Appl. Phys. B **95**, 69–73 (2009)
14. M. Koyama, T. Hirose, M. Okida, K. Miyamoto, T. Omatsu, Power scaling of a picosecond vortex laser based on a stressed Yb-doped fiber amplifier. Opt. Express **19**, 994–999 (2011)
15. M. Koyama, T. Hirose, M. Okida, K. Miyamoto, T. Omatsu, Nanosecond vortex laser pulses with millijoule pulse energies from a Yb-doped double-clad fiber power amplifier. Opt. Express **19**, 14420–14425 (2011)
16. J. Hamazaki, R. Morita, K. Chujo, Y. Kobayashi, S. Tanda, T. Omatsu, Optical vortex laser ablation. Opt. Express **18**, 2144–2151 (2010)
17. T. Omatsu, K. Chujo, K. Miyamoto, M. Okida, K. Nakamura, N. Aoki, R. Morita, Metal microneedle fabrication using twisted light with spin. Opt. Express **18**, 17967–17973 (2010)
18. K. Toyoda, K. Miyamoto, N. Aoki, R. Morita, T. Omatsu, Using optical vortex to control the chirality of twisted metal nanostructures. Nano Lett. **12**, 3645–3649 (2012)
19. K. Toyoda, F. Takahashi, S. Takizawa, Y. Tokizane, K. Miyamoto, R. Morita, T. Omatsu, Transfer of light helicity to nanostructures. Phys. Rev. Lett. **110**, 143603–1-5 (2013)
20. N.K. Viswanathan, D.Y. Kim, S. Bian, J. Williams, W. Liu, L. Li, L. Samuelson, J. Kumar, S.K. Tripathy, Surface relief structures on azo polymer films. J. Mater. Chem. **9**, 1941 (1999)
21. M. Watabe, G. Juman, K. Miyamoto, T. Omatsu, Light induced conch-shaped relief in an azo-polymer film. Sci. Rep. **4**, 4281 (2014)
22. C.J. Barrett, A.L. Natansohn, P.L. Rochon, Mechanism of optically inscribed high-efficiency diffraction gratings in azo polymer films. Phys. Chem. **100**, 8836 (1996)
23. H. Ishitobi, M. Tanabe, Z. Sekkat, S. Kawata, The anisotropic nanomovement of azo-polymers. Opt. Express **15**, 652 (2007)

# Index

**A**
Acceptor, 298
Adenine (A), 249
Adiabatic potential, 300–303
Angle-integrated UPS, 17
Angle-resolved photoemission spectroscopy (ARPES), 315, 316
Angle-resolved ultraviolet photoelectron spectroscopy (ARUPS), 12–13
Anisotropy, 19, 22
ARPES *See* Angle-resolved photoemission spectroscopy (ARPES)
Arrhenius formula, 301
Au(111), 57, 95

**B**
Band bending, 4
Band gap states, 4, 49–63
Band structure, 33, 319
Band transport, 15
Bandwidth, 4, 16
Bandwidth-narrowing, 316
Binding energy, 10
Binding energy of polaron state, 332–333
BiOTFT, 252, 259–260
BiOTFT memory, 261
Bis-(1,2,5-thiadiazolo)-p-quinobis (1,3-dithiole) (BTQBT), 14
Boltzmann distribution, 302
Born–Oppenheimer adiabatic approximation, 301
Bragg condition, 93
Bragg-reflected wave, 91
Bragg reflection, 93

**C**
Canonical equation of motion, 329
Carrier density, 321
Carrier traps, 309
Charge density, 322
Charge mobilities, 313
Charge transfer complex (CTC), 90
Charge transfer interaction, 345
Charge transport gap, 63
Charging effects, 52
Clustering of impurity atoms, 309
Coherent fraction, 94
Coherent length, 22
Coherent position, 94
Color tunable OLED, 270–274
Commensurate structure, 360
Conductance, 317
Conductivity, 321, 322
Contact potential difference, 10
Continuum medium, 4
Covalent bonding, 298
Crystal phonons, 5
Cu(111), 95
Cytosine (C), 249

**D**
Deep-level states, 298
Degradation, 297
Delocalization, 22
Density of gap states (DOGS), 56
Density of states (DOS), 17–18, 54, 322
Diffusion, 297
  coefficient, 323
  constant, 303


H. Ishii et al. (eds.), *Electronic Processes in Organic Electronics*,
Springer Series in Materials Science 209, DOI 10.1007/978-4-431-55206-2

Diindenoperylene (DIP), 94
Dipoles, 101
Dirac point, 304
Disorder, 4, 316
DNA, 249
DNA BiOLED, 266–270
DNA-cetyltrimethylammonium (CTMA), 252
DNA-surfactant complexes, 259–260
Donor, 298
Doping, 4
Drift mobility, 15
Dynamical diffraction theory, 92
Dynamical lattice distortions, 328

**E**
Effective mass, 12, 15, 320
Einstein energy, 19
Einstein relation, 303, 322
Elastic energy loss, 305
Electron affinity (EA), 58, 87, 357
Electron density, 108
Electronegativity, 298
Electronic coupling, 38
Electronic polaron, 5
Electron-phonon coupling, 19, 25, 37, 44, 330
Electron spin resonance (ESR), 315
Electron transfer, 304
Encapsulation, 365
Energy-level alignment (ELA), 87
Energy-momentum (E-k) dispersion relation, 12
Excitation energy dependent UPS, 14

**F**
Fermi distribution function, 322
Fermi level (EF), 2, 10
Fermi level pinning, 4
Fermi's Golden Rule, 37
Field-effect transistors (FETs), 313
First-principles calculation, 297
Flood gun, 13
Franck-Condon, 37
Franck-Condon principle, 42
$F_{16}$ZnPc, 97

**G**
Gas exposure effects, 54–59
Gate voltage, 321
Graphene, 303–306, 343
Graphite, 29
Group velocity, 320
Guanine (G), 249

**H**
He Iα, 53
He IIα, 53
Helmholtz equation, 89
Hierarchical polarons, 5
Highest occupied molecular orbital (HOMO), 14, 90, 298, 317
  band, 14
Hole-vibration coupling, 26, 44
Holstein (small polaron), 316
Holstein model, 315, 323–324
Holstein–peierls model, 325–326
Holstein polaron, 37
HOMO *See* Highest occupied molecular orbital (HOMO)
Hopping mobility, 45
Hopping transport, 38
Huang-Rhys factors, 39
Hubbard U, 5
Hybridization, 305

**I**
Impurity, 297
Inner potential, 12
In-pore high pressure effect, 355
Interaction energy, 309
Interaction potential, 344
Intercalation, 345
Intercalation of $C_{60}$, 357
Interface(s), 4
  dipole, 51
  electric dipole, 50
  state, 17
Intermolecular interaction, 4
Inter-molecule diffusion, 308
Interstitial, 298
Intra-molecular band dispersion, 14
Intramolecular vibrations, 329
Intra-molecule diffusion, 307
Inverse photoemission spectroscopy (IPES), 58

Ionization, 308
Ionization energy (IE), 20, 87
Ionization potential (IP), 55, 357

**K**
Kubo formula, 316, 321–323

**L**
Laplace's formula, 351
Large polaron, 5, 316
Lifetime, 33, 34
Light-emitting diodes (LEDs), 313
Liquid phase adsorption, 345
Local electric dipoles, 4
Local phonons, 5
Lowest unoccupied molecular orbital (LUMO), 2, 90, 298

**M**
Marcus electron-transfer theory, 37
Marcus equation, 315
Mean free path, 320–321, 323, 335
Metallic bonding, 310
Methylene blue (MB), 365
Mobility, 1, 315
Mobility edge, 63
Molecular distortions, 101
Molecular orientation, 4
Molecular size, 4
Molecular-vibration, 5
Momentum relaxation time, 323
Monolayer site, 344
Morphological defects, 349
Mott-Hubbard insulator, 4
Mulliken's resonance theory, 356
Multimode analysis (MMA), 42

**N**
Naphthalene, 298, 362
Naphthalene derivatives, 362
NO dimerization, 352
Non-adiabatic (diabatic) processes, 301
Nonlocal phonons, 5

**O**
Occupied molecular orbital (HOMO), 2
Oligoacenes, 45, 306–309
1D sulfur chain, 353
One-dimensional tight-binding (1D-TB) approximation, 14
Organic field effect transistor (OFET), 1
Organic light emitting diode (OLED), 1
Organic-metal interfaces, 87
Organic semiconductors, 25
Organic single crystals, 14
Organic solar cell (OSC), 1

**P**
Pauli repulsion, 90, 94–97
Peierls model, 316, 325
Peierls transition, 306
Pentacene (PEN), 16, 27, 96, 298, 317
6,13-pentacenequinone (P2O), 98
5,7,12,14-pentacenetetrone (P4O), 98
Perfluoropentacene (PFP), 39, 96
3,4,9,10-perylene-tetracarboxylic-dianhydride (PTCDA), 98
Phase of the complex amplitude, 91
Phonon, 16, 301
Phonon-band structure, 330
Photoconductivity, 14, 19
Photoconductors, 1
Photoelectron spectroscopy, 9
Photoelectron yield, 93–94
Photoelectron yield spectroscopy (PYS), 20
Photoemission electron microscopy (PEEM), 30
Phthalocyanines (Pcs), 26, 97
Physisorption, 94–97
π-conjugated organic systems, 297
π–conjugated surface, 344
Polarization, 26, 60
Polarization/screening, 5
Polaron, 29, 60–63, 316
Polaron binding energy, 25
Polaronic relaxation, 329
Polyaromatic hydrocarbons (PAH), 343
Polymorphs, 17
Poly(acetylene)/Polyacetylene, 261, 306
Potential barrier, 302
Push-back effect, 89

**Q**
Quasi-interface dipole, 51

**R**
Radiation damages, 52

Reflection high-energy electron diffraction (RHEED), 18
Relaxation(s), 60, 101
energies, 25, 27, 39, 45
time, 15, 323
Reorganization energy, 25, 38, 43, 324
$Ru(bpy)_3^{2+}$, 262
Rubrene, 14, 20
Rubrene single crystals, 1

S
Sample charging, 12, 19
Secondary electron cut-off (SECO), 11
Self-synthesizing function, 6
Semiconductors, 297
Sensitivity, 52
Single mode analysis (SMA), 27
Single-wall carbon nanotube (SWCNT), 343
Single wall carbon nanohorn (SWCNH), 346
Small polaron, 5
Spatial distribution, 4
$sp^2$ carbon atoms, 344
$sp^3$ carbon atoms, 344
Spread, 36
$S_8$ ring structure, 353
Static disorder, 335–337
Stone-Thrower-Wales (STW) defects, 350
Strain energy, 308
Strongly correlated systems, 5
Substitutional, 298
Sulfur-encapsulated carbon nanotubes, 355
Surface Brillouin zone (SBZ), 18
Surface-enhanced Raman scattering (SERS), 350
Surface-induced aromatic stabilization, 90
Surface sensitivity, 52
Surface solid, 344
Su-Schrieffer-Heeger model, 325
SWCNT bundles, 345
Symmetry, 4

T
Tailing states, 60
TB approximation, 19
2,3,5,6-tetrafluoro-7,7,8,8-tetracyanoquinodimethane (F4-TCNQ), 98
Thermal carrier, 59
Thin-film transistor (TFT), 315
Thymine (T), 249
Time-dependent wave-packet diffusion (TD-WPD), 326
Time-energy uncertainty relationship, 16
Time-evolution operator, 328
Transfer energy, 333–334
Transfer integral, 14, 25, 34, 38
Two-dimensional (2D), 19

U
Ultraviolet photoelectron spectroscopy (UPS), 26, 88
Universal features of organic crystals, 2

V
Vacuum levels, 10, 87
Valence band, 14
van der Waals (vdW) attraction, 90
van der Waals interaction, 298
Velocity of a crystal electron, 319
Velocity operator, 322
Vertical field effect transistor, 16
Very small polaron, 5
Vibration effects, 323–326
Voigt function, 33, 42

W
Watson–Crick model, 249
Wave-function, 36
Work function (WF), 10, 59, 87

X
Xe I, 53
X-ray diffraction, 57
X-ray photoelectron spectroscopy (XPS), 88
X-ray reflectivity (XRR), 108
X-ray standing wave, 90–94

Z
ZnPc, 97

MIX
Papier aus verantwortungsvollen Quellen
Paper from responsible sources
FSC® C105338

If you have any concerns about our products,
you can contact us on
**ProductSafety@springernature.com**

In case Publisher is established outside the EU,
the EU authorized representative is:
**Springer Nature Customer Service Center GmbH**
**Europaplatz 3, 69115 Heidelberg, Germany**

Printed by Libri Plureos GmbH
in Hamburg, Germany